Lecture Notes in Computer Science 4802

Commenced Publication in 1973
Founding and Former Series Editors:
Gerhard Goos, Juris Hartmanis, and Jan van Leeuw

Jean-Luc Hainaut   Elke A. Rundensteiner
Markus Kirchberg   Michela Bertolotto
Mathias Brochhausen   Yi-Ping Phoebe Chen
Samira Si-Saïd Cherfi   Martin Doerr   Hyoil Han
Sven Hartmann   Jeffrey Parsons   Geert Poels
Colette Rolland   Juan Trujillo   Eric Yu
Esteban Zimányi (Eds.)

Advances in Conceptual Modeling – Foundations and Applications

ER 2007 Workshops CMLSA, FP-UML,
ONISW, QoIS, RIGiM, SeCoGIS
Auckland, New Zealand, November 5-9, 2007
Proceedings

 Springer

Volume Editors

Jean-Luc Hainaut E-mail: jean-luc.hainaut@fundp.ac.be
Elke A. Rundensteiner E-mail: rundenst@cs.wpi.edu
Markus Kirchberg E-mail: M.Kirchberg@massey.ac.nz
Michela Bertolotto E-mail: michela.bertolotto@ucd.ie
Mathias Brochhausen E-mail: mathias.brochhausen@ifomis.uni-saarland.de
Yi-Ping Phoebe Chen E-mail: phoebe@deakin.edu.au
Samira Si-Saïd Cerfi E-mail: sisaid@cnam.fr
Martin Doerr E-mail: martin@ics.forth.gr
Hyoil Han E-mail: hhan@ischool.drexel.edu
Sven Hartmann E-mail: s.hartmann@massey.ac.nz
Jeffrey Parsons E-mail: jeffreyp@mun.ca
Geert Poels E-mail: geert.poels@ugent.be
Colette Rolland E-mail: rolland@univ-paris1.fr
Juan Trujillo E-mail: jtrujillo@dlsi.ua.es
Eric Yu E-mail: eric.yu@utoronto.ca
Esteban Zimányi E-mail: ezimanyi@ulb.ac.be

Library of Congress Control Number: 2007938044

CR Subject Classification (1998): H.2, H.4, H.3, F.4.1, D.2, C.2.4, I.2, J.1

LNCS Sublibrary: SL 3 – Information Systems and Application, incl. Internet/Web and HCI

ISSN   0302-9743
ISBN-10   3-540-76291-4 Springer Berlin Heidelberg New York
ISBN-13   978-3-540-76291-1 Springer Berlin Heidelberg New York

Springer is a part of Springer Science+Business Media

springer.com

© Springer-Verlag Berlin Heidelberg 2007
Printed in Germany

Typesetting: Camera-ready by author, data conversion by Scientific Publishing Services, Chennai, India
Printed on acid-free paper   SPIN: 12181322   06/3180   5 4 3 2 1 0

Preface

The 26th International Conference on Conceptual Modeling in Auckland, New Zealand, hosted six workshops which allowed participants to focus their presentations and discussions on advanced topics that cannot easily fit the general conference scope.

Thirteen good quality proposals were received and nine were selected. Due to the similarity of their scope, two pairs were suggested to merge, leading to seven proposals. One workshop attracted fewer submissions than expected, so that its selected papers were integrated into the conference. Finally, six workshops were kept. Interestingly, four of them (FP-UML, ONISW, QoIS, SeCoGIS) were a sequel of workshops that were held in the last few years, while two were new (CMLSA, RIGiM), exhibiting both the maturity and the innovation of the workshops.

Following the call for papers, we received 114 complete submissions, from which 40 quality papers were selected, giving an acceptance rate of 35% (a fairly standard score for workshops).

The following six workshops were organized:

- Conceptual Modelling for Life Sciences Applications (CMLSA 2007), chaired by Yi-Ping Phoebe Chen and Sven Hartmann. This workshop addressed the specific challenges posed by the large data volumes, the complexity and the data and software heterogeneity involved by life science applications.
- Foundations and Practices of UML (FP-UML 2007), chaired by Juan Trujillo and Jeffrey Parsons. The third edition of this workshop gathered researchers and practitioners on topics related to data warehouses, security, model transformation, state diagrams development and model quality.
- Ontologies and Information Systems for the Semantic Web (ONISW 2007), chaired by Mathias Brochhausen, Martin Doerr and Hyoil Han. The second edition of this workshop focused on the potential and actual roles of ontologies in Web information systems. In particular, the papers addressed ontology extraction from texts, ontology interoperability, automatic semantic annotation and domain modeling.
- Quality of Information Systems (QoIS 2007), chaired by Samira Si-Saïd Cherfi and Geert Poels. The papers of the third edition of this workshop dealt with quality models, model understandability and (at last!) quality assessment of scientific conferences.
- Requirements, Intentions and Goals in Conceptual Modeling (RIGiM 2007), chaired by Colette Rolland and Eric Yu. The workshop goal was to explore the relations between requirements engineering and conceptual modeling. The selected papers particularly addressed such topics as model-driven approaches, the use of i*, refinement of goal-based methodologies, elicitation techniques and visualization.

– Semantic and Conceptual Issues in Geographic Information Systems (SeCo-
  GIS 2007), chaired by Esteban Zimányi and Michela Bertolotto. The fourth
  edition of this workshop included presentations on data interoperability, spatial data warehouses, spatial data querying, data quality, conceptual modeling, spatio-temporal aspects of moving objects as well as special spatial
  applications.

We are particularly grateful to all the workshop organizers, the Program Committee members and the authors for the time and effort spent to guarantee the
high quality of the programs. Special thanks are due to Markus Kirchberg who
took charge of most of the editorial work of these proceedings.

November 2007   Jean-Luc Hainaut
                                                    Elke A. Rundensteiner

ER 2007 Conference Organization

General Chair

Bernhard Thalheim (Christian Albrechts University Kiel, Germany)

Program Committee Co-chairs

Christine Parent (University of Lausanne, Switzerland)
Klaus-Dieter Schewe (Massey University, New Zealand)
Veda C. Storey (Georgia State University, USA)

Organization Chair

Gillian Dobbie (University of Auckland, New Zealand)

Steering Committee Liaison

Tok Wang Ling (National University of Singapore, Singapore)

Publicity chair

Markus Kirchberg (Massey University, New Zealand)

Workshop Co-chairs

Jean-Luc Hainaut (University of Namur, Belgium)
Elke A. Rundensteiner (Worcester Polytechnic Institute, USA)

Tutorial Co-chairs

Sven Hartmann (Massey University, New Zealand)
Alberto H.F. Laender (UFMG, Brazil)

Panel Chair

John F. Roddick (Flinders University, Australia)

Industrial Chair

John Grundy (University of Auckland, New Zealand)

Demonstration and Poster Chair

Leszek Maciaszek (Macquarie University, Australia)

Treasurers

Patricia Rood (University of Auckland, New Zealand)
Stephen W. Liddle (Brigham Young University, USA)

Organized By

Massey University, New Zealand
The University of Auckland, New Zealand

Sponsored By

The ER Institute

In Cooperation With

ACM SIGMIS
ACM SIGMOD

ER 2007 Workshop Organization

ER 2007 Workshop Co-chairs

Jean-Luc Hainaut, University of Namur, Belgium
Elke A. Rundensteiner Worcester Polytechnic Institute, USA

CMLSA 2007 – International Workshop on Conceptual Modelling for Life Sciences Applications

Workshop Chairs

Yi-Ping Phoebe Chen, Deakin University, Australia
Sven Hartmann, Massey University, New Zealand

Publicity Chair

Markus Kirchberg, Massey University, New Zealand

Program Committee

Isabelle Bichindaritz, USA
Leonard N. Bloksberg, New Zealand
Jake Chen, USA
Alexei Drummond, New Zealand
Silke Eckstein, Germany
Amarnath Gupta, USA
David Hansen, Australia
Peter Hunter, New Zealand
Nikola Kasabov, New Zealand
Dirk Labudde, Germany
Sebastian Link, New Zealand
Huiqing Liu, USA

Jingchu Luo, China
Victor Maojo, Spain
Sudha Ram, USA
Keun Ho Ryu, Korea
Amandeep S. Sidhu, Australia
Tanya Soboleva, New Zealand
Thodoros Topaloglou, Canada
Haixun Wang, USA
Jing Wang, New Zealand
Xiaofang Zhou, Australia
Esteban Zimányi, Belgium

External Referees

Claudia Täubner, Germany
Henning Köhler, New Zealand

FP-UML 2007 – International Workshop on Foundations and Practices of UML

FP-UML 2007 was organized within the framework of the following projects: META-SIGN (TIN2004-00779) from the Spanish Ministry of Education and Science and DADS (PBC-05-012-2) from the Castilla-La Mancha Ministry of Science and Technology (Spain).

Workshop Co-chairs

Juan Trujillo, University of Alicante, Spain
Jeffrey Parsons, Memorial University of Newfoundland, Canada

Program Committee

Doo-Hwan Bae, South Korea
Michael Blaha, USA
Cristina Cachero, Spain
Tharam Dillon, Australia
Gillian Dobbie, New Zealand
Brian Dobing, Canada
Dirk Draheim, Germany
Joerg Evermann, New Zealand
Eduardo Fernández, Spain
Jens Lechtenbörger, Germany
Tok Wang Ling, Singapore
Pericles Loucopoulos, UK
Hui Ma, New Zealand

Andreas L. Opdahl, Norway
Oscar Pastor, Spain
Witold Pedrycz, Canada
Mario Piattini, Spain
Ivan Porres, Finland
Colette Rolland, France
Matti Rossi, Finland
Manuel Serrano, Spain
Keng Siau, USA
Il-Yeol Song, USA
Ambrosio Toval, Spain
Antonio Vallecillo, Spain
Panos Vassiliadis, Greece

External Referees

M. Kirchberg
A. Tretiakov

O. Thonggoom
G. Abraham

ONISW 2007 – International Workshop on Ontologies and Information Systems for the Semantic Web

Workshop Co-chairs

Mathias Brochhausen, Saarland University, Germany
Martin Doerr, Foundation for Research and Technology - Hellas, Greece
Hyoil Han, Drexel University, USA

Program Committee

Yuan An, Canada
Sofia J. Athenikos, USA
Boualem Benatallah, Australia
Mathias Brochhausen, Germany
Namyeon Choi, USA
Crisitan Cocos, Germany
Martin Doerr, Greece
Ramez Elmasri, USA
Fabien Gandon, France
Raul Garcia-Castro, Spain

Hyoil Han, USA
Mary Elizabeth Jones, USA
Aneesh Krishna, Australia
Werner Kuhn, Germany
Sang-Koo Lee, Korea
SeungJin Lim, USA
Dimitris Plexousakis, Greece
Barry Smith, USA and Germany
Krishnaprasad Thirunarayan, USA

QoIS 2007 – International Workshop on Quality of Information Systems

Workshop Co-chairs

Samira Si-Saïd Cherfi, CEDRIC-CNAM, France
Geert Poels, Ghent University, Belgium

Steering Committee

Jacky Akoka, CEDRIC-CNAM & INT, France
Mokrane Bouzeghoub, PRISM, University of Versailles, France
Isabelle Comyn-Wattiau, Conservatoire National des Arts et Métiers
   and ESSEC, France
Marcela Genero, Universidad de Castilla-La Mancha, Spain
Jeffrey Parsons, Memorial University of Newfoundland, Canada
Geert Poels, Ghent University, Belgium
Keng Siau, University of Nebraska, USA
Bernhard Thalheim, University of Kiel, Germany

Program Committee

Jacky Akoka, France
Laure Berti-Equille, France
Andrew Burton-Jones, Canada
Tiziana Catarci, Italy
Corinne Cauvet, France
Isabelle Comyn-Wattiau, France
Marcela Genero, Spain
Paul Johannesson, Sweden
Jacques Le Maître, France
Jim Nelson, USA
Jeffrey Parsons, Canada
Oscar Pastor, Spain

Houari Sahraoui, Canada
Farida Semmak, France
Keng Siau, USA
Guttorm Sindre, Norway
Samira Si-Saïd Cherfi, France
Monique Snoeck, Belgium
Il-Yeol Song, USA
David Tegarden, USA
Bernhard Thalheim, Germany
Dimitri Theodoratos, USA
Juan Trujillo, Spain

External Referees

Ki Jung Lee, USA

RIGiM 2007 – International Workshop on Requirements, Intentions and Goals in Conceptual Modeling

Workshop Co-Chairs

Colette Rolland, Université Paris1 Panthéon Sorbonne, France
Eric Yu, University of Toronto, Canada

Program Committee

Daniel Amyot, Canada
Mikio Aoyoma, Japan
Ian Alexander, UK
Aybuke Arum, Australia
Franck Barbier, France
Daniel Berry, Canada
Sjaak Brinkkemper, The Netherlands
Lawrence Chung, USA
Luiz Cysneiros, Canada
Eric Dubois, Luxembourg
Vincenzo Gervasi, Italy
Aditya K. Ghose, Australia
Peter Haumer, USA
Zhi Jin, China
Aneesh Krishna, Australia
John Krogstie, Norway
Lin Liu, China

Peri Loucopoulos, UK
John Mylopoulos, Canada
Selmin Nurcan, France
Bashar Nuseibeh, UK
Andreas Opdahl, Norway
Barbara Pernici, Italy
Klaus Pohl, Germany
Jolita Ralyte, Switzerland
Bjorn Regnell, Sweden
Camille Salinesi, France
Motoshi Saeki, Japan
Pnina Soffer, Israel
Carine Souveyet, France
Leon Sterling, Australia
Yair Wand, Canada
Roel Wieringa, The Netherlands

External Referees

Arosha Bandara, UK
Chiara Francalanci, Italy
Joy Garfield, UK

Kim Lauenroth, Germany
Thorsten Weyer, Germany

SeCoGIS 2007 – International Workshop on Semantic and Conceptual Issues in Geographic Information Systems

Workshop Co-chairs

Esteban Zimányi, Université Libre de Bruxelles, Belgium
Michela Bertolotto, University College Dublin, Ireland

Program Committee

Gennady Andrienko, Germany
Natalia Andrienko, Germany
Yvan Bédard, Canada
David Bennett, USA
Alex Borgida, USA
Patrice Boursier, France
Bénédicte Bucher, France
James Carswell, Ireland
Christophe Claramunt, France
Eliseo Clementini, Italy
Maria Luisa Damiani, Italy
Clodoveu Davis, Brazil
Max Egenhofer, USA
Fernando Ferri, Italy
Andrew Frank, Austria
Anders Friis-Christensen, Italy
Antony Galton, UK
Bo Huang, Canada
Marinos Kavouras, Greece

Werner Kuhn, Germany
Sergei Levashkin, Mexico
Ki-Joune Li, South Korea
Thérèse Libourel, France
Jose Macedo, Switzerland
Peter van Oosterom, The Netherlands
Dimitris Papadias, China
Fabio Porto, Switzerland
Dieter Pfoser, Greece
Ricardo Rodrigues Ciferri, Brazil
Andrea Rodriguez, Chile
Sylvie Servigne-Martin, France
Stefano Spaccapietra, Switzerland
Emmanuel Stefanakis, Greece
Kerry Taylor, Australia
Christelle Vangenot, Switzerland
Antonio Miguel Vieira Monteiro, Brazil
Agnès Voisard, Germany

Table of Contents

FP-UML 2007 – International Workshop on Foundations and Practices of UML

Improving the Use of UML Diagrams

Model Transformations and Extensions

ONISW 2007 – International Workshop on Ontologies and Information Systems for the Semantic Web

QoIS 2007 – International Workshop on Quality of Information Systems

RIGiM 2007 – International Workshop on Requirements, Intentions and Goals in Conceptual Modelling

Keynote

Requirements and Goals – Methods

Requirements and Goals – Concepts

SeCoGIS 2007 – International Workshop on Semantic and Conceptual Issues in Geographic Information Systems

Moving Objects

Advances in Conceptual Modelling for GIS

Integrity Constraints and Approximate Reasoning

Preface to CMLSA 2007

Yi-Ping Phoebe Chen[1] and Sven Hartmann[2]

[1] Deakin University, Australia
[2] Massey University, New Zealand

Life sciences applications typically involve large volumes of data of various kinds and a multiplicity of software tools for managing, analyzing and interpreting them. There are many challenging problems in the processing of life sciences data that require effective support by novel theories, methods and technologies. Conceptual modelling is the key for developing high-performance information systems that put these theories, methods and technologies into practice. The fast-growing interest in life sciences applications calls for special attention on resource integration and collaborative efforts in information systems development.

This volume contains the papers presented at the 1st International Workshop on Conceptual Modelling for Life Sciences Applications (CMLSA 2007) which was held in Auckland, New Zealand, November 5-9, 2007 in conjunction with the 26th International Conference on Conceptual Modeling (ER 2007). On behalf of the Program Committee we commend these papers to you and hope you find them useful.

The primary objective of the workshop is to share research experiences in conceptual modelling for applications in life sciences and to identify new issues and directions for future research in relevant areas, including bioinformatics, health informatics, medical and veterinary informatics. The workshop invited original papers exploring the usage of conceptual modelling ideas and techniques for developing and improving life sciences databases and information systems. The scope of CMLSA 2007 included topics such as:

- Conceptual modelling for biological and laboratory information systems
- Conceptual modelling for analysis tools in bioinformatics
- Conceptual modelling for health information systems
- Conceptual modelling for medical and veterinary information systems
- Life sciences databases, data exchange and data distribution
- Data representation and visualization
- Life sciences ontologies and taxonomies
- Data, process and application integration in life sciences
- Data warehousing and data mining
- Privacy and security in life sciences applications
- Electronic health records
- Reuse and reengineering of life sciences applications
- Collaborative and open source software development in life sciences

Following the call for papers which yielded 26 submissions, there was a rigorous refereeing process that saw each paper refereed by three international experts. The eight papers judged best by the Program Committee were accepted and are included in this volume.

We are grateful to Allen Rodrigo, from the University of Auckland, who kindly agreed to present the CMLSA keynote address on "Computational Challenges in the Metagenomic Age."

We wish to thank all authors who submitted papers and all workshop participants for the fruitful discussions. We would also like to thank the members of the Program Committee for their timely expertise in carefully reviewing the submissions, and Markus Kirchberg for his excellent work as CMLSA Publicity Chair. Finally, we wish to express our appreciation to the local organizers at the University of Auckland for the wonderful days in New Zealand.

Pattern Recognition of Single-Molecule Force Spectroscopy Data

Dirk Labudde[1,2], Annalisa Marsico[1], K. Tanuj Sapra[2], and Michael Schroeder[1]

[1] Department of Bioinformatics, Center of Biotechnology, TU Dresden,
Tatzberg 47-51, 01307 Dresden, Germany
[2] Department of Cellular Machines, Center of Biotechnology, TU Dresden,
Tatzberg 47-51, 01307 Dresden, Germany
{dirk.labudde, annalisa.marsico, tanuj.sapra,
michael.schroeder}@biotec.tu-dresden.de

Abstract. Motivation: Misfolding of membrane proteins plays an important role in many human diseases such as retinitis pigmentosa, hereditary deafness, and diabetes insipidus. Little is known about membrane proteins as there are only a very few high-resolution structures. Single-molecule force spectroscopy is a novel technique which measures the force necessary to pull a protein out of a membrane. Such force curves contain valuable information about the protein's structure, conformation, and inter- and intra-molecular forces. High-throughput force spectroscopy experiments generate hundreds of force curves including spurious and good curves, which correspond to different unfolding pathways and to different functional states of an investigated membrane protein.

Results: In the present work we propose a novel application of automated unfolding pattern recognition routines. We apply our method to datasets from unfolding experiments of bacteriorhodopsin (bR) and bovine rhodopsin (Rho). As a result, we discuss the different unfolding pathways of bR, and two functional states for Rho could be observed . Overall, the algorithm tackles the force spectroscopy bottleneck and leads to more consistent and reproducible results paving the way for high-throughput analysis of structural features of membrane proteins.

Keywords: unfolding, force-distance pattern, single-molecule force spectroscopy, pattern recognition.

1 Introduction

Integral membrane proteins play essential roles in cellular processes, including photosynthesis, transport of ions and small molecules, signal transduction and light harvesting. Despite the central importance of transmembrane proteins, the number of high-resolution structures remains small due to the practical difficulties in crystallising them [1]. Many human disease-linked point mutations occur in transmembrane proteins [2]. These mutations cause structural instabilities in a transmembrane protein leading it to unfold or misfold in an alternative conformation [2, 3].

J.-L. Hainaut et al. (Eds.): ER Workshops 2007, LNCS 4802, pp. 3–13, 2007.

Protein folding is described by multidimensional energy landscapes or folding funnels resulting from the interplay of complex inter- and intra-molecular interactions [4]. Atomic force microscopy (AFM) is mostly known for its imaging capabilities [5, 6, 7]. Recently, single-molecule force spectroscopy (SMFS) has proved to be a novel tool for detecting and locating inter- and intra-molecular forces on a single molecule level [8, 9, 10]. SMFS experiments allow measuring the stability of membrane proteins and also probing the energy landscapes [11]. In Fig. 1A a schematic representation of the force spectroscopy instrumentation is shown. Molecules with complex three-dimensional structures, such as proteins, can be unfolded in a controlled way. Titin and bacteriorhodopsin (bR) are examples of proteins that have been intensively studied [11, 12, 13]. When transmembrane proteins are unfolded in force spectroscopy experiments, during continuous stretching of the molecule the applied force is measured by the deflection of the cantilever and plotted against extension (tip-sample separation), yielding a characteristic force-distance curve (F-D curve) (see Fig. 1). From the analysis of single molecule force spectra it is possible to associate the peaks to individual stable structural segments within membrane proteins.

For a given protein under study, the F-D curves exhibit certain patterns, which contain information about the strength and location of molecular forces established within the molecule, stable intermediates and reaction pathways, and the probability with which they occur. For membrane proteins the sequence of the unfolding peaks follows the amino acid sequence of the protein [14]. Fitting each peak to a hyperbolic function, the worm-like chain model (WLC), gives the number of already unfolded amino acids [15]. Consequently, with the peaks and the predicted secondary structure, it is possible to associate the peaks to the structural domains (see Fig. 1) [14]. To discriminate force curves showing specific and unspecific interactions and different unfolding pathways, classification and pattern recognition algorithms have to be applied, as the manual annotation is slow and subject to human mistakes. Even if some off-line software packages have been developed to analyze single-molecule force spectroscopy data [41, 42], there is an increasing demand for data analysis techniques and suitable pattern recognition algorithms that offer fully automated processing of force spectroscopy data sets on the basis of scientific criteria. Here, we develop an algorithm for high-throughput classification and statistical analysis of force spectra.

We apply our method to datasets from unfolding experiments performed on bR and bovine rhodopsin (Rho). The light-driven proton pump bR was chosen as a model system for this study because it represents one of the most extensively studied membrane protein [16, 17]. bR converts light energy ($\lambda \nabla 500$–650 nm) into an electrochemical proton gradient, which in turn is used for ATP production by the cellular ATP synthase. Its structural analysis has revealed the photoactive retinal embedded in seven closely packed transmembrane α-helices [18-22], which builds a common structural motif among a large class of related G protein-coupled receptors (GPCRs) [23-27]. The helices in bR are lettered A, B, C, D, E, F, and G, to which the C-terminal end is connected. With increasing knowledge of its structural and functional properties, bR has become a paradigm for α-helical membrane proteins [28]. Together with the adjacent lipids bR molecules assemble into trimers, which are packed into a two-dimensional hexagonal lattice of the purple membrane as a chemically distinct domain of the cell membrane.

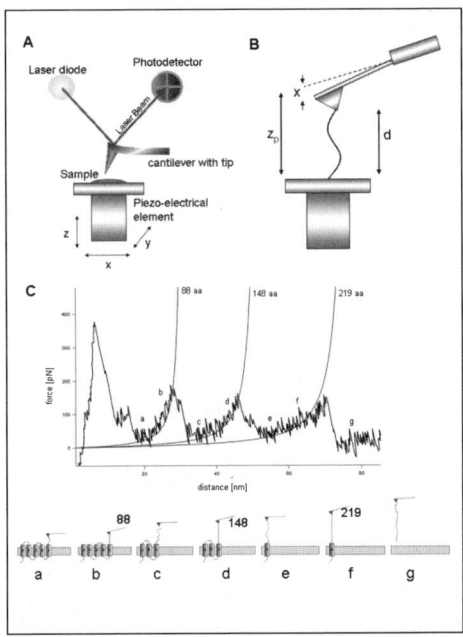

Fig. 1. A) Schematic representation of AFM. The sample is mounted on a piezo-electric element and scanned under a sharp tip attached to the cantilever. The voltage difference of the photodetector is proportional to the deflection of the cantilever. **B) Unfolding of a transmembrane protein.** A single molecule is attached between the tip and the sample while the distance between tip and sample is continuously increased. **C)** Typical spectrum obtained from an unfolding experiment of bR with the main peaks fitted by a hyperbolic function (WLC model) and correlated to the unfolding of secondary structure elements (cartoon at the bottom).

The main unfolding pathway of bR shown in Fig. 1C is characterized by the presence of three main peaks that suggest a pairwise unfolding of the transmembrane helices. Moreover, on unfolding and analyzing many single bR molecules it was found that besides these main three peaks that occur in every F-D curve, other peaks referred to as side peaks occur with much smaller probabilities indicating that bR exhibits sometimes different unfolding intermediates [13,14].

Rho, one of the most widely studied GPCRs, is a light activated receptor that initiates phototransduction in the rod outer segments (ROS) of the retinal photoreceptor cells. Rho has a molecular mass of 42,002 Da, is composed of a seven transmembrane apoprotein, opsin, with a polypeptide chain of 348 amino acids, and a chromophore, 11-*cis*-retinal, covalently bound via a protonated Schiff base to Lys296 in helix VII [29]. The counter ion for the protonated Schiff base, Glu113 in helix III, is highly conserved among all known vertebrate visual pigments [3]. The specific localization of Rho in the internal discs of the ROS, its high expression level in the retina (constituting > 90% of all proteins in disc membranes), and the lack of other highly abundant membrane proteins have facilitated studies in this system with a number of biochemical, biophysical and molecular biological methods that are difficult to be carried out with any other GPCR [29]. Since the elucidation of its X-ray crystal structure, Rho has served as a structural basis and template for studying and understanding this family of receptors and the signaling systems that they regulate [30, 31]. A number of diseases are associated with mutations that cause destabilization and misfolding of GPCRs [32]. Understanding the molecular interactions that stabilize or destabilize GPCRs is therefore fundamental to our understanding of their function. However, little is known about the underlying molecular mechanisms. The majority of these types of mutations in Rho lead to the neurodegenerative disease *retinitis pigmentosa* [33, 34]. Most mutations target the transmembrane and extracellular domains of Rho [3], and induce misfolding by the replacement of the conserved Cys110-Cys187 disulfide (S-S) bond with an abnormal disulfide bond Cys185-Cys187 or Cys110-Cys185 [35].

2 Force Distance Pattern Recognition Method

To draw biologically relevant conclusions on molecular interactions, about how strong they are and where they occur, or whether they are independent or occur only in presence with other events, one must analyze many F-D curves by identical objective procedures. Thus, there is an increasing demand for data analysis techniques that offer fully automated processing of many datasets with identical analysis procedures. To discriminate force spectra showing specific and non-specific interactions and different unfolding/unbinding pathways, classification and pattern recognition algorithms are urgently needed. Our unfolding pattern recognition method is based on dynamic programming procedures [36]. The method can separated in three steps:

1. pre-processing of the data and reduction of the noise by applying dimension reduction
2. alignment of curves with dynamic programming and computation of pair-wise distances; The distance is defined as 1-sim(FD1,FD2), were sim(FD1,FD2) is the similarity of two curves.
3. clustering of the curves based on their distances

In the second step of our method the curves are aligned using global multiple sequence alignment with dynamic programming [37]. For the detailed description of the overall spectra alignment procedure see [36]. The key reason for using a sequence alignment technique is the meaningful definition of matches/mismatches, insertions, and deletions. Matches and mismatches reward/penalize more or less fitting parts of the force curves. Insertions and deletions are important, as peaks in the curves may vary by up to six residues and as peaks may be missing completely between two curves. As result we get the Z-score.

Finally, we cluster the curves using hierarchical clustering with average linkage. We define the distance of two FD curves as 1-sim(FD1, FD2). Only curve pairs with a Z-score of sim(FD1, FD2) better than 0.65 are considered for clustering, as a lower value indicates an outlier, which belongs to the class of spurious curves. As a general result we get a hierarchical clustering tree. Each cluster corresponds to a different unfolding pattern. We use a WLC fitting routine to correlate the curve alignment of one cluster with unfolding process of the investigated membrane protein.

The WLC model is the standard approach for describing force peaks revealed from stretching a polypeptide. Unfolded proteins behave almost like random coils whose elasticity is captured by the WLC with a persistence length l_p = 4 nm [21, 14]. The gradual, nonlinear increase in the extension traces can be fitted using the WLC model with only one free parameter, namely the contour length, L, of the stretched portion of the molecule. The following equation describes the increasing slope of the F-D trace at low forces (few hundreds pN) in good approximation:

$$F(x) = \frac{k_b T}{l_p} \left(\frac{1}{4} \left(1 - \frac{x}{L} \right)^{-2} + \frac{x}{L} - \frac{1}{4} \right)$$

Herein, k_b denotes the Boltzmann constant, and T the temperature. The WLC fit of a peak thus provides the contour length L of the unfolded portion of the protein, which is the position of the corresponding barrier against unfolding. From knowledge of the attachment point of the protein to the cantilever tip, the position of an unfolding barrier with respect to the amino acid sequence of the protein can be determined. Consequently, with the peaks and the predicted secondary structure, it is possible to associate the peaks with structural domains.

3 Results and Discussion

3.1 Analyses of SMFS Experiments of bR

One aim of the analyses of experimental data from SMFS measurements is the detection of possible unfolding pathways. For the demonstration of the performance of our method we used a SMFS data set of the membrane protein bR. The data set included 53 F-D curves. Fig. 2 shows only a subset of all F-D curves (n=30). In this example we will show the correlation between a sub-cluster and possible unfolding pathways. The evaluation was based on the work described in [13,14].

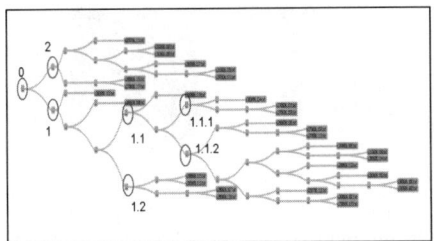 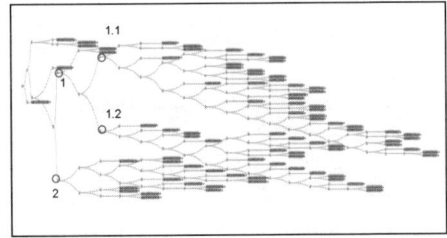

Fig. 2. Left: The illustrated hierarchical cluster tree includes 30 of the 53 analyzed F-D curves of bR. Highlighted are different sub-clusters. The number of each sub-cluster is connected with the level inside of the whole cluster, e.g., sub-clusters 1.1 and 1.2 are derived from sub-cluster 1. **Right:** The illustrated hierarchical cluster tree of Rho includes 84 F-D curves. Highlighted are different sub-clusters. The sub-cluster 2 is connected with the erroneous curves, and the sub-cluster 1 can be separated in two further sub-clusters (1.1 and 1.2). The sub-cluster 1.1 contains the F-D curves of the S-S bridge intact state and sub-cluster 1.2 contains the curves from the S-S bridge broken state. 7 F-D curves are not located in the sub-clusters 1 or 2.

Our alignment algorithm generates a hierarchical clustering tree where each cluster corresponds to a class of unfolding events. We compare it against a manual annotation of the curves based on manual peak detection with the WLC described above. A hierarchical clustering based on curve alignment reflects the similarity of clustered F-D curves.

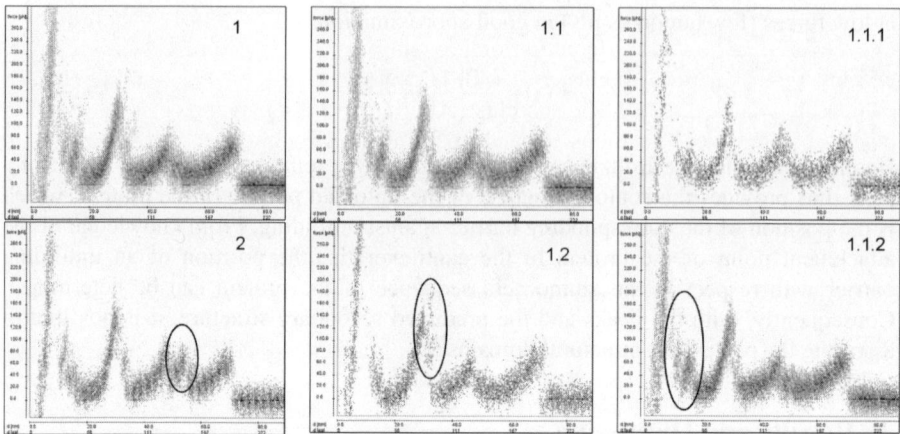

Fig. 3. Alignment of F-D curves of all highlighted sub-clusters in fig. 2. The main unfolding steps are present in each sub-cluster. On the first separation level the sub-clusters, 1 and 2, differ in the region highlighted blue. Inside sub-cluster 1 we found different sub-clusters, 1.1 and 1.2. The difference in the alignment is visible in the blue circle. Sub-clusters 1.1.1 and 1.1.2 differ in the beginning of the alignment.

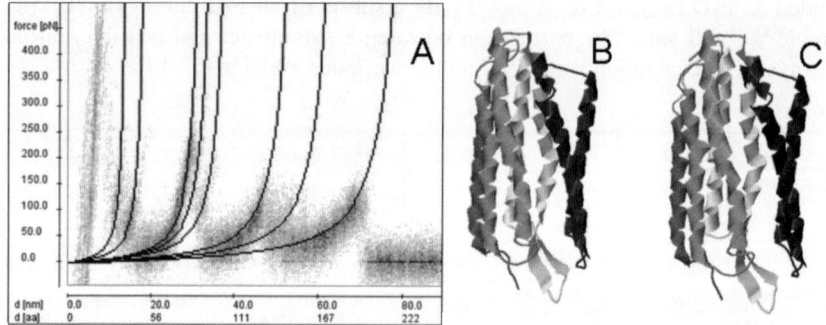

Fig. 4. A- Alignment of all 30 F-D curves of the illustrated tree. The blue lines are WLC fits and correspond to unfolding events of the membrane protein bR. Mapping the unfolding events of sub-cluster 1.1 (**B**) and 1.2 (**C**) on the structure of bR (1brr). The pairwise unfolding of the transmembrane helices are colored in yellow, blue and green. In figure **B** is illustrated the unfolding pathway of sub-cluster 1.1. In figure **C** is shown the possible unfolding pathway of sub-cluster 1.2. Red and grey are additional events derived from the sub-cluster 1.2.

Clusters in the tree correspond to classes of unfolding events at different levels of granularity. We find that all the spectra in the dataset share three main peaks (88 ± 5, 148 ± 5 and 219 ± 5 amino acids) indicating the unfolding of two transmembrane helices and their connecting loop in a single step (fig. 3). Furthermore, we can identify sub-clusters in the hierarchical tree relating to different unfolding events showing the presence of side peaks besides the main unfolding pathway, indicating that helices not always unfold pairwise but exhibit more unfolding intermediates.

This result agrees with previous studies that analyse individual unfolding pathways of bR. Fig. 4 shows the alignment of all F-D curves in the tree "0" with the WLC fits. A comparison of the sub-clusters 1, 2, 1.1, 1.2, 1.1.1 and 1.1.2 clarify the correlation of each sub-cluster with different unfolding pathways. We could detect an additional unfolding step in the sub-cluster 2 with a contour length of 175 amino acids on comparing to sub-cluster 1.The sub-cluster 1.1 and 1.2 differ in the occurrence of the peaks at amino acid positions 95 and 105. In the sub-cluster 1.1.1 the peak with contour length 38 amino acids is missing compared to the sub-cluster 1.1.2. Fig. 4 (B, C) shows the mapping of possible unfolding pathways on the know structure of bR (PDB_ID: 1brr) in the sub-cluster 1.1 and 1.2.

3.2 Analyses of SMFS Experiments of Rho

Using SMFS it was possible to detect molecular interactions within native bovine Rho and discover structural segments of well-defined mechanical stability [38]. Two main classes of force curves were observed. One class corresponded to the unfolding of Rho with the highly conserved Cys110–Cys187 disulfide (S-S) bond intact and the other class corresponded to the unfolding of the entire Rho polypeptide chain when the Cys110-Cys187 bridge is broken [38]. The changes in unfolding barriers detected with and without the S-S bridge highlight the structural importance of this disulfide bond and may form the basis of dysfunctions associated with its absence. To evaluate our routine we analyzed a set of 84 measured F-D curves from Rho. The annotation and interpretation are based on the work of Sapra et al. [38].

Form the manual annotation it was known that 30 F-D curves corresponded to the S-S intact state, 30 were correlated to the S-S broken state and 24 F-D curves were erroneous curves. For a discussion of a filter characteristic of our method we calculated the recalls and precisions for each separated sub-cluster (Fig. 2). The recall are calculated by the equations:

$$recall = \frac{N_{cluster\_i}^{correct}}{N_{cluster\_i}^{incorrect} + N^{total}} \qquad precision = \frac{N_{cluster\_i}^{correct}}{N_{cluster\_i}^{incorrect} + N_{cluster\_i}^{correct}}$$

Herein, $N_{cluster\_i}^{correct}$ denotes the number of the correct classified F-D curves in the sub-cluster i, $N_{cluster\_i}^{incorrect}$ is the number of the incorrect classified F-D curves in the sub-cluster i and N^{total} is the total number of F-D curves of the different states.

Figure 5 illustrates the location of the S-S bridge in the topology of Rho. Both sates of the S-S bridge lead to different unfolding pathways of the protein. With our method it was possible to distinguish both pathways. Additionally we sorted out erroneous F-D curves. The values for recall and precision of the sub-cluster from tree are listed in Table 1. The highest precision (96.6%) was calculated for the S-S bridge intact state. The best recall value (76.0%) correlated with the classification of the erroneous F-D curves.

Table 1. Confusion matrix of the cluster calculation for the tree with F-D curves from the membrane protein Rho

	S-S bridge intact	S-S bridge broken	erroneous curves	recall
sub-cluster 1.1	29	5	5	72,5%
sub-cluster 1.2	0	22	0	73.3%
sub-cluster 2	1	0	19	76.0%
Total curve number	30	30	24	
precision	96.6%	81.5%	79.2%	

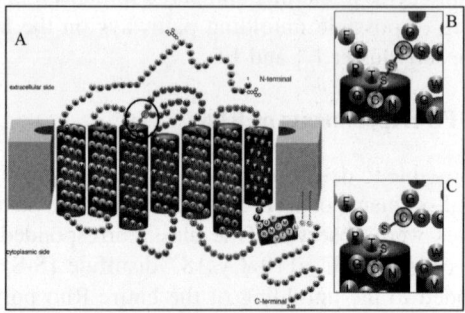

Fig. 5. A) Topology of rhodopsin. Encircled is the highly conserved S-S bridge among family A GPCRs. The intact S-S bridge state is illustrated in **B**. The broken S-S bridge is shown in **C**.

4 Conclusions

The sensitivity of SMFS with AFM provides a unique opportunity to study membrane protein (un)folding at a high resolution (sub-Å) in the native lipid bilayer environment [39, 40]. During continuous stretching of the molecule the applied forces are measured by the deflection of the cantilever and plotted against extension yielding a characteristic F-D curve that is formed by subsequent events of molecular interactions. These interactions can represent unbinding or unfolding events, or ligand binding of the protein. In the near future, force spectroscopy assays will permit screening of physiologically relevant parameters (e.g. pH, electrolyte concentration, temperature, mutations etc.), which alter the inter- and intramolecular interactions that determine, for example, the folding, structure, stability, function and assembly of biological macromolecules. In order to obtain statistically relevant results, several hundreds to thousands single-molecule experiments have to be performed, each resulting in a unique F-D curve. To draw reliable conclusions on molecular interactions one must be able to analyse a statistically relevant number of F-D curves by identical objective procedures. Thus, there is an increasing demand for data analysis techniques that offer fully automated processing of many data sets applying identical scientific criteria. Manual analysis of force spectroscopy data is a time consuming process and requires expert knowledge. But suitable algorithms to process the data still have to be developed. The solution we propose greatly simplifies and accelerates the data processing step in specific force measurements compared to a manual selection and annotation. With our automated approach the recognition of unfolding events is no more considered subjective as for manual recognition but has the advantage of being reproducible and quantitative. The hierarchical tree, as output from the described procedure, can be helpful in the interpretation of the experimental data, in discriminating different possible unfolding pathways and calculating their probability of occurrence. Our dynamic programming approach was used successfully for classification problems. The novel methods can bridge the gap between fast device development and requirements for efficient analysis and interpretation of unfolding experiments. We propose a novel algorithm for high-throughput classification of F-D curves. On the example of the membrane protein bR we separated the investigated curves in sub-clusters, which are correlated with different unfolding pathways of the protein. On Rho we demonstrated the filter quality of our method. The two functional states of rhodopsin (broken and intact S-S bridge) were separated in different sub-clusters. The recall and precision values were calculated and yield to recall (S-S intact) of 72.5% and a precision of 96.6%. For the S-S bridge broken we determine a recall of 73.3% and a precision of 81.5%.

Acknowledgements

Funding through the EFRE project CODI and FoldUnfold is kindly acknowledged.

References

1. Bowie, J.U.: Solving the membrane protein folding problem. Nature 438(7068), 581–589 (2003)
2. Sanders, C.R., Myers, J.K.: Disease-related misassembly of membrane proteins. Annu. Rev. Biophys. Biomol. Struct. 33, 25–51 (2004)
3. Mirzadegan, T., Benko, G., Filipek, S., Palczewski, K.: Sequence analyses of G-protein coupled receptors: similarities to rhodopsin. Biochemistry 42(10), 2759–2767 (2003)
4. Onuchic, J.N., Wolynes, P.G.: Theory of protein folding. Current Opinion in Structural Biology 14, 70–75 (2004)
5. Müller, D.J., Engel, A.: Voltage and pH-induced channel closure of porin OmpF visualized by atomic force microscopy. J. Mol. Biol. 285, 1347–1351 (1999)
6. Müller, D.J., Sass, H.J., Muller, S.A., Buldt, G., Engel, A.: Surface structures of native bacteriorhodopsin depend on the molecular packing arrangement in the membrane. J. Mol. Biol. 285, 1903–1909 (1999)
7. Seelert, H., Dencher, N.A., Muller, D.J.: Fourteen protomers compose the oligomer III of the proton-rotor in spinach chloroplast ATP synthase. J. Mol. Biol. 333, 337–344 (2003)
8. Park, P.S., Sapra, K.T., Kolinski, M., Filipek, S., Palczewski, K., Muller, D.J.: Stabilizing effect of Zn2+ in native bovine rhodopsin. J. Biol. Chem. (2007)
9. Rief, M., Gautel, M., Schemmel, A., Gaub, H.E.: The mechanical stability of immunoglobulin and fibronectin III domains in the muscle protein titin measured by atomic force microscopy. Biophys. J. 75, 3008–3014 (1998)
10. Janshoff, A., Neitzert, M., Oberdorfer, Y., Fuchs, H.: Force spectroscopy of molecular systems-single molecule spectroscopy of polymers and biomolecules. Angew Chem. Int. Ed. Engl. 39(18), 3212–3237 (2000)

11. Janovjak, H., Struckmeier, J., Hubain, M., Kedrov, A., Kessler, M., Muller, D.J.: Probing the energy landscape of the membrane protein bR. Structure 12(5), 871–879 (2004)

12. Oesterhelt, F., Oesterhelt, D., Pfeiffer, M., Engel, A., Gaub, H., Muller, D.J.: Unfolding pathways of individual bacteriorhodopsins. Science 288(5463), 143–146 (2000)

13. Sapra, K.T., Besir, H., Oesterhelt, D., Muller, D.J.: Characterizing molecular interactions in different bacteriorhodopsin assemblies by single-molecule force spectroscopy. J. Mol. Biol. 355(4), 640–650 (2006)

14. Muller, D.J., Kessler, M., Oesterhelt, F., Moller, C., Oesterhelt, D., Gaub, H.: Stability of bacteriorhodopsin alpha-helices and loops analyzed by single-molecule force spectroscopy. Biophys. J. 83(6), 3578–3588 (2002)

15. Rief, M., Gautel, M., Oesterhelt, F., Fernandez, J.M., Gaub, H.E.: Reversible unfolding of individual titin immunoglobulin domains by afm. Science 276(5315), 1109–1112 (1997)

16. Haupts, U., Tittor, J., Oesterhelt, D.: Closing in on bacteriorhodopsin: progress in understanding the molecule. Annu. Rev. Biophys. Biomol. Struct. 28, 367–399 (1999)

17. Oesterhelt, D.: The structure and mechanism of the family of retinal proteins from halophilic archaea. Curr. Opin. Struct. Biol. 8, 489–500 (1998)

18. Belrhali, H., Nollert, P., Royant, A., Menzel, C., Rosenbusch, J.P., Landau, E.M., Pebay-Peyroula, E.: Protein, lipid and water organization in bacteriorhodopsin crystals: a molecular view of the purple membrane at 1.9. A resolution. Struct. Fold. Des. 7, 909–917 (1999)

19. Essen, L-O., Siegert, R., Lehmann, W.D., Oesterhelt, D.: Lipid patches in membrane protein oligomers: Crystal structure of the bacterorhodopsin-lipid complex. Proc. Natl. Acad. Sci. 95, 11673–11678 (1998)

20. Grigorieff, N., Ceska, T.A., Downing, K.H., Baldwin, J.M., Henderson, R.: Electron-crystallographic refinement of the structure of bacteriorhodopsin. J. Mol. Biol. 259, 393–421 (1996)

21. Luecke, H., Schobert, B., Richter, H.T., Cartailler, J.P., Lanyi, J.K.: Structure of bacteriorhodopsin at 1.55 Å resolution. J. Mol. Biol. 291, 899–911 (1999)

22. Mitsuoka, K., Hirai, T.T., Murata, K., Miyazawa, A., Kidera, A., Kimura, Y., Fujiyoshi, Y.: The structure of bacteriorhodopsin at 3.0 Å resolution based on electron crystallography: implication of the charge distribution. J. Mol. Biol. 286, 861–882 (1999)

23. Baldwin, J.M.: The probable arrangement of the helices in G protein-coupled receptors. EMBO J. 12, 1693–1703 (1993)

24. Helmreich, E.J.M., Hofmann, K-P.: Structure and function of proteins in G-protein coupled signal transfer. Biochem. Biophys. Acta. 1286, 285–322 (1996)

25. Kolbe, M., Besir, H., Essen, L.O., Oesterhelt, D.: Structure of the light-driven chloride pump halorhodopsin at 1.8 Å resolution. Science 288, 1390–1396 (2000)

26. Palczewski, K., Kumasaka, T., Hori, T., Behnke, C.A., Motoshima, H., Fox, B.A., Le Trong, I., Teller, D.C., Okada, T., Stenkamp, R.E., Yamamoto, M., Miyano, M.: Crystal structure of rhodopsin: a G protein-coupled receptor. Science 289, 739–745 (2000)

27. Royant, A., Nollert, P., Edman, K., Neutze, R., Landau, E.M., Pebay-Peyroula, E., Navarro, J.: X-ray structure of sensory rhodopsin II at 2.1-Å resolution. Proc. Natl. Acad. Sci. 98, 10131–10136 (2001)

28. Lanyi, J.K.: Progress toward an explicit mechanistic model for the light-driven pump, bacteriorhodopsin. FEBS Lett. 464, 103–107 (1999)

29. Filipek, S., Teller, D.C., Palczewski, K., Stenkamp, R.: The crystallographic model of rhodopsin and its use in studies of other G protein-coupled receptors. Annu. Rev. Biophys. Biomol. Struct. 32, 375–397 (2003)

30. Filipek, S., Stenkamp, R.E., Teller, D.C., Palczewski, K.: G protein-coupled receptor rhodopsin: a prospectus. Annu. Rev. Physiol. 65, 851–879 (2003)
31. Ridge, K.D., Abdulaev, N.G., Sousa, M., Palczewski, K.: Phototransduction: crystal clear. Trends Biochem. Sci. 28, 479–487 (2003)
32. Dryja, T.P., Li, T.: Molecular genetics of retinitis pigmentosa. Hum. Mol. Genet. 4, 1739–1743 (1995)
33. Liu, X., Garriga, P., Khorana, H.G.: Structure and function in rhodopsin: correct folding and misfolding in two point mutants in the intradiscal domain of rhodopsin identified in retinitis pigmentosa. Proc. Natl Acad. Sci. 93, 4554–4559 (1996)
34. Rader, A.J., Anderson, G., Isin, B., Khorana, H.G., Bahar, I., Klein-Seetharaman, J.: Identification of core amino acids stabilizing rhodopsin. Proc. Natl Acad. Sci. 101, 7246–7251 (2004)
35. Hwa, J., Klein-Seetharaman, J., Khorana, H.G.: Structure and function in rhodopsin: mass spectrometric identification of the abnormal intradiscal disulfide bond in misfolded retinitis pigmentosa mutants. Proc. Natl Acad. Sci. 98, 4872–4876 (2001)
36. Marsico, A., Labudde, D., Sapra, T., Muller, D.J., Schroeder, M.: A novel pattern recognition algorithm to classify membrane protein unfolding pathways with high-throughput single-molecule force spectroscopy. Bioinformatics 23, 231–236 (2007)
37. Eddy, S.R.: What is dynamic programming? Nature Biotechnology 22, 909–910 (2004)
38. Tanuj Sapra, K., Park, P.S., Filipek, S., Engel, A., Muller, D.J., Palczewski, K.: Detecting molecular interactions that stabilize native bovine rhodopsin. J. Mol. Biol. 358, 255–269 (2006)
39. Janovjak, H., Kedrov, A., Cisneros, D.A., Sapra, K.T., Müller, D.J.: Imaging and detecting molecular interactions of single transmembrane proteins. Neurobiol of Aging 27, 546–561 (2006)
40. Muller, D.J., Sapra, K.T., Scheuring, S., Kedrov, A., Frederix, P.L., Fotiadis, D., Engel, A.: Single-molecule studies of membrane proteins. Curr. Opin. Struct. Biol. 16, 489–495 (2006)
41. Kasas, S., Rieder, B.M., Catsicas, S., Cappella, B., Dietler, G.: Fuzzy logic algorithm to extract specific interaction forces from atomic force microscopy data.
42. Review of Scientific Instruments 71(5), 2082–2086 (2000)
43. Kuhn, M., Janovjak, H., Hubain, M., Muller, D.J.: Automated alignment and pattern recognition of single-molecule force spectroscopy data. J. Microsc. 218, 125–132 (2005)

Massive Protein Structural Property Explorations Using New Indexing Mechanism

Yu-Feng Huang[1], Chia-Chen Chang[2], and Chien-Kang Huang[2,*]

[1] Department of Computer Science and Information Engineering, National Taiwan University,
Taipei, Taiwan 106
[2] Department of Engineering Science and Ocean Engineering, National Taiwan University,
Taipei, Taiwan 106
* Tel.: +886 2 3366 5736, Fax: +886 2 2932 9885
yfhuang@csie.ntu.edu.tw, {r95525051, ckhuang}@ntu.edu.tw

Abstract. In order to comprehend residue environment, we use residue environmental sphere which is a sphere with 10 Å of radius, to describe environment information surrounding a residue. For the purpose of detecting residue-residue contacts more quickly and efficiently, we decompose a protein structure into lots of spheres, and it is a great challenge to store protein structure and sphere information in database. Therefore, we build a database for protein structure, ligand/substrate, and DNA/RNA information for quick search and mining to observe residue environment of protein structure. In each residue environmental sphere, we can easily identify neighbor residues and their properties, including secondary structure, physicochemical property, and b-factor, could be considered. In this paper, we focus on disulfide bond which stabilizes protein folding. Furthermore, we detect all possible residue contacts of cysteine pairs in three-dimensional space and disulfide bonds between two cysteines annotated in Protein Data Bank to analyze how disulfide bond affects protein structures. We use a sphere to represent a protein structure and build a database for protein structure and structure representation for further analysis.

Keywords: Residue environmental sphere (RES), protein structural property mining, residue contact, disulfide bond.

1 Introduction

As of July 3, 2007, there are 44,476 determined protein structures examined by X-ray or nuclear magnetic resonance (NMR) in Protein Data Bank (PDB) [4]. They include proteins, protein complexes, nucleic acids and protein nucleic acid complexes. Applying mining technique on protein structures is an interesting issue to discover residue environmental information inside protein structure [12, 13, 14]. Residue environment has been studied for many years and applied on protein threading and protein binding site characterization [2, 15]. In the protein structure, a residue is the essential element for conformation, and residue-residue contacts will affect the overall

---

* Corresponding author.

J.-L. Hainaut et al. (Eds.): ER Workshops 2007, LNCS 4802, pp. 14–23, 2007.
© Springer-Verlag Berlin Heidelberg 2007

framework of a protein structure. Therefore, residue environment can help us to comprehend protein structure conformation. In addition, binding site environment analysis is also a good starting point to understand how residue contacts affect protein binding and protein function [6, 18].

In previous researches, residue-residue contact is an important issue to be investigated for protein structure fold, protein structure conservation, and protein function [3, 7, 9, 17, 21]. With the fast growth of protein structure, it provides more materials on the study of discovering local residue environment with/without chemical bond information. Furthermore, protein conformation is highly correlated to residue contact with chemical bonds such as covalent bonds, ionic bonds, hydrogen bonds, Van der Waals attractions, or disulfide bonds. For quick searching of residue environment, we use residue environmental sphere to describe environment information surrounding a residue. On the purpose of protein structural property exploration, we have to trace residue neighborhood on whole protein structure collection. Furthermore, to handle huge protein structure collection is also a great challenge to store entire structure and sphere information in database.

2 Review of Protein Structural Property Exploration

In sequence based prediction, the position-specific scoring matrix (PSSM) is used to improve their prediction accuracy for protein sequence analysis. The PSSM gives the log-odds score for finding a particular matching amino acid against to a target sequence. Therefore, the prediction tools treat PSSM as sequence property for each amino acid. In protein structure prediction, amino acid property, secondary structure information, b-factor, accessible surface area (ASA), or relative solvent accessibility (RSA) are structural properties. In 1992, Singh and Thornton [19] discovered the atlas of protein side-chain interaction to understand sidechain-sidechain interactions. In this research, they revealed interactions for 20 * 20 amino acids, and counted the frequency for each amino acid pairs. In addition, Glaser et. al. [10] also studied structural property of residues at protein-protein interfaces. In order to realize the inside of protein structure conformation, protein structural property exploration is very important such as amino acid interactions or residue-residue contact.

3 Proposed Indexing Mechanism for Massive Structural Property Exploration

3.1 Residue Environmental Sphere and Indexing Mechanism

In order to describe residue environment of protein local structure, our original idea comes from the neighbor string (NS_r) developed by Jonassen et al. for mining structure motif [11]. This string encodes all residues in the structure that are with a distance of d Å from r (d=10, as default), including r itself from N-terminal to C-terminal. The protein structure is folded by the interactions between amino acids to connect with each other; therefore, amino acid plays an important role on protein folding. Therefore, each 10 Å sphere representation, residue environmental sphere (RES), can describe environmental information inside a protein. This distance cut-off of 10 Å [8] is Van der Waals contribution and it dominates for less then 3 Å but is insignificant at 10 Å. And we know that residue-residue interaction will affect protein structure conformation so that the residue environmental sphere should be a good candidate to extract residue environment to understand residue-residue contact for each protein structure. Fig. 1 is an example to illustrate residue environmental sphere as indexing unit.   Now, we use RES to identify each local structure surrounding a residue, and it is also a index unit to index protein structure residue by residue for quick database search, and this sphere is the essential/abstract form to record environmental information such as nearest neighbor residues, secondary structure information, biochemical property, and so on. With the great help of database, we store all structure information and index entire residue environmental sphere for analyzing residue-residue contacts.

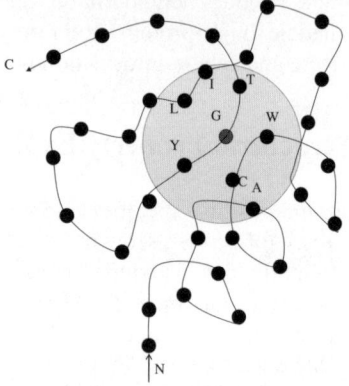

Fig. 1. Residue environmental sphere. The area in gray is the area with 10 Å of radius surrounding the central residue G.

3.2 Materials

In this work, we analyze entire protein structures in Protein Data Bank, and all structure information will be considered, such as coordinate information, connectivity annotation, heterogen information, physicochemical properties, and secondary structure information. In coordinate information, both ATOM and HETATM will be considered for protein structures, DNA/RNA structures, and hetero-atom structures respectively. The heterogen information is extracted from pdb file with HET and HETATM tags, which describe non-standard residues, such as prosthetic groups, inhibitors, solvent molecules, and ions for which coordinates are supplied. In our database implementation, DNA/RNA structures could be viewed as special chemical components. In connectivity annotation, SSBOND is the most important information to observe disulfide bonds both intra-molecularly and inter-molecularly.   The fundamental physicochemical properties will be also concerned include hydrophobic, hydrophilic, charge (negative and positive), polar, etc. Currently, we select whole protein structures of 43427 as our data collection from Protein Data Bank in early 2007. In this collection, there are 40303 protein structures, 1152 protein/DNA

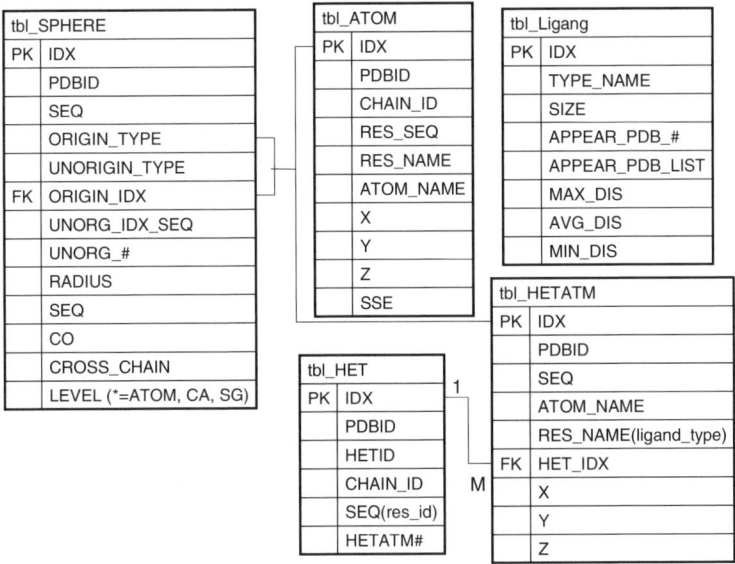

Fig. 2. Database table schema for structural property exploration

complexes, 465 protein/RNA complexes, 28 DNA/RNA hybrid structures, 43 protein/DNA/RNA complexes, 892 DNA structures, and 544 RNA structures.

3.3 Database Design

For the purpose of quick search on residue environment, we use residue environmental sphere as indexing unit to speed up table lookup and mine residue-residue contacts. Cooperating with atom coordinate table, and ligand/substrate table, it can be easy to mine residue environment surrounding a residue. In Fig 2, we illustrate database table schema for atom, hetatom, ligand, and residue environmental sphere. In database design, the great challenge is to put huge scale of protein structure into tables includes residue environmental sphere, coordinate information, substrate/ligand/DNA/RNA information, and bone connectivity. As we know, each PDB ID has 4-character code that uniquely defines an entry in the Protein Data Bank. The first character must be a digit from 1 to 9, and the remaining three characters can be letters or numbers. Therefore, we use middle two characters as table identifier; for example, if the PDB IDs are 4hhb, 2hhb, and 3hhb, their atom coordinates will be stored together in the database with table identifier "hh". At last, we have 4 kinds of database tables to store protein structure information, and they are atom coordinate table, ligand/substrate table, and residue environmental sphere table. Unlike data cube structure, we don't use grid structure to describe a protein structure, and a residue environmental sphere is used to describe neighborhood information surrounding a residue.

4   Statistical Analysis of Structural Properties on Protein Data Bank

4.1   Residue-Residue Contacts

In protein structure, residue-residue interactions make a protein to fold as a stable conformation. If two residues are considered to be in contact with each other provided the distance between their alpha carbon atom (C_α) below a certain cutoff. Therefore, we collect residue-residue contacts from whole protein structures and extract all residue pairs and its neighbor residues to understand how interactions help protein folding. Moreover, each residue can have multiple properties on it such as biochemical property (hydrophobic, hydrophilic, charge, etc), physicochemical property, and secondary structure element type (α-helix, β-sheet, or coil). Inside the residue environmental sphere, we first use C_α in backbone to represent geometry information, but in order to describe detail residue contact with chemical bond, therefore, atom level residue-residue contacts will be also considered.

4.2   Chemical Component Contacts

In this sub-section, we try to observe residue environment surrounding a chemical component to understand the interaction environment between protein and ligand or substrate. We also use residue environmental sphere to observe chemical component close to a residue contacts. According to PDB format, HET records are used to describe chemical components or non-standard residues, such as prosthetic groups, inhibitors, solvent molecules, and ions for which coordinates are supplied. Groups are considered HET if they are not part of a biological polymer described in SEQRES and considered to be a molecule bound to the polymer, or they are a chemical species that constitutes part of a biological polymer that is not one of the following: (a) not one of the standard amino acids, and (b) not one of the nucleic acids (C, G, A, T, U, and I), and (c) not an unknown amino acid or nucleic acid where UNK is used to indicate the unknown residue name. Because we focus on residue-residue contacts to realize how they interacts with chemical component, and chemical component information is used to understand how interaction begins.

4.3   Property Analysis on Disulfide Bond

4.3.1   Disulfide Bond

In general, disulfide bonds are suggested to stabilize protein folding which has been reviewed [1, 5, 16, 20]. In biochemistry, disulfide bond or disulfide bridge is connected between C_β-S_γ-S_γ-C_β (S_γ is a SG atom in PDB, and C_β is a beta carbon) which can occur intra-molecularly (i.e within a single polypeptide chain) and inter-molecularly (i.e. between two polypeptide chains). Disulfide bond in intra-molecular stabilize the tertiary structures of proteins while those that occur inter-molecularly are involved in stabilizing quaternary structure. In this paper, we focus on SSBOND section which identifies each disulfide bond in protein and polypeptide structures by identifying the two residues involved in the bond. Furthermore, we also use residue environmental sphere to detect residue-residue contacts of cysteine pairs intra-molecularly.

4.3.2 SSBOND

In PDB, the connectivity annotation section is used to allow the depositors to specify the existence and location of disulfide bonds and other linkages. The bond between two S_γ atoms is disulfide bond annotated as SSBOND by Protein Data Bank. We separate this collection into two groups, intra-molecular and inter-molecular; therefore, we have 48152 pairs in intra-molecular group and 2115 pairs in inter-molecular group. While applying secondary structure information, we observe that SSBOND tends to grasp at β-sheets and coils.

4.3.3 Residue-Residue Contacts of Cysteine Pairs

Unlike SSBOND discovery, not all protein structures contain disulfide bonds; therefore, we observe all cysteine pairs in whole PDB to distinguish the difference between SSBOND and residue-residue contacts of cysteine pair. In this work, we only collect all cysteine pairs in both C_α and atom level (S_γ) intra-molecularly to observe their environment. The reason to use atom level discovery is that we will miss some cysteine pairs if we only count C_α atom level. Therefore, we have 114,777 residue-residue contacts intra-molecularly for further analysis.

4.4 Results

Although we detect all possible residue-residue contacts among whole protein structures in PDB; according to previous studies, we select SSBOND annotation in PDB and residue-residue contacts of cysteine pair as example to explore protein structural property because of well-studied topic on disulfide bond.

4.4.1 Residue-Residue Contacts and Chemical Component Contacts

We detect all pairs of amino acid combination to discuss relationship among residue interaction and secondary structure property. In our experimental result, the top-10 residue-residue contacts contain Glycine, and the pairs are Gly-Gly, Gly-Ala, Gly-Ser, Gly-Pro, Gly-Asp, Gly-Glu, Gly-Lys, Gly-Leu, Gly-Thr, and Gly-Val ranked by their occurrence frequency. According to amino acid property, the amino acid glycine tends to contact with small or tiny amino acids such as Ala, Ser, Asp, Thr, and Pro. Focusing on cysteine paris, we observe that Cys-Cys occurs in β-sheet and loop frequently. Moreover, the chemical component is defined as hetID in PDB; thus we totally extract about 6827 different hetIDs from PDB. The top-5 hetIDs are SO4, _CA, _ZN, _MG, and MSE.

Table 1. Statistical result of SSBOND and Cysteine pair

		Intra-molecular	Inter-molecular	Total
SSBOND	(A)	48152	2115	50267
	(B)	3333	95	3429
Cysteine Pairs	(A)	114777	-	114777
	(B)	12847	-	12847

(A) Number of pairs; (B) Chemical component contacts.

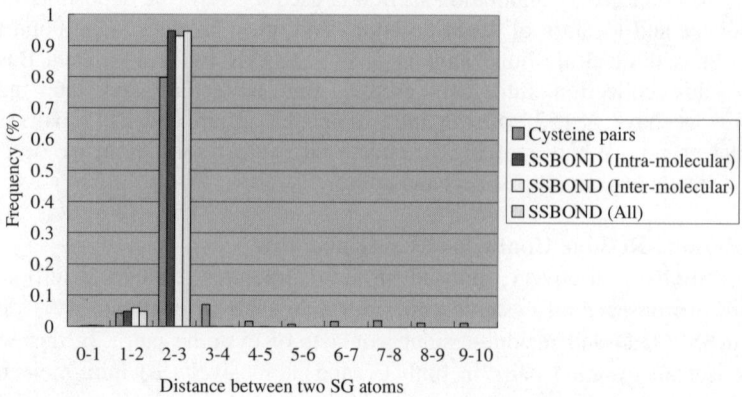

Fig. 3. Distribution between distance and its frequent. In X-axis, for example, the annotation of 0-1 represents the measured distance is larger or equal to 0 Å and smaller than 1 Å. Most frequent distance between two SG falls in 2-3 Å.

4.4.2 Disulfide Bond

In Table 1, number of pairs and chemical component contacts are listed in both intra-molecular and inter-molecular for SSBOND and cysteine pair. We also measure min, max and average distance between two S_γ atoms of SSBOND and cysteine pairs. In Fig. 4, we also report distance distribution for SSBOND and cysteine pair. We collect 50627 SSBOND entries to analyze the connection between two amino acids of cysteine from PDB. In our discovered collection, we find the following problematic points: (1) extreme long bond length between two S_γ atoms exists intra-molecularly or inter-molecularly (e.g. > 10 Å); (2) a residue in SSBOND would be a missing residue; (3) a residue in SSBOND would be heterogen, and most of them are modified residues. According to Protein Data Bank content guide, if S_γ of cysteine is disordered then there are possible alternate linkages. PDB's practice is to put together all possible SSBOND records. This is problematic because the alternate location identifier is not specified in the SSBOND record.

5 Discussion

5.1 Difference of SSBOND and Cysteine Pairs

Based on disulfide bond analysis between SSBOND and residue-residue contact of cysteine pairs, the most frequent bond length between two cysteines ranges from 2 to 3 Å, and in general the disulfide bond length is around 2.8 Å. Due to disulfide bond conformed by two S_γ atoms, atom level analysis within a sphere is necessary rather than C_α atom. The problematic points we detected are minimum distance and maximum distance between two S_γ atoms in SSBOND. The condition of zero distance between two S_γ atoms comes from the same coordinates of Cysteines annotated in SSBOND. Furthermore, we also find that the distance between two S_γ

atoms larger than 10 Å (e.g. 149.663 Å in intra-molecular of 1RHG:C 64 and 74, and 77.881 Å in inter-molecular of 1uMR:B 135 and 1UMR:D 203), and it's might be incorrect annotation of bond connectivity. In addition, because the size of chemical component will affect the result, we only select large-size chemical component for structure similarity evaluation. In whole residue environmental spheres containing SSBOND, we have 16 residue environmental spheres containing BEN and 66 residue environmental spheres containing FAD. We find that residue environmental spheres of SSBOND surrounded by two chemical components of FAD and BEN respectively have highly conserved region of spheres. In Fig. 4, atoms in yellow are S_γ atoms and the chemical component in CPK mode is FAD. Unlike previous researches, we try to index whole PDB dataset to analysis all residue-residue contacts and bond connectivity inside a protein structure while previous researches focus on only analyze special pair preference and residue frequencies [10].

5.2   File Parsing and Efficiency of Database Query

To extract atom coordinate information, we have to parse pdb file to obtain structure information, but file parsing is the worst way for data mining because of information reusable. Besides, the use of sphere can gain the advantage of neighborhood information. Thus, for the purpose of structure mining on PDB, we try to simplify our mining procedure, and then we parse PDB raw files, index whole protein structures with residue environmental sphere and deposit all information into database. To avoid database connection I/O, we use database dump technique to prepare dump file for database restore instead of row-by-row insertion. Comparing with file parsing and database query, without consideration of preprocessing, we spend about 1 hour to select sphere information from database for detect residue-residue contact of cysteine pairs via database query while spending 17 hours via file parsing without database utilization. Therefore, we can gain more benefit from indexing mechanism and database query.

Fig. 4. Atoms in yellow are S_γ atoms that build the disulfide bond annotated as SSBOND in PDB ID 1BHY and the chemical component is FAD (FLAVIN-ADENINE DINUCLEOTIDE) in CPK mode

6 Conclusions

In summary, sphere-based neighborhood searching is an appropriate local structure representation for structure mining on PDB. Consequently, we obtain huge scale collection of residue environmental sphere for describing protein local environment based on the believed assumption of protein function interacted with local structure. In order to searching and mining among this collection, indexing mechanism is very important; therefore, the residue environmental sphere is local structure representation and indexing unit for the reason of information reuse. Focusing on disulfide bond, the observation can be put on both SSBOND and cysteine pairs. Although there is some problematic information in SSBOND, they still provide useful information to compare with SSBOND and cysteine pair. In the future, further analysis on different residue-residue contacts and discussion on structural property should be scrutinized.

Reference

1. Aitken, A., Learmonth, M.: Quantification and location of disulfide bonds in proteins. Methods in molecular biology 64, 317–328 (1997)
2. Alton, D., Adab, P., Roberts, L., Barrett, T.: Relationship between walking levels and perceptions of the local neighbourhood environment. Archives of disease in childhood 92, 29–33 (2007)
3. Bagley, S.C., Altman, R.B.: Characterizing the microenvironment surrounding protein sites. Protein Sci. 4, 622–635 (1995)
4. Berman, H.M., Battistuz, T., Bhat, T.N., Bluhm, W.F., Bourne, P.E., Burkhardt, K., Feng, Z., Gilliland, G.L., Iype, L., Jain, S., Fagan, P., Marvin, J., Padilla, D., Ravichandran, V., Schneider, B., Thanki, N., Weissig, H., Westbrook, J.D., Zardecki, C.: The Protein Data Bank. Acta crystallographica 58, 899–907 (2002)
5. Betz, S.F.: Disulfide bonds and the stability of globular proteins. Protein Sci. 2, 1551–1558 (1993)
6. Chalk, A.J., Worth, C.L., Overington, J.P., Chan, A.W.: PDBLIG: classification of small molecular protein binding in the Protein Data Bank. Journal of medicinal chemistry 47, 3807–3816 (2004)
7. Cheng, J., Baldi, P.: Improved residue contact prediction using support vector machines and a large feature set. BMC bioinformatics 8, 113 (2007)
8. Crowley, M., Darden, T., Cheatham, T., Deerfield, D.: Adventures in Improving the Scaling and Accuracy of a Parallel Molecular Dynamics Program. The Journal of Supercomputing 11, 255–278 (1997)
9. Fan, S.C., Zhang, X.G.: Characterizing the microenvironment surrounding phosphorylated protein sites. Genomics, proteomics & bioinformatics / Beijing Genomics Institute 3, 213–217 (2005)
10. Glaser, F., Steinberg, D.M., Vakser, I.A., Ben-Tal, N.: Residue frequencies and pairing preferences at protein-protein interfaces. Proteins 43, 89–102 (2001)
11. Jonassen, I., Eidhammer, I., Conklin, D., Taylor, W.R.: Structure motif discovery and mining the PDB. Bioinformatics 18, 362–367 (2002)
12. Lutteke, T., Frank, M., von der Lieth, C.W.: Data mining the protein data bank: automatic detection and assignment of carbohydrate structures. Carbohydrate research 339, 1015–1020 (2004)

13. Oldfield, T.J.: Creating structure features by data mining the PDB to use as molecular-replacement models. Acta crystallographica 57, 1421–1427 (2001)
14. Oldfield, T.J.: Data mining the protein data bank: residue interactions. Proteins 49, 510–528 (2002)
15. Plochocka, D., Kosinski, J., Rabczenko, A.: Formation of the local secondary structure of proteins: local sequence or environment. Acta biochimica Polonica 33, 109–118 (1986)
16. Raina, S., Missiakas, D.: Making and breaking disulfide bonds. Annual review of microbiology 51, 179–202 (1997)
17. Rodionov, M.A., Johnson, M.S.: Residue-residue contact substitution probabilities derived from aligned three-dimensional structures and the identification of common folds. Protein Sci. 3, 2366–2377 (1994)
18. Shin, J.M., Cho, D.H.: PDB-Ligand: a ligand database based on PDB for the automated and customized classification of ligand-binding structures. Nucleic acids research 33, D238–241 (2005)
19. Singh, J., Thornton, J.M.: Atlas of Protein Side-Chain Interactions, vol. I, II. IRL press, Oxford (1992)
20. Wedemeyer, W.J., Welker, E., Narayan, M., Scheraga, H.A.: Disulfide bonds and protein folding. Biochemistry 39, 4207–4216 (2000)
21. Zhang, C., Kim, S.H.: Environment-dependent residue contact energies for proteins. Proceedings of the National Academy of Sciences of the United States of America 97, 2550–2555 (2000)

Data Access and Management in ACGT: Tools to Solve Syntactic and Semantic Heterogeneities Between Clinical and Image Databases

Luis Martín[1], Erwin Bonsma[2], Alberto Anguita[1], Jeroen Vrijnsen[2],
Miguel García-Remesal[1], José Crespo[1], Manolis Tsiknakis[3], and Víctor Maojo[1]

[1] Biomedical Informatics Group, Artificial Intelligence Laboratory,
School of Computer Science, Universidad Politécnica de Madrid
Campus de Montegancedo S/N, 28660 Boadilla del Monte, Madrid, Spain
{lmartin, aanguita, mgarcia, vmaojo}@infomed.dia.fi.upm.es
[2] Phillips Research, Healthcare System Architecture
High Tech Campus 37, 5656 AE Eindhoven, The Netherlands
{erwin.bonsma, jeroen.vrijnsen}@philips.com
[3] Center of eHealth Technologies, Institute of Computer Science,
Foundation for Research and Technology - Hellas
GR-71110 Heraklion, Crete, Greece
tsiknaki@ics.forth.gr

Abstract. Recent changes in data management within post-genomic clinical trials have emphasized the need for novel methods and tools to solve semantic and syntactic heterogeneities among distributed sources of information. ACGT is an Integrated Project funded by the European Commission that aims at building a GRID-based platform comprised by a set of tools to support multi-centric post-genomic clinical trials on cancer. The main goal of ACGT is to provide seamless access to heterogeneous sources of information. For this purpose, two core tools were developed and included in the ACGT architecture: the ACGT Semantic Mediator (ACGT-SM), and the Data Access Wrappers (ACGT-DAWs). The ACGT-SM addresses semantics and schema integration, while the ACGT-DAWs cope with syntactic heterogeneities. Once the sources are bridged together, they can be seamlessly accessed using the RDQL query language. We tested our tools using a set of three relational and DICOM based image sources obtaining promising results.

Keywords: Database Integration, Semantic Mediation, Ontologies, Clinical Trial Databases, DICOM.

1 Introduction

During the last years, health research and care institutions have generated a large amount of relevant to cancer treatment and research. This wealth of information emphasizes the need of new methods and tools targeted to integrate all these data that can be used to design novel therapeutic procedures.

J.-L. Hainaut et al. (Eds.): ER Workshops 2007, LNCS 4802, pp. 24–33, 2007.

ACGT (Advancing Clinico-Genomic Trials on Cancer) is an Integrated Project funded by the European Commission that aims at providing a framework to support multicentric post-genomic clinical trials on cancer. The ultimate goal of ACGT is to provide a GRID-enabled platform comprised by a set of software tools to be used within proper workflows defined for specific clinical trials on cancer.

One of the main challenges in ACGT is to provide users with transparent access to heterogeneous, disparate sources of information. These sources include clinical, genetic, and image databases related to patients involved in clinical trials.

The different types of heterogeneity that arise when integrating heterogeneous sources can be classified into two categories: syntactic heterogeneities and semantic heterogeneities. Syntactic heterogeneities include aspects such as differences in hardware platforms, operating systems, database management systems, and query languages. Conversely, semantic heterogeneities appear when sources related to a common domain represent the same information in different manners.

In this work, we present two core components in the ACGT platform namely the ACGT Semantic Mediator (ACGT-SM) and the ACGT Data Access Wrappers (ACGT-DAWs). The role of the ACGT-SM is to solve semantic conflicts, while the ACGT-DAWs deal with syntactic heterogeneity. The ACGT relies on the ACGT Master Ontology on Cancer (ACGT-MOC) —developed by IFOMIS, a member of the ACGT consortium—that covers the domain of clinical trials on cancer.

This paper is organized as follows. Section 2 reviews recent related work. Section 3 describes the ACGT-DAWs and the ACGT-SM tools. Section 4 presents the results of the evaluation experiments. Finally, in section 5 we draw the conclusions and we point out future research lines.

2 Background

The main goal of the database integration task is to facilitate users to query a set of disparate and heterogeneous data sources in a transparent and uniform way.

During the last fifteen years, different approaches to perform database integration were proposed, including *Data Warehousing* [1], *Federated Database Systems* [2], *Mediator-based approaches* [3], and other hybrid approaches. However, all existing database integration methods can be classified into three different categories: 1) information linkage (IL), 2) data translation (DT), and 3) query translation (QT) approaches.

IL is the simplest method to perform database integration. This approach establishes links among different sources by using cross references. IL is used by many public online sources such as MEDLINE, GENBANK, OMIM, Prosite, etc. Furthermore, the World Wide Web is a huge IL-based repository.

By contrast, the goal of DT based methods is creating a centralized repository containing all the data from the different sources to be integrated. This centralized repository is normally equipped with a unified and normalized schema. Data coming from different sources have to be translated to match the normalized schema. Data warehousing can be regarded as the most representative DT-based approach.

Conversely, QT approaches do not translate any data, but queries. Queries are launched against a mediator that divides them into different sub-queries—one sub-query for each physical database. Then, the sub-queries are translated into the native query language of their corresponding database management system. After the execution of all sub-queries, the mediator gathers, unifies, and presents the results to the user. QT approaches include federated database systems, mediated systems, and other hybrid approaches.

We can distinguish two different categories when dealing with QT based approaches: 1) Global as View (GaV), and 2) Local as View (LaV). Both of them are equipped with a global schema to which user queries are launched. In GaV [4], the global schema is defined by means of views to the local schemas (mappings). This approach supports an straightforward translation of queries, but implies revising the global schema and the existing mappings each time a new source is added. On the other hand, when using LaV based approaches, the local schemas are defined as views from the global schema. This solves the problems that may arise when using GaV, but turns the query translation process into a difficult task [5][6].

In the next section we describe in detail the database integration architecture that we adopted to integrate clinical trials databases including genomic information. This architecture follows a QT-LaV based approach.

3 The ACGT Data Access Tools

Data Sources

When dealing with clinical trials involving genomic information, biomedical researchers have to use different types of data sources. The most important types include:

- Relational databases. These, amongst others, typically store the data gathered using the Clinical Report Forms.
- Medical image databases. Virtually all hospitals nowadays use Picture Archiving Communication System (PACS) for digitally storing medical images.
- Public web databases. The most important public genome and protein sequence databases are freely available through a web interface.
- Files in various formats. Investigators of trials may, for example, collect and maintain laboratory results in simple spreadsheet documents.

We have currently created data access services for the two first types of sources: relational databases and medical image databases. The reasons for initially choosing these two are three fold. First, these types of databases are the most relevant in clinical trials. Second, both kinds of sources are equipped with established standards for accessing the data—i.e. SQL/JDBC for relational databases, and Digital Imaging and Communication in Medicine (DICOM) [7] for medical image databases. By using these standards, the data access services can be used to access many different data sources from different vendors, resulting in a high pay-off given the invested development effort. Finally, the expressive power of the queries supported by both kinds of sources is completely different. Thus, it is an interesting challenge to create a uniform data access interface for both kinds of sources.

In ACGT, syntactic heterogeneities are solved by the ACGT-DAWs, which provide a seamless way to access different kinds of databases. By contrast, semantic heterogeneities are solved by the ACGT-SM, which uses the ACGT-MOC, a global model describing the cancer domain. Syntactic and semantic heterogeneities can be removed by our tools, providing homogeneous views for all sources to be integrated. These views can be automatically bridged together by the ACGTM, providing a virtual view that can be queried using the RDF Data Query Language (RDQL) [8].

We tested our tools using a set of three clinic-genomic databases. This includes two relational clinical trials sources— SIOP (International Society of Pediatric Oncology trial database, related to nephroblastoma), and TOP (Trial of Principle study database, related to breast cancer)— and a DICOM-based images database. We performed two pairwise integration experiments, each involving one of the relational clinical trials sources, and the DICOM database.

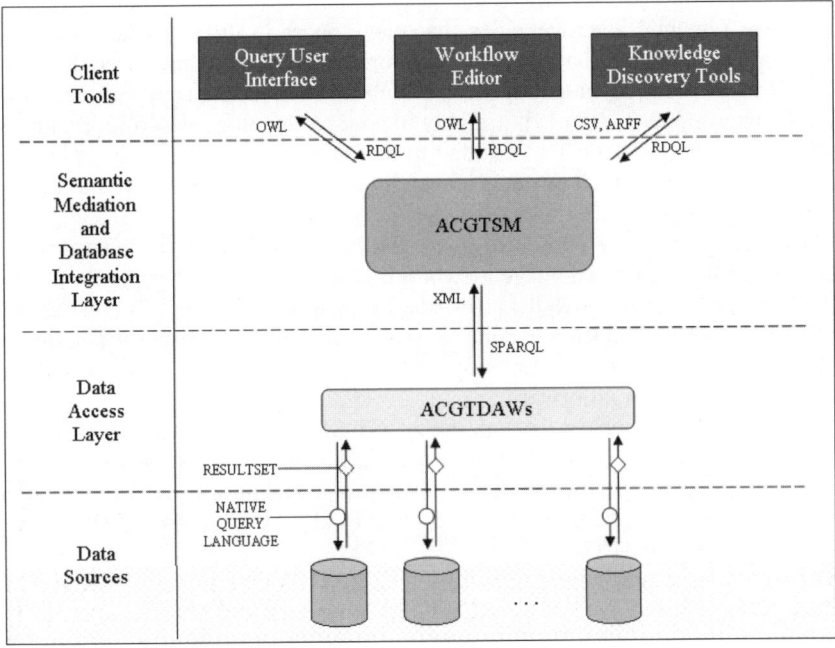

Fig. 1. Architecture of the Data Access Tools in ACGT

Figure 1 shows an architecture comprised by the different components involved in the ACGT data access. The upper layer includes the different kinds of client tools that need to access the mediation service. The Semantic Mediation and Database Integration Layer main component is the ACGT-SM, which gives access to virtual views of integrated databases solving the semantic heterogeneities. The Data Access Layer is comprised by the ACGT-DAWs, which provide transparent access to different types of data sources. Finally, all the underlying physical sources of information are located in the bottom layer. The ACGT-SM and the ACGT-DAWs are described in detail in the next paragraphs.

The ACGT Semantic Mediator

The role of the ACGT-SM is to solve semantic heterogeneities among different data sources, and to perform database integration. These functionalities are provided via a web service interface. The interface can be used by different tools within the ACGT platform.

We have adopted an approach based in QT-LaV to achieve database integration. As stated before, LaV normally presents poor performance in the query translation process—in the worst case, all the data sources need to be explored. To deal with this issue, we have constrained the mappings and the views representing the integrated set of databases. The ACGT-SM has been designed to cover only the kind of queries most frequently used by users—i.e. a user requirements driven restriction of the domain.

To solve semantic heterogeneities that may appear in different data sources we have adopted an ontology-based approach. More concretely, we have used the ACGT-MOC. The latter was developed using the Web Ontology Language with Description Logic Language (OWL-DL) [9], a highly expressive ontology description language. The reason to choose OWL-DL was that future integration requirements may require advanced mappings, which strongly depend on the expressiveness of the ontology representation language.

The language we have chosen to query the ACGT-SM is RDQL. The reason for selecting RDQL as query language is twofold: 1) it is a subset of Simple Protocol and RDF Query Language (SPARQL) [10]—the language used by the ACGT-DAWs—so the translation process does not imply a loss of information, and 2) it is expressive enough to meet the user requirements, as well as simple enough to make the query translation process straightforward.

When a query is launched through the ACGT-SM, it is automatically divided into different dedicated queries in SPARQL for the different wrappers that provide access to the underlying databases. We have developed a middleware to translate RDQL queries into SPARQL (the query language used by the ACGT-DAWs). This interface is included within the ACGT-SM.

The results of a query are returned by the ACGT-DAWs in the SPARQL Query Results XML Format [11]. The ACGT-SM integrates different result sets from different data sources into a set of instances of classes belonging to the ACGT-MOC. These instances are represented using the OWL ontology description language [9]. The ACGT-SM can also export the results into the Comma Separated Variables (CSV) and Attribute-Relation File Format (ARFF) formats—suitable for knowledge discovery tools such as Weka [12].

Regarding efficiency when a new database needs to be included in the system, or the schema of an existing one changes, the LaV-based approach chosen allows its integration by only creating or changing the view describing this single source.

The ACGT Data Access Wrappers

The ACGT-DAWs are responsible for solving syntactic heterogeneities. I.e. they need to *provide a uniform data access interface* (This includes uniformity of transport protocol, message syntax, and query language and data format). Additionally, the

ACGT-DAWs also need to *export the data model of data sources* (Clients need this to construct queries), enforce *access policies of data source,* and *audit access to data sources* (For clinical data, there are strict legal and ethical requirements that need to be adhered to).

The data access services were implemented as web services. More specifically, the data access services are implemented as Open Grid Services Architecture Data Access and Integration (OGSA-DAI) services [13]. OGSA-DAI is the standard web services framework for providing data access and its activity framework enables efficient and flexible services invocation.

SPARQL was chosen as the query language for the data access services. It is a modern RDF query language that is more expressive than its predecessor RDQL, the language used by an early version of the mediator. SPARQL has an intermediate level of expressiveness. It is less expressive than Structured Query Language (SQL) [14], mainly because it does not support any form of aggregation. It can only return the values that are in the underlying database, not derived values obtained through counting, averaging, summation, etc. On the other hand, it is more expressive than DICOM. Note that SQL cannot be used a the common query language for the data access services, as it is specific to relational databases. As we are not using a data translation approach, we need a query language that is more generic.

```
PREFIX xsd: <http://www.w3.org/2001/XMLSchema#>
PREFIX dicom: <http://example.philips.com/dicom#>
SELECT ?name ?dob ?studyId ?studyDescr
WHERE {
   ?patient dicom:PatientsName ?name ;
            dicom:PatientsBirthDate ?dob .
   ?study dicom:Patient ?patient ;
          dicom:StudyID ?studyId .
   OPTIONAL {
      study dicom:StudyDescription ?studyDescr .
   }
   FILTER ( ?dob >= "1970-01-01"^^xsd:date ||
            ?dob <  "1970-02-01"^^xsd:date )
}
```

Fig. 2. An example DICOM query expressed using SPARQL

Implementation of the SPARQL-enabled web services for querying relational databases was straightforward, as we could use D2RQMap [15] for handling the required query transformation. The implementation of the data access services for querying DICOM image databases required more work. Here we had to develop the functionality to transform SPARQL queries into DICOM queries ourselves, as it did not yet exist. Furthermore, we had to implement functionalities for retrieving DICOM images. DICOM uses a four-level hierarchical information model. At the top-level is the patient. A patient can have one or more studies associated with her. A study may contain one or more series, and a series may include one or more images. Each level has specific attributes associated with it. A few are required, but most are optional. For example, Patient Name is a required attribute at the patient level but Patient's Sex is an optional attribute.

DICOM queries consist of a list of attributes, together with a required match. Various types of matching are supported. Single value matching is the most basic, and requires that the value specified in the request is exactly matched. For certain data types, slightly more powerful matching is supported— e.g. basic wild card matching for string values, and range matching for data values.

Figure 2 shows an example of a DICOM query expressed using SPARQL as supported by the image data access service. For all patients born in January 1970, it retrieves the patient's name, as well as some details about the patient's studies.

4 Results

We used our tools to conduct some integration experiments in the context of the ACGT project. For this purpose, a set of three different sources was used. This set includes two relational clinical trials databases—SIOP and TOP—and a DICOM images repository. SIOP is a nephroblastoma database, provided by the International Society of Pediatric Oncology, including clinical patient data collected from clinical report forms. The TOP repository is a breast cancer related database, available at the Jules Bordet Institute (Brussels), also containing clinical data. The DICOM database contains several thousands of images including magnetic resonance, computed tomography, and radiography imaging.

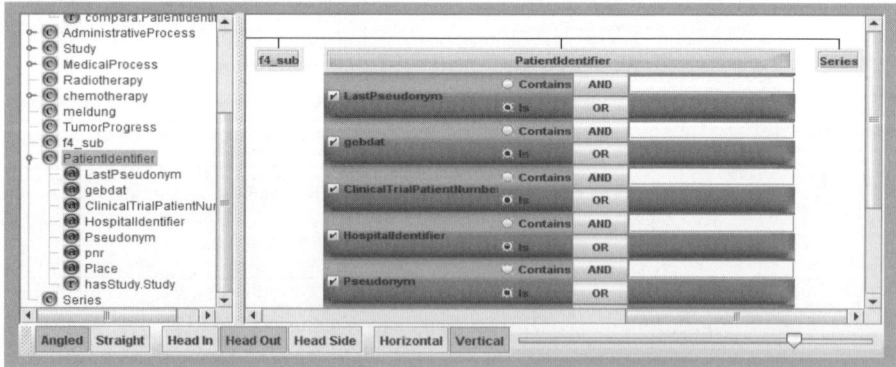

Fig. 3. Schema of the SIOP and DICOM integration

We performed two integration experiments between the DICOM source and each of the clinical trials databases. We did not bridge together the clinical databases since they belong to different domains—i.e. nephroblastoma and breast cancer clinical trials. Our tool integrated the sources successfully, and the generated schemas were validated by experts in the domain.

Figure 3 shows the integrated schema for the SIOP and DICOM databases. The leftmost window shows the subset of the MOC that describes the portion of the domain covered by both sources. As it can be seen, the schema is composed of concepts, attributes, and relationships. Once the user has selected the concept she is interested in, the right window shows a detailed view of the targer concept. This window facilitates users to query the integrated repository. Users can constraint the searches by filling in the textboxes—i.e. desired values for attributes. Besides, logical operators such as AND or OR can also be used.

For instance, let us assume that the user has chosen the concept *PatientIdentifier* and she has requested all the instances of such concept whose attribute *HospitalIdentifier* happen to be "*3*". Once the query has been submitted and processed, the user is presented with the set of instances shown in figure 4.

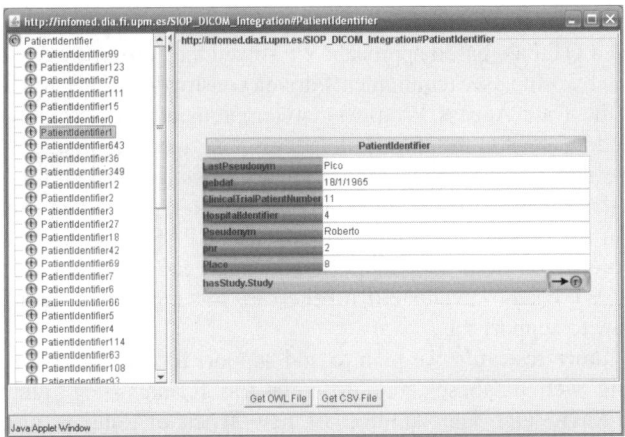

Fig. 4. Result set retrieved by the system after the execution of a query

The left window shows the retrieved set of instances, while the right window presents the available information related to the currently selected instance. As can be seen it is possible to export all the retrieved instances to other formats, either OWL or formatted text file—including CSV or ARFF formats—just by clicking on the corresponding button.

The instances shown in figure 4 contain clinical information about patients—i.e. data comes from the SIOP source. It is possible to retrieve instances from other concepts related to the currently selected instance just by clicking on the relationships buttons. In figure 4, the concept *PatientIdentifier* is related to the concept *Study*, whose instances are DICOM metadata—i.e. data comes from the DICOM database. Once the user clicks on the relationship *hasStudy.Study*, the system presents the user with all the available instances of the concept *Study* related to the selected instance of the concept *PatientIdentifier* using the same interface shown in figure 4.

The web interface has been designed for demonstration purposes in the ACGT project, providing an interactive and intuitive querying service. Therefore, it only supports queries involving a single concept. However, the underlying web service can handle multi-concept queries. Thus it is possible to design and use other interfaces handling more complex queries.

This prototype was developed using Java and HTML languages.

5 Conclusions and Future Work

We have contributed to the ACGT infrastructure developing two core components to deal with semantic and syntactic heterogeneities in data sources.

To solve semantic heterogeneities, we have created a semantic mediator that can carry out the database integration task. This mediator uses the ACGT-MOC—i.e. the global model—to build the local schemas that represent the semantics of the underlying databases. Furthermore, these local schemas are actually subsets of the ACGT-MOC.

The results of the conducted experiments prove that our approach can seamlessly integrate relational and image databases within a single view.

We selected a QT-LaV based approach. We solved the performance issues associated to this approach by using user requirements-driven constriction of the domain.

Regarding the Data Access Wrappers, we implemented OGSA-DAI services for two types of data source: relational databases and medical image databases. For uniformity, we decided to use a common query language. We chose SPARQL and demonstrated that it can be successfully applied to both types of data sources.

However, in the case of the relational database, SPARQL is less expressive than SQL. For example, it does not support data aggregation (averaging, summation, counting, etc). We need to evaluate if it necessary to support data aggregation, and if so, find out how to support this.

Regarding future research, we plan to add support for new types of sources. This includes public web databases and different file formats—e.g. plain text, Excel spreadsheets, XML, etc. The addition of new types of sources requires a more expressive query language than RDQL. Thus, we are working towards replacing the RDQL query language used in the ACGSTM with SPARQL.

Acknowledgments. This research has been supported by the Advanced Clinico-Genomic Trials in Cancer (ACGT) project (FP6 IST-2005-026996) funded by the European Commission. The group at the UPM is also supported by the ONTOMINEBASE project, Ministry of Science and Education, Spain.

References

1. Kimball, R.: The Data Warehouse Toolkit: Practical Techniques for Building Dimensional Data Warehouses. John Wiley, New York (1996)
2. Sheth, A.P., Larson, J.A.: Federated Database Systems for Managing Distributed, Heterogeneous, and Autonomous Databases. ACM Computing Surveys 22(3), 183–236 (1990)

3. Wiederhold, G.: Mediators in the Architecture of Future Information Systems. IEEE Computer 25(3), 38–49 (1992)
4. Cali, A., De Giacomo, G., Lenzerini, M.: Models for information integration: Turning local-as-view into global-as-view. In: Proc. of Int. Workshop on Foundations of Models for Information Integration (10th Workshop in the series Foundations of Models and Languages for Data and Objects) (2001)
5. Abiteboul, S., Duschka, O.: Complexity of answering queries using materialized views. In: PODS 1998. Proc. of the 17th ACM SIGACT SIGMOD SIGART Symp. on Principles of Database Systems, pp. 254–265. ACM Press, New York (1998)
6. Ullman, J.D.: Information integration using logical views. In: Afrati, F.N., Kolaitis, P.G. (eds.) ICDT 1997. LNCS, vol. 1186, pp. 19–40. Springer, Heidelberg (1996)
7. National Electrical Manufacturers Association: Digital Imaging and Communications in Medicine (DICOM) (2004) (last accessed in May 2007), http://medical.nema.org/dicom/2004.html
8. http://www.w3.org/Submission/RDQL/ (last accessed in May 2007)
9. http://www.w3.org/2004/OWL/ (last accessed in May 2007)
10. Prud'hommeaux, E., Seaborne A.: SPARQL Query Language for RDF, W3C Working Draft (March 26, 2007) (last accessed in May 2007), http://www.w3.org/TR/2007/WD-rdf-sparql-query-20070326/
11. http://www.w3.org/TR/rdf-sparql-XMLres/ (last accessed in May 2007)
12. http://www.cs.waikato.ac.nz/ml/weka/ (last accessed in May 2007)
13. Antonioletti, M., et al.: The Design and Implementation of Grid Database Services in OGSA-DAI. Concurrency and Computation: Practice and Experience 17(2-4), 357–376 (2005)
14. Eisenberg, A., Melton, J., Kulkarni, K., Michels, J., Zemke, F.: SQL:2003 has been published. SIGMOD Rec. 33(1), 119–126 (2004)
15. Bizer, C., Seaborne, A.: D2RQ - Treating Non-RDF Databases as Virtual RDF Graphs. In: McIlraith, S.A., Plexousakis, D., van Harmelen, F. (eds.) ISWC 2004. LNCS, vol. 3298, Springer, Heidelberg (2004)

Ontology-Based Data Integration in Data Logistics Workflows

Olivier Curé[1] and Stefan Jablonski[2]

[1] Université Paris-Est, S3IS, Marne-la-Vallée, France
ocure@univ-mlv.fr
[2] University of Bayreuth, Bayreuth, Germany
Stefan.Jablonski@uni-bayreuth.de

Abstract. Data integration has become an essential issue in the development of information systems, especially in the biomedical and healthcare domain. Thereby it is required to have methods and tools available to support this effort in a systematic and structured manner, i.e. to have a conceptual, model based approach. We present such an approach; it is constituted by two components which are coping with the two main challenges of data integration: Data Logistics copes with the technical task of data transmission and data exchange; an ontology-based transformation copes with the semantic issues of data integration by dealing with the heterogeneity of formats, terminologies and ontologies. This paper shows how the synergetic combination of these concepts provides an adequate solution to data integration in healthcare applications.

1 Introduction

Nowadays, data integration is one of the most frequent and one of the must crucial tasks in data management. In [8] [9] we describe the data integration task in a medical/clinical application. There, different data sources belonging to separated clinical applications have to be integrated into a comprehensive data base in order to support specialized medical applications; in this example, the applications are stemming from the area of ophthalmology. Integrating data and analyzing them together enables new kinds of ophthalmologic analysis which would not be possible if data are investigated separately.

Another data integration task is examined in [5]. There, a medical application XIMSA (eXtended Interactive Multimedia System for Auto-medication) [3] is investigated. It offers medical services to patients enabling safe self-medication. One of the main problems here is that drug information is scattered over distributed and heterogeneous data bases and ontologies. This information has to be integrated to provide higher quality information to patient through inferences, e.g. consistent drug prescription and drug contra-indication detections.

Although the two application areas seem to be very diverse they are quite well comparable from a data integration point of view. In principle data of different source data bases have to be integrated. We summarize the main requirements of

J.-L. Hainaut et al. (Eds.): ER Workshops 2007, LNCS 4802, pp. 34–43, 2007.
© Springer-Verlag Berlin Heidelberg 2007

these data integration tasks in the following with the goal to derive a conceptual approach from that:

(Req1) Data that must be integrated often use different data formats, data models, and notions. Data integration therefore means to homogenize these heterogeneous issues. Thus, a set of transformations is required for data integration, namely format, terminology and ontology transformations.

(Req2) Data integration is a complex and complicated task that will happen frequently (e.g. whenever new data are generated). Thus, it is required to apply a model based, conceptual approach that illustrates its complexity and makes it possible to systematically enact it. Process modelling has proven to be an adequate technology for this purpose.

In [8] and [9] we proposed a principal solution for the requirements identified above which is called Data Logistics (DaLo). One of the main issues of data integration is to sustain consistency among the integrated data. In [8] also a (primitive) mechanism is proposed that allows the adaptation of different formats, terminologies, and ontologies. The scope of this mechanism is rather limited and needs replacement through a systematic and portable concept. Conversely, in [5] a general and comprehensive framework for ontology integration is presented named DBOM (Data Base Ontology Mapping). This system solves the integration problem for heterogeneous ontologies. However, the data integration task is not focussed by the DBOM approach. Due to this situation, our goal is to combine the two approaches DaLo and DBOM in a synergetic way: we intend to take the data integration capabilities of DaLo and integrate the ontology management capabilities of DBOM. Thus, the two requirements Req1 and Req2 revealed above are covered: Req1 is completely covered by the DBOM approach, Req2 is completely covered by the DaLo system. We consider the format, terminology, and ontology transformation as one of the most crucial issues in data integration. So the replacement of our simple method with a systematic, comprehensive and scalable method is a mandatory prerequisite for the acceptance and portability of our entire approach.

In Section 2, we present DaLo. In Section 3, we propose an architecture for data, terminology and ontology transformations and in Section 4, we analyze related works in the ontology merging domain.

2   Data Logistics Approach

The DaLo method is based on process management. However, to cope with the requirements identified in Section 1, process management must be tailored specifically. In a first phase of the DaLo method 'normal' workflows (or processes; we use both terms synonymously in this paper) describing the application must be specified [7]. For instance, in an antibiotic drug prescription application, workflows describe how patients are being prescribed antibiotic drugs in a safe and efficient manner. Prescription data will be stored in a data base.

Fig. 1 depicts the principle concept of DaLo. In the upper part of the figure a 'normal' workflow is described. In a first step a general practitioner (GP)

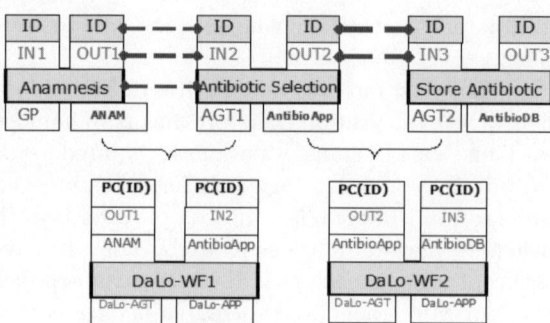

Fig. 1. The Data Logistic approach

has to perform the anamnesis of a patient (work step 'Anamnesis'), using the ANAM program; as input to that work step the patient ID is provided and some more data which are shortly named IN1 here. The outcome of the work step is some output drug related data, here called OUT1. This output data have to be transmitted to the next work step 'Antibiotic Selection', where it becomes the input data IN2. At that work step a computerized agent (AGT1) has to select antibiotic drugs for this patient - identified by the input parameter ID - with the AntibioApp application. The outcome of this examination is the output data OUT2. This data is input (IN3) to the work step 'Store Antibiotic' in which the data are stored in a data base using the application 'AntibioDB'. The data contained in this data base are used for statistical evaluation.

The main idea of DaLo is now to derive DaLo workflows from 'normal' process descriptions which focus on data exchange between client applications, i.e. between the different data bases associated with these applications. The lower part of Fig. 1 shows the result of such a derivation. In our example, the exchange of data between the applications ANAM, AntibioApp and AntibioDB are considered. One DaLo workflow (e.g. DaLo-WF1) is just dealing with one data transmission between different data bases. For each data transmission task one DaLo workflow is created.

When the actual data transmission is performed, a so-called DaLo-AGT is triggering a so-called DaLo-APP to transfer data between the involved data bases. In our example, the data OUT1 are taken from the data base of the ANAM application and are transmitted to the data base of the AntibioApp system in the DaLo workflow (DaLo-WF1). In the second DaLo workflow (DaLo-WF2) the OUT2 data of the AntibioApp data base are transmitted to the IN3 data on the AntibioDB application, i.e. to the data base associated AntibioDB application.

Due to different notions and name spaces of participating applications of a DaLo workflow, format, terminology and ontology transformation is needed. This transformations are already implemented in the DaLo system, however in a kind of 'handcrafted' and not generalizable manner. Concretely, each transformation (format, terminology, and ontology) is coded in a (Java) program. Even this is effective and is proven to be feasible in a productive clinical application this form of implementation is not applicable. A first drawback is that each change and/or enhancement of transformations has to be enacted by an application programmer (triggered by a domain expert). Another weakness is that it does not provide a conceptual model for all format, terminology, and ontology transformations. Such a model is especially of great interest in complex application scenarios since it could improve besides understanding also consistency. Thus we must replace this program based transformation with a model based one to downsize the drawbacks of the current implementation and to gain a clear conceptual approach.

3   Ontology-Based Transformations

In general the applications used in a workflow are based on different data models and so the data has to be transformed in order to be interpretable among the different applications. Before presenting the architecture of our solution, we first provide a concrete example of these transformations.

Example 1. This example deals with a single DaLo workflow (DaLo-WF1) related to medical prescriptions (MedicalPrescription) where the output of APP1 is a CSV (Comma Separated Values) document about allopathy drug prescriptions. The schema is defined in terms of a citizen's social security number (SSN), a french drug identifier (CIP) and an international identifier for molecules (ATC). The input of APP2 is an XML document concerned with antibiotic drug prescriptions for a given patient (Patient). The data contained in this document require a filtering of the data stored in the output of APP1, i.e. all non-antibiotic drug from APP1's output are not integrated in APP2's input. This input contains the following elements: SSN of the patient, drug name and molecule name.

  Example 1 emphasizes the following three transformations: (i) at the format level, to transform a CSV document into an XML document. (ii) at the ontological level, the antibiotic drugs of APP1's output have to be identified. This process is supported by a classification of the involved drugs into a general classification (ontology). (iii) at the terminological level, to transform the CIP and ATC codes to drug and molecule names in DaLo-WF1. This task is easy to accomplish by just selecting the drug name and molecule name, respectively, out of the CIP code and the ATC code.

  To enable efficient transformations in DaLo workflows, DaLo must cope with semantics of the output and input of involved applications. Thus we propose an ontology-based approach which deals with the problems of ontology alignment and merging of local ontologies via the use of a reference ontology. This enables to perform the required transformations through the use of inferences.

3.1   Basic Notions

In information technology, an ontology provides a shareable and reusable piece of knowledge about a specific domain. In such a context, an ontology is a set of concepts and their relationships, that are specified more or less formally in order to create an agreed-upon vocabulary. In our solution, we have selected Description Logics (DL) [2] as a mean to represent ontologies. This family of knowledge representation formalisms allows to represent and reason over domain knowledge in a formally and well-understood way. Central DL notions are concepts (unary predicates) and relationships, also called roles (binary predicates). In Example 1, notions like 'MedicalPrescription' and 'Patient' are concepts while 'prescribeTo' is a role relating a patient to his medical prescriptions.

A standard DL knowledge base is usually defined as $\mathcal{K} = \langle \text{ T,A } \rangle$ where T (or TBox) and A (or ABox) consist respectively of a set of concept descriptions (resp. concept assertions and role assertions). A concept description is $AlloDrug \doteq \forall madeOf.ChemicalSubstance \sqcap \exists madeOf.ChemicalSubstance \sqcap Drug$. This description states that a concept AlloDrug (allopathy drugs) describes those drugs that are made of at least one chemical substance and contains only chemical substances. Concept assertions, for instance, are *Patient(Bob)* and *MedicalPrescription(X101)*; an example of a role assertion is *prescribeTo(X101,Bob)*. Based on $\mathcal{K}$, we are using standard DL reasoning tasks such as concept satisfiability, concept subsumption and instance checking which are detailed in [2].

In our approach, the transformation problems are related to the ontology alignment problem. This is usually defined as the problem of discovering similarities between ontologies. Ontology alignment is generally described in terms of the Match operator. In [12], this operator is defined as a function that takes two arbitrary schemas (ontology TBoxes, etc.) S1 and S2 as input and returns a mapping between S1 and S2 as output, called the match result. In our solution, the match result supports the merging of the local ontologies with the reference ontology. The outcome of this operation is a 'merged' ontology, which is a kind of union of the source ontologies based on the correspondences between them.

3.2   General Architecture

In order to support the ontology-based transformations, we need to change the model of a DaLo workflow, e.g. DaLo-WF1 of Fig.1. This new model requires the introduction of new parameters Mapping, LocalOnt and RefOnt: (i) The reference ontology (RefOnt) is added to the DaLo step and proposes a common vocabulary for the domain of the DaLo workflow. (ii) A local ontology (LocalOnt) is introduced for each of the applications involved in a DaLo workflow. A local ontology describes the concepts and roles required to explain the domain of the input, respectively ouput, of a DaLo workflow. The concept descriptions of these local ontologies are defined in terms of the reference ontology. (iii) A set of mappings (Mapping) for each element of the output and input of a DaLo workflow. These annotations use the concepts and roles of their respective local ontologies and enable relationships between the concrete parameters of the input, respectively output, of a DaLo workflow with the corresponding concepts and roles of the local ontologies. For example, the value '3572151' in the output corresponds to the CIP of the drug named 'Biocalyptol'. The adoption of this model simplifies considerably the alignment of both local ontologies. This is due to the role of the reference ontology which proposes a common vocabulary to compute the subsumption relations between concepts from the local ontologies.

Based on these relations, standard reasoning services, and the mappings, the terminology and ontology transformations are effective and efficient.

According to this model, DaLo transformations, including format transformation, can be formally described with the following function:

$$\text{DaLoT: (OUT, F, } \langle \text{ R, } L_1, M_1 \rangle) \rightarrow (\text{IN, F, } \langle \text{ R, } L_2, M_2 \rangle)$$

where OUT and IN describe respectively the output and input data of a DaLo-WF, F stands for Format, R stands for the reference ontology, L_x stands for a local ontology and M_x stands for a set of mappings of application x (APPx). The function DaLoT therefore maps an output data OUT to an input data IN whereas the output and the input data bring their individual formats, local ontologies and annotations (mappings).

The reference ontology regroups concepts and roles of the general domain of the DaLo workflow. The expressive power of this ontology is restricted: (i) for atomic concepts C and D, terminological axioms are limited to concept inclusion $(C \sqsubseteq D)$ and equality $(C \equiv D)$. (ii) for roles, considered as binary predicates, we limit ourself to specify the parameters (domain and range).

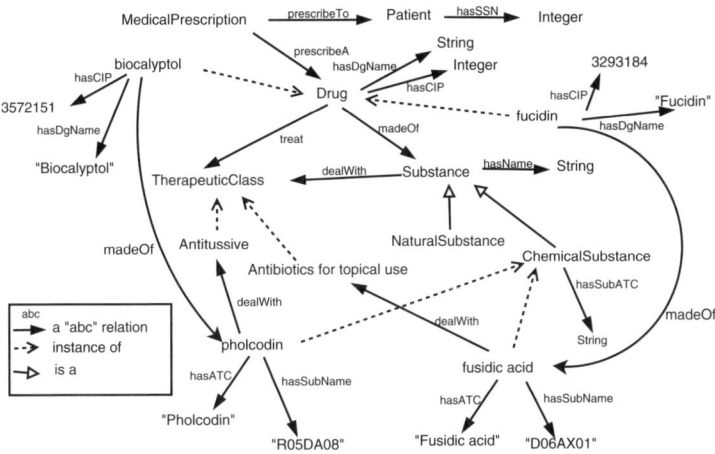

Fig. 2. A reference ontology for Example 1

In Fig.2, we propose a reference ontology which supports the transformations for DaLo-WF1 in Example 1. This reference ontology contains two types of properties (or roles). In Fig.2, the 'prescribeTo' property is called an object property as its range is defined in terms of a concept, i.e. 'Patient'. But the 'hasCIP' role is called a data type property because its range is a type, i.e. an integer. In order to support terminological transformations, the reference ontology has to be equipped with instances for some of its concepts. In Fig.2, instances are defined for the concept 'ChemicalSubstance', i.e. 'pholcodin' and 'fucidic acid', as well as the Drug concept, i.e. 'fucidin' and 'biocalyptol'. Both of these instances have some data type properties defined, e.g. 'hasATC(pholcodin,

R05DA08)'. The integration of instances in the reference ontology is facilitated by procedures supported by the DBOM system [4,5].

A local ontology is defined for each application involved in a DaLo workflow. The TBox of a local ontology assigns names to (possibly complex) descriptions using the terms of the reference ontology and/or its own set of terms. Instances are not required for the knowledge base associated with a local ontology, instead, they are stored in the data bases related to each DaLo workflow application.

Example 2. The local ontologies L_1 (1-3) for APP1 and L_2 (4-6) of APP2 of Example 1 are now defined:
(1) $Citizen \doteq Patient$. The concept 'Citizen' is equal to the concept 'Patient'.
(2) $AlloDrug \doteq \forall\ madeOf.ChemicalSubstance \sqcap \exists\ madeOf.ChemicalSubstance \sqcap Drug$. Already explained in Section 3.1.
(3) $AlloMedicalPrescription \doteq MedicalPrescription \sqcap \exists\ prescribeTo.Person \sqcap \forall prescribeA.AlloDrug \sqcap \exists\ prescribeA.AlloDrug$. An allopathy drug prescription is defined as a medical prescription (MedicalPrescription) which is prescribed to a patient and prescribes at least one allopathy drug and only allopathy drugs.
(4) $AntibioticSubstance \doteq ChemicalSubstance \sqcap dealWith \in AntibioticForTopicalUse$
(5) $AntibioticDrug \doteq Drug \sqcap \exists\ madeOf.AntibioticSubstance$
(6) $AntibioticPrescription \doteq \forall\ prescribeA.AntibioticDrug \sqcap MedicalPrescription \sqcap \exists\ prescribeTo.Patient \sqcap \exists\ prescribeA.AntibioticDrug$

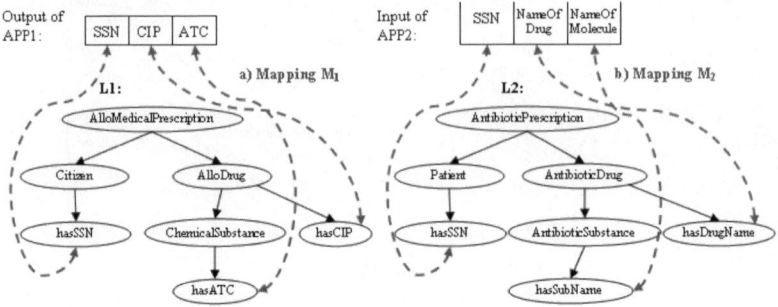

Fig. 3. Mapping M_1 (a) and M_2 (b) for Example 1

The DaLoT function requires mappings on the output and input schemas of a DaLo workflow. These mappings are expressed in terms of the application local ontology and possibly the reference ontology. Intuitively, the user responsible for DaLo workflow application designs the related local ontology and defines annotations. These annotations aim to define correspondences between the elements of the output/input of a DaLo workflow to concepts and roles of the local or reference ontologies. In Example 1, the output of APP1 is a CSV record. First, the users need to associate a concept to this type of record, e.g. AlloMedicalPrescription from L_1, then they study all fields of the record and relate concepts to each field or aggregation of fields. These aggregations may also be associated to concepts. Finally, the fields themselves have to be associated to data type properties, e.g. the field containing a CIP drug identifier is associated to the role 'hasCIP'. These mappings can be represented using a tree representation. This tree representation needs to satisfy the fact that leafs and nodes of a mapping tree for APPx are defined by data type properties, respectively concepts, defined in R (reference ontology) or its local ontology L_x. In Fig.3, we present trees for M_1 and M_2 (resp. mapping for local ontology L_1 and L_2) of DaLo-WF1.

3.3   Terminological and Ontological Transformations

Terminology and ontology transformations are processed using a new ontology (called merged ontology) which corresponds to a merging of the source ontologies, i.e. the reference and both local ontologies. The idea is that all concepts and roles associated to the mappings are present within a single ontology.

Transformations are based on the exploitation of correspondences between the mapping trees of M_1 and M_2 and reasoning services. The correspondences between M_1 and M_2 can be considered as a morphism between the trees and are defined in terms of equivalence ($\equiv$), generalization ($\sqsubseteq$), specialization ($\sqsupseteq$), overlapping ($\sqcap$) and disjoints ($\perp$) relations between concepts. The relationships between elements of M_1 and M_2 based on concept hierarchy of the merged ontology are: (1) $M_1.AlloMedicalPrescription \sqsupseteq M_2.AntibiocPrescription$. The AntibioticPrescription concept is a subconcept of the AlloMedicalPrescription concept. (2) $M_1.Citizen \equiv M_2.Patient$. The Citizen and Patient concepts are equivalent. (3) $M_1.AlloDrug \sqsupseteq M_2.AntibioticDrug$. (4) $M_1.ChemicalSubstance \sqsupseteq M_2.AntibioticSubstance$

Although there are just two mappings used in our example, we discuss the five possible correspondences between nodes. Here, C_x is a node of M_x as well as a concept in L_x: (1) In a case of equivalence between C_1 and C_2, i.e. $C_1 \equiv C_2$, the ontology transformation is straightforward as all instances of C_1 are also instances of C_2, e.g. in Example 1 the concepts Citizen and Patient are equivalent, thus in the context of this ontology, all citizens are also patients. (2) In case of a specialization relation from C_1 to C_2, i.e. $C_1 \sqsubseteq C_2$, it will always be the case that instances of C_1 are also instances of C_2. (3) In the case of a generalization relation from C_1 to C_2, $C_1 \sqsupseteq C_2$, it may not always be the case that instances of C_1 are also instances of C_2, e.g. a ChemicalSubstance is not always an AntibioticSubstance. In such a situation, some records from APP1's output may not be present in APP2's input. (4) In the case of an overlapping between C_1 and C_2, i.e. $C_1 \sqcap C_2$, there might be instances that belong to both C_1 and C_2. If an instance corresponding to an entry in APP_1's output lies in $C_1 \sqcap C_2$, then this is a candidate for APP2's input otherwise it is ignored. (5) A case of disjointness between C_1 and C_2, i.e. $C_1 \sqcap C_2 \sqsubseteq \perp$, implies that instances of C_1 coming from the output of APP1 cannot be mapped to the concept C_2.

In the cases 3 and 4, we need to check if each instance of C_1 is also an instance of C_2. This operation is perfomed using the instance checking reasoning service of DL.

Example 3. In the context of Example 1, we can consider the following output: 181xx corresponds to a social security number (SSN), 3293184 and 3572151 are CIP codes for respectively the Fucidin and Biocalyptol drugs, ATC codes

D06AX01 and R05DA08 correspond respectively to the Fucidic acid and Pholcodin substances. So the entries in the CSV document of APP1 are: 181xx; 3293184; D06AX01 and 181xx; 3572151; R05DA08.

Using the mappings on this output, we are able to create two AlloDrug individuals, on for the Fucidin drug and one for the Biocalyptol drug. Then we have to check in the merged ontology if these individuals are also instances of the AntibioticDrug concept. This is the case for Fucidin as it is a drug that contains an antibiotic chemical substance, but it is not the case for the Biocalyptol drug, i.e. pholcodin is an antitussive. The last step is to proceed the terminological transformations on candidates. This task uses a form of functional dependency defined on data type properties of $\mathcal{T}$.

4   Related Approaches

Solutions proposed in the field of ontology alignment and merging usually adopt one or a combination of the following analysis methods: lexical, structural, semantical and matching heuristics. Surveys about these systems are proposed in [13]. In this section, we briefly present two solutions that use a mediated approach, meaning that a reference ontology is exploited to align and merge the input ontologies. In the S-match [6] system, information sources are graph-like structures containing terms and their inter-relationships. The main principle is to translate labels at the nodes of these trees into propositional formulas and thus to codify the matching problem into a propositional unsatisfiability problem. In HCONE [10], a reference ontology is designed from mapping input ontologies to external domain sources like Wordnet. The merging method then translates concept definitions of the source ontologies to this common vocabulary and finally, merges these translated definitions by means of specific merging rules and DL reasoning procedures. The main difference with our approach is that HCONE designs the intermediate ontology on-the-fly and does not require instances.

A number of workflow systems have emerged to facilitate scientific applications. Most of them like Kepler [1] and Taverna [11] follow a bottom up approach. They are based on component models with a functional procedural-call interface. Their composition graphs are then mapped to the predefined component models. Data models are expressed in form of fixed and embedded ports and are not provided as separate models. Moreover, these system cannot adapt external data models, especially they cannot incorporate external ontologies. There are some more deficiencies of scientific workflow systems like lacking interaction support and specification of organizational issues (e.g. responsibilities). Nevertheless, these powerful tools integrate scientific applications but they do not have clear conceptual foundations and are more built upon handcrafted implementations.

5   Conclusion

This paper presents a mechanism for data integration that copes with the technical issues of data transmission and exchange and with semantic issues of format, terminology and ontology adaptation. One of the main advantages of our mechanism is its adaptability to different application areas, in this paper we presented as an example an application from the pharmaceutical realm. This application is already implemented and makes use of the proposed concepts. This implementation uses SemanticWeb technologies such as RDF/S and OWL to represent the ontologies and HP's Jena Java framework. A first set of evaluations has been conducted on the XIMSA project where several self-medication services need to exchange data, e.g. services related to drug proposition and diagnosis, management of the personal health care electronic record, etc.

References

1. Altintas, I., Berkley, C., Jaeger, E., Jones, M., Ludaescher, B., Mock, S.: Kepler: Towards a Grid-Enabled system for scientific workflows. In: Workflow in Grid Systems Workshop in GGF10, Berlin (March 2004)
2. Baader, F., Calvanese, D., McGuinness, D., Nardi, D., Patel-Schneider, P.: The Description Logic Handbook: Theory, Implementation and Applications. Cambridge University Press, Cambridge (2003)
3. Curé, O.: XIMSA: eXtended Interactive Multimedia System for Auto-medication. In: Proceedings of CBMS 2004, Bethesda, USA, pp. 570–575 (2004)
4. Curé, O.: Semi-automatic data Migration in a Self-medication Knowledge-Based System. In: Proceedings of Wissensmanagement 2005, Kaiserslautern, Germany, pp. 373–383 (2005)
5. Curé, O., Squelbut, R.: Data Integration Targeting a Drug Related Knowledge Base. In: Proceedings of EDBT Workshops, Munich, Germany, pp. 411–422 (2006)
6. Giunchiglia, F., Shvaiko, P., Yatskevich, M.: S-match: an algorithm and an implementation of semantic matching. In: Bussler, C.J., Davies, J., Fensel, D., Studer, R. (eds.) ESWS 2004. LNCS, vol. 3053, pp. 61–75. Springer, Heidelberg (2004)
7. Jablonski, S., Bussler, C.: Workflow management - modeling concepts, architecture and implementation. International Thomson Computer Press, London (1996)
8. Jablonski, S., Lay, R., Meiler, C., Müller, S., Hümmer, W.: Data Logistics as a Means of Integration in Healthcare Applications. In: Preneel, B., Tavares, S. (eds.) SAC 2005. LNCS, vol. 3897, pp. 236–241. Springer, Heidelberg (2006)
9. Jablonski, S.: Process Based Data Logistics: Data Integration for Healthcare Applications. In: proceedings of European Conference on eHealth 2006 (2006) GI-Edition
10. Kotis, K., Vouros, G., Stergiou, K.: Towards automatic merging of domain ontologies. The HCONE-merge approach Journal of Web Semantics 4(1), 60–79 (2006)
11. Oinn, T., et al.: Taverna: Lessons in creating a workflow environment for the life sciences. Concurrency Computation: Pract. Exper., 1–7 (2000)
12. Rahm, E., Bernstein, P.: A survey of approaches to automatic schema matching. VLDB Journal 10(4), 334–350 (2001)
13. Shvaiko, P., Euzenat, J.: A Survey of Schema-based Matching Approaches. Journal on Data Semantics IV, 146–171 (2005)

Model-Driven Development Based Transformation of Stereotyped Class Diagrams to XML Schemas in a Healthcare Context*

Eladio Domínguez[1], Jorge Lloret[1], Beatriz Pérez[1],
Áurea Rodríguez[1], Ángel L. Rubio[2], and María A. Zapata[1]

[1] Dpto. de Informática e Ingeniería de Sistemas.
Facultad de Ciencias. Edificio de Matemáticas.
Universidad de Zaragoza. 50009 Zaragoza. Spain
noesis,jlloret,beaperez,arv852,mazapata@unizar.es
[2] Dpto. de Matemáticas y Computación. Edificio Vives.
Universidad de La Rioja. 26004 Logroño. Spain
arubio@unirioja.es

Abstract. The health sector uses clinical guidelines as instruments for helping decision making. We are interested in the development of a ubiquitous decision support system (UDSS) for clinical guidelines in order to help the medical staff in their decisions and in order to record a trace of the application of the guidelines. For the development of such a system in a Model–Driven Development (MDD) setting, we propose the use of class diagrams of the Unified Modeling Language (UML) with stereotypes and eXtensible Markup Language (XML) schemas. When both languages, UML and XML, have to share a common modelling space, the necessity of transforming UML models into XML schemas arises. However, to our knowledge, previous transformation proposals do not consider the case in which some profiles have been applied to the UML model. For this reason, in this paper we propose a set of rules for translating stereotyped UML class diagrams into XML schemas, storing a trace of the application of the guideline.

Keywords: Clinical Guideline, UML Class Diagram, XML Schema.

1 Introduction

The daily work of physicians involves the obligation of taking difficult decisions since they have implications for patients' lives. Specifically, guidelines and protocols are the most common way of assisting doctors in decision making. In general, guidelines and protocols incorporate procedures and analyses for different aspects of medical practice and management [14], but we focus our attention on guidelines used for diagnosis and prognosis. Within this context one of the

* This work has been partially supported by DGI (project TIN2005-05534 and FPU grant AP2003-2713), by the Government of La Rioja (project ANGI 2005/19), by the Government of Aragon and by the European Social Fund.

J.-L. Hainaut et al. (Eds.): ER Workshops 2007, LNCS 4802, pp. 44–53, 2007.

most important challenges is to provide the practitioner with computer assistance during the application of a guideline [16].

As a way of achieving this, we are studying, as a long term research goal, the automatic development of ubiquitous decision support systems (UDSS) for clinical guidelines [17]. We propose to tackle this goal by using a model-driven development approach (MDD) [18] so that, given a particular guideline, this is expressed by means of a model from which we aim at generating a UDSS for the guideline in question. There are several advantages that such a system provides to the medical staff (see [17] for a detailed study) but in this paper we want to focus our attention on the fact that a trace of the guideline application can be automatically recorded (physician's decisions, history of patient's states,...).

For the development of the UDSS in an MDD setting, we propose to use the Unified Modeling Language (UML) and the eXtensible Markup Language (XML) as modelling languages. In particular, UML class diagrams are used, as platform independent models (PIM), for modelling the trace of the execution of guidelines. But, whenever UML is applied to different domains [21], the necessity of adapting the UML proposal to a particular context arises. This adaptation could be achieved by using UML lightweight extension mechanisms, that is, by defining a UML profile [15]. We have employed this possibility so that stereotyped UML class diagrams are used to store the information that is generated while the guideline is being applied. Furthermore, the ubiquitous nature of this system makes data interchange a critical issue and therefore the use of XML Schema, as a platform specific model (PSM), appears to be a natural choice.

When both languages, UML and XML, are used within an MDD approach, a transformation between them must be provided. In our particular case, stereotyped UML class diagrams have to be translated into XML schemas. However, although there are many works in the literature proposing a transformation of UML class diagrams into XML schemas, after an exhaustive revision and comparison of such transformations, we conclude in [4] that none of the analyzed approaches specifies transformation rules for the general application of UML profiles. For this reason, as the main contribution of this paper, we propose a set of rules for translating stereotyped UML class diagrams into XML schemas. This set of rules is applied to our case of stereotyped UML class diagrams representing a trace of the application of some guideline.

The remainder of this paper is structured as follows. In the following section we discuss related work. In Section 3 we explain the case study and our proposal for automatic generation of a UDSS for a given guideline in an MDD setting. In section 4 we describe the transformation of stereotyped UML class diagrams into XML schemas. Finally, conclusions and future work are presented.

2   Related Work

Several proposals for providing computer assistance for the application of clinical guidelines can be found in the literature [7,16]. Many of them propose their own set of representation primitives to capture the structure of guidelines. Among them, it is worth noting the Guideline Interchange Format (GLIF) and PRO-forma [7,14] which define specific ontologies for representing guidelines. Nevertheless, as is claimed in [8], the use of mainstream modelling languages such as UML and XML has potential benefits. In particular, both UML [2,11] and XML [3,19] are widely accepted as practical and beneficial formalisms within the healthcare realm. For this reason, we have chosen these modelling languages for the development of a UDSS [17].

A noteworthy characteristic of UML is the possibility of being adapted for solving certain types of problems in a particular domain by way of UML profiles. This feature is important within the medical context [21] since different specific necessities must be tackled in this domain such as data confidentiality [9], knowledge modelling [1] or the business aspects of hospitals [20], among others.

The problem is that although several authors propose algorithms for translating UML models into XML Schema, to our knowledge, none of them specifies transformation rules for the general application of UML profiles. In [4] we present a survey comparing different UML to XML schema transformation approaches (a total of 22 papers are analyzed) and, among them, only one [12] takes into account UML stereotypes, but without considering their properties.

Given the necessity of translating UML stereotypes into XML Schemas for the development of medical systems, we consider it suitable to have a general mechanism available for such a transformation.

As for the development strategy of the decision support systems proposed in the literature [7,16], to our knowledge, none of these uses an MDD approach, so that this characteristic distinguishes our UDSS proposal [17] from others.

3   Case Study

Recently we have proposed an MDA–based approach to automatically develop ubiquitous decision support systems (UDSS) for clinical guidelines [17]. One of the goals of a UDSS is to guide the physician during the application of a guideline in a very specific way in order to help in her decision making. Another goal of such decision support systems is to record automatically the trace of the application of the guideline. That is, since the application of a guideline can vary depending on the specific characteristics of the patient, our aim is to record all the information resulting from its application, such as the patient's clinical conditions, the physician's decision, the antibiotic treatment given to the patient and the actions or tests that are carried out regarding the patient's clinical state. All this information will be stored as part of the patient's clinical history. This could be used for obtaining a summary of the application of the guideline, the purposes of physicians' future decisions, legal support, etc [6]. From now on, we will focus our attention on this second goal of our UDSS proposal.

In order to store the trace of the application of a guideline, first of all we propose to represent the dynamics of the given guideline by means of a state-chart [15]. Next, a stereotyped UML class diagram is obtained starting from this

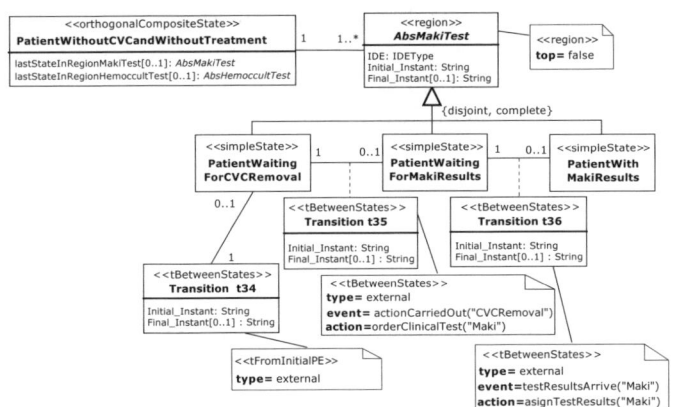

Fig. 1. Part of the stereotyped UML class diagram of the IRC guideline

statechart. The process we propose for generating the class diagram is based on some ideas proposed in [13], although our objective is different (in [13] the goal is to generate Java code). In addition to the ideas described in [13], we propose to annotate the class diagram with information about the UML statechart from which it comes, resulting in a stereotyped UML class diagram. The resulting stereotyped UML class diagram allows us to register information concerning the application of the guideline (patient's states, physician's decisions and clinical actions). Finally, in order to implement the physical storage, the stereotyped UML class diagram is transformed into an XML schema. So, in this context, a general mechanism that allows the transformation between these models is essential. The general mechanism we propose is described in the next section.

As an example of application, we will use a particular clinical guideline presented in [6]. This guideline is used by doctors in the Intensive Care Unit (ICU) when it is suspected that a patient has an infection related to an intravenous catheter (IRC). This guideline is necessary because the use of catheters is more and more frequent and they may be associated with complications, especially in Intensive Care Units (ICU) where patients in critical states suffer from a wide range of serious illness and injury. In particular, this guideline for the prevention of catheter-related infections (IRC guideline) establishes the criterion for the prevention, diagnosis and antibiotic treatment of infections caused by catheters.

For example, we can see a part of the stereotyped UML class diagram of the IRC guideline in Fig. 1. This UML diagram is stereotyped making use of a *profile for statechart execution persistence* (SEP profile) we have defined. This part of the diagram will allow the trace of the patient's states to be stored when the medical staff thinks the patient has an infection related to an intravenous catheter but the doctors do not yet know if the sick person has an infection. Moreover, the information relating to activities regarding Maki tests (such as clinical actions that are carried out, and timestamps in which the Maki test is ordered and its results are received) will be recorded. In particular, we can observe the class *Transition T35* which corresponds to the change in the patient's state when the catheter is removed. In these circumstances, the patient's state while waiting for the catheter removal (*PatientWaitingForCVCRemoval*) changes to another state in which she is waiting for the results of the Maki Test (*PatientWaitingForMakiResults*). The *TransitionT35* class is described by means of two tag values, one indicating the event which has triggered the transition (*actionCarriedOut('CVCRemoval')*) and the other tag value representing the action which has been carried out when the previous event has taken place (*orderClinicalTest('Maki')*).

4   Translation of Stereotyped Models to XML Schemas

UML profiles have been used in different medical information systems with different aims, such as for data confidentiality [9] or for historical records as in our case. In the medical context they can be applied to different types of UML diagrams (such as use case, class or sequence diagrams). Taking this into account, our proposal aims to be generally applicable and therefore allows the translation of any stereotyped UML model into XML Schema. Nevertheless, since in our specific case we start from a stereotyped UML class diagram model, in this section, we show our proposal applied to this type of UML diagram.

Our approach consists of two translation steps. In the first step we propose to create an XML schema for each UML profile, which makes the XML schema reusable for any UML model to which the profile is being applied. In the second step, we translate the stereotyped UML model into an XML schema, which will make use of the XML schema resulting from the previous step.

One of the strengths of our approach is that we do not present 'yet another approach' to the mapping of UML models into XML Schemas, since any previously published mapping can be taken as a starting point. For the translation of UML class diagrams, we have used the approach of [12], since we consider that this approach is one of the most complete proposals for these transformations [4].

Next, we explain in detail our approach using as an example the transformation of the stereotyped UML class model for trace recording of Fig. 1.

4.1   Step 1: Translation of a UML Profile

As we have pointed out, we create an XML schema for each UML profile. In order to define it, we have taken into account not only several peculiarities of the clinical context as far as conceptual modelling is concerned, but also specific semantics of UML Profiles. Considering these factors, the first decision that we have taken is to define this XML schema without declaring a `targetNamespace`, that is, to define a Chameleon XML schema [10]. The main reason for defining such a type of schema is the following. Since several profiles can be applied to different UML models in the medical environment, it is of interest for the XML schemas created from those UML models to provide their own application–specific namespace no matter what the profile or profiles applied to it. So, we prefer not to declare a namespace for each profile XML schema; rather, we want

```
 1  <?xml version="1.0"?>
 2  <xs:schema xmlns:xs="http://www.w3.org/2001/XMLSchema" elementFormDefault="qualified">
 3  <xs:complexType name="TransitionBetweenStatesDefinitionType">
 4    <xs:sequence>
 5      <xs:element name="type" type="TransitionKindEnume" minOccurs="1" maxOccurs="1"/>
 6      <xs:element name="event" type="xs:string" minOccurs="0" maxOccurs="unbounded"/>
 7      <xs:element name="guard" type="xs:string" minOccurs="0" maxOccurs="1"/>
 8      <xs:element name="action" type="xs:string" minOccurs="0" maxOccurs="1"/>
 9    </xs:sequence>
10    <xs:attribute name="profile" type="xs:string" fixed="SEPProfile"/>
11  </xs:complexType>
12  <xs:complexType name="StateDefinitionType">
13    <xs:sequence>
14      <xs:element name="entry" type="xs:string" minOccurs="0" maxOccurs="1"/>
15      <xs:element name="exit" type="xs:string" minOccurs="0" maxOccurs="1"/>
16      <xs:element name="do" type="xs:string" minOccurs="0" maxOccurs="1"/>
17    </xs:sequence>
18    <xs:attribute name="profile" type="xs:string" fixed="SEPProfile"/>
19  </xs:complexType>
20  <xs:complexType name="SimpleStateDefinitionType">
21    <xs:complexContent>
22      <xs:extension base="StateDefinitionType"/>
23    </xs:complexContent>
24  </xs:complexType>
25  ...
26  </xs:schema>
```

Fig. 2. Part of the SEP Profile XML schema

containing schemas to provide their own namespace. In this sense, chameleon schemas provide a lot of flexibility since they take on the `targetNamespace` attribute of the XML schema that includes them [10].

From now on we will refer to the obtained chameleon XML schema as the *profile XML schema*. It will be defined as follows. In each chameleon XML schema, every stereotype is represented as a global `complex type` element whose name is the name of the stereotype plus the word `DefinitionType`. This complex type will be defined in accordance with the following:

- when the stereotype has properties, this complex type will contain an XML `sequence` element including one XML `element` for each property of the stereotype. Each of these elements is named as the property. In addition, the multiplicity of the property is represented by using the `minOccurs` and `maxOccurs` attributes.
- whether or not the stereotype has properties, this complex type will have an XML `attribute` named `profile` of the type `string`. The value of this attribute will be the name of the profile to which the stereotype belongs.

For example, part of the profile XML schema obtained from the created SEP Profile is shown in Fig. 2. We show the translation of stereotypes «TransitionBetweenStates» (see lines from 3 to 11) and «SimpleState» (see lines from 20 to 24). For instance, in the translation of the definition of the «TransitionBetweenStates» stereotype, we have defined a complex type `Transition BetweenStatesDefinitionType` with a sequence of elements, one for each property (see lines from 5 to 8) and an attribute named `profile` with the name of the profile (see line 10).

```
1  <?xml version="1.0"?>                               33 <complexType name="TransitionT35Type">
2  <schema                                             34   <sequence>
3  xmlns="http://www.w3.org/2001/XMLSchema"            35     <element name="initial_instant" type="string"/>
4  xmlns:cd="http://www.XMLproyect.com/UMLCD"          36     <element name="final_instant" type="string"
5  targetNamespace="http://www.XMLproyect.com/UMLCD"   37       minOccurs="0"/>
6  elementFormDefault="qualified">                     38     <element name="TransitionBetweenStates"
7    <include schemaLocation="SEPProfileDefinition.xsd"/> 39         type="cd:TransitionBetweenStatesType"/>
8    <element name="TraceClassDiagram">                40   </sequence>
9      <complexType>                                   41 </complexType>
10       <sequence>                                    42 <complexType name="TransitionBetweenStatesType">
11         <element name="PatientWaitingForMakiResults" 43   <complexContent>
12           type="cd:PatientWaitingForMakiResultsType" 44     <restriction base="cd:TransitionBetween
13           minOccurs="0" maxOccurs="unbounded"/>     45       StatesDefinitionType">
14         <element name="Transition35"                46       <sequence>
15           type="cd:Transition35Type"                47         <element name="type" type="cd:TransitionKindType"
16           minOccurs="0" maxOccurs="unbounded"/>     48           fixed="external"/>
17           ...                                       49         <element name="event" type="string"
18       </sequence>                                   50           fixed="actionCarriedOut('CVCRemoval')"/>
19     </complexType>                                  51         <element name="action" type="string"
20     <key ...>...</key>                              52           fixed="orderClinicalTest('Maki')"/>
21   </element>                                        53       </sequence>
22   <complexType name="PatientWaitingForMakiResultsType"> 54     </restriction>
23     <complexContent>                                55   </complexContent>
24       <extension base="cd:AbsMakiTestType">         56 </complexType>
25         <sequence>                                  57   ....
26           ...                                       58 </schema>
27           <element name="simpleState"
28             type="cd:StateDefinitionType"/>
29         </sequence>
30       </extension>
31     </complexContent>
32   </complexType>
```

Fig. 3. The class diagram XML schema

4.2   Step 2: Transformation of the Stereotyped Class Model

Once an XML schema is defined for each profile, the second step deals with the creation of the XML schema for the stereotyped class model, called *stereotyped XML schema*. The idea consists of extending the approach of [12] by adding new transformation rules concerning the stereotypes applied to the class model.

As we will see in this section, the complex types created in the profile XML schemas will be used for the translation of the stereotypes applied to the class model. Therefore, the final XML schema created for the stereotyped class model will include, for each profile applied to the class model, an XML include element, representing schema dependencies. For example, an extract of the XML schema obtained from the stereotyped class diagram for trace recording of Fig. 1 is shown in Fig. 3. In line 7 of this figure we include the SEP profile XML schema in order to reuse its defined complex types along the XML schema.

In order to explain our approach for step 2, we will show how to translate stereotyped UML classes, attributes and binary associations by using the skeletons presented in Table 1. We have distinguished two translation options, depending on whether the stereotype (or stereotypes) applied to the UML element has tagged values. In each option, we indicate in bold the XML code which corresponds to the translation of a stereotype of the UML class model. In addition, italics describe code which has to appear in the XML schema.

As we show in Table 1(see (a)(c)(e)), the translation of applied stereotypes without tagged values consists of defining an XML element whose Type is the complex type defined for that stereotype in the profile XML Schema. If the applied stereotype has tagged values (see (b)(d)(f)), an XML element and an additional global XML complex type are defined. This complex type is declared as a derivation by restriction [22] of the complex type defined for the stereotype in the profile XML schema. Therefore, each XML element which corresponds with a tag value of the applied stereotype is redeclared. Also, the XML `fixed` attribute with the corresponding tag value as value is added to the redeclared XML element.

Table 1. Our approach to translate several stereotyped UML elements

<table>
<tr>
<th>Stereotyped UML class
in which the stereotype has no tagged values</th>
<th>Stereotyped UML class
in which the stereotype has tagged values</th>
</tr>
<tr>
<td>

```
1  <element name=className type=classNameType/>
2  <complexType name=classNameType>
3      <sequence>
4          XML elements which translate
             attributes of the UML class
5          <element name=stereotypeName
6              type=stereotypeNameDefinitionType/>
7      </sequence>
8      <attribute name="id" type="ID"
9          use="required"/>
10     XML attributes which translate
           attributes of the UML class
11 </complexType>
```

</td>
<td>

```
1  <element name=className type=classNameType/>
2  <complexType name=classNameType>
3      <sequence>
4          XML elements which translate
             attributes of the UML class
5          <element name=stereotypeName
6              type=stereotypeNameType/>
7      </sequence>
8      <attribute name="id" type="ID"
9          use="required"/>
10     XML attributes which translate
           attributes of the UML class
11 </complexType>

12 <complexType name=stereotypeNameType>
13     <complexContent>
14         <restriction
15             base=stereotypeNameDefinitionType>
16             <sequence>
17                 Redeclared elements from the
                     elements of the base type
18             </sequence>
19         </restriction>
20     </complexContent>
21 </complexType>
```

</td>
</tr>
<tr>
<td align="center">(a)</td>
<td align="center">(b)</td>
</tr>
<tr>
<th>Stereotyped UML attribute
in which the stereotype has no tagged values</th>
<th>Stereotyped UML attribute
in which the stereotype has tagged values</th>
</tr>
<tr>
<td>

```
1  <element name=attributeName
2      type=attributeNameType/>
3  <complexType name=attributeNameType>
4      <sequence>
5          XML element (with name "value")
             which translates the UML attribute
6          <element name=stereotypeName
7              type=stereotypeNameDefinitionType/>
8      </sequence>
9  </complexType>
```

</td>
<td>

```
1  <element name=attributeName
2      type=attributeNameType/>
3  <complexType name=attributeNameType>
4      <sequence>
5          XML element (with name "value") which
             translates the UML attribute
6          <element name=stereotypeName
7              type=stereotypeNameType/>
8      </sequence>
9  </complexType>

10 Include a complex type as the stereotypeNameType
     defined in (b) of this figure.
```

</td>
</tr>
<tr>
<td align="center">(c)</td>
<td align="center">(d)</td>
</tr>
<tr>
<th>Stereotyped UML binary association
in which the stereotype has no tagged values</th>
<th>Stereotyped UML binary association
in which the stereotype has tagged values</th>
</tr>
<tr>
<td>

```
1  <element name=associationEndRolName
2      type=associationEndRolNameType/>
3  <complexType name=associationEndRolNameType>
4      <sequence>
5          XML element (with name "id") which
             translates the UML association
6          <element name=stereotypeName
7              type=stereotypeNameDefinitionType/>
8      </sequence>
9  </complexType>
```

</td>
<td>

```
1  <element name=associationEndRolName
2      type=associationEndRolNameType/>
3  <complexType name=associationEndRolNameType>
4      <sequence>
5          XML element (with name "id") which
             translates the UML association
6          <element name=stereotypeName
7              type=stereotypeNameType/>
8      </sequence>
9  </complexType>

10 Include a complex type as the stereotypeNameType
     defined in (b) of this figure.
```

</td>
</tr>
<tr>
<td align="center">(e)</td>
<td align="center">(f)</td>
</tr>
</table>

Finally, the XML element generated for each applied stereotype is included as part of the translation of the stereotyped class element in the following way:

- if the transformation of the stereotyped UML element involves the creation of an XML `complex type` according to the chosen transformation (in our case the approach proposed in [12]), then the XML element generated for the stereotype is nested inside the last child of the sequence of the global complex type.
- otherwise, we propose to modify the translation approach of the stereotyped UML element so that it involves the creation of an XML `complex type`. In this way, the XML element created for the stereotype can be nested inside that complex type.

An example of our approach is shown in Fig. 3. In this XML schema, among other things, the translation of class *TransitionT35* with the stereotype «transitionBetweenStates» previously commented in Section 3 is depicted. For the translation of this stereotyped class, since its applied stereotype has tagged values, we have followed the approach of Table 1(b). We have created the XML element `transitionBetweenStates` of type `transitionBetweenStatesType` (see lines 38 and 39). Additionally, we have defined the complex type `transitionBetween-StatesType` as a derivation by restriction of the type `transitionBetweenStatesDefinitionType` (see lines from 42 to 56). In this case, we have added to that complex type three new elements corresponding to three of the tagged values of the applied stereotype. Each of those elements has the XML element `fixed`, indicating the specific value of the corresponding property. For instance, to translate the event that has triggered the transition (*actionCarriedOut('CVCRemoval')*), we use the XML element `event` defined with the fixed value `actionCarriedOut('CVCRemoval')` (see lines 49 and 50).

5   Conclusions and Future Work

In this paper we propose the use of stereotyped UML class diagrams to store a guideline application in an MDD setting. We also provide a general mechanism for the transformation of stereotyped UML class models into XML schemas.

There are several lines of future work. The main issue is the incorporation of our transformation proposal into our general framework of evolution of UML/XML models [5]. In particular, the development of an evolution tool is an ongoing project. As for the transformation of UML profiles, the definition of a procedure for the translation of UML constraints is a goal for further work.

References

1. Abdullah, M.S., Paige, R., Benest, I., Kimble, C.: Knowledge Modelling Using The UML Profile. In: AIAI 2006, vol. 204, pp. 70–77. Springer, Heidelberg (2006)
2. Aggarwal, V.: The Application of the Unified Modeling Language in Object-Oriented Analysis of Healthcare Information Systems. Systems Journal of Medical Systems 26(5), 383–397 (2002)

3. Dart, T., Xu, Y., Chatellier, G., Degoulet, P.: Computerization of guidelines: towards a "guideline markup language". Medinfo. 10, 186–190 (2001)
4. Domínguez, E., Lloret, J., Pérez, B., Rodríguez, A., Rubio, A.L., Zapata, M.A.: A Survey of UML Models to XML Schemas Transformations (submitted for publication)
5. Domínguez, E., Lloret, J., Rubio, A.L., Zapata, M.A.: Evolving XML Schemas and Documents Using UML Class Diagrams. In: Andersen, K.V., Debenham, J., Wagner, R. (eds.) DEXA 2005. LNCS, vol. 3588, pp. 343–352. Springer, Heidelberg (2005)
6. Domínguez, E., Pérez, B., Rodríguez, A., Zapata, M.A.: Medical protocols for taking decisions, in an ubiquitous computation context. Novática 177, 38–41 (2005)
7. Elkin, P.L., Peleg, M., Lacson, R., Bernstam, E., Tu, S., Boxwala, A., Greenes, R.A., Shortliffe, E.H.: Toward Standardization of Electronic Guideline Representation. MD Computing 17(6), 39–44 (2000)
8. Hederman, L., Smutek, D., Wade, V., Knape, T.: Representing Clinical Guidelines in UML: A Comparative Study. In: Medical informatics in Europe 2002, Amsterdam, Netherlands, pp. 471–477. IOS Press, Amsterdam (2002)
9. Heldal, R., Schlager, S., Bende, J.: Supporting Confidentiality in UML: A Profile for the Decentralized Label. In: Third International Workshop on Critical Systems Development with UML (2004)
10. Hunter, D., et al.: Beginning XML, 3rd edn. Wrox (September 2004)
11. Jones, V.M., Rensink, A., Brinksma, H.: Modelling mobile health systems: an application of augmented MDA for the extended healthcare enterprise. In: Proceedings of the EDOC 2005, pp. 58–69. IEEE Computer Society Press, Los Alamitos (2005)
12. Krumbein, T., Kudrass, T.: Rule-Based Generation of XML Schemas from UML Class Diagrams. In: WebDB. Proceedings of the XML Days at Berlin, Workshop on Web Databases, pp. 213–227 (2003)
13. Niaz, I.A., Tanaka, J.: Code Generation From UML Statecharts. In: SEA 2003 (2003)
14. Ohno-Machado, L., Gennari, J.H., Murphy, S.Ñ., et al.: The GuideLine Interchange Format: A Model for Representing Guidelines. Journal of the American Medical Informatics Association 5(4), 357–372 (1998)
15. OMG. UML 2.0 Superstructure Specification (August 2005), Document formal/05-07-04, Available at http://www.omg.org/
16. OpenClinical. Methods and tools for representing computerised clinical guidelines. http://www.openclinical.org/gmmsummaries.html
17. Porres, I., Dominguez, E., Perez, B., Rodriguez, A., Zapata, M.A.: Development of an Ubiquitous Decision Support System for Clinical Guidelines using MDA (submitted for publication)
18. Selic, B.: The pragmatics of model-driven development. IEEE Software 20(5), 19–25 (2003)
19. Shabo, A., Rabinovici-Cohen, S., Vortman, P.: Revolutionary impact of XML on biomedical information interoperability. IBM Systems Journal 45(2), 361–372 (2006)
20. Tanaka, A., Ñagase, Y., Kiryu, Y., Ñakai, K.: Applying ODP Enterprise Viewpoint Language to Hospital Information System. In: EDOC 2001. Proceedings of the Fifth IEEE International, pp. 188–192. IEEE Computer Society, Los Alamitos (2001)
21. van der Maas, A.A.F., Ter Hofstede, A.H.M., Hoopen, A.J.T.: Requirements for Medical Modeling Languages. Journal of the American Medical Informatics Association 8, 146–162 (2001)
22. W3C. XML Schema Part 0: Primer Second Edition (October 2004), Available at http://www.w3.org/TR/xmlschema-0/

An Extendable System for Conceptual Modeling and Simulation of Signal Transduction Pathways

Silke Eckstein and Claudia Täubner*

Technical University of Braunschweig, Germany
Institute of Information Systems
eckstein@ifis.cs.tu-bs.de, taeubner@ifis.cs.tu-bs.de
http://www.ifis.cs.tu-bs.de/welcome_e.html

Abstract. In this paper we present an extendable system to generate models of signal transduction pathways in different modeling languages and to simulate these models with the belonging simulation tools. Systems biology is an emerging research area in Life Sciences with the aim "to develop a system level understanding of biological systems" (Kitano). One step in this direction – besides a lot of other necessary steps – is to model and simulate signal transduction pathways, which describe the flow of signals inside the cell. This can very well be done by using discrete modeling languages such as Petri Nets or Life Sequence Charts. Another step forward is to generate such models out of databases for signal transduction pathways.

1 Introduction

Systems biology is an emerging research area in Life Sciences with the aim "to develop a system level understanding of biological systems" [Kit00]. This aim of developing a somehow complete model of e.g. a living cell will possibly be achieved by the development of models describing single aspects of a cell, which can then be composed together to a comprehensive model. One of these single aspects is the transduction of signals inside the cell. In few words, a signal transduction pathway describes, how a signal received by a receptor is processed inside the cell via biochemical reactions. Usually, it is transfered into the nucleus, where it causes a change in the currently active genetic program. Modeling and simulating this information flow is beneficial for several reasons: It helps to understand the flow of signals in a complex network, to test hypotheses in silico before validating them with experiments and last but not least to validate the data collected about a certain pathway.

The fact that a flow of information shall be described has led to the idea of applying conceptual modeling languages in this area, even if these were originally developed to assist the construction or engineering of systems and not the description of already existing systems [FHH+04]. A couple of modeling languages, e.g. Petri Nets, Life Sequence Charts etc. qualify themselves for this task,

---

* Partially supported by the German BMBF, "Intergenomics" (031U210C).

J.-L. Hainaut et al. (Eds.): ER Workshops 2007, LNCS 4802, pp. 54–63, 2007.

all bringing different advantages and drawbacks with them. In the same way the belonging simulation tools have different strengths and weaknesses. Hence, it would be beneficial to develop different models for one signal transduction pathway using different modeling languages and to be thereby able to apply different simulation tools.

There exist a number of pathway databases compiling the current knowledge about biological pathways. It would be a great advantage to be able to generate different pathway models directly out of those databases.

Thus, we developed an extendable system to generate different models, expressed with different modeling languages, out of the data stored in pathway databases in order to simulate these models with the belonging simulation tools. At the moment we operate on one of the major pathway databases, TRANS-PATH® [KPV+06], and use Colored Petri Nets (CPNs) and Life Sequence Charts (LSCs) as modeling languages. The belonging simulation tools are CPN Tools [JKW07] and Play-Engine [HM03], respectively.

The rest of the paper is organized as follows: We present some background knowledge about signal transduction pathways in the second section. Afterwards we discuss the conceptual modeling of signal transduction pathways with Colored Petri Nets and Life Sequence Charts, respectively. In section five we present the architecture and the main workflow of our systems. Related work is discussed in section six, before we end up with some conclusions and an outlook on future work.

2   Signal Transduction Pathways

In organisms proteins have a wide variety of functions and they interact with each other in likewise multifaceted ways. These interactions of proteins are being described by means of pathways or networks, which are typically represented as certain kinds of maps. A distinction is drawn between metabolic and regulatory pathways, whereas metabolic pathways describe the conversion of classes of substances into other classes of substances. Regulatory pathways on the other hand describe, how the function of something is being regulated. Here, not the conversion of classes of substances into other classes is significant but the transduction of signals. That is why they are also called signal transduction pathways.

Put the other way round, a signal transduction pathway describes, how a cell responses to an extracellular signal, e.g. a signaling molecule excreted by a bacterium. The signal is being received at a receptor protein and then transfered via biochemical reactions into the nucleus, where it changes the currently active genetic program. Figure 1 shows an example of a signal transduction pathway, where the gray area marks the inside of a cell and the light-colored area stands for the outside. The nucleus is represented as a gray ellipse.

In this map molecules are represented with different shapes and colors, which encode the role a certain molecule plays in the pathway under consideration. Examples for such roles are extracellular signals, represented as stars, receptors, represented as rectangles across the cell membrane, and adapter proteins,

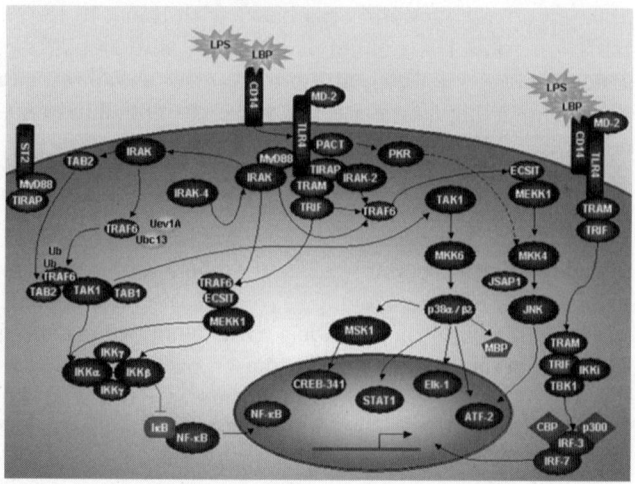

Fig. 1. TLR4 signal transduction pathway in the TRANSPATH® database [KPV+06]

represented as blue ellipses. Interactions of the molecules appear as lines and arrows, whereas their different shapes stand for different kinds of interactions, e.g. direct or indirect activation or inhibition. Molecules also interact by building molecule complexes, which are represented through narrow cumulations of molecules.

Such pathways are being composed by experts, who study the relevant literature and accumulate small pieces to a big picture. This is, because worldwide there are various groups doing research in very small parts of a signal transduction pathway, e.g. research about short sequences of chemical reactions. These information pieces are then composed bottom-up to a signal transduction pathway by other groups, who collect these information into databases and provide an integrated view on the whole pathway.

Examples for such signal transduction pathway databases are TRANSPATH® [KPV+06], KEGG [KGH+06], Reactome [JTGV+05] and BioCyc [KOMK+05]. They usually provide a web interface for interactive searches and also make their data available as text files in flat file or XML format. Some of the databases already use a more or less standardized exchange format on XML basis, e.g. SBML (Systems Biology Markup Language).

Our database to start with was TRANSPATH®, which provides the accession methods described above but uses its own XML format. In January 2007 TRANSPATH® contained entries about 60,000 molecules, 100,000 chemical reactions, 20,000 genes and 57 pathways. The information was based on 30,000 publications. The web interface provides access to all these entries and in addition contains interactive maps, which give an overview over a certain pathway (cf. Fig. 1). The XML version of the database is divided into 6 files containing data about molecules, genes, reactions, pathways, annotations and references, respectively. They are accompanied by a DTD describing the structure of the files.

3 Conceptually Modeling Pathways with CPNs

In this section we give an overview over our representation of signal transduction pathways with Colored Petri Nets (CPNs). In contrast to regular Petri Nets [GV03] in Colored Petri Nets the token are distinguishable [Jen97]. Each token has a certain value belonging to a certain data type. The values of the tokens are called colors and the data types color sets. Each place must carry the information, which kind of tokens it accepts, i. e. tokens of which color (data type).

In our approach, the tokens represent molecules and molecule complexes. Each place can therefore carry certain molecules or molecule complexes. Transitions represent chemical reactions. The incoming and outgoing arcs of a transition specify together with their weights, how many molecules of which kind function as educts, enzymes, inhibitors or products, respectively.

In Fig. 2 a class diagram is shown, which represents pathways, as they are stored in TRANSPATH®, in an abstract way. Each pathway consists of a number of chains of reactions. Molecules participate in reactions as reactants (or educts), enzymes, inhibitors and products. Molecules can be members of molecule complexes, which are treated as molecules themselves. Molecules can also be in different states, e.g. hydroxylated or not. In TRANSPATH® these different states are represented with two entries in class molecule and a reference (state of) between the two instances.

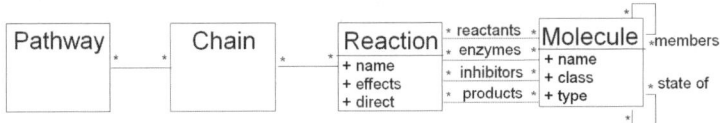

Fig. 2. Representation of pathways as a class diagram

When mapping a pathway into a Petri Net the breakdown into chains is abandoned with the consequence, that only the reactions belonging to a pathway and the molecules participating in the reactions have to be translated. The general assignment of elements of the pathway class diagram to Petri Net concepts is shown in Tab. 1.

An example for a transformation of a certain reaction – the binding of three molecules to a complex – into a Petri Net is shown in Fig. 3. Figure 4 contains a subnet of the Petri Net representation of the TLR4 pathway, which was shown in Fig. 1.

To gain some experiences in how to model a signal transduction pathway with CPN we first did a transformation by hand [TMK+06, Fle05] and derived out of it the assignment of concepts presented above. That transformation was based on numerous discussions with experts from biology, among them one member of the TRANSPATH® team.

There is some post- and preprocessing necessary in order to build a Colored Petri Net, which can be simulated using CPN Tools [JKW07], out of the

Table 1. General assignment of elements of the pathway class diagram to PN concepts

Pathway	CPN
Reaction	transition
Reaction.name	-
Reaction.effects	transition name
Reaction.direct	arrow color (gray)
Molecule	places/token
Molecule.name	initial marking
Molecule.class	color-set
Molecule.type	identification of molecule groups

Pathway	CPN
reactants	input places
enzymes	input-/output places
inhibitors	input places
products	output places
members	group color-sets
state_of	complex color-sets
	color highlighting

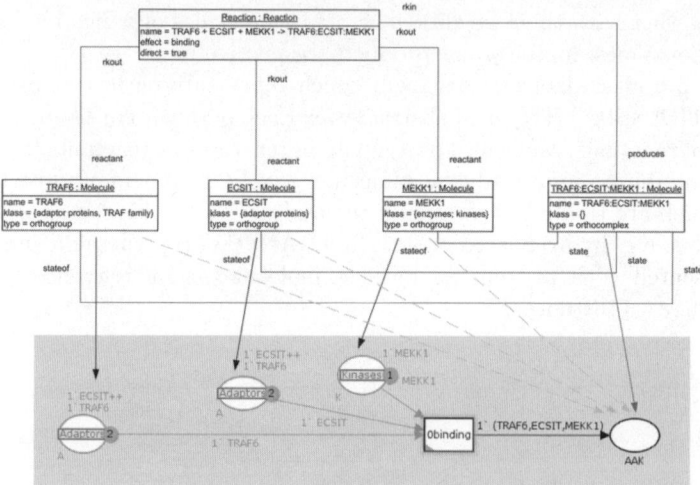

Fig. 3. Exemplary transformation of a reaction into a CPN

TRANSPATH® XML data. First of all, the data belonging to a certain pathway has to be collected out of a number of XML files. After this preprocessing step the data is available in the form described by the class diagram in Fig. 2.

CPN Tools stores its Petri Nets as XML files and a DTD describing the structure is also available. Therefore, we decided to generate valid XML files with respect to this DTD according to the transformation rules sketched above. One problem that arises in this context is, that we can not generate any layout information for the Petri Net out of our pathway data. This means, that we have to use a placing algorithm to constitute the layout information. This postprocessing step is still work in progress. Thus, the Petri Net of the TLR4 pathway shown above in Fig. 4 was generated out of the TRANSPATH® data, but the layout was later done by hand. First simulations already revealed some data inconsistencies in TRANSPATH® like e.g. missing links and missing molecules.

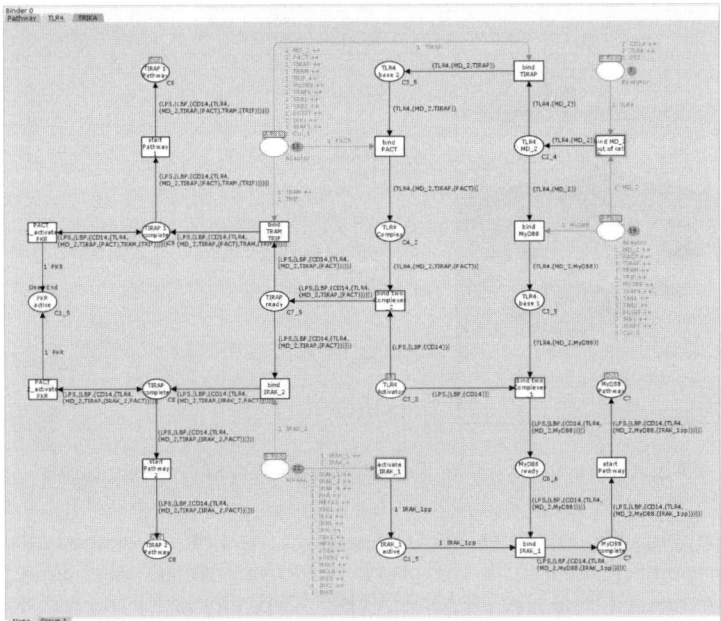

Fig. 4. Subnet of the TLR4 CPN

4   Conceptually Modeling Pathways with LSCs

The second formalism we use to represent signal transduction pathways are Life Sequence Charts [DH01], which are an extension of Message Sequence Charts and allow to describe the behavior of a concurrent system. The belonging simulation tool is called Play-Engine [HM03]. This simulation tool is especially interesting in our context, not only because it implements another discrete modeling language, which enables us to generate different models for a certain pathway and to compare these models with each other. But it is also interesting, because it allows to provide a graphical user interface (GUI) representing the panel of the system being developed. In the simulation phase this GUI is being animated according to the events occurring in the LSCs. Think of a control panel, where some settings can be undertaken. Subsequently the system is simulated according to these settings and the resulting values for some monitored functions are displayed.

These GUIs can be developed completely according to the users needs and their constituents can be connected to the LSC concepts. Our idea in this context is to use the pathway maps provided by the TRANSPATH® (cf.´Fig. 1) as GUIs. Consequently, we will be able to animate these maps according to the simulation of the pathway. This means, that the experts from biology will be able to use a simulation tool, which displays the simulation run in their own graphical language.

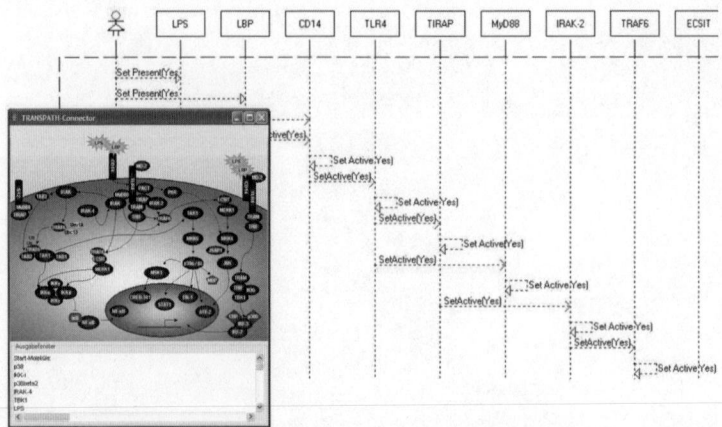

Fig. 5. Fragment of the TLR4 pathway in LSC representation

The mapping of signal transduction pathways to LSCs is defined analogously to the mapping presented in the previous section. We use the same pathway instance extracted from the TRANSPATH® XML files as for the transformation into Petri Nets (cf. Fig. 2). In Fig. 5 you can see a fragment of the TLR4 pathway in LSC representation.

At the moment, we are able to to generate LSCs out of the TRANSPATH® data and to simulate them with the Play-Engine. The animation of the pathway maps is still work in progress.

5 Architecture and Workflow

The main architecture of our system "Pathway Modeler" is presented in Fig. 6. We will explain it by discussing the main workflow.

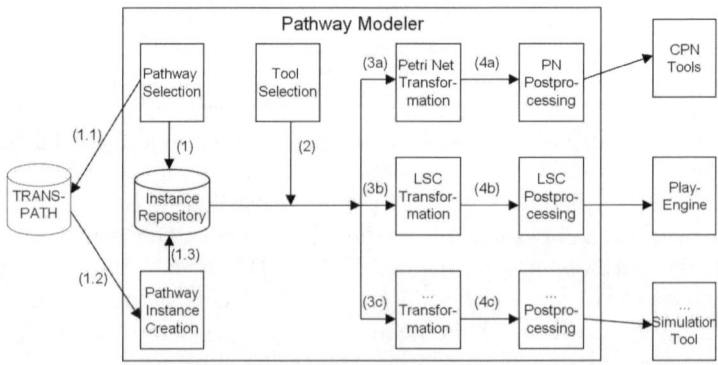

Fig. 6. Architecture

Initially, the user has to select the pathway he wants to simulate. The system checks, if it is already available in the Instance Repository (1). Otherwise, it collects all data belonging to the pathway out of the TRANSPATH® XML files and creates a pathway instance (1.1 - 1.3). Afterwards, the user selects the simulation tool he wants to generate a pathway model for (2). Presently, CPN Tools and the Play-Engine are available. According to the choice, a Petri Net representation or a LSC representation of the selected pathway are constructed (3a/3b). The system is prepared for further simulation tools to be integrated (3c). After the transformation is done, some postprocessing steps follow. In case of the Petri Net representation e.g. the layout algorithm we discussed in sec. 6 has to be applied. Finally, an input file for the selected simulation tool is the result of the transformation process.

6   Related Work

In systems biology several modeling and simulation approaches are being applied [KHK+05]. Most of them can be classified according to the following three dimensions [UDZ05]: discrete and continuous, qualitative and quantitative as well as deterministic and stochastic. When examining biological pathways, the considered events are discrete but the other two dimensions are not determined. Especially various kinds of Petri Nets have been applied in this area [HR04], ranging from Colored Petri Nets describing metabolic pathways qualitativly [HK04] to Stochastic Petri Nets, which allow a quantitative modeling, and Hybrid Functional Petri Nets quantitatively representing metabolic pathways [NDMM05].

Life Sequence Charts have been applied to model a certain part in the development of the worm *C.elegans* [KHK+04] and the π-calculus [RS04] to model gene regulatory networks.

What most of the approaches have in common is, that the models are constructed "by hand" and not generated out of a database. And in those approaches, where the models are derived out of one or more databases (e.g. [NDMM05]), only one kind of models is generated. This is in contrast to our approach, where we offer the possibility to generate different representations for a certain pathway.

7   Conclusion

In this paper we presented an extendable system for the modeling and simulation of signal transduction pathways. As a source for the pathways to be simulated we use the TRANSPATH® database and as simulation tools CPN Tools and the Play-Engine for Coloured Petri Nets and Life Sequence Charts, respectively.

In the future we plan to support other modeling languages and simulation tools, as e.g. State Charts and the Rhapsody tool, for which first preparations already have begun. Furthermore, we will adapt our system to accept CSML (Cell System Markup Language, [NJD+06]) as an input format for signal transduction pathways. CSML is an XML based exchange format for biological pathways.

The developers of CSML provide tools to translate pathways described with other XML based languages, as e.g. CellML [LHN04] and SBML [HFB+04] (Systems Biology Markup Language), into CSML. Since most of the major pathway databases make their data available using one of these exchange formats, supporting CSML will enable us to process pathways not only from TRANSPATH® but from the other major databases as well.

Acknowledgments

We would like to thank C. Choi, E. Shelest, B. Störmann and C. Rio Bartulos for providing us insights in signal transduction pathways, our colleagues from the Intergenomics project (especially R. Münch) for fruitful discussions and last but not least all the students that contributed – directly or indirectly – through their diploma theses to this project: N. Fleischer, H. Langhorst, M. Pirsch, F. Rudolph, T. Springmann, J.-C. Treusch, O. Witthöft and M. Ziegler.

References

[DH01]   Damm, W., Harel, D.: LSCs: Breathing Life into Message Sequence Charts. Formal Methods in System Design 19(1), 45–80 (2001)

[FHH+04]   Fisher, J., Harel, D., Hubbard, E.J.A., Piterman, N., Stern, M.J., Swerdlin, N.: Combining State-Based and Scenario-Based Approaches in Modeling Biological Systems. In: Danos, V., Schachter, V. (eds.) CMSB 2004. LNCS (LNBI), vol. 3082, pp. 236–241. Springer, Heidelberg (2005)

[Fle05]   Fleischer, N.: Modellierung und Simulation der P. aeruginosa Infektion mit Petri-Netzen. Master's thesis, TU Braunschweig (2005)

[GV03]   Girault, C., Valk, R.: Petri Nets for Systems Engineering - A Guide to Modeling, Verification, and Applications. Springer, Berlin (2003)

[HFB+04]   Hucka, M., Finney, A., Bornstein, B.J., Keating, S.M., Shapiro, B.E., Matthews, J., Kovitz, B.L., Schilstra, M.J., Funahashi, A., Doyle, J.C., Kitano, H.: Evolving a Lingua Franca and Associated Software Infrastructure for Computational Systems Biology: The Systems Biology Markup Language (SBML) Project. Systems Biology 1(1), 41–53 (2004)

[HK04]   Heiner, M., Koch, I.: Petri Net Based Model Validation in Systems Biology. In: Cortadella, J., Reisig, W. (eds.) ICATPN 2004. LNCS, vol. 3099, pp. 216–237. Springer, Heidelberg (2004)

[HM03]   Harel, D., Marelly, R.: Come, Let's Play - Scenario Based Programming Using LSCs and the Play-Engine. Springer, Berlin (2003)

[HR04]   Hardy, S., Robillard, P.N.: Modeling and Simulation of Molecular Biology Systems Using Petri Nets: Modeling Goals of Various Approaches. J. Bioinform Comput. Biol. 2(4), 595–613 (2004)

[Jen97]   Jensen, K.: Coloured Petri Nets - Basic Concepts, Analysis Methods and Practical Use, 2nd edn. Springer, Berlin (1997)

[JKW07]   Jensen, K., Kristensen, L.M., Wells, L.: Coloured Petri Nets and CPN Tools for modelling and validation of concurrent systems. In: STTT. Int. J. on Software Tools for Technology Transfer, Sp. Sec. CPN 04/05 (2007)

[JTGV+05]   Joshi-Tope, G., Gillespie, M., Vastrik, I., D'Eustachio, P., Schmidt, E., de
            Bono, B., Jassal, B., Gopinath, G.R., Wu, G.R., Matthews, L., Lewis, S.,
            Birney, E., Stein, L.: Reactome: a knowledgebase of biological pathways.
            Nucleic Acids Research 33(suppl_1), D428–432 (2005)
[KGH+06]   Kanehisa, M., Goto, S., Hattori, M., Aoki-Kinoshita, K.F., Itoh, M.,
            Kawashima, S., Katayama, T., Araki, M., Hirakawa, M.: From genomics
            to chemical genomics: new developments in KEGG. Nucleic Acids Re-
            search 34(suppl_1), D354–357 (2006)
[KHK+04]   Kam, N., Harel, D., Kugler, H., Marelly, R., Pnueli, A., Hubbert, E.J.A.,
            Stern, M.J.: Formal Modelling of C. elegans Development - A Scenario-
            Based Approach. In: Ciobanu, G., Rozenberg, G. (eds.) Modelling in
            Molecular Biology, pp. 151–173. Springer, Berlin (2004)
[KHK+05]   Klipp, E., Herwig, R., Kowald, A., Wierling, C., Lehrach, H.: Systems
            Biology in Practice - Concepts, Implementation and Application. Wiley-
            VCH, Weinheim (2005)
[Kit00]   Kitano, H.: Perspectives on systems biology. New Generation Comput-
            ing 18(3), 199–216 (2000)
[KOMK+05]   Karp, P.D., Ouzounis, C.A., Moore-Kochlacs, C., Goldovsky, L., Kaipa,
            P., Ahren, D., Tsoka, S., Darzentas, N., Kunin, V., Lopez-Bigas, N.: Ex-
            pansion of the BioCyc collection of pathway/genome databases to 160
            genomes. Nucleic Acids Research 33(19), 6083–6089 (2005)
[KPV+06]   Krull, M., Pistor, S., Voss, N., Kel, A., Reuter, I., Kronenberg, D.,
            Michael, H., Schwarzer, K., Potapov, A., Choi, C., Kel-Margoulis, O.,
            Wingender, E.: TRANSPATH(R): an information resource for storing
            and visualizing signaling pathways and their pathological aberrations.
            Nucleic Acids Research 34(suppl_1), D546–551 (2006)
[LHN04]   Lloyd, C.M., Halstead, M.D.B., Nielsen, P.F.: CellML: its future, present
            and past. Progress in Biophysics and Molecular Biology 85(2-3), 433–450
            (2004)
[NDMM05]   Nagasaki, M., Doi, A., Matsuno, H., Miyano, S.: Petri Net Based Descrip-
            tion and Modeling of Biological Pathways. Algebraic Biology - Computer
            Algebra in Biology, 19–31 (2005)
[NJD+06]   Nagasaki, M., Jeong, E., Doi, A., Saito, A., Miyano, S.: CSML 3.0 - Cell
            System Markup Language - Basic Concept and Specification (2006)
[RS04]   Regev, A., Shapiro, E.: The pi-calculus as an Abstraction for Biomolecu-
            lar Systems. In: Ciobanu, G., Rozenberg, G. (eds.) Modelling in Molecular
            Biology, pp. 219–266. Springer, Berlin (2004)
[TMK+06]   Täubner, C., Mathiak, B., Kupfer, A., Fleischer, N., Eckstein, S.: Mod-
            elling and Simulation of the TLR4 Pathway with Coloured Petri Nets.
            In: Dhawan, A.P., Laine, F.A., Akay, M., Chon, K.H. (eds.) 28th Annual
            Int. Conf. of IEEE Engineering in Medicine and Biology Society, New
            York, 2006, p. 221. IEEE Computer Society Press, Los Alamitos (2006)
[UDZ05]   Uhrmacher, A.M., Degenring, D., Zeigler, B.: Discrete Event Multi-level
            Models for Systems Biology. In: Priami, C. (ed.) Transactions on Compu-
            tational Systems Biology I. LNCS (LNBI), vol. 3380, pp. 66–89. Springer,
            Heidelberg (2005)

Toward an Ontological Database for Subcellular Neuroanatomy*

Amarnath Gupta, Stephen D. Larson, Christopher Condit,
Sandeep Gupta, Lisa Fong, Li Chen, and Maryann E. Martone

University of California San Diego, La Jolla, CA 92093, USA
gupta@sdsc.edu, slarson@ucsd.edu, condit@sdsc.edu, sandeep@sdsc.edu,
lfong@ncmir.ucsd.edu, lichen@sdsc.edu, mmartone@ncmir.ucsd.edu

Abstract. We present the semantic data model for an ontological database for subcellular anatomy for Neurosciences. The data model builds upon the foundations of OWL and the Basic Formal Ontology, but extends them to include novel constructs that address several unresolved challenges encountered by biologists in using ontological models in their databases. The model addresses the interplay between models of space and objects located in the space, objects that are defined by constrained spatial arrangements of other objects, the interactions among multiple transitive relationships over the same set of concepts and so on. We propose the notion of parametric relationships to denote different multiple ways of parcellating the same space. We also introduce the notion of phantom instances to address the mismatches between the ontological properties of a conceptual object and the actual recorded instance of that object in cases where the observed object is partially visible.

1 Introduction

An Ontology-based database (OBDB) is a new class of information systems that consists of a domain ontology and a database content that references the ontology, such that the system can be queried both as a database and through the ontology. In recent work, [1,2] have developed an object-oriented framework for managing and querying OBDBs. We consider this class of systems to be especially important for life science applications and develop elements of a semantic data model to address a number of specific modeling issues we have encountered in developing ontology-based databases for neuroscience. A primary difference between the work in [1,2] and ours is that we consider the ontology to be much richer, specified in terms of OWL-DL or any other $\mathcal{SHOIN}$-compliant description logic, supplemented by constructs and rules of entailment from the Basic Formal Ontology (http://www.ifomis.uni-saarland.de/bfo) that describes, among other things, a formalization of space and objects in space.

This paper investigates how the semantic data model of an ODBMS is influenced when the application domain contains information about interacting

* This work is supported by NIH BIRN-CC Award No. 8P41 RR08605-08S1, NIH Human Brain Project Award No. 5RO1DC03192.

J.-L. Hainaut et al. (Eds.): ER Workshops 2007, LNCS 4802, pp. 64–73, 2007.
© Springer-Verlag Berlin Heidelberg 2007

extended objects in space. To focus the issue around a life science problem, we consider the domain of neuroanatomy as our application context.

Application Domain. To situate the application context of our problem domain, consider that one of the fundamental organizing principles of biological information is anatomical, i.e., based on the physical structure of living beings, and how these entites relate to each other. Anatomy provides one of the *covering* disciplines that spans every single entity in any biological organism. Further, a number of biological functions can be associated with anatomical entities, the functional relationship between two entities is often effected through their anatomical relationship. Anatomy presents a number of interesting issues. As a simple example, any anatomical structure (at least for multicellular organisms) can be recursively broken down into increasingly finer substructures – starting from substructures that are at the same scale (i.e., measured by the same spatial units) like forearm and phallanges, and extending to substructures that span scales such as tissues and cellular ultrastructures. This meronymic continuum is complemented by other relationships among anatomical enities such as physical and chemical connectivity, functional groups, and cytoarchitectural associations. It is the task of a conceptual model to capture the intricacies of their relationships to various biological functions. Because of the central role of anatomy in structuring biological information, it may also be used to serve as a common backbone for problems in biological information integration. For example, an organ-level biological database and a cell-level biological database can be semantically integrated through the multi-granular meronymy of an anatomical ontology.

The central role of anatomy in modeling and understanding biological systems has inspired a number of formal modeling efforts. For example, the cellular component fragment of the well-known Gene Ontology (http://www.geneontology.org/) captures is-a and part-of relationships among intracellular structures of generic cells. The Foundational Model of Anatomy (FMA), on the other hand, captures gross anatomical object including spatial relationships like has_boundary. These two ontologies overlap in that the cell-level terms of the FMA is a subset of that in Gene Ontology. And yet, their combined collection of terms and relationships leave some holes in the eyes of the neuroanatomist. The present work is an extension and re-formalization of *SAO* [3], the ontology for subcellular anatomy arises from our recognition that (a) these ontologies do not reflect the complexity needed to express cellular and subcellular neuroanatomical information, and (b) there is no available ontology that formally reflects *mesoscale* information, where those structures that sit between more gross anatomical scales and the level of individual protein and other macromolecules. In our example area of the nervous system, it comprises the dimensional range of nanometers to microns, encompassing cellular networks, subcellular microdomains, and their macromolecular constituents. These spatially extended structures lie at the heart of information processing in the nervous system, providing the adaptive spatial framework in which molecular and biochemical processes occur. In order to uncover the complex, detailed structural and dynamic inter-relations critical to the functional properties of nervous system, entities at this scale must be observed and described, and inferred using a combination of experimental observations and formal treatment. The purpose of the semantic modeling framework proposed in this paper is to create an OBDB that serves this goal.

Contributions. In this paper we propose the notion of parametric relationships to denote different multiple ways of parcellating the same space. We also introduce the notion of phantom instances to address the mismatches between the ontological properties of a conceptual object and the actual recorded instance of that object in cases where the observed object is partially visible.

2   Data Model

2.1   Preliminaries

An OBDB consists of the triple $(\mathcal{S}, \mathcal{O}, \mu)$, where $\mathcal{S}$ is a schema, $\mathcal{O}$ is an ontology and μ defines a set of mappings between $\mathcal{O}$ and $\mathcal{S}$. For simplicity, we start with the assumption that $\mathcal{S}$ is a relational schema containing relations $R_1, R_2 \ldots$. We also make the simplifying assumption at this point that we have a single ontology $\mathcal{O}$, and that the ontology roughly corresponds to the $\mathcal{SHIN}$ description logic. The focus of this paper is to define a variety of mapping relationships μ that can be concretely used for anatomical information.

Any relation $R_i(\mathbf{A}_o, \mathbf{A}_d)$ may have two kinds of attributes – $\mathbf{A}_o$ is the subset of attributes that are *ontology-mapped*, while $\mathbf{A}_d$ is the subset of attributes having data domains. If $R_i.A_j$ is an ontology-mapped attribute of relation R_i, the domain of $R_i.A_j$ is defined to be an ontology expression that specifies which part of the ontology will provide values for the attribute. In our example domain, consider the relation neuron-image containing images of traced neurons.

neuron-image(image-id, image-type, neuron-name, x-size, y-size)

where neuron-name is mapped to the subcellular anatomy ontology (SAO) and the mapping μ_{dom} is described as:

dom(neuron-image.neuron-name) = SAO:subclass*('neuron')

where 'neuron' is a term of the SAO (Subcellular Anatomy Ontology) [3] ontology and subclass* refers to the transitive closure of the subclass relationship of SAO. We will present more examples of ontology-expressions as we proceed. Suppose t (e.g., 'basket cell') is a term belonging to SAO:subclass*('neuron') and r_i is a tuple in neuron-image such that r_i.neuron-name $= t$. We say tuple r_i is an *evidence-of* t. The set of all evidences of a term t in a relation R is called the *range* of t in R. Why should one invent yet another relationship like *evidence-of*? Semantically, it will be incorrect to say that tuple r_{20} of neuron-image is an instance of 'basket cell', for example, because r_{20} is actually an instance of class image which happens to *depict* a basket cell. Thus this image is like a container of an instance of an ontological object. We capture this semantic relationship through the distiguished relationship *evidence-of.*

Now consider the relation that stores information about all structures that have been segmented from images:

segments(<u>structure-name</u>, <u>cell-name</u>, image-id, length, volume)

where the domain of structure-name is SAO:component-of*('neuron') and the domain of cell-name is SAO:subclass*('neuron'). Note that has-component (hc) is a kind of partonomic relationship such that the component has a specific function within the cell. Clearly, every tuple of the relation segments is an evidence of some 'neuron' and some 'component'. But in our model this depicts a stronger semantic association. Since the primary key of the segments relation consists of two attributes having domains in two parts of the SAO ontology, this implies that every tuple in segments is an *instance-of the direct or inferred path connecting the corresponding neuron and the structure of the ontology*. Let us say the SAO ontology has the following entries:

Purkinje-neuron $\sqsubseteq$ neuron
dendrite $\sqsubseteq$ neural-component
neuron $\equiv \forall$ has-component.neural-component
neuron $\equiv \exists$ has-component.dendrite

Here Purkinje-neuron $\xrightarrow{\sqsubseteq}$ neuron $\xrightarrow{hc}$ dendrite is a path p_1. Consequently, Purkinje-neuron $\xrightarrow{hc}$ dendrite is a path p'_1 in SAO. If the relation segments has tuples:

r_1: segments(dendrite, Purkinje-cell, 12, 4.67, 2.11)
r_2: segments(dendrite, Purkinje-cell, 39, 3.93, 1.52)

we can say r_1, r_2 are instances of path p'_1 (or p_1 for that matter). However, it is not meaningful to create instances of all paths in an ontology. For example, while it is possible to have an instance of $\exists R.C_1$ it is not possible to have an instance of $C_2 \sqcup \exists R.C_1$. In general, instances of paths containing $\sqcap$ or $\sqcup$ are disallowed. Instantiations of more complex concepts are possible with more specialized constructs. We will return to this issue when we present our ontology model for aggregates in Section 2.2.

2.2   A Semantic Model for Extended Objects

Anatomical objects are spatial objects – they are located in a spatial context, occupy space, and maintain functional relationships with other objects often by virtue of their (relative) positions in space. It is therefore imperative that both the ontology component and the schema component of an OBDB has an adequate semantic model of extended objects and their spatial context. The fact that spaces and objects are conceptually distinct but are related intricately through relationships of *spatial inclusion*, *meronymy* and *topology* has been a topic of philosophical research [4,5,6]. In our work, we encapsulate all this development through our adoption of the Basic Formal Ontology (BFO) [7], described next.

Basics. BFO serves as an upper-ontology for biomedical science. BFO uses a concept called continuant for everything that is not dependent on time. Two subcategories of continuant relevant to us are spatial-region, that represents space, and independent-continuant, that represents material objects or named portions of objects. According to the BFO catergorization, spatial-region represents space that is not relativized, i.e., not specified with respect to any object (e.g., extra-cellular space surrounding the post-synaptic density). In contrast, independent-continuants have some characteristic shape (that may be specified with respect to other continuants) in which they are completely enclosed, and are further subcategorized into object, object-boundary, site, object-aggregate and fiat-object-part. Briefly, sites are entities that can be occupied by other continuant entities (e.g., a post-synaptic-density is also occupied by the dendrite of which it is a part); objects are independent, spatially extended, maximally self-connected and self-contained entities, and possesses an internal unity; object-boundary (e.g., the outer membrane of a mitochondrion) is an independent continuant entity that is a lower-dimensional part of some other continuant entity; an object-aggregate is a named independent continuant entity that is a mereological sum of sepa-rate objects (e.g., a gap junction consists of the two membranes of two different neurons apposed to each other); finally, a fiat-object-part is a named part of an object that does not have an identifiable boundary (e.g., the distal dendrites of a neuron). In addition to these entities, BFO also admits a number of relation-ships – the two relevant ones are part-of that holds between two continuants or between two spatial-regions and located-in, that holds between an independent continuant and a spatial-region.

Recently [8] has investigated the problem of representing space and extended objects in the light of biomedical ontologies. They introduced the primitive func-tion region-of, which given an extended object instance e, returns the spatial-region s that e fully occupies, such that

$$\forall x, y : \text{located-in}(x, y) =_{def} \text{part-of}(\text{region-of}(x), \text{region-of}(y))$$

For extended objects they also distinguish between two forms of located-in, namely, location due to parthood and that due to spatial containment. If o is the instance of a spatial extended object, they define:

$$\forall x, y : \text{contained-in}(x, y) =_{def} \text{object}(x) \wedge \text{object}(y) \wedge \text{located-in}(x, y) \wedge \neg \text{part-of}(x, y)$$

Our semantic data model adopts this ontological framework and extends it to provide a more realistic ontology to instance mapping constructs that can model extended data objects observed in images and videos and stored in the database.

Multiple Paritioning. As in BFO, in our model, a spatial-region, extended object, or site can be partitioned using part-of relationship. However, our anatomy-motivated viewpoint persuades us to model spatial partitioning in a finer detail. The following snippet shows an example of our partition specification:

```
s1 ⊑ SAO:site
p1 ≡ ∃ complete-partition-of.s1
p2 ≡ ∃ partial-partition-of.s1
s11 ≡ ∀ part-of(p1).s1
s12 ≡ ∀ part-of(p1).s1
s21 ≡ ∀ part-of(p2).s1
disjoint(s11,s12)
```

The first line establishes s1 as a SAO:site (which is the same as a BFO:site). The next two lines show that our role hierarchy for the partition-of relationship has two values complete-partition-of and partial-partition-of. The first sub-role means that all the spatial parts that constitute the particular partition-of relationship are completely enumerated in the ontology, while the latter sub-role states that this is not the case. The next three lines illustrate a *parametric role* or a *parametric relationship*, our extension to the standard OWL-DL. It simply states that s11 and s12 are parts of site s1 according to the partitioning scheme p1, whereas s21 is part of site s1 according to the partitioning scheme p2. The last line implies that our partition-of relationship does not assume spatial disjointedness of the parts – otherwise this constraint must be explicitly asserted as in this example. Also note that although p1 is a complete partitions11 s12 are not the only parts of s1 because we don't claim the OBDB is completely known.

Multiple Inheritance of BFO Categories. Our model permits multiple inheritance from the basic BFO categories. For example, the class synapse inherits from both an aggregate-object and from a fiat object part because a synapse is a *junction*, i.e., it is a portion of extracellular space (hence it has no demarcated boundaries) where axon terminals and dendritic processes are situated (hence it is an aggregate) closely enough such that chemical neurotransmitters can pass from the axon terminals to the neurotransmitter receptor portions (e.g., post-synaptic density) of those dendrites. We show the ontological definition of synapse in Figure 1. Other neuroanatomical entities like the *gap junction* or the *node of Ranvier* share the same characteristics – they are aggregates of fiat object parts such that certain predicates over selected properties of the participating objects hold.

Implicit Subclasses. Our model also supports the creation of *implicit subclasses* but considering all possible values of a class based on its data properties. Consider a class called post-synaptic-density that has a data property called *morphological-type* whose value is a 2-tuple (m_1, m_2) where m_1 can be 'symmetric' or 'asymmetric' and m_2 can be 'macular' or 'perforated'. When implicit subclasses are used, the system implicitly creates a cross-product of 4 subclasses of post-synaptic-density. If one of the system-generated classes is not valid, the ontologist has to specify a constraint like invalid-class('asymmetric_perforated_post-synaptic-density'). One might argue that this creates a huge proliferation of possible classes. We view this as a necessary evil because thanks to the bottom-up nature of biological discoveries, very often a scientist would discover a new characteristic of an biological entity that had not been materialized as a separate class before because there was no prior evidence that such a class would be biologically significant.

2.3   Bridging the *Instantiation Gap*

We return to the issue of mapping between the ontology part and the database part of an OBDB in the context of our semantic model.

An implicit assumption in almost all ontology literature is that the primary technical problems lies in the expressive power and computational complexity

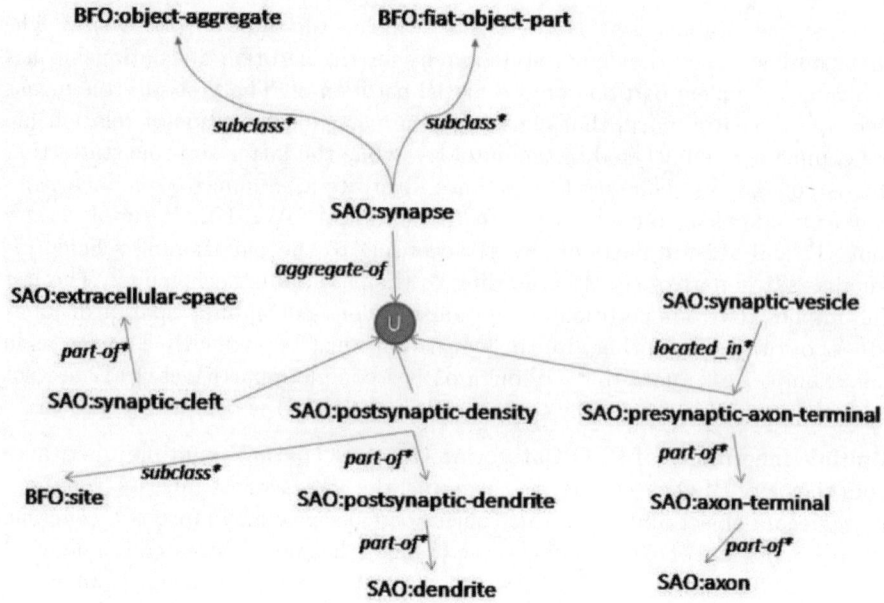

Fig. 1. A rough ontological definition for the concept synapse

related to the specification of the ontology and the reasoning over instances. We introduce a problem called the *instantiation gap* problem between the ontology and its instance repository, which in an OBDB, has a schema constructed independently of the ontology. To our knowledge, this problem has not been considered by prior research. Simply, an *instantiation gap* occurs when either of the following cases hold:

1. an instance store has a data object d marked as instance of a concept c from the ontology, but one or more of the following occurs:
   (a) d does not have all the inherited properties of c
   (b) d does not have all the local properties of c
   (c) d's extended object properties violate the expected object properties of c
2. an instance store has instantiations of all properties that an instance of concept c from the ontology is supposed to have, but there is no data object d in the instance store that can be assigned as an instance of c.

It is easy to see that all three conditions under case (1) may occur when an image depicts one or more extended object that are partially visible. We show that case (2) also occurs due to partially visible structures. For example, consider an image (image-id = 23) showing a chemical synapse from a granule cell to a Purkinje cell as part of its content. As a constituent of the synapse, the presynaptic axon terminal exists in the image but is partly visible, the postsynaptic density of the synapsing dendrite is also partly visible, other dendritic processes of the same postsynaptic neuron are also partial. Let us imagine the image also shows a number of synaptic vesicles, some in the synaptic cleft, some in the presynaptic axon, and the rest in the extracellular space as well as parts of other neural structures such as the endoplasmic reticulum. Now consider Table 1, a slightly extended version of the segments relation we saw earlier, with the contents as recorded in from the scientist's annotation of image 23.

Table 1. A portion of the segments table showing the experimenter's data about the synapse

Obj-ID	Structure-name	Cell-name	Image-id	Visibility	length	volume	class
1001	axon-terminal:02	granule-cell:21	23	partial			
1002	synaptic-vesicle:45	granule-cell:21	23	partial			fused
1003	psd:15	Purkinje-cell:19	23	partial			
1004	dendritic-process:25	Purkinje-cell:19	23	partial			
1005	dendritic-process:26	Purkinje-cell:19	23	partial			
1006	dendritic-spine:58	Purkinje-cell:19	23	partial			thin
1007	cell-membrane:34	granule-cell:21	23	partial			
1008	synaptic-vesicle:46		23	complete		0.10	normal
1009	synaptic-vesicle:47		23	complete		0.11	
1010	synaptic-vesicle:48		23	partial			
1011	synaptic-vesicle:48		23	partial			
1012	synaptic-cleft:03		23	complete			
1013	ER:09	granule-cell:21	23	partial			
1014	extracellular-space:28		23	partial			none

Notice that since the database has been developed to support experimental documentation and ontology has been developed to model the scientist's perception of biological reality, there are a number of disconnects between the ontology's depiction and the database's depiction of the objects. First, for oid = 1003, the instance of the post-synaptic-density could not be assigned any classification (e.g., 'symmetric_macular') due to the partial observation, although the ontological data property of the class posits this as a mandatory property[1] A different mismatch occurs for oid = 1001. We know from the ontology that an axon-terminal is a part of an axon and can't really exist without the axon. However, the database does not indicate the presence of the axon. In absence of the axon however, oid = 1007 can't be considered to be an instance of axonal-membrane, which is defined as the part of the cell-membrane covering the axon, and is a mandatory part of axon. Along the same lines, the fiat aggregate object called synapse as defined in the ontology (Figure 1) does not exactly exist, although pieces that consitute the synapse do.

To bridge this instantiation gap, our system computes an additional concept-to-instance mapping called *unobserved-instance-map* that first copies partially observed extended objects, and then recursively fills-in *phantom instances* of

---

[1] Notice that in contrast, the classification for the dendritic spine could be filled in even if the spine is not completely visible.

objects for which there is sufficient evidence (more than $x\%$ of mandatory properties). This would create an axon instance and an neuron instance given the axon terminal. The expected properties of these phantom instances are then matched against the observed properties in the database to derive further correspondences between the concepts and properties in the ontology to the data records. This will instantiate the synapse when most of its expected components are located. When the process terminates, the unobserved-instance-map is committed to the database, that can be verified by the domain expert.

3   Querying the OBDB with OntoQuest 2.0

In [9], we presented OntoQuest 1.0, the first version of our ontology exploration software. OntoQuest 1.0 is built upon the IODT framework from IBM (http://www.alphaworks.ibm.com/tech/semanticstk) and allows navigational queries over the ontology content and SPARQL+aggregate queries over instances in distributed stores. In general however, one should be able to pose arbitrary ad hoc queries to an OBDB that permit ontology navigation and instance fetching together, and utilize all the concept-to-instance mappings that are available.

OntoQuest 2.0 uses a storage model for the concept graph that is similar to that of IODT. In this model, all distinguished relationships permitted by OWL (e.g., subclass, allValuesFrom, disjoint ...) are stored in separate tables, while all user-defined relation names are stored in a quad-store. However, the instances (both observed and phantom) are stored in a graph-like manner. The properties of the nodes and the edges are stored in relational column stores, while the connectivity of the nodes and edges are stored in triple stores with additional path indexes. The mappings between the concepts and the relations are maintained using standard primary key foreign key relationships between the concept tables and the node/edge property tables.

OntoQueL, the query language of OntoQuest 2.0 is influenced by SPARQL, OntoQL [1], OWL-QL [10]; since a full treatment of the language is beyond the scope of this paper, we illustrate it with examples.

The query   select succ*(concept:'neuron', part-of, 3) from MyOnto as graph returns a 3-level deep part-of hierarchy from the concept-graph starting with the term 'neuron'. If we want to get the part-of hierarchy of all *observed instances* of the concept 'neuron', we will write the query as   select succ*(instance-of('neuron') $X, part-of, 3) from MyOnto as graph where observed($X). Due to the built-in predicate observed, the query does not return phantom instances, but does return both complete and partially visible instances. If partially visible instances are not desirable, we add the condition not(partial($X)).

OntoQueL also permits *instance navigation* queries. The query
select neighborhood(instance:'Purkinje-cell' $X) as $G from MyOnto
where $X.length > 4 stop before (exists $Y and phantom-instance($Y) and $Y in $G) starts with each instance of Purkinje-cell satisfying the length predicate and follows graph edges until it touches a phantom instance of any kind. The condition stop before ensures that the phantom instance does not get included

in the result graph; if want to stop the graph navigation after the phantom node is included, we will use the construct stop when instead of stop before.

4   Conclusion

In this paper, we presented our model of capturing the semantics of an ontology-based database system, and the extensions we made to the standard OWL and BFO semantics. We introduced the problem of instantiation gap and presented our current solution. However, we believe more elegant solutions to the problem should be possible. We briefly discussed how the OntoQueL, the query language of the OntoQuest 2.0 system treats concept-grapha and instance graph queries, and allows the formulation of more complex queries that current ontology-based query languages do not support.

References

1. Jean, S., At-Ameur, Y., Pierra, G.: Querying Ontology Based Database Using OntoQL (an Ontology Query Language). In: Ontologies, DataBases, and Applications of Semantics (ODBASE'2006), pp. 704–721 (2006)
2. Jean, S., Dehainsala, H., Xuan, D.N., Pierra, G., Bellatreche, L., At-Ameur, Y.: OntoDB: It is Time to Embed your Domain Ontology in your Database. In: Database Systems for Advanced Applications (DASFAA 2007) (2007) (demo paper)
3. Fong, L.L., Larson, S.D., Gupta, A., Condit, C., Bug, W., Chen, L., West, R., Lamont, S., Terada, M., Martone, M.E.: An ontology-driven knowledge environment for subcellular neuroanatomy. In: 3rd Int. Workshop on OWL: Experiences and Directions, Innsbruck, Austria (2007)
4. Casati, R., Varzi, A.: The structure of spatial localization. Philosophical Studies 82, 205–239 (1996)
5. Casati, R., Smith, B., Varzi, A.C.: Ontological tools for geographic representation. In: Guarino, N. (ed.) Formal Ontology in Information Systems, pp. 77–85. IOS Press, Amsterdam (1998)
6. Smith, B.: Mereotopology: a theory of parts and boundaries. Data Knowl. Eng. 20, 287–303 (1996)
7. Spear, A.D.: Ontology for the Twenty First Century: An Introduction with Recommendations. The Institute for Formal Ontology and Medical Information Science (2006), http://www.ifomis.uni-saarland.de/bfo/manual
8. Schulz, S., Marko, K., Hahn, U.: Spatial location and its relevance for terminological inferences in bio-ontologies. BMC Bioinformatics 8, 134 (2007)
9. Chen, L., Martone, M.E., Gupta, A., Fong, L., Wong-Barnum, M.: Ontoquest: Exploring ontological data made easy. In: VLDB. Proc. 31st Int. Conf. on Very Large Databases, pp. 1183–1186 (2006)
10. Fikes, R., Hayes, P.J., Horrocks, I.: OWL-QL - a language for deductive query answering on the semantic web. Journal on Web Semantics 2, 19–29 (2004)

Seed-Based Generation of Personalized Bio-ontologies for Information Extraction

Cui Tao* and David W. Embley*

Department of Computer Science,
Brigham Young University, Provo, Utah 84602, U.S.A

Abstract. Biologists usually focus on only a small, individualized, sub-domain of the huge domain of biology. With respect to their sub-domain, they often need data collected from various different web resources. In this research, we provide a tool with which biologists can generate a sub-domain-size, user-specific ontology that can extract data from web resources. The central idea is to let a user provide a seed, which consists of a single data instance embedded within the concepts of interest. Given a seed, the system can generate an extraction ontology, match information with the user's view based on the seed, and collect information from online repositories. Our initial experimentations indicate that our prototype system can successfully match source data with an ontology seed and gather information from different sources with respect to user-specific, personalized views.

1 Introduction

To do activities such as performing background research for a field of study, gaining insights into relationships and interactions among different research discoveries, or building up research strategies inspired by other's hypotheses, biologists often need to search several online databases and gather information of interest. Biologists usually have to traverse different web sources and collect the data of interest manually. This task is a tedious and time-consuming.

It would be beneficial if we could generate a data-extraction ontology specifically for each individual user that would automatically collect the information of interest. But generating an ontology, especially an ontological description for an information repository, is non-trivial; it not only requires domain expertise, but also requires knowledge of specific ontology language. Data heterogeneity and different user objectives makes the task even more daunting.

To illustrate the difficulties biologists encounter in gathering information from a variety of sources and also to illustrate the challenges involved in building an extraction ontology to automatically collect data, consider some examples. For chromosome location of a gene, some users might only care about the chromosome on which this gene is located. Other users might care about a more detailed location like the start and end base pairs. Sources, not knowing user objectives,

---

* Supported in part by the National Science Foundation under Grant #0414644.

J.-L. Hainaut et al. (Eds.): ER Workshops 2007, LNCS 4802, pp. 74–84, 2007.

provide their own view of the data. One source could describe a chromosome location of a gene as one concept. Others could describe the location in terms of multiple concepts such as chromosome number, start location, end location, orientation, and size. The size is actually an implicative value which is equal to end minus start, therefore some designer could also choose to omit this concept. As another example, consider the use of different units for the same concept. For example, one site could use kilo-dalton as an unit for molecular weight, one could use dalton, and another could provide both. Still other problems arise because different sites might provide information directly or indirectly. A protein database, for example, could use Gene Ontology (GO) terms to describe molecular functions. In order to obtain information for the description of the definitions of the terms, a user usually needs to go to the GO database.

In this paper, we introduce a system that can automatically build a data-extraction ontology given a user's seed. We call this system SIH (Seed-based Information Harvester, pronounced "sigh"), because once built, we can use the ontology to harvest information from web repositories. A seed consists of a single sample data instance embedded within the concepts of interest. Based on the ontology seed, SIH matches information with the user's view, builds a personalized ontology, and collects information of interest from online repositories. The advantages of this system are (1) it does not require knowledge of conceptual modeling or ontology languages to build ontologies, and (2) it can automatically harvest information from multiple sites and present the information according to a user-specified view.

We present the details of SIH and our contribution to user-specified ontology generation and subsequent information harvesting as follows. Section 2 positions our work within recent work on ontology creation. Section 3 introduces OSM ontologies, the ontology framework we use in this research. Section 4 describes the interface used to create a seed ontology for SIH. Section 5 explains how SIH maps site labels to seed ontology labels and how the generated extraction ontology collects information from various sources. Section 6 reports the results of some initial experiments we conducted with our SIH implementation, makes concluding remarks, and mentions some future directions we wish to pursue in this research.

2   Related Work

In recent years, many researches have tried to facilitate ontology generation. Manual editing tools such as Protege [6] and OntoWeb [8] have been developed to help users create and edit ontologies. It is not trivial, however, to learn ontology modeling languages and complex tools in order to manually create ontological description for information repositories.

Because of the difficulties involved in manual creation, researchers have developed semi-automatic ontology generation tools. Most efforts so far have been devoted to automatic generation of ontologies from text files. Tools such as OntoLT [1], Text2Onto [2], OntoLearn [5], and KASO [11] use machine learning methods to generate an ontology from arbitrary text files. These tools usually require a large training corpus and use various natural language processing algorithms to derive features to learn ontologies. The results, however, are not very satisfactory [7]. Tools such as TANGO [10] and the one developed by Pivk [7] use structured information (HTML tables) as a source for learning ontologies. Structured information makes it easier to interpret new items and relations. These approaches, however, derive concepts and relationships among concepts from source data, not from users. SIH, on the other hand, allows users to provide their own views and generate user-specified ontologies.

Potentially, it should be possible to derive biologist-specific view ontologies from large biological ontologies such as the Gene Ontology, the NCI Thesaurus, and the SNOMED Ontology. Our own experience in this direction [3], however, has not been very successful, mostly because the existing large biological ontologies are usually more like hierarchial vocabulary lists than the conceptual-model-based ontologies we need for information extraction.

3   OSM Ontologies

We use OSM [4] as the semantic data model for an extraction ontology. Figure 1 shows a graphical view of a sample ontology. The structural components of OSM include object sets, relationship sets, and constraints over these object and relationship sets. An object set in an OSM ontology represents a set of objects which may either be lexical or non-lexical. A dashed box represents a lexical object set and a solid box represents a non-lexical object set. A lexical object set contains concrete values. For example, "T-complex protein 1 subunit theta" is a possible value of the *Name* object set in Figure 1. A non-lexical object describes an abstract concept, such as *Protein* in Figure 1. Lines among object sets represent the relationship sets among them. A small circle at one end of a line indicates optional. For example, a *Protein* can have zero or more *GO Function Definition*s. An arrow indicates functional from domain to range. For example, a *Protein* can only have at most one *Molecular Weight*; the relationship set is therefore functional from *Protein* to *Molecular Weight*. OSM also supports n-ary relationships with multiple lines connecting the object sets involved.

We have found OSM to be more expressive than other standard ontology representations, such as RDF and OWL, which, for instance, only supports binary relationships [10]. In addition, and more important, an OSM ontology can support data extraction from source documents [4].

4   Seed Ontology Creation

In this section, we explain how SIH generates an ontology based on a user's seed. We provide our users with a graphical user interface (GUI) where they can create a seed easily. We adapt the user interface proposed by Zhou [12]. Through this GUI, a user can generate an ontology seed by creating a form and then provides the seed values by filling out the form. The form tells SIH

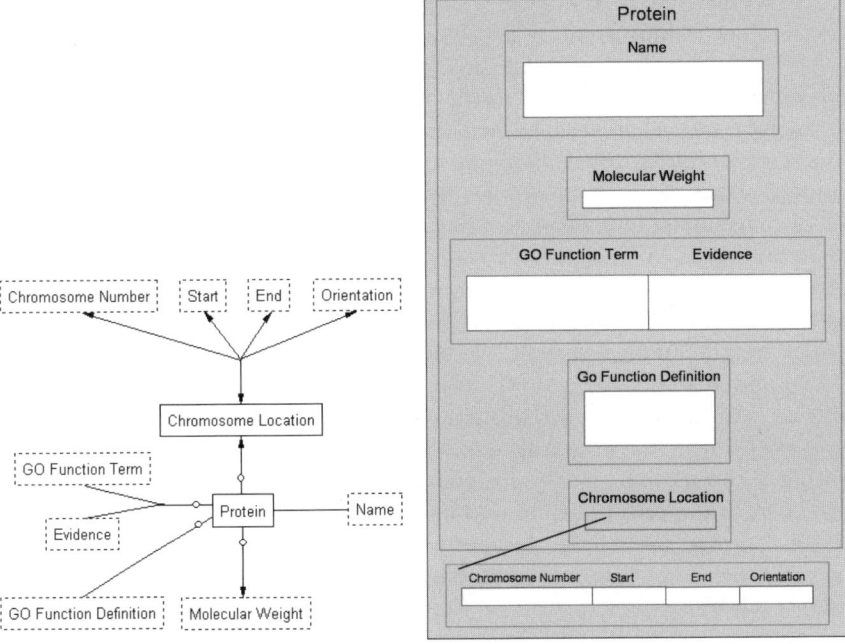

Fig. 1. The Graphical View of a Sample Ontology

Fig. 2. A Sample Form for Generating the Ontology in Figure 1

how to generate the ontology structure, and the seed values tell SIH something about how to collect information from different sources. We discuss information collection in Section 5.

The GUI provides users with an intuitive method for defining different kinds of form features. There are four basic form fields from which users can choose: *single-label/single-entry* fields, *single-label/multiple-entry* fields, *multiple-label/-single-entry* fields, and *multiple-label/multiple-entry* fields. Users can also recursively nest a form inside any basic form field. Nested forms allow users to describe their interests in more structured and meaningful ways.

For each new ontology to be built, a user creates a form and gives the form a meaningful title. Based on this title, SIH generates a new ontology and a non-lexical object set with this title as the name. Every label represents an object set in the corresponding ontology; the label is the name for the object set. If the label is for a field containing a nested form, its object is non-lexical; otherwise its object set is lexical. SIH generates relationship sets among the object sets as follows. Between the form-title object set and each single-label field, it generates a binary relationship set; between the form-title object set and each multiple-label field, it generates an n-ary relationship set; between each field and a single-label object set nested in side of it, SIH generates a binary relationship set; between each field and a multiple-label object set nested inside of it, SIH generates an n-ary relationship set. Cardinalities for relationship sets depend on whether the form field is single-entry or multiple-entry, which respectively indicates that the relationship set is functional or non-functional from form title or nested form title to field or fields. In the reverse direction, the cardinalities are non-functional except when there is exactly one single-entry field. Thus, for example, we have a one-to-one relationship set between *Chromosome Location* and the quadruple *Chromosome Location, Start, End,* and *Orientation.*

Figure 2 shows an example of form generation. Suppose we are interested in basic information about human proteins (their names, locations, functions, and sizes). In our example, we choose "Protein" as the base-form title. We know each protein can have one or more names, so we choose to add a *single-label/multiple-entry* field to the form and label it *Name.* Since we know there is only one molecular weight for one protein, we choose to use a *single-label/single-entry* form field and label it *Molecular Weight.* We are also interested in the locations of proteins. We know that each protein can have only one location, so generate a *single-label/single-entry* form field and label it *Chromosome Location.* We also know that a chromosome location is composed of four parts: chromosome number, start location, end location, and orientation. In this situation, we choose to create a nested form field. A nested form field is defined in separate panels in the same way as users define basic form fields. Here we choose to use a *multiple-label/single-entry* field as Figure 2 shows. A line between *Chromosome Location* form field and this *multiple-label/single-entry* field indicates the nesting. We use the GO (Gene Ontology) to describe a protein function. Each protein has a set of GO function terms to describe their functions. Each GO function term also has an evidence designator associated with it. We thus create a *multiple-label/multiple-entry* field as Figure 2 shows. For each protein, we also want to include the GO function definitions and use a *single-label/multiple-entry* to define it. Overall, SIH generates the ontology in Figure 1.[1]

We complete the creation of a seed by filling in a created form. SIH provides users with a GUI where they can copy values from source pages and paste them into generated forms. Users can browse the online databases with which they are familiar or from which they want to collect information and copy and paste values for one instance into the form. Figure 3 shows an example of copying values for chromosome location and molecular weight from a source page to the corresponding form fields in the form in Figure 2. The highlighted values in the document are the copied values.

5   Data Collection

We explain in this section how we generate data-extraction specifications from an ontology seed. SIH collects information from source repositories that present their information in structured/semi-structured ways. SIH first interprets source

---

[1] The optional constraints in Figure 1 are not defined by the user's form. Instead, they come from observing source data as explained in Section 5. We may also adjust reverse cardinalities according to our observations of source data.

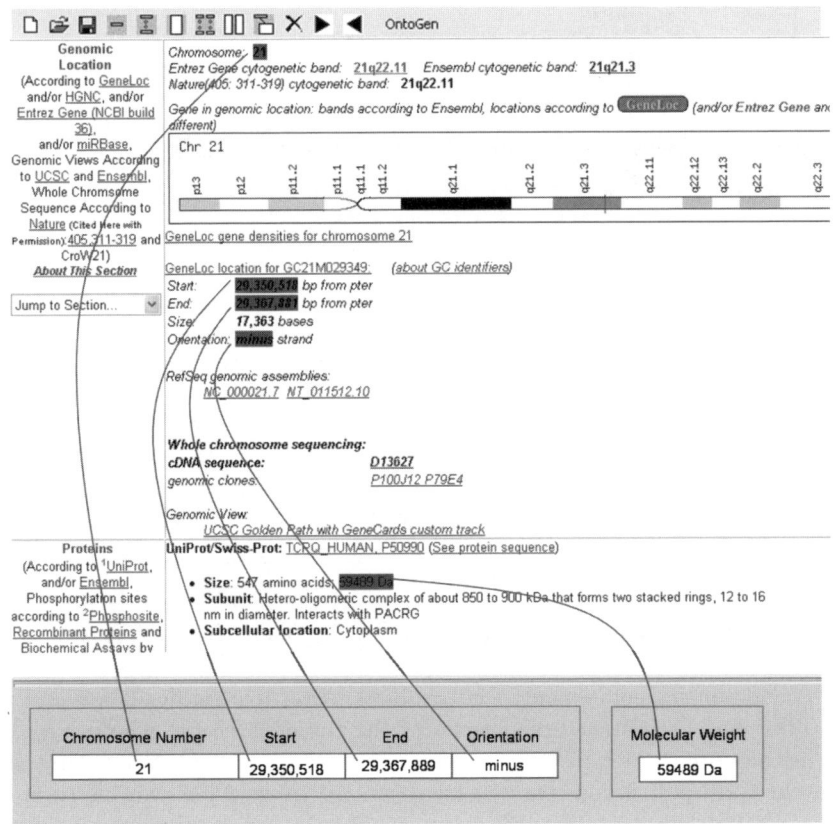

Fig. 3. An Sample Ontology Seed with a Source Data Page (Partial)

pages from these online resources. It then maps the labels in a generated user-specific ontology to the labels in the source pages. Once the mappings are defined, SIH can collect source data for the user.

5.1   Source Page Interpretation

Many online repositories present their data in dynamically generated pre-defined templates in response to submitted queries. Pages from this kind of repository usually have the same or similar structure. We call pages that are from the same web site and have similar structures *sibling pages* and the corresponding tables in sibling pages *sibling tables*. Figure 4 shows a sibling page for the page with *Molecular Weight* seed values in Figure 3.

We have developed a system called *TISP* (*Table Interpretation with Sibling Pages*) [9], which can automatically interpret the structure of sibling pages and find the association between labels and values. TISP first decomposes source sibling pages, unnests all the HTML tables, and finds sibling tables. TISP then compares a pair of sibling tables to identify nonvarying components (category

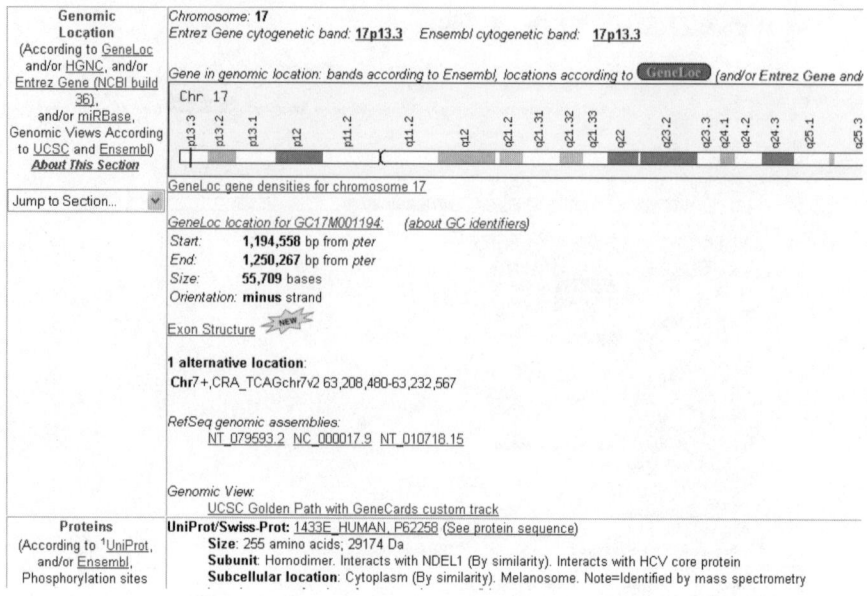

Fig. 4. An Example of a Sibling Page for the Source Page in Figure 3

labels) and varying components (data values). After it identifies labels and values, TISP finds the structure pattern of the table. It checks whether a table matches any pre-defined pattern template by testing each template until it finds a match. With a structure pattern for a specific table, TISP can interpret the table and all its sibling tables. We assume that values under the same label from different sibling tables are for the same or same set of concepts. Using TISP, we can collect all the values under the same label from a source repository.

5.2 Source-Target Mapping and Data Collection

The next step for SIH is to map source labels to concepts in the generated ontology. Seed values provide the main means of determining these mappings. SIH knows both the label for a seed value and the value's source-page label, and can therefore link the two labels. The basic idea is that source values for source-page labels are values that can fill in the form field for the form labels.

Unfortunately, mapping generation is not quite so simple because the labels may not have exactly the same meaning. Size in Figure 3 and 4, for example, does not have the same meaning as *Molecular Weight* in Figure 1 and 2. Size values include both the number of amino acids and the molecular weight. Thus, SIH must "split" size values and pick up only the part giving the molecular weight. In general, we must handle five different cases we encountered during the mapping process: direct mappings, unions, selections, splits, and merges.

Direct Mapping. When a seed value matches a source value exactly, SIH infers a direct mapping. For example, the highlighted value "21" under the source label

Protein description	
Protein name	**T-complex protein 1 subunit theta**
Synonyms	TCP-1-theta CCT-theta Renal carcinoma antigen NY-REN-15

Fig. 5. An Example of Union

Annotations				
	Ontology	**Annotation**	**Evidence**	**Source**
Gene Ontologies	**Molecular Function**	Unfolded protein binding	IEA	GOA/IPI
		ATP binding	IEA	GOA/IPI
		ATPase activity, coupled	TAS	GOA/IPI
	Biological Process	Protein folding	IEA	GOA/IPI
	Cellular Component	Cytosol	NR	GOA/IPI

Fig. 6. An Example of Selection

chromosome	21
base	from 29350518to 2937881
strand	+

Fig. 7. An Example of Split

Chromosome in Figure 3 matches the seed value "21" in the form under seed label *Chromosome Number*; thus SIH infers a direct mapping from *Chromosome* to *Chromosome Number*. For information harvesting, SIH just collects all the information under the same label from all the sibling pages.

Union/Selection. When individual source values under different source labels match individual seed values under one label, SIH infers a union mapping. For example, suppose a user creates seed values for *Name* (of *Protein*) by copy-and-paste of "T-complex protein 1 subunit theta", "TCP-1-theta", "CCT-theta", and "Renal carcinoma antigen NY-REN-15" from Figure 5. Then, in this case, SIH detects a union mapping from *Protein name* and *Synonyms* to *Name*. When individual seed values under a seed label match a subset of individual source values under only one source label, SIH infers a selection mapping. For example, in Figure 6, suppose that the seed values only include the first three values under *Annotation* because the user only cares about protein functions. SIH then can infer a selection mapping from the source label *Annotation* for only *Molecular Function* annotations. For information harvesting for a union mapping, SIH collects information from all source fields. For selection, if the desired values are labeled in the source pages as they are in Figure 6, SIH collects the information under the proper, restricted label. If not, the user needs to provide a selection expression to filter the values (see future work).

Split/Merge. When part of a seed value matches an individual source value, SIH detects a split mapping. For example, in Figure 3, only part of the source value "29,350,518 bp from pter" under the source label *Start* matches the seed value "29,350,518". In this case, SIH detects a split mapping. Sometimes, one source value could be split into multiple seed values. For example, the value under the label *base* in Figure 7 matches seed values for labels *Start* and *End* and thus should be split and mapped to two ontology concepts. To do the split, SIH stores and uses patterns. The pattern "<start> bp from pter" works for extracting *Start* values for the site of Figures 3 and 4, and the pattern "from <Start>to <End>" works for extracting *Start* and *End* values from the site from which Figure 7 was taken. A merge mapping is the opposite of a split mapping. If the user's form had a *single-label/single-entry* field for base that expected values like the base value in Figure 7, then for the site of Figure 3 and 4, SIH would need to merge *Start* and *End* values. For information harvesting, SIH collects values to be split by filtering them through generated patterns and collects values to be merged by obtaining all of them and concatenating them, separated by a delimiter (e.g. " - ").

To harvest information from multiple sites, the user specifies multiple seeds. The user does not specify another form, but does fill in the form with seed values from each new site. If, however, SIH can find a match in a new site with either the original seed values, or with any seed values it has already collected, the user need not even specify new seed values for the new site.

After SIH collects information, it checks cardinality constraints. For example, if SIH finds that each value under a single-label/single-entry field is unique, it marks this concept as a unique identifier for the base form or subform concept. If SIH observes that values for some fields are not available, it marks the field as optional. For example, if SIH finds that some web site does not provide information about *Molecular Weight* for a protein, it marks this concept as optional. If data collected contradicts user-provided constraints, SIH warns the user and allows the user to determine if any adjustment needs to be made.

6   Experimental Results, Conclusion, and Future Work

We have conducted some preliminary tests for SIH by creating the sample ontology in Figure 1 and extracting information from seven different web sites. Although there were seven web sites, we only needed to create three sets of seed values. Some web site had various values or values in different formats for the same concept. For example, some web sites used "minus" for *Orientation*, whereas other sites used "-". There were a total of 31 concept mappings. Among these 31 mappings, 11 were direct mappings, which SIH was able to handle 100% correctly. SIH also successfully detected and correctly processed all 4 union mappings it encountered. There were 15 split mappings, SIH detected and found the correct patterns for 12 of them. One error was due to the use of different delimiters in the same site, and two errors were due to the seed value being only a small subset of the source field. (Additional work on finding delimiters is needed.)

SIH did not encounter any merge mapping—our specified fields were always at the lowest granularities. For selection mappings, we only tested the part where a source page uses a label to mark the selection. SIH encountered one selection mapping and was able to detect it successfully.

As a conclusion, we can say that seed-based harvesting of information via bio-ontologies appears to be both possible and reasonable. SIH can match source data with an ontology seed and gather information from different sources with respect to user-specific, personalized views.

Several directions remain to be pursued. First, we would like to support additional form features such as allowing the user to specify filter functions or desired units for form fields. Second, we want to integrate the data, not just harvest it. Finally, we want to improve SIH, so that the users do not need to create and fill our forms. We plan to have the users just highlight the values they want from sample pages; our future system would generate ontologies directly from these highlighted values.

References

1. Buitelaar, P., Olejnik, D., Sintek, M.: Ontolt: A protege plug-in for ontology extraction from text. In: Fensel, D., Sycara, K.P., Mylopoulos, J. (eds.) ISWC 2003. LNCS, vol. 2870, Springer, Heidelberg (2003)
2. Cimiano, P., Völker, J.: Text2Onto—a framework for ontology learning and data-driven change discovery. In: Montoyo, A., Muñoz, R., Métais, E. (eds.) NLDB 2005. LNCS, vol. 3513, pp. 227–238. Springer, Heidelberg (2005)
3. Ding, Y., Lonsdale, D.W., Embley, D.W., Hepp, M., Xu, L.: Generating ontologies via language components and ontology reuse. In: NLDB 2007. Proceedings of 12th International Conference on Applications of Natural Language to Information Systems, Paris, France, pp. 131–142 (June 2007)
4. Embley, D.W., Campbell, D.M., Jiang, Y.S., Liddle, S.W., Lonsdale, D.W., Ng, Y.-K., Smith, R.D.: Conceptual-model-based data extraction from multiple-record Web pages. Data & Knowledge Engineering 31(3), 227–251 (1999)
5. Navigli, R., Velardi, P., Cucchiarelli, A., Neri, F.: Quantitative and qualitative evaluation of the OntoLearn ontology learning system. In: Proceedings of the 20th International Conference on Computational Linguistics, Geneva, Switzerland, pp. 1043–1050 (August 2004)
6. Noy, N.F., Sintek, M., Decker, S., Crubezy, M., Fergerson, R.W., Musen, M.: Creating semantic web contents with Protege-2000. IEEE Intelligent Systems 16(2), 60–71 (2001)
7. Pivk, A.: Automatic ontology generation from web tabular structures. AI Communnications 19(1), 83–85 (2006)
8. Spyns, P., Oberle, D., Volz, R., Zheng, J., Jarrar, M., Sure, Y., Studer, R., Meersman, R.: OntoWeb—a semantic web community portal. In: Karagiannis, D., Reimer, U. (eds.) PAKM 2002. LNCS (LNAI), vol. 2569, pp. 189–200. Springer, Heidelberg (2002)
9. Tao, C., Embley, D.W.: Automatic hidden-web table interpretation by sibling page comparison. In: Afrati, F.N., Kolaitis, P. (eds.) ER 2007. LNCS, vol. 4801, pp. 566–581. Springer, Heidelberg (2007)

10. Tijerino, Y.A., Embley, D.W., Lonsdale, D.W., Ding, Y., Nagy, G.: Toward ontology generation from tables. World Wide Web: Internet and Web Information Systems 8(3), 251–285 (2004)
11. Wang, Y., Völker, J., Haase, P.: Towards semi-automatic ontology building supported by large-scale knowledge acquisition. In: AAAI Fall Symposium On Semantic Web for Collaborative Knowledge Acquisition, Arlington, Virginia, vol. FS-06-06, pp. 70–77 (October 2006)
12. Zhou, Y.: Generating data-extraction ontologies by example. Master's thesis, Brigham Young University (December 2005)

Preface to FP-UML 2007

Juan Trujillo[1] and Jeffrey Parsons[2]

[1] University of Alicante, Spain
[2] Memorial University of Newfoundland, Canada

The unified modeling language (UML) has been widely accepted as the standard object-oriented (OO) modeling language for modeling various aspects of software and information systems. The UML is an extensible language, in the sense that it provides mechanisms to introduce new elements for specific domains if necessary, such as database applications, business modeling, software development processes, data warehouses and so on. Furthermore, the latest version of UML 2.0 got even bigger and became more complicated with a larger number of diagrams, with some good reasons. Therefore, heuristics, design guidelines, and lessons learned from experiences are extremely important for the effective use of UML 2.0 and so to avoid unnecessary complication. Moreover, UML plays a central role in model transformations in the software development life-cycle, and therefore novel proposals on UML model transformations are also needed.

The 3rd International Workshop on Best Practices of UML was renamed as Foundations and Practices of UML (FP-UML 2007) and is a sequel to the successful BP-UML 2005 and BP-UML 2006 workshops held in conjunction with ER 2005 and ER 2006 respectively. FP-UML 2007 was held with the 26th International Conference on Conceptual Modeling (ER 2007) and was intended to be an international forum for exchanging ideas on the best and new practices of UML in modeling and system developments, and on the foundations and model transformations on UML. To keep the high quality of the workshops held in conjunction with ER, a strong International Program Committee was organized.

The workshop attracted papers from nine different countries from all over the world: Austria, Chile, China, France, Hong Kong, Malaysia, Spain, Tunisia and USA. We received 18 abstracts and 15 papers were finally submitted. The Program Committee only selected six papers, making an acceptance rate of 40%.

The accepted papers were organized into two sessions. In the first one, two papers presented techniques for improving the use of different UML diagrams, and the other one focused on the use of UML 2.0 for business process modeling. In the second session, two papers focused on model transformations, and the other one presented a UML profile.

We would like to express our gratitude to the Program Committee members and the external referees for their hard work in reviewing papers, the authors for submitting their papers, and the ER 2007 Organizing Committee for all their support. We would also like to thank Miguel Ángel Varó, Jose-Norberto Mazón, and Jesus Pardillo for their support in the organization of this workshop.

Developing State Diagrams Using a State Specialization Technique

Il-Yeol Song and Ki Jung Lee

College of Information Science and Technology, Drexel University
Philadelphia, PA 19104, USA
{song, kijung.lee}@drexel.edu

Abstract. A state diagram models the dynamic aspect of a class or a system, representing its behavior over its lifetime. Our survey shows that there are few systematic methods of developing state diagrams. In this paper, we present a systematic method of developing state diagrams using a state specialization technique. Our method identifies meaningful states of a target class before a state diagram is constructed. We use three state-identification rules - state-valued attributes, association paths, and conditions stated in integrity constraints. By using our technique, system designers will be able to systematically identify interesting and meaningful states.

Keywords: Unified Modeling Language, State Diagram, Statecharts, State Specialization, State Identification Methods.

1 Introduction

The state diagram represents the dynamic aspect of an object, visualizing its behavior over its lifetime. The behavior of an object is represented by a set of states and a set of events that causes transitions between those states.

The target of a state diagram can be a class, a use case, a subsystem, or a complete system. In this paper, however, we concentrate on state diagrams for a class. A state diagram for a class visualizes the behavior of the class, showing possible states that objects of the class can go through and the events that possibly cause transitions between the states. The state diagram identifies models of meaningful moments for the target class in its lifecycle.

Developing a state diagram is a difficult process for most novice designers. From our experience of teaching system analysis and design classes for almost 20 years, we have found that many students have trouble identifying meaningful states and detailed transitions. Most UML books explain the notations and terminologies of pre-built state diagrams and expect designers to be able to learn to develop a state diagram by understanding examples and a few simple guidelines. Although some authors provide simple steps for creating state diagrams, the steps are not usually detailed enough in identifying all meaningful states and transitions.

In this paper, we present a systematic method of developing state diagrams using a state specialization technique. The underlying principle in our method is the notion of

J.-L. Hainaut et al. (Eds.): ER Workshops 2007, LNCS 4802, pp. 86–95, 2007.

the state specialization. This technique identifies all the meaningful states of a target class before the state diagram is constructed. We use three state identification rules – state-valued attributes, association paths, and constraints. These three components represent semantic components that affect the behaviors of the target class. Then, rules for constructing the state diagrams are assigned. By using our method, unnecessary trial and error during a system design process can be avoided as much as possible. We illustrate the method using a case study.

The rest of the paper is organized as follows: In Section 2, we review related studies. We present the specialization method in Section 3 and a case study in Section 4. We then conclude our paper in Section 5.

2 Related Studies

UML state diagram was adapted from the statecharts designed by Harel [5] to represent the complex nature of objects and the way they evolve. For core concepts of state diagrams on states, events, and transitions, we refer Booch, Rumbaugh, and Jacobson [1], Shlaer & Mellor [11], Cook & Daniels [2], and Harel [5, 6].

Most research on state diagram has been focused on formalism and semantics issues [5, 12], model validation issues [10], and implementation issues [4]. In this section, we briefly review existing approaches for developing state diagrams.

Booch et al. [1] sketches a method of developing a state diagram in five steps. Roques [9] and Reed [8] suggest a similar simple procedure for creating a state diagram for classes. All these methods are not detailed enough for novice designers.

Some studies have discussed the use of specialization in *understanding* state diagrams [2, 7]. To our knowledge, however, there have been few studies that suggested the state specialization as part of a systematic method for *developing* state diagrams.

Using specialization can offer potential advantage by clustering processes and enabling methodological guidance of step by step procedure. Moreover, an empirical study shows that the use of composite states in the state diagram significantly improves understandability of different stakeholders [3].

Some researchers discuss the state specialization as a design guideline for state diagram. Cook and Daniels [2] use the notion of the state type view, treating state as a type. Here, state nesting is depicted by showing the nested states as subtypes of the enclosing state type. Shlaer and Mellor [11] as well as Wyner and Lee [13] define a set of sequential states along with related events and transitions for each subtype object, instead of specializing states directly.

We find that these guidelines are too simple to use and not sufficient in developing a detailed sate diagram. Existing methods are limited in that they are simple procedural guidelines without deeper discussion on how to identify meaningful states, events, and transitions systematically. Our method differs from the other research on state diagrams in that we present a sequence of steps that enable novice designers to find meaningful states in an organized manner.

3 Developing State Diagrams Using State Specialization

In this section, we discuss the use of specialization in the state modeling and how the principle can be applied to state diagram development.

3.1 Basic Convention in State Specialization Approach

We first present the naming convention used in our discussion. *A state specialization tree* is a diagram that shows a state specialization of a class in multiple levels. Fig. 1 shows a typical state specialization diagram showing three levels of specialization. The state at the top of the tree is called *the root state*. The root state represents a class we want to develop into a state diagram. State A in Fig. 1 is the root state. A state located at the bottom of a tree is called *a leaf state*. States D, E, F, and G in Fig. 1 are leaf states. *The depth of a tree* is called the level of the tree. In Fig. 1, A is on level 1; B and C are on level 2; and D, E, F, and G are on Level 3. Level 1 always represents the class that we want to design the state diagram. A is the parent of both B and C. B and C are *sibling states*. Each sibling produces a different *subtree*. State B has a subtree whose leaf states are D and E. State C has its own subtree whose leaf states are F and G.

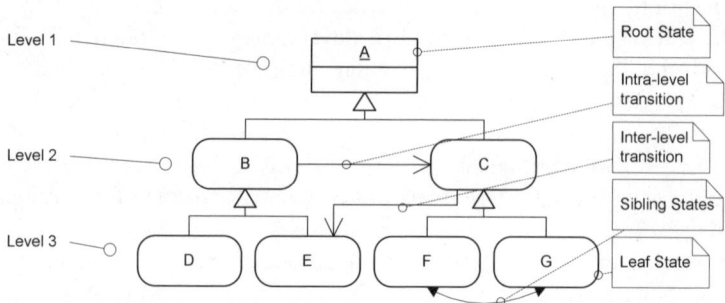

Fig. 1. A Typical Form of a State Specialization Tree Diagram of Level 3

We now define the terminology for transitions. When there is a transition from state B to state C, we call B the *source state* and C the *target state*. Any transition between two states at the same level is called an *intra-level transition*. For example, any transition between D, E, F, and G is an intra-level transition. Any transition between two states each at a different level is called an *inter-level transition*. For example, a transition between C and E is an inter-level transition.

3.2 The Procedure of Using the State Specialization Approach

One of the key factors in designing a state diagram is to ensure that all meaningful states of a class are included in the diagram. However, this can be a challenge for inexperienced designers since state diagrams, in the conventional development approach, are created through trial and error. An inexperienced designer is likely to miss many states in the first attempt at designing a state diagram. With specialization, designers would start by looking at the class and breaking it into substates recursively. A state specialization tree diagram breaks down the class so that all meaningful states are included. The procedure for developing a state diagram using our state specialization is presented below and illustrated in an activity diagram in Fig. 2.

1. Identify the target class that we want to develop into a state diagram
2. Identify possible states based on the three state identification (SI) methods
   a. Using State-Valued Attributes (SI-SVA)
       i. If an attribute has values that can classify the object into different states, it is called a state-valued attribute.
   b. Using Association-Paths (SI-AP)
       i. Identify states of the target class by traversing all the association paths emanating from the target class.
   c. Using Constraints (SI-CB)
       i. Identify states by examining a set of business rules based on the class diagram.
3. Using the heuristics of creating substates, organize the identified states in a state specialization tree diagram
4. Identify meaningful transitions in conjunction with triggering events
   a. Identify intra-level transitions
   b. Identify inter-level transitions
5. Identify initial and final states if necessary
6. Draw a state diagram based on identified states, composite state(s), events, and transitions.

4 A Case Study Using an Example of a Movie Rental System

In this section, we demonstrate our state specialization method using the case of a movie rental system. Fig. 3 represents a domain class diagram of a movie rental system.

4.1 Step 1

In Step 1, we identify the target class, which is the class that we want to develop into a state diagram. We have general consensus with Reed [8] in that the following types of classes usually have interesting behaviors, and would like to visualize their dynamic behaviors in the form of a state diagram:

1. A class that has a role of controller
2. A class that send/receive a high volume of messages
3. A class that has a role of interface, and
4. A class that has many domains and/or context relevance

In our example, we choose the *Item* class since it is the most relevant in the context of a movie rental system, while at the same time, it handles substantial communication messages.

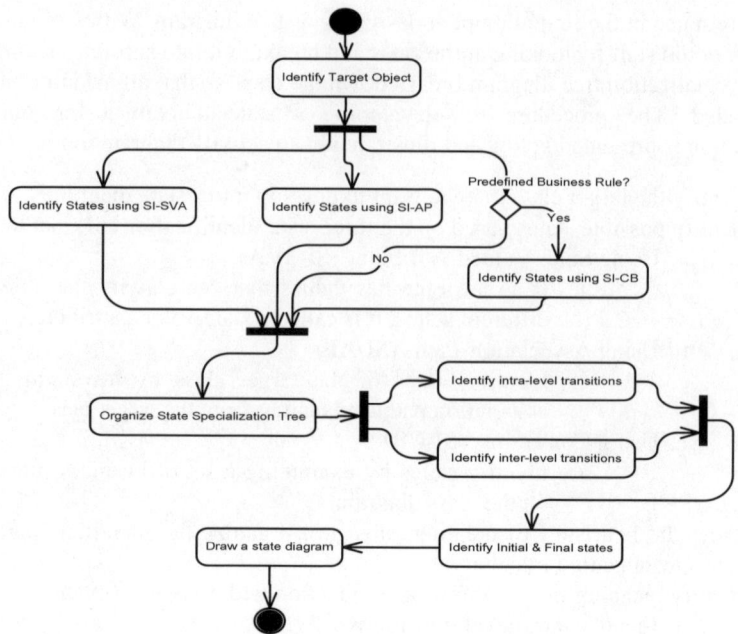

Fig. 2. An Activity Diagram Describing the Procedure of a State Specialization Method

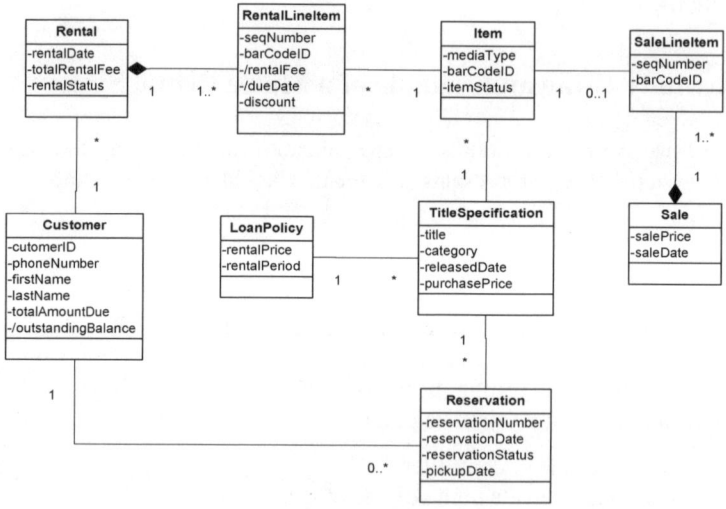

Fig. 3. A Domain Class Model of a Movie Rental System

4.2 Step 2

In Step 2, we identify possible states based on the state identification methods. We propose three state identification methods –state-valued attributes, association-paths, and constraints. First, if an attribute has values that can denote different states of the target class, the attribute is called a *state-valued attribute*. For example, values of *itemStatus* attribute (rented/not-rented) can be used in identifying states of a target class. Therefore, we call *itemStatus* a state-valued attribute. Second, a state can be identified by traversing association paths emanating from the target class. Each association path is examined to identify a possible state of the target class. For example, the existence of the association between *Item* and *RentalLineItem* implies that *Item* class has a state called "Rented" or "Off Shelf". The third state identification method uses integrity constraints. Examination of conditions stated in business rules can be used for identifying states. An example is given below.

1. **Using state-valued attribute**
   a. *The itemStatus* attribute can have "In Store", "Out of Store", "Damaged" or "Lost" states.
2. **Using association-paths.** We traverse each association path emanating from the *Item* class.
   a. In the association path of *Item-SaleLineItem-Sale*, we can identify the item status of "Sold" or "Not Sold". Since "Not Sold" does not add meaning to our system design, we can disregard it.
   b. In the association path of *Item-TitleSpecification-Reservation*, we can identify the state of "Reserved". We can specialize this state into "Canceled" or "To be picked-up".
   c. In the association path of *Item-RentalLineItem-Rental*, we can identify item the status of "Rented" or "Not Rented".
3. **Using constraints:** Suppose we have the following constraint:
   a. *The Overdue charge rule*: The overdue charge is the same as the rental fee for every rental period. If the total overdue charge exceeds the purchased price, the purchased price is charged.
      overDueCharge = MIN (TitltSpecification.purchasePrice,
      INTEGER (NOD/RP)*RentalLineItem.rentalFee), where
         RP = LoanPolicy.rentalPeriod and NOD = Number of overdue days,
   Based on this integrity constraint, the "Overdue" or "Not Overdue" states are identified.

Note that only meaningful states are used in state diagram representation below.

4.3 Step 3

In Step 3, using the heuristics of creating substates, we organize the identified states in a State Specialization Tree diagram. Some of the heuristics are:

1. A super state is an abstraction of its substates
2. Substates (from a single superstate) in the same level are mutually exclusive
3. Every instance of a substate is also an instance of the super state of the substate.
4. The specialization is transitive (When C is a substate of B and B is a substate of A, then C is a substate of A).

As the first step of the decomposition, a movie item is represented in two states: "In Store" and "Out of Store" (see Fig. 4 for illustration.). The "In Store" state can be specialized into "On Shelf" and "Off Shelf". A movie that is "Out of Store" can be "Sold", "Rented" or "Lost". "Off Shelf" can be further divided into the following states "On Sale", "Reserved", "Lost" or "Damaged", which are placed on level 4. The state "On shelf" which is under the state "In Store" cannot be broken into any lower states. The "Sold" state is at its final state because a sold item will not be returned to the store. The "Rented" state can be specialized into "Overdue" and "Not Overdue" states which are placed on level 4.

4.4 Steps 4 and 5

In Step 4, we identify meaningful transitions in conjunction with triggering events. We identify all the transitions by examining transitions between nodes at different levels (i.e., inter-level transitions) and those between leaf states in the same level (i.e., intra-level transitions). Transitions are defined beginning from the lowest level in the hierarchy.

- "Canceled", a substate of "Reserved", has one inter-level transition. The other substate of "Reserved", "To be picked up" has one inter-level transition and one intra-level transition. A "Reserved" movie can be re-shelved to "On Shelf" when the customer "Cancel" the reservation, or can be "Rented" when the customer "Picked up" the reserved movie. Therefore, we impose the transitions from "Canceled" to "On Shelf" and from "To be picked up" to "Rented". Also, the sequence between "Canceled" and "To be picked up" is represented as a transition from the latter to the former.
- "Overdue" has one inter-level transition and one intra-level transition; it can become "Sold" when the customer is charged the purchased price for the overdue charge, as indicated in the business rule, "Not Overdue" when the rental is renewed or "In Store" when returned with payment.
- "Not Overdue" has an intra-level transition but no inter-level transitions. When a due date is over, the item transits from "Not Overdue" to "Overdue".
- The transition from "Off Shelf" is covered by its substates. "Rented" can transit to "In Store" when the item is returned to the store and to "Lost" when the item is missing during customer's rental period.
- An "On Shelf" movie has two intra-level transitions but no inter-level transitions.
- A "Sold" item is at the lowest level once a movie is sold; it makes a transition to a final state.
- "On Sale" state becomes "Sold" when a customer buys the movie.
- In level 4, "Damaged" and "Lost" states make transition to a final state since they represent the end of an object lifecycle.

4.5 Step 6

We first design states, including composite states, and then design transitions later.

Once the state specialization tree diagram is complete with transitions, we can design the state diagram. We should first identify the initiating state. For this example, the initiating state is "In Store".

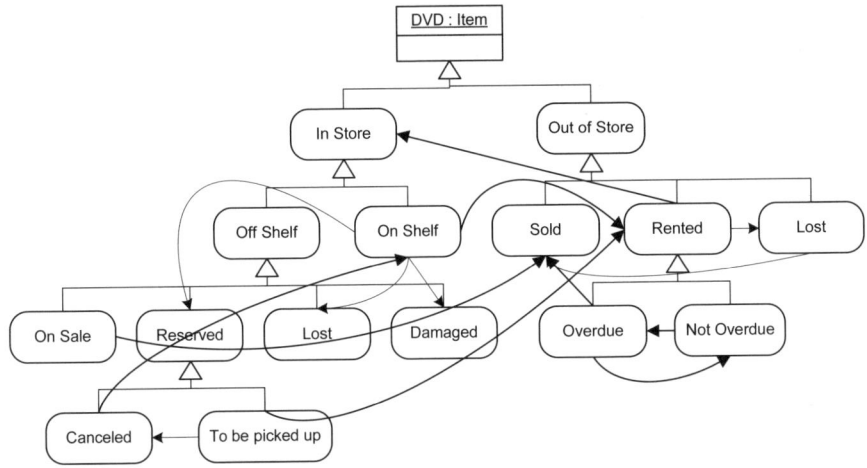

Fig. 4. A State Specialization Tree diagram with Transitions for a Movie Rental System

Next, we place all substates of the "In Store" state in the composite representation. "In Store" state has two child states: "On Shelf" and "Off Shelf". We include those child states inside the "In Store" state. "On Shelf" state does not have any child state; so we can leave it as is. "Off Shelf" state has child states; so we can include its child states inside "Off Shelf" state. We include "On Sale", "Reserved", "Lost", and "Damaged" child states into "Off Shelf". Since both "Lost" and "Damaged" do not evolve to other states, they can be combined and make transitions to a final state. We classified them as a "Deactivated" state. "Canceled" and "To be picked up" states can be placed into the "Reserved" state. This completes the representation process of the "In Store" state.

We now design the "Out of Store" state. The "Out of Store" state has three child states: "Sold", "Rented" and "Lost". The state "Sold" does not have a child state and means the end of the lifecycle in our system. Therefore, it transitions into a final state. "Rented" is specialized into its child states: "Overdue" and "Not overdue". By adding those states within the "Rented" state, we have completed the "Out of Store" state.

Our next step is to add transitions to the state diagram using the ones we identified in our State Specialization Tree diagram. In order to add them at the appropriate level of generalization, we look from top down to identify transitions. Therefore, level 2 of the specialization diagram is examined first. "On Shelf" can become "Reserved", "Lost" or "Damaged"; so we impose a transition from "On Shelf" to "Reserved", "Lost" and "Damaged". "On Shelf" can also become "Rented" when a customer rents the movie.

Next we look at level 3 of the specialization diagram and identify transitions for this level of generalization. An item can be in a state "To be picked up" when a "Reservation" is made. When the item is "Picked up" it is "Rented". On the other hand, when the "Reservation" is canceled, the item becomes "Canceled" state, then re-shelved. Therefore, we can impose the transitions from "To be picked up" to "Canceled", then to "On Shelf". Notice that "Rented" has two child states and thus we do not need to draw a transition to both states but only to the parent state. The next state in level 3 is "Overdue". An item that is "Overdue" can be "Renewed" or not. If the item is returned, it is "In Store". If it is not returned, it is "Sold" to the customer based on the business rule. We can impose a transition to "Sold" and "Not Overdue" from the "Overdue" state. The last state on level 3 is "Not Overdue". An item that is "Not Overdue" can make a transition to "Overdue" when the time of the rental has expired. And a renewed item can indicate transition from "Overdue" to "Not Overdue". A completed state diagram is shown in Fig. 5.

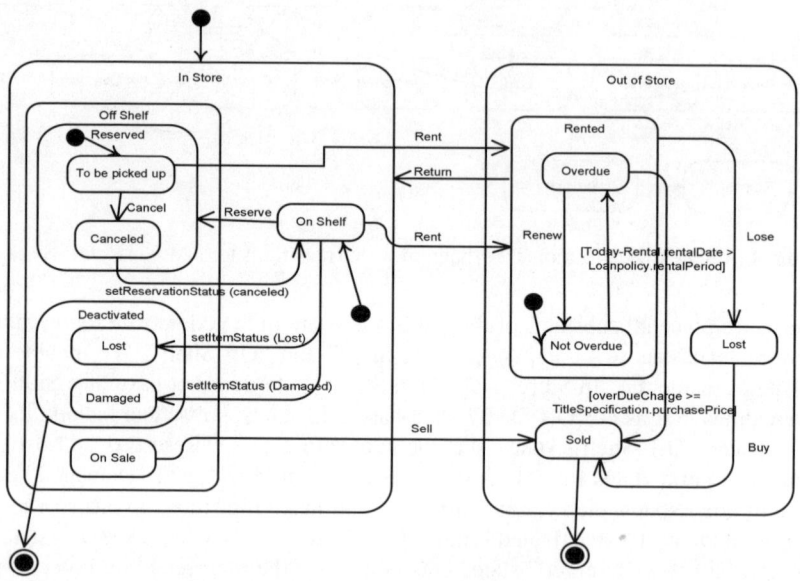

Fig. 5. A Completed State Diagram of *Item* Class of a Movie Rental System

5 Conclusion

In this paper, we have presented a systematic method for developing state diagrams using a state specialization technique. Our technique focused on identifying all the states using the three state identification methods –state-valued attributes, traversing association paths, and constraints. The important characteristic of our method is that it identifies meaningful states using the semantics captured in a class diagram and business rules. In identifying transitions, we identify intra-level and inter-level transitions. We have illustrated our method using an example of a movie rental system. Our step-by-step process is significant in developing state diagrams since unnecessary trial and error can be avoided as much as possible.

Our technique will be helpful to many novice designers and will enable them to create a complete state diagram for their systems systematically. Future work includes extending our method to design concurrent state diagrams, developing detailed rules for transitions, and performing experiments for completeness and usability.

References

1. Booch, G., Rumbaugh, J., Jacobson, I.: The Unified Modeling Language User Guide Addison-Wesley. Upper Saddle River, NJ (2005)
2. Cook, S., Daniels, J.: Designing Object Systems: Object-Oriented Modeling with Syntropy. Prentice-Hall, Englewood Cliffs (1994)
3. Cruz-Lemus, J.A., Genero, M., Piattini, M., Toval, A.: An Empirical Study of the Nesting Level of Composite States within UML Statechart Diagrams. In: Akoka, J. (ed.) ER Workshops 2005. LNCS, vol. 3770, pp. 12–22. Springer, Heidelberg (2005)
4. Fernandez-Aleman, J.L., Toval-Alvarez, A.: Can Intuition Become Rigorous?: Foundations for UML Model Verification Tool. In: ISSRE 2000. The Eleventh International Symposium on Software Reliability Engineering, San José, California (2000)
5. Harel, D.: Statecharts: A Visual Formalism for Complex Systems. Science of Computer Programming 8, 231–274 (1987)
6. Harel, D.: On Visual Formalisms. Communications of the ACM 31, 514–530 (1988)
7. Hu, Z., Shatz, S.M.: Explicit Modeling of Semantics Associated with Composite States in UML Statecharts. Automated Software Engineering 13, 423–467 (2006)
8. Reed, P.R.: Developing Applications with Visual Basic and UML. Addison-Wesley Publishing Company, Reading, MA (2000)
9. Roques, P.: UML in Practice: The Art of Modeling Software Systems Demonstrated through Worked Examples and Solutions. Wiley, Hoboken, NJ (2004)
10. Schafer, T., Knapp, A., Merz, S.: Model Checking UML State Machines and Collaborations. Electronic Notes in Theoretical Computer Science 47, 1–13 (2001)
11. Shlaer, S., Mellor, S.J.: Object Lifecycles Modeling the World in States. Yourdon Press, Englewood Cliffs, NJ (1992)
12. von der Beeck, M.: A Structured Operational Semantics for UML-Statecharts. Software and Systems Modeling 1, 130–141 (2002)
13. Wyner, G.M., Lee, J.: Process Specialization: Defining Specialization for State Diagrams. Computational & Mathematical Organization Theory 8, 133–155 (2002)

Quality Dependencies Among Use Case Models and Sequence Diagrams Developed by Novice Systems Analysts

Narasimha Bolloju[1] and Vijayan Sugumaran[2]

[1] Department of Information Systems
City University of Hong Kong,
Kowloon Tong, Hong Kong
narsi.bolloju@cityu.edu.hk
[2] Department of Decision and Information Sciences
School of Business Administration
Oakland University, Rochester, MI 48309
sugumara@oakland.edu

Abstract. Capturing and representing information systems requirements using quality artifacts is an important step in successful systems development process. Better quality artifacts in the early stages of systems development help in early detection and correction of errors. Although UML is widely used for modeling systems requirements, it is often difficult for novice systems analysts to develop quality UML artifacts. This paper mines a set of errors frequently observed in use case models and sequence diagrams created by novice analysts, and develops propositions identifying intra- and inter-artifacts quality. The data mining tool, AlphaMiner, was used to discover dependencies as association rules between frequently committed errors. Propositions have been developed based on these dependencies identified from student projects.

Keywords: UML artifacts, novice systems analysts, quality dependencies, association rules.

1 Introduction

Capturing information systems requirements and representing those requirements using appropriate models is an important part of information systems development process. Many information systems development project failures have been linked to problems associated with the requirements capturing. Further, only half of the required features and functions make it to the released products [6]. Highlighting the importance of quality conceptual modeling, Wand and Weber [20] state that better quality models facilitate early detection and correction of errors.

Nowadays, different techniques available in the Unified Modeling Language are being widely used for modeling systems requirements. However, many researchers (e.g., [2, 16]) observe difficulties and complexities associated with using UML.

J.-L. Hainaut et al. (Eds.): ER Workshops 2007, LNCS 4802, pp. 96–105, 2007.
© Springer-Verlag Berlin Heidelberg 2007

Despite the availability of guidance and recommendations for developing quality UML artifacts, typical novice systems analysts fail to get maximum benefit due to the cognitive overload associated with those guidelines and recommendations. Bolloju & Leung [4] investigated specific difficulties frequently encountered by novice analysts using the conceptual model quality framework proposed by Lindland et al. [13]. Such investigations can help develop better training procedures and enhanced support in the modeling process.

Use case diagrams and descriptions have become the corner stone of requirements elicitation and specification in systems development, especially after its introduction into UML. Client involvement with these artifacts has been found greatest compared to other UML models [8]. It is the most commonly used approach for describing system requirements from a user's perspective. Use case diagrams and the associated use case descriptions serve as the basis for identifying objects [9] and deriving other UML models such as class diagrams, sequence diagrams, etc. Since use case models are supposed to effectively capture system requirements, ensuring the quality of the use case diagrams and descriptions is of utmost importance. The quality of the remaining models and artifacts generated are dependent on the quality of the use case models. However, there is little empirical research identifying the factors that impact the quality of use case models. Early work in use case modeling has focused on creating semantically correct diagrams and more recently, researchers have focused on methods to create good use case descriptions. While it is essential to support the development of good use case models, it is equally important to study and understand the factors that affect the quality of these models.

By understanding the types of errors or mistakes that modelers make and how they get propagated from one set of artifacts to the next, one can design tools and environments to help users in their use case modeling activity and thus create quality use case models and sequence diagrams. Thus, the objectives of this research are to:

a) *identify intra- and inter-artifact quality dependencies using frequently observed errors in the models created by novice analysts, and*

b) *develop a set of propositions that can be empirically tested, which can lead to recommendations for analysts in creating quality models.*

The contributions of this research are three fold. First, the results from this research will help in providing intelligent support to novice systems analysts in addressing difficulties associated with requirements capturing using use cases and other associated diagrams. This can also be used in training new recruits such as analysts/programmers. Second, findings from this research can help develop facilities for enforcing syntactic conventions in CASE tools. Third, this work will result in identifying different ways of enhancing the quality of the resulting models.

The following section briefly reviews a conceptual model quality framework and its suitability for evaluating quality of UML models. Section 3 presents an overview of the approach employed to identify quality dependencies among use case models and sequence diagrams. Section 4 analyzes the association rules for identifying quality dependencies among different types of errors.

2 Background

Quality of the UML artifacts can be evaluated following the approaches used for evaluating conceptual model quality. A wide range of studies on conceptual model quality have been surveyed by Wand and Weber [20]. Frameworks for conceptual model quality provide a systematic structure for evaluation. Genero and Piattini [10] review four major frameworks for this purpose and describe the key elements of each framework. Many of these frameworks offer mostly subjective ways of evaluating the quality of conceptual models.

Lindland et al. [13] discuss the need for a framework addressing both process and product for the treatment of quality. They proposed a framework using three linguistic concepts viz. syntax, semantics and pragmatics, suitable for analyzing the quality of conceptual models. Syntactic correctness of a model implies that all statements in the model are according to the syntax of the language. Semantic quality captures the quality of a model in terms what the model lacks that is present in the domain and what the model includes that is not present in the domain. Semantic quality is described in terms of validity and completeness goals. The validity goal specifies that all statements in the model are correct and relevant to the problem domain. The completeness goal specifies that the model contains all statements about the problem domain that are correct and relevant. Pragmatic quality addresses the comprehension aspect of the model from the stakeholders' perspective. This framework has been applied by researchers in different situations such as evaluating the quality of information models [14] and comparing the quality of different ontology languages and tools [18]. Extensions to this framework have also been proposed (e.g. [10]).

Since UML artifacts produced in early phases are more conceptual in nature, this framework can be applied effectively to evaluate the quality of such artifacts. In addition, many of the UML artifacts model the system requirements from different perspectives and hence would be helpful to study and analyze the dependencies in intra- and inter-artifact quality. Results from such analysis are expected to contribute towards better understanding of quality problems and thereby help in developing methods and tools for modeling systems requirements effectively and efficiently.

Most of the existing studies that assess the quality of UML models treat each of the models independently. Several studies have examined the quality of use case models in isolation and provide suggestions for how to improve them [1, 7, 17]. For example, Phalp et al. [15] examine the quality of use case descriptions and provide ways to improve the quality of use case descriptions. Similarly, Lange [12] discusses issues related to improving the quality of UML models in practice and develops some research hypotheses to be tested. Several authors have also examined the quality of sequence diagrams separately and articulate some of the common mistakes made and the pitfalls in creating sequence diagrams [5, 19].

Since a set of existing UML artifacts are used to develop the next set of artifacts, it is important to ensure the quality of artifacts at each and every stage of development, especially in the initial stages. Since artifacts such as use case models, domain models and dynamic models are commonly used in the initial stages, we focus our study only on some relevant UML artifacts (i.e., use case diagrams and descriptions, class diagrams and sequence diagrams). Very few studies have looked at the quality of UML models, particularly, the use case model and the sequence diagram together.

Bolloju and Leung [4] present the results of an empirical study aimed at identifying the most typical set of errors frequently committed by novice systems analysts in four commonly used UML artifacts—use case diagrams, use case descriptions, class diagrams, and sequence diagrams. They also discuss how these errors affect the quality of the artifacts developed. They conclude that by ensuring the artifacts to be free of such errors, novice analysts would develop better-quality UML artifacts. They have developed a typology of errors for the syntactic, semantic, and pragmatic quality categories suggested by Lindland et al. [13] and also a coding scheme. In their study, Bolloju and Leung [4] again assess the quality of the models individually and suggest recommendations for improvement. Our current research builds on this work and advances it further by studying the quality dependencies between use case models and sequence diagrams. Understanding these dependencies and their impact is essential in ensuring the quality of the UML artifacts that get generated downstream in the life cycle. The following section discusses our proposed approach for investigating the quality dependencies between use case models and sequence diagrams.

3   Approach

This section presents an overview of the approach employed in identifying quality dependencies among use case models and sequence diagrams. This research makes use of part of the data, collected in prior research, pertaining to errors committed frequently by 14 project teams of novice systems analysts while developing UML artifacts viz. use case diagrams & descriptions, class diagrams and sequence diagrams (see [4] for details). The Appendix contains the list of error codes and their descriptions corresponding to use case models and sequence diagrams.

Three data sets were prepared for mining association rules among the errors observed in each of the 14 projects: (i) errors in use case diagrams and descriptions, (ii) errors in sequence diagrams, and (iii) combination of data sets (i) and (ii) for finding quality dependencies within use case models, within sequence diagrams and between use case models and sequence diagrams respectively. AlphaMiner [3], an open source data mining platform, was used for generating association rules within each of these data sets. Table 1 lists some of the association rules generated for use case models.

A two-step procedure was employed for pruning the large number of rules generated. First, a support cut-off at 40% and a confidence cut-off at 80% were used to select rules with good support and confidence[1]. Then, in order to identify

---

[1] For a given association rule X → Y, support is the percentage of transactions in the entire data set that contain both X and Y, and confidence is the percentage of occurrences of Y in transactions that contain X.

meaningful and useful quality dependencies from the remaining set of rules, only the rules affecting one or more errors belonging to the semantic quality category were selected. By focusing on the semantic quality, it is expected that any dependencies identified will be helpful in addressing validity and completeness of requirements. The remaining rules were used to identify specific sets of errors that affect intra-artifact semantic data quality for use case models and sequence diagrams. The third set of rules, after pruning, was used to identify use case model errors that affect semantic quality of sequence diagrams (inter-artifact semantic quality).

Table 1. Sample set of association rules within use case models

Rule No.	Rule	Items Size	Support (%)	Confidence (%)
72	UPu3e, USy2b => USc1c	3	42.857	100
55	USc1c, USv1b, USy2b => UPu3e	4	35.714	100
62	UPu3e, USv1b, USy2b => USc1c	4	35.714	100
10	UPu3d, USc1c => UPu3e	3	35.714	100
16	UPu3d, USc2a => UPu3e	3	35.714	100
25	UPu3e, USv1e => UPu3d	3	35.714	100
28	UPu3d, USv1e => UPu3e	3	35.714	100
87	USc2a, USv1c => UPu3f	3	35.714	100
90	UPu3f, USv1c => USc2a	3	35.714	100
96	UPu3f, USv1f => USc2a	3	35.714	100
105	USc2a, USy2a => USc1c	3	35.714	100
108	USc1c, USy2a => USc2a	3	35.714	100
49	USc1c, USv1b => UPu3e	3	42.857	85.714

4 Analysis of Association Rules

This section analyzes the pruned sets of association rules for identifying dependencies among different types of errors and then developing the propositions.

4.1 Use Case Diagrams and Descriptions

Frequently observed errors in use case descriptions are largely related to step descriptions. Errors observed in use case diagrams are related to identification of proper relationships (viz. includes, extends and generalization) between use cases.

The semantic quality component of use case models, which includes invalid step extensions, missing steps and ambiguous step descriptions in use case descriptions, is considered to be of importance in this analysis because it deals with validity and completeness of functional requirements of the system under development. Figure 1 depicts dependencies identified using the association rules by focusing on the rules

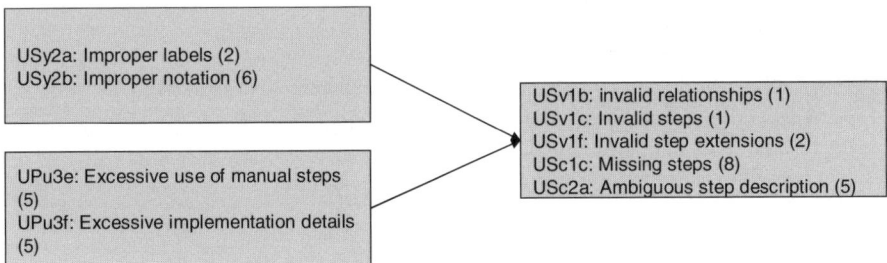

Fig. 1. Quality dependencies identified in use case models

having errors belonging to the semantic quality on their right hand side. It can be noted that errors belonging to the other two quality components have approximately equal contributions to this quality component.

Since use case names are expected to start with strong action verbs indicating its purpose, wrong selection of a name may not convey the expected functionality that is normally described in the use cases. Similarly, by not adhering to the notation for diagrams and description (e.g., <subject-verb-direct object> format for step description), one is likely to end up with ambiguous step descriptions. Ambiguity in step descriptions could also be the result of invalid steps' description (e.g., steps that cannot be performed by the system) and invalid step extensions. Excessive use of manual steps obviously implies that majority of the tasks are performed by the actor(s), which may lead to missing steps that can be delegated to the system. Finally, excessive use of implementation details at the stage of determining logical requirements shift the focus away from assigning responsibility to the system as captured in step description. Based on these observations, the following proposition can be established.

Proposition 1: Syntactic quality problems related to use case naming and not *adhering to notation, and pragmatic quality problems related to excessive use of manual steps and implementation details in step descriptions are associated with semantic quality problems in the form of invalid step extensions and missing or ambiguous use case step descriptions.*

4.2 Sequence Diagrams

One of the major purposes of sequence diagrams is to identify responsibilities of various classes or objects of the problem domain in terms of methods and method signatures (or interfaces). Messages included in sequence diagrams clearly help in achieving this purpose. The majority of errors found in the development of sequence diagrams can be attributed to inexperience of novice analysts, lack of problem-structuring skills such as decomposition and to difficulties in applying object-oriented concepts related to distribution of responsibilities across participating objects. Many syntactic errors are related to message flow control such as missing initial trigger

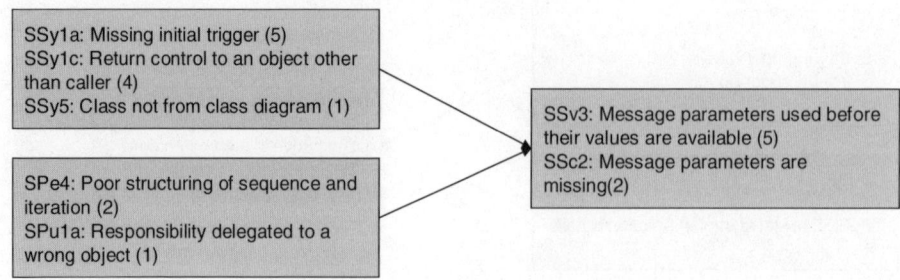

Fig. 2. Quality dependencies identified in sequence diagrams

messages, returning control to objects other than the calling object, etc. Pragmatic errors included improper delegation of responsibility (often to a wrong object/class) and making a class/object perform tasks that could be delegated to other objects.

The semantic quality of sequence diagrams, as identified based on the analysis of association rules, includes usage of message parameters before their values are made available (e.g., findCustomer(ID) using a known value of customer ID), and missing parameters in messages. Figure 2 lists various errors identified using the association rules containing these two types of errors on the right hand side. Since the relative contributions of SSy5 and SPu1a are quite low, these errors are excluded from further analysis.

Often, input parameter values to the process represented by a sequence diagram are provided through an initial trigger message. Absence of an initial trigger is likely to result in missing parameters. In a similar way, returning control to objects other than the caller object causes some parameter values undefined for subsequent use by the caller object. Poor structuring of actions can also contribute to missing parameters or undefined parameters in one or more messages. The above observations lead to the following proposition.

Proposition 2: *Syntactic problems with flow control and poor structuring of sequence and iterations in sequence diagrams are associated with semantic quality problems via missing parameters and undefined parameter usage.*

4.3   Use Case Models and Sequence Diagrams

Association rules generated using the combined set of errors in use case models and sequence diagrams were analyzed to identify possible errors in use case models that affect semantic quality of the corresponding sequence diagrams. Considering the frequencies of associations found (see Figure 3), it is apparent that pragmatic quality problems associated with step descriptions have a significant effect on the quality of sequence diagrams. It is, however, possible that this effect could be indirect via the structure of actions in sequence diagrams. Proposition 3 is derived from this analysis.

Proposition 3: *Pragmatic quality problems related to step descriptions in use cases are associated with semantic quality problems of sequence diagrams related to message descriptions.*

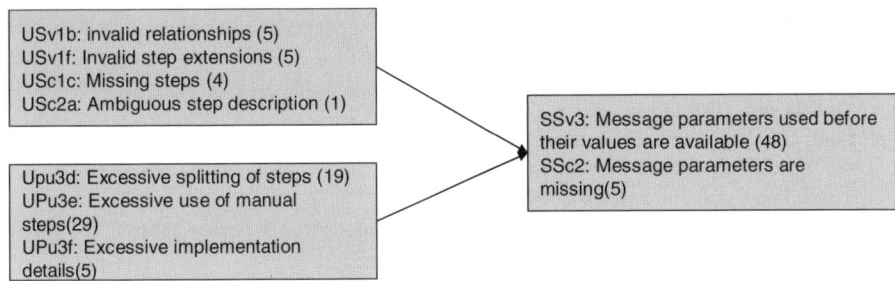

Fig. 3. Quality dependencies identified between use case models & sequence diagrams

5 Discussion and Conclusion

Using the data collected through an empirical investigation involving an application of Lindland et al's conceptual model quality framework for analyzing the quality of UML artifacts, three propositions related to quality dependencies among use case models and sequence diagrams developed by novice systems analysts have been presented. These dependencies among different types of errors frequently observed in use case models and sequence diagrams, could be exploited in designing training programs for systems analysts and in providing intelligent support in modeling tools.

A major limitation of this study is the use of data collected based on project work completed by teams of students, though these students had prior experience with structured systems analysis and design method. Considering difficulties in accessing the work produced by experienced systems analysts in real life projects, we think the limited experience of subjects and semester-long duration of team projects address this limitation to a large extent. Another limitation of the study is related to types of artifacts considered. Though some of the widely used UML artifacts are considered in this study, other artifacts such as activity diagrams, class diagrams and collaboration diagrams may be relevant for such studies.

Further research investigations and studies are required for validating the propositions developed in this research, incorporating some of these findings into CASE tools to provide intelligent and adaptive support to novice systems analysts, and identifying the effect of quality of analysis phase artifacts on that of design phase artifacts.

Acknowledgement. Funding for this research was received from the Strategic Research Grant # 7001871 of City University of Hong Kong.

References

1. Adolph, S., Bramble, P., Cockburn, A., Pols, A.: Patterns for Effective Use Cases. Addison-Wesley, Reading (2002)
2. Agarwal, R., Sinha, A.P.: Object-oriented modeling with UML: A study of developers' perceptions. Communications of the ACM 46(9), 248–256 (2003)

3. Alphaminer, E-Business Technology Institute, The University of Hong Kong (2005), http://www.eti.hku.hk/alphaminer/
4. Bolloju, N., Leung, F.: Assisting Novice Analysts in Developing Quality Conceptual Models with UML. Communications for the ACM 49(7), 108–112 (2006)
5. Carr, J.T., Balci, O.: Verification and Validation of Object-Oriented Artifacts throughout the Simulation Model Development Life Cycle. In: Proceedings of the 2000 Winter Simulation Conference, pp. 866–871 (2000)
6. Chaos Report (2003) Standish Group Chaos Report - Press Releases (March 25, 2003), http://www.standishgroup.com/press/article.php?id=2
7. Cockburn, A.: Writing Effective Use Cases. Addison-Wesley, Reading (2001)
8. Dobing, B., Parsons, J.: How UML is Used. Communications of the ACM 49(5), 109–113 (2006)
9. Dobing, B., Parsons, J.: Understanding the Role of Use Cases in UML: A Review and Research Agenda. Journal of Database Management 11(4), 28–36 (2000)
10. Genero, M., Piattini, M.G.: Quality of Conceptual Modelling in Information and Database Quality. In: Piattini, M.G., Calero, C., Genero, M. (eds.) Quality of Conceptual Modelling in Information and Database Quality, pp. 13–44. Kluwer, Dordrecht (2002)
11. Krogstie, J., Lindland, O.I., Sindre, G.: Towards a Deeper Understanding of Quality in Requirements Engineering. In: Proceedings of the 7th International Conference on Advanced Information Systems Engineering, Finland, pp. 82–95 (1995)
12. Lange, C.F.J.: Improving the Quality of UML Models in Practice. In: ICSE 2006, Shanghai, China, May 20-28, pp. 993–996 (2006)
13. Lindland, O.I., Sindre, G., Sølvberg, A.: Understanding quality in conceptual modeling. IEEE Software 11(2), 42–49 (1994)
14. Moody, D.L., Sindre, G., Brasethvik, T., Sølvberg, A.: Evaluating the Quality of Information Models: Empirical Testing of a Conceptual Model Quality Framework. In: Proceedings of the 25th International Conference on Software Engineering, pp. 295–305 (2003)
15. Phalp, K.T., Vincent, J., Cox, K.: Assessing the Quality of Use Case Descriptions. Software Quality Journal 15(1), 69–97 (2007)
16. Siau, K., Cao, Q.: Unified Modeling Language: A Complexity Analysis. Journal of Database Management 12(1), 26–34 (2001)
17. Somé, S.: Supporting use case based requirements engineering. Information and Software Technology 48(1), 43–58 (2006)
18. Su, X., Ilebrekke, L.: Using a Semiotic Framework for a Comparative Study of Ontology Languages and Tools, Information Modeling Methods and Methodologies. In: Krogstie, J., Halpin, T.A., Siau, K. (eds.) IDEA Group, pp. 278–299 (2005)
19. Unhelkar, B.: Verification and Validation for Quality of UML 2.0 Models. John Wiley & Sons, Inc., Hoboken, NJ, U.S.A (2005)
20. Wand, Y., Weber, R.: Research commentary: Information systems and conceptual modeling - A research agenda. Information Systems Research 13(4), 363–376 (2002)

Appendix: Errors Categorized Using Conceptual Model Quality Framework [2]

Use Case Diagram and Descriptions

Error Category	Error Code	Error Description
Syntactic	USy2a	Improper labels (especially for use case names)
	USy2b	Improper notation used in diagrams and descriptions
Semantic - validity	USv1b	Invalid relationship between use cases
	USv1c	Invalid step in a use case
	USv1e	Invalid actor either in use case model or description
	USv1f	Invalid extension(s) to a step
Semantic - completeness	USc1c	Missing an important use case step
	USc2a	Ambiguous use case step description
Pragmatic	UPu3d	Excessive splitting of step(s)
	UPu3e	Excessive use of manual step(s)
	UPu3f	Excessive implementation details (user interface; database; programming style)

Sequence Diagrams

Error Category	Error Code	Error Description
Syntactic	SSy1a	Improper flow control - missing initial trigger
	SSy1c	Improper flow control - return to an object different from caller or control not transferred to caller
	SSy5	Class in sequence diagram does not belong to class diagram
Semantic - validity	SSv3	Message parameters are used before their values are available
Semantic - completeness	SSc1	Essential or key classes/objects left out of sequence diagram
	SSc2	Message parameters are missing (completely or occasionally)
Pragmatic	SPe4	Poor structuring of actions especially sequence & iteration
	SPu1a	Responsibility delegated to a wrong object

---

[2] Only the error codes that were part of the association rules generated by AlphaMiner are shown in these tables.

M-BPSec: A Method for Security Requirement Elicitation from a UML 2.0 Business Process Specification

Alfonso Rodríguez[1], Eduardo Fernández-Medina[2], and Mario Piattini[2]

[1] Departamento de Auditoría e Informática, Universidad del Bio Bio,
Chillán, Chile
alfonso@ubiobio.cl

[2] ALARCOS Research Group, Information Systems and Technologies Department
UCLM-Indra Research and Development Institute
University of Castilla-La Mancha
Ciudad Real, Spain
{Eduardo.FdezMedina,Mario.Piattini}@uclm.es

Abstract. The early attainment of requirements in a software development process allows us to improve the quality of the product. Although many methods through which to elicit requirements exist, few of them are specifically designed for security requirements. This paper describes a method - M-BPSec - which permits the elicitation of security requirements which form part of a business process description carried out with a UML 2.0 Activity Diagram. M-BPSec is made up of stages, actors, tools and artifacts which, when applied in a coordinated manner, allow us to specify security requirements in business processes and to obtain class and use cases from this specification.

1 Introduction

The elicitation of requirements is perhaps the activity which is most often regarded as the first step in the requirement engineering process [8]. Since many elicitation techniques are available to the requirements engineer, some guidance on their use is needed. *Methods* provide one way of delivering such guidance. Each method itself has its strengths and weaknesses, and is normally best suited for use in particular application domains [8]. Elicitation methods related to security requirement are scarce. One reason for this is that few elicitation methods are specifically directed towards security requirements. Another reason is that organizations seldom specifically address the elicitation of security requirements and instead lump them in with other traditional requirement elicitation methods [6].

Furthermore, business processes (BP), which are defined as a set of procedures or activities which collectively pursue a business objective or policy goal [15], are not only an important resource in enterprise performance and in maintaining competitiveness, but are also an important requirement source.

Moreover, the business scene, in which there are many participants and an intensive use of communications and information technologies, implies that

J.-L. Hainaut et al. (Eds.): ER Workshops 2007, LNCS 4802, pp. 106–115, 2007.

enterprises not only expand their businesses but also increase their vulnerability. Despite the fact that the importance of business processes security is widely accepted, the business analyst perspective in relation to security has hardly been dealt with until now. At the present it is possible to capture high level security requirements which are easily identifiable by those who model business processes, because: (i) the business process representation has improved in the UML 2.0 version, (ii) the security requirement will tend to have the same basic kinds of valuable and potentially vulnerable assets [3], and (iii) empirical studies show that customers and end users are able to express their security needs [5]. Consequently, in [12], we have extended the UML 2.0 Activity Diagram (UML 2.0-AD) [10] by creating the BPSec-Profile. This profile allows us to capture security requirements and create a Secure Business Process (SBP) specification.

In this paper, we demonstrate a method of security requirement elicitation which uses SBP specification as a starting point. Our method considers stages, workers, tools, models and artifacts which, if grouped together, permit (i) the design of an SBP (ii) the attainment of analysis-level classes and use cases which include security aspects and (iii) the storage of information related to the specification of the business process.

The structure of the remainder of the paper is as follows: in Section 2, we will present our proposal in brief and related works, in Section 3 we will present M-BPSec. Finally, in Section 4 our conclusions will be drawn.

2   Our Proposal and Related Work

Our proposal is a method by which to elicit security requirements which are captured through the application of an extension of a UML 2.0-AD. By using BPSec-Profile, it is possible to represent the business analyst's perspective with regard to security. In this way, a new point of view with a high level of abstraction is obtained, and this complements existing perspectives concerning security.

Given that M-BPSec includes models (at different levels of abstraction) and artifacts (pieces of information that are produced, modified, or used in a method [11]), the Model-Driven Architecture (MDA) [9] approach for the models and the Unified Process (UP) [4] for the use of the artifacts have been used as a reference framework. MDA is composed of the following perspectives: (i) the computation independent viewpoint which focuses on the environment of the system, (ii) the platform independent viewpoint which focuses on the operation of a system whilst concealing the details necessary for a particular platform, and (iii) the platform specific viewpoint which combines the platform independent viewpoint with an additional focus on the details of the use of a specific platform by a system. We have also used the UP, which is composed of a set of activities which are necessary for the transformation of user requirements into a software system.

The basic aspects of our proposal are shown in Figure 1. The first column (on the left) shows three types of models which conform to the MDA. The last column shows the UP disciplines. The central part shows our proposal and the artifacts which are derived from the M-BPSec application. The SBP specification is created by using the UML 2.0-AD and the BPSec-Profile [12]. The Analysis-level Classes and Use cases are obtained by a transformation which takes the SBP model as its starting point and

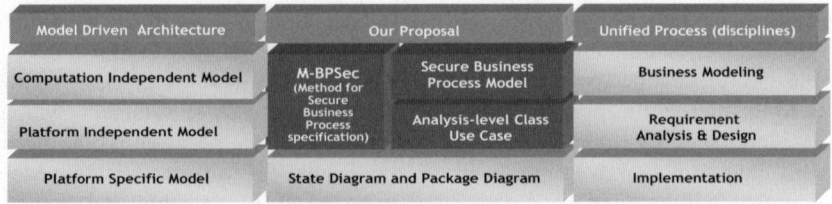

Fig. 1. Overview of our proposal

uses QVT rules, refinement rules and checklists. If Figure 1 is observed horizontally it will be noted that there is a correspondence between the elements presented in our proposal and the MDA models and UP disciplines. Thus, an SBP description corresponds with a Computation Independent Model (CIM). Said model can be used as a complement in the "Business Modeling" discipline of the UP. The Use Cases, which form a part of a Platform Independent Model (PIM), will complement the "Requirement" and "Analysis & Design" disciplines.

In related work dealing with security requirement elicitation methods we found in [6] a comparative analysis of nine elicitation methods. The author has used the following comparison criteria: adaptability, a computer-aided software engineering (CASE) tool, stakeholder acceptance, easy implementation, graphical output, quick implementation, a shallow learning curve, high maturity and scalability. In [7] the authors analyzed seven proposals orientated towards establishing security requirements in the development of an information system scope. The comparative criteria are: degree of agility, help support, degree of integration with other software requirements, user friendliness and contributions of the proposal as regards security. And finally, a comparative evaluation of three approaches: common criteria, misuse cases and attack trees which take into account learnability, usability, solution inclusiveness, clarity of output and analyzability criteria is carried out by [2].

However, none of the proposals for the acquisition of security requirements which we have reviewed consider the acquisition of these requirements by using a business process described with UML 2.0-AD as a starting point. Our proposal considers this situation, and also the possibility of automatically obtaining other UML artifacts which contain security requirements and which complement the traditional stages in a software creation process.

3   M-BPSec: A Security Requirement Elicitation Method

In this section we show our proposal, a security requirements elicitation method (see Figure 2). In Section 3.1 we shall describe the workers who participate in M-BPSec, the tools and models which are used and the artifacts which are generated. The *Construction, Security requirement incorporation, Refining* and *Transformations* stages are described in Sections 3.2, 3.3, 3.4, and 3.5 respectively. Besides having defined each stage of the M-BPSec, we have also used an example which allows us to relate the *workers*, *tools, models*, and *artifacts* which form a part of each stage.

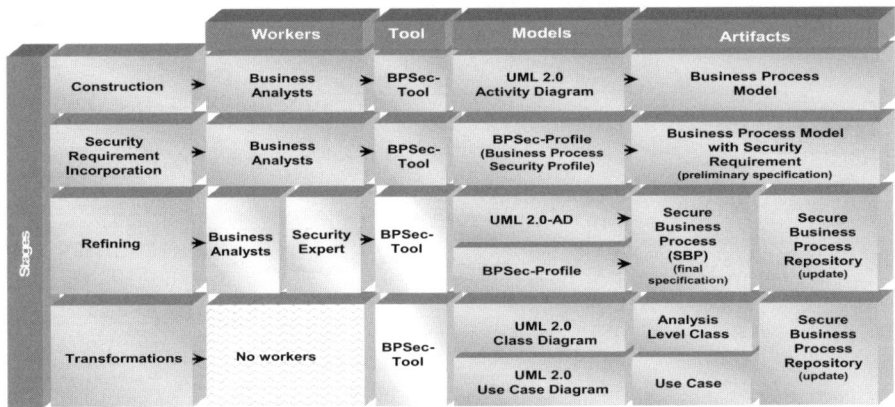

Fig. 2. M-BPSec: A method for security requirement elicitation

3.1 Workers, Tool, Models and Artifacts

Two types of workers are involved in M-BPSec: business analysts and security experts. The *business analyst* is responsible for the specifications related to the business itself as well as for incorporating (from his/her point of view) security requirements into the specifications by considering a high level of abstraction. The *security expert*: is the person responsible for refining the security specifications indicated by the business analyst. Such refinement considers the verification of the validity and the completeness of the specifications.

The Tool is the *BPSec-Tool*, which is used to design the SBP, to automatically transform models and to update the data contained in the secure business process repository. The BPSec-Tool was built by using a 3-tiered architecture to separate the presentation, application, and storage components, using MS-Visio, C#, and MS-Access technology respectively.

The Models employed are: (i) *UML 2.0-AD*, used for business process specification; (ii) *BPSec-Profile* used for security requirement specification; (iii) a *Class Diagram* and a *Use Case Diagram*, which both contain the specifications which will be automatically obtained from the SBP specification.

The artifacts are: (i) a *Business Process Model* described with UML 2.0-AD in which the business process is modelled without security requirements; (ii) a *Business Process Model with Security Requirement* and a *Secure Business Process*, both of which are described with UML 2.0-AD and BPSec-Profile. The first of these corresponds with a preliminary specification and the second corresponds with the final specification; (iii) *Secure Business Process Repository* (see Figure 3) in which the SBP, analysis-level class and use case are stored, and (iv) *Analysis-Level Class* and *Use Case*, both of which are automatically created and contain security specifications.

The repository contains information about the secure business process specification. Information concerning the activity diagram, security requirements specification, analysis-level classes and use cases are stored here. The repository is

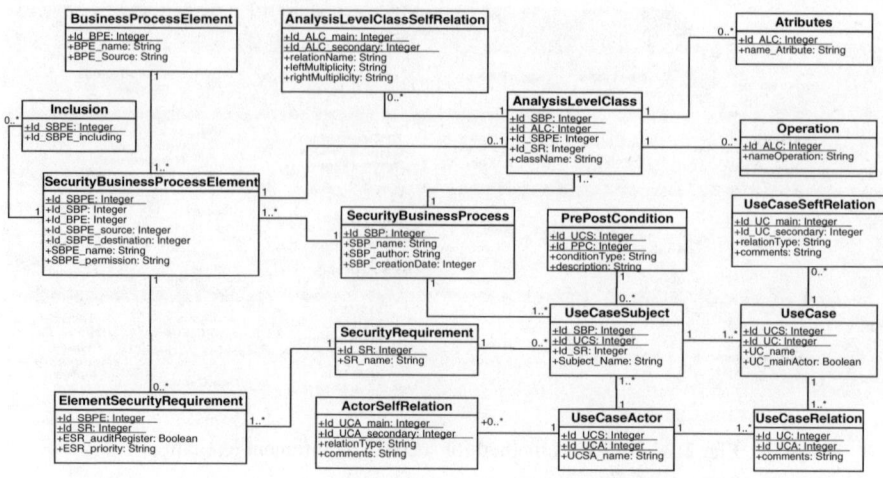

Fig. 3. Secure Business Process Repository

automatically created from the SBP specification. It is completed with the information from the analysis-level classes and the use cases which are obtained automatically. The repository can additionally be used to obtain historical information in order to make the business process specifications secure.

In the following sections we shall explain each stage in the M-BPSec. The stages will be described through the worker, tool, model and output artifacts that are used in each stage. We shall additionally use the example of a business process which is related to the admission of patients to a medical institution (see Figure 4).

3.2 *Construction* Stage

In this stage the objective is to build the business process model. To attain this objective, the UML 2.0-AD must be used. The *Construction* stage is carried out by the business analyst, who sets the time and place of the work activities, ensures that they have a beginning and an end, and who clearly explains the business view point. The BPSec-Tool is used to make a business process design with UML 2.0-AD. The final result of this stage is a Business Process Model.

In our example the business process is initiated with an admission request which is filled in by the *Patient*. This document, called an *Admission Request*, is sent to the *Administration Area*. In this area, information related to insurance is captured, and the existence of the patient's medical file is verified.

Once the patient's documentation has been validated and completed, it is sent to the *Medical Area*. The *Medical Evaluation* area uses a set of pre-admission tests to determine the patient's medical condition. If necessary, additional examinations are carried out, and these must be registered from both the clinical and the economic point of view. Finally, the *Medical Evaluation* document is filled in with information about the patient, and this is then sent to him/her. The business process is completed when the patient receives his/her *Medical Evaluation*.

3.3   Security Requirement Incorporation Stage

In this stage the security requirements can be added to the business process description from the business analyst's viewpoint. The business analyst must able to identify the potential threats in the business process model. The security requirements must subsequently be incorporated, using BPSec-Profile (supported by the BPSec-Tool). The output artifact in this stage is a Business Process Model with Security Requirement.

In [12] we have proposed the following types of security requirements: Access Control, Attack Harm Detection, Security Auditing, Non Repudiation, Integrity and Privacy. In our proposal we have used a padlock, standard *de facto*, to represent security. The same symbol, the padlock, but with a twisted corner is used to represent a Security Requirement with Audit Register.

In the "Admission of Patients" business process specification, the business analyst has specified «Privacy» (anonymity) for the *Patient* ActivityPartition, with the aim of preventing the disclosure and storage of sensitive information about Patients. «Nonrepudiation» has been defined for the control flow that goes from the *Fill Admission Request* action to the *Capture Insurance Information* and *Check Clinical Data* actions with the aim of avoiding the denial of the *Admission Request* reception. «AccessControl» and «Privacy» (confidentiality) have been defined for the Interruptible Activity Region. A «SecurityRole» can be derived from this specification. *Admission/Accounting* will be another role. All objects in an interruptible region must be considered for permission specification. The Access Control specification has been complemented with an audit requirement. This implies that it must register information about the security role and security permissions. An «Integrity» (high) requirement has been specified for the *Clinical Information* DataStore and finally, the business analyst has specified «AttackHarmDetection» for the *Medical Evaluation* DataStore, so that, all events related to the attempt or success of attacks or damages are registered. The graphic specification of the business process which has been carried out by using the BPSec-Tool, is shown in Figure 4.

Additionally, the priority attribute, according to that established in [1], and the permission for the actions, data store and object flows in access control scope, must be specified.

3.4   *Refining* Stage

In the *Refining* stage the security requirement specified in the business process description must be reviewed and complemented. The workers, the business analyst and the security expert, work together and the specifications that will finally be incorporated into the business process are agreed. The BPSec-Tool must be used to achieve a Secure Business Process (final specification). The repository which contains information about the secure business process specification is also automatically generated.

The priority specification of each security requirement and security permissions associated with the access control specification are also refined in this stage. The *must be* priority was assigned to Privacy (anonymity) and AuditRegister, and to

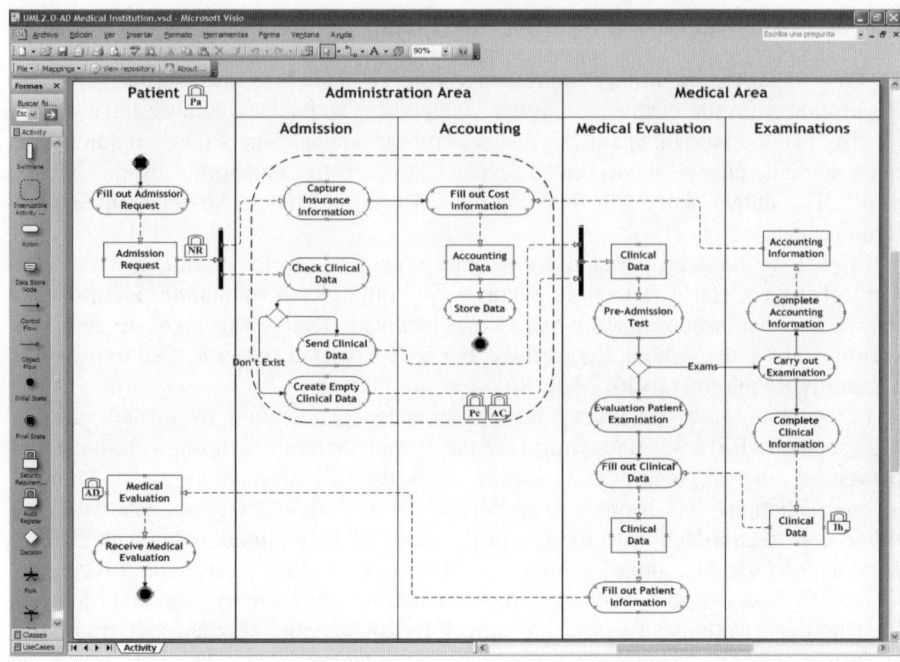

Fig. 4. Admission of Patients to a Medical Institution with security requirement

Nonrepudiation, *should have* was assigned to Access Control, and to Privacy (confidentiality), and *could have* was assigned to Integrity (high).

The e*xecution* permission was assigned to *Capture Insurance Information*, *Check Clinical Data*, *Store Data*, and *Create Empty Clinical Data*. The *CheckExecution* permission was assigned to *Fill out Cost information*. And the *Update* permission was assigned to *Accounting Data*.

3.5 Transformation Stage

Finally, in the *Transformation* stage, the analysis-level classes and use cases are obtained and are stored in the repository. This stage does not require workers because the artifacts and the repository are generated automatically.

Transformations to analysis-level classes, from CIM to PIM (C2P), require a set of rules which have been specified in QVT language and also a set of refinement rules [13]. For transformations to use cases (C2P) it is necessary to specify a set of QVT rules, refinement rules and a checklist [14].

The result of the application of the QVT rules and the refinement rules in relation to the attainment of analysis-level classes can be seen in Figure 5.

The use case model derived from the business process specifications for the admission of patients (see Figure 6 on the left-hand) and the use case model derived from the Access Control and Privacy specification security requirements (see Figure 6 on the righ-hand) are graphically shown.

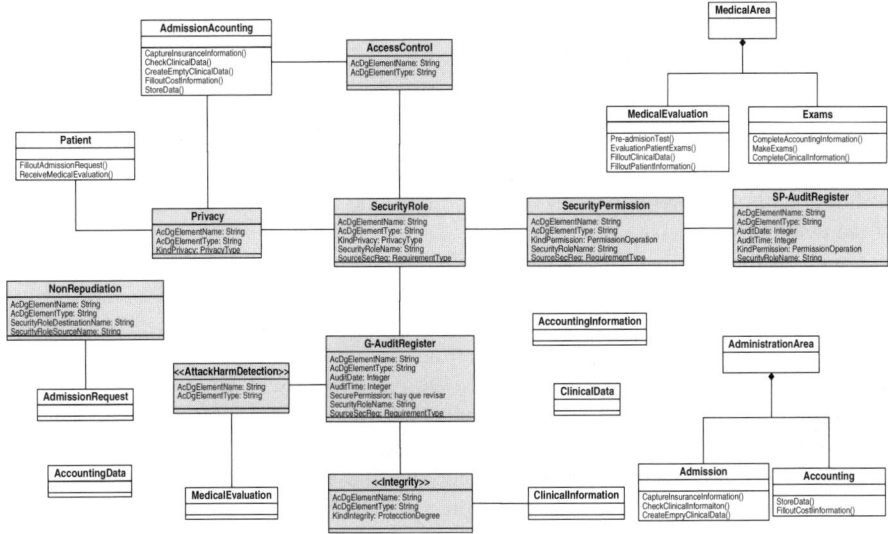

Fig. 5. Analysis-Level Class from Patient Admission

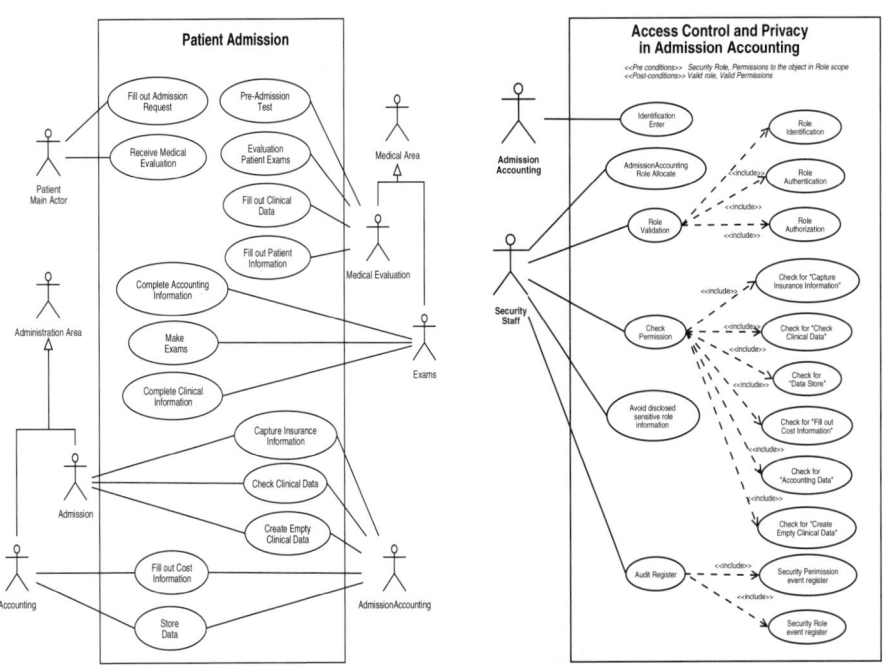

Fig. 6. Patient Admission and Access Control/Privacy use case specification

4 Conclusions

A business process specified with a UML 2.0 Activity Diagram containing security requirement allows the incorporation of a new perspective with regard to the security in a software creation process. However, this specification alone is not sufficient. The attainment of security requirements at this level of abstraction should be framed within a method which allows us to ensure both the acquisition and the adequate use of the resulting artifacts.

Owing to the fact that the elicitation of requirements is perhaps the most important activity in requirement engineering, in this paper we have defined a method for security requirement elicitation from a UML 2.0 Business Process specification. M-BPSec is a regular and systematic method by which to accomplish (i) security requirement specification in a business process described with UML 2.0-AD, and (ii) the attainment of the UML artifact, analysis-level classes and use cases, through which it is possible to achieve most concrete models which include security.

In M-BPSec the stages, workers, tool-models and artifacts are clearly defined and, when grouped together, allow us to elicit security requirements by using a secure business process as a starting point. We believe that M-BPSec satisfactorily fulfills the evaluation criteria of elicitation methods for requirements such as: the computer-aided software tool, easy implementation, graphical output, quick implementation, and the shallow learning curve.

The next steps in our research are orientated towards applying M-BPSec to a real case orientated towards enriching the method and the artifacts. We shall also improve the BPSEc-Tool in order to allow us to include other notations for the specification of SBP.

Acknowledgements. This research is part of the following projects: DIMENSIONS (PBC-05-012-1), and MISTICO (PBC06-0082) both partiality supported by the FEDER and the "Consejería de Ciencia y Tecnología de la Junta de Comunidades de Castilla-La Mancha", Spain, COMPETISOFT (506AC087), granted by CYTED and ESFINGE (TIN2006-15175-C05-05/) granted by the "Dirección General de Investigación del Ministerio de Ciencia y Tecnología", Spain.

References

1. Arlow, J., Neustadt, I.: UML 2 and the Unified Process: Practical Object-Oriented Analysis and Design, 2nd edn., p. 592. Addison-Wesley, Reading (2005)
2. Diallo, M.H., Romero-Mariona, J., Sim, S.E., Alspaugh, T.A., Richardson, D.J.: A Comparative Evaluation of Three Approaches to Specifying Security Requirements. In: REFSQ. 12th International Working Conference on Requirements Engineering: Foundation for Software Quality, Luxembourg (2006)
3. Firesmith, D.: Specifying Reusable Security Requirements. Journal of Object Technology 3(1), 61–75 (2004)
4. Jacobson, I., Booch, G., Rumbaugh, J.: The Unified Software Development Process, p. 463 (1999)
5. Lopez, J., Montenegro, J.A., Vivas, J.L., Okamoto, E., Dawson, E.: Specification and design of advanced authentication and authorization services. Computer Standards & Interfaces 27(5), 467–478 (2005)

6. Mead, N.R.: Experiences in Eliciting Security Requirements, CrossTalk. The Journal of Defense Software Engineering 19(12) (2006)
7. Mellado, D., Fernández-Medina, E., Piattini, M.: A Comparative Study of Proposals for Establishing Security Requirements for the Development of Secure Information Systems. In: Computational Science and Its Applications (ICCSA), Glasgow, UK, pp. 1044–1053 (2006)
8. Nuseibeh, B., Easterbrook, S.M.: Requirements Engineering: A Roadmap. In: ICSE 2000. 22nd International Conference on on Software Engineering, Future of Software Engineering Track, Limerick Ireland, pp. 35–46. ACM Press, New York (2000)
9. Object Management Group; MDA Guide Version 1.0.1(2003), http://www.omg.org/docs/omg/ 03-06-01.pdf
10. Object Management Group; Unified Modeling Language: Superstructure Version 2.1.1 (formal/2007-02-05) (2007), http://www.omg.org/docs/formal/07-02-05.pdf
11. Rational Software, Rational Unified Process, Best Practices for Software Development Teams, p.21 (2001)
12. Rodríguez, A., Fernández-Medina, E., Piattini, M.: Towards a UML 2.0 Extension for the Modeling of Security Requirements in Business Processes. In: Fischer-Hübner, S., Furnell, S., Lambrinoudakis, C. (eds.) TrustBus 2006. LNCS, vol. 4083, pp. 51–61. Springer, Heidelberg (2006)
13. Rodríguez, A., Fernández-Medina, E., Piattini, M.: Analysis-Level Classes from Secure Business Processes through Models Transformations. In: Lambrinoudakis, C., Pernul, G., Tjoa, A.M. (eds.) TrustBus 2007. LNCS, vol. 4657, pp. 104–114. Springer, Heidelberg (2007)
14. Rodríguez, A., de Guzmán, I.G.-R.: Obtaining Use Cases and Security Use Cases from Secure Business Process through the MDA Approach. In: Workshop on Security in Information Systems (WOSIS), Funchal, Madeira - Portugal (2007)
15. WfMC, Workflow Management Coalition: Terminology & Glossary, p.65 (1999)

Applying Model Transformation By-Example on Business Process Modeling Languages*

Michael Strommer[1], Marion Murzek[2,**], and Manuel Wimmer[1]

[1] Business Informatics Group
Institute of Software Technology and Interactive Systems
Vienna University of Technology, Austria
{strommer,wimmer}@big.tuwien.ac.at
[2] Womens Postgraduate College of Internet Technologies
Institute for Software and Interactive Systems
Vienna University of Technology, Austria
murzek@wit.tuwien.ac.at

Abstract. Model transformations are playing a vital role in the field of model engineering. However, for non-trivial transformation issues most approaches require imperative definitions, which are cumbersome and error-prone to create. Therefore, Model Transformation By Example (MTBE) approaches have been proposed as user-friendly alternative that simplifies the definition of model transformations. Up to now, MTBE approaches have been applied to structural models, only. In this work we apply MTBE to the domain of business process modeling languages, i.e., Event-driven Process Chains and UML activity diagrams. Compared to structural languages, business process modeling languages cover static semantic constraints, which are not specified in the metamodel. As a consequence, reasoning on the abstract syntax level is not sufficient. The contribution of this paper is to extend our existing MTBE approach by new alignment operators on the user level, which further improves the transparency of model transformation code. Concrete syntax and the knowledge about mapping operators are to be the only requisite artifacts.

Keywords: Business Process Models, Model Transformation, MTBE.

1 Introduction

With the rise of model engineering, model transformations have been steadily put in the limelight in the past five years. Growing tool support indicates, that model transformations are not only attractive for researchers, but also for industrial parties accommodating to their customers needs. Model transformation

* This work has been partly funded by the Austrian Federal Ministry of Transport, Innovation and Technology (BMVIT) and FFG under grant FIT-IT-810806.0.
** This research has been partly funded by the Austrian Federal Ministry for Education, Science, and Culture, and the European Social Fund (ESF) under grant 31.963/46-VII/9/2002.

J.-L. Hainaut et al. (Eds.): ER Workshops 2007, LNCS 4802, pp. 116–125, 2007.

scenarios include transformations between different refinement levels of models (*vertical transformations*) as well as transformations between different modeling languages which rely on the same abstraction level (*horizontal transformations*). Languages used for defining model transformations are ATL [4], QVT [8], Triple Graph Grammars [2], XSLT, and even general purpose languages such as Java [1]. Many model transformation languages are hybrid, meaning that besides a declarative style, an imperative style is provided for problems that cannot be solved by the provided declarative features. Nevertheless, the development of imperative fragments is often a tedious and error-prone task. In contrast, declarative solutions are compact descriptions but not always intuitive to find. We believe Model Transformation By-Example (MTBE) can help overcome the difficulties arising in the production of declarative and imperative transformation code.

Instead of focusing on the domain of *structural modeling languages*, what has been done in previous investigations [11], [10], in this paper we concentrate on *behavioral modeling languages*. More specifically, we apply MTBE on the domain of business process modeling (BPM), which, up to our best knowledge, has not yet been subject to the MTBE approach. The definition of requirements for MTBE in the context of business process modeling and the specification of proper mapping operators comprise the main contribution of this paper. Therefore, we present *main challenges encountered in business process (BP) model transformations*, and how these challenges can be tackled by extending already proposed MTBE *mapping operators*. The proposed extensions are explained by a running example in which two prominent BP modeling languages are used, namely the UML Activity Diagram and Event Driven Process Chains.

2   Motivation for MTBE

In this section we give a brief outline of our MTBE approach we introduced in [11]. We have recognized two main issues in conjunction with the task of defining model transformations. The first one is about the gap between the way a person thinks about models and the way a computer represents those models internally. And the second issue is about the way concepts are represented in the metamodel (MM), i.e., whether one needs to have expert knowledge to identify those concepts or not. We call the phenomenon of hidden concepts in a metamodel *concept hiding* [5]. Having those issues in mind one can easily accept the fact, that the task of creating model transformation rules is not a user-friendly one. This is why we have come up with the idea of MTBE, that can be seen as a semi-automatic approach for the generation of model transformation rules. One of the main benefits of MTBE is the shift in abstraction. Mostly all of the proposed model transformation approaches operate on the abstract syntax (AS), although modelers might not be familiar with the abstract syntax. Therefore, we intend to make the transformation task more concrete and operate on a level the modelers or designers are familiar with, i.e, on the concrete syntax (CS). Hence, with the application of MTBE one does not need any programming language experience or knowledge of the underlying metamodel to have some model transformation defined and executed. Furthermore we envision MTBE as agile and iterative process, in which the user can consequently improve the transformation outcome by adjusting the mappings and mapping operators applied on the CS.

3   Models for Business Processes

Business process models are in use for quite a long time and continue to gain importance as support from the software engineering field is improving significantly. Particularly model engineering fosters research in the area of BPM. There exist several metamodels for existing languages in order to raise their acceptance and tool interoperability. Due to this growing interest in BPM and proper tool support, we believe MTBE can be advantageous for specifying model transformations between BP models. As real world scenario consider the case in which two companies , that use different BPM techniques, merge and have to integrate their BP models. In this work we use the two BPM languages UML 2.1 AD [9] and EPC [6] shown in Figure 1 for our running examples.

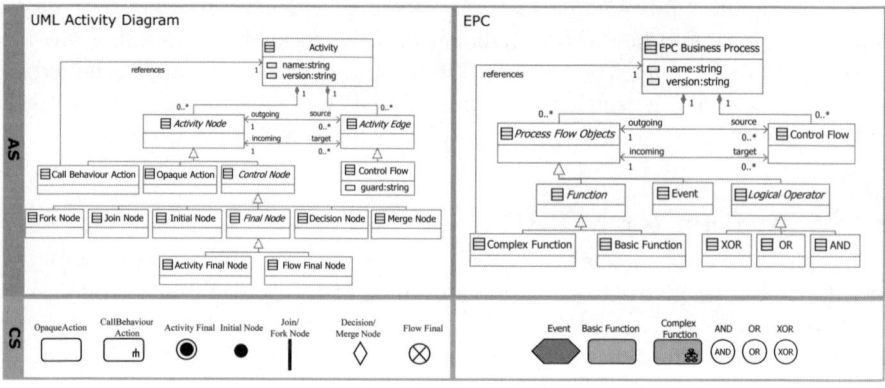

Fig. 1. Parts of the UML 2.1 AD and EPC meta models and their concrete syntax

4   Mapping Operators for MTBE in the Light of BPM

During our investigation of BP models we discovered, that there are considerable differences compared to structural models concerning the requirements for MTBE. To transform structural models, one has to be familiar with the notation and hidden concepts in the metamodels, especially when dealing with UML diagrams. Resulting ambiguities on the metamodel layer have to be solved either by reasoning algorithms or user input, as we described in detail in [11]. Now, with the task of transforming BP models we have to deal with quite different issues, in order to apply our MTBE approach. A lot of interesting aspects concerning the heterogeneity of BP models have been identified by Murzek et al. in [7]. The

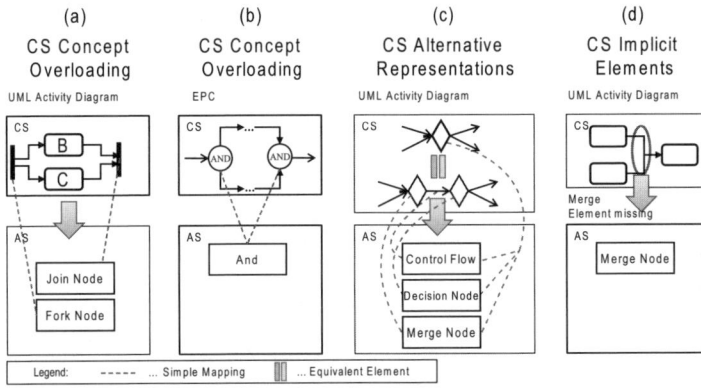

Fig. 2. Overview of BP models heterogeneities

heterogeneities shown in Figure 2 are partly based on their work. One of the special requirements coming along with BP models has its root in the mapping from concrete to abstract syntax layer (notation) and the number of modeling elements involved on each layer. As we now allow for zero or more elements on each layer in the CS-AS mapping, the notation can be now defined as

$$Triple :=< as\_E^*, cs\_E^*, const(as\_E)? >$$ (1)

Note that the original definition of the notation defined in [11] has been extended in this work. In UML AD we have for example the notation:

$$< \{MergeNode, ControlFlow, DecisionNode\}, \{DecisionMergeFigure\}, \{\} >$$

as is illustrated in Figure 2 c for the CS modeling element on the very top. Note that the used modeling construct is here just an abbreviation on the CS layer and could be equivalently expressed by the following pattern of notation triples:

$$< \{DecisionNode\}, \{DecisionFigure\}, \{\} >$$
$$< \{ControlFlow\}, \{ConnectionFigure\}, \{\} >$$
$$< \{MergeNode\}, \{MergeFigure\}, \{\} >$$

We also observed several heterogeneities between modeling languages, which pose further requirements for MTBE. Figure 2 gives four examples for the peculiarities we found in the two BP modeling languages we introduced in Section 3. Examples *a* and *b* in Figure 2 depict the case of so called CS overloading in UML AD and EPC. In example *a* we encounter no problems because with the help of the notation we can distinguish between the two concepts join and fork despite the CS overloading. In example *b* CS overloading represents a real challenge for MTBE as two equal CS elements, but in fact featuring two different meanings, are mapped to the same AS element.

When we have to deal with alternative representations in the CS, see Figure 2c, we can use the notation in MTBE to find them. The challenge arises not until we have to map two languages, where one consists of such variation points in the CS. Example *d* in Figure 2 shows the possibility in UML AD to merge

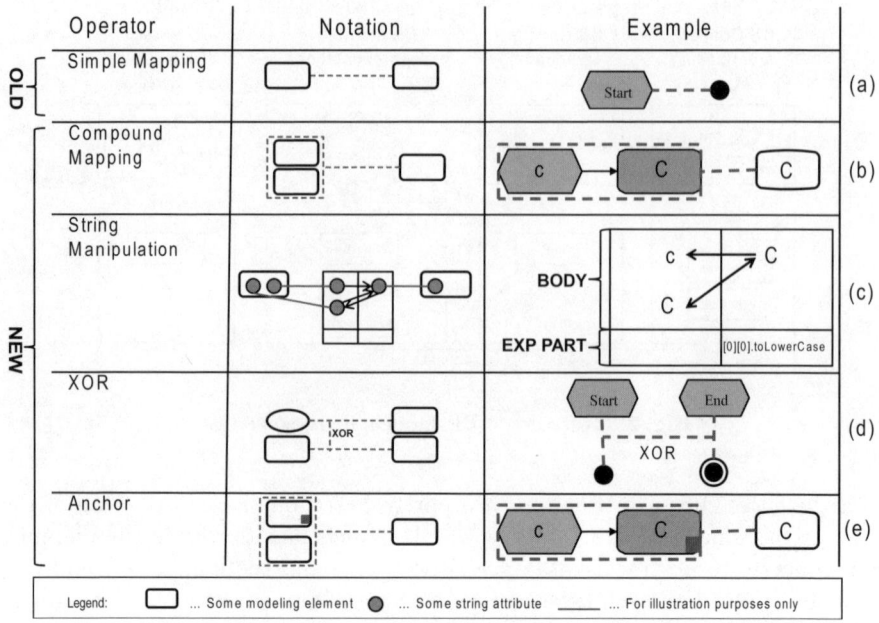

Fig. 3. Overview of MTBE Mapping Operators

parallel flows implicitly by omitting a merge/join node, i.e., we have no mapping from the AS to the CS.

In the following we present new mapping operators, which resolve heterogeneities, as expressed in the examples *a*, *b*, and *c*, in Figure 2. Unfortunately, up to now we are not able to cope in MTBE with implicit elements as shown in example *d*. The problem here is twofold. First we have to address the question how to map these implicit elements on the concrete syntax layer. And second we have to adjust the code generation process accordingly. One could now argue, that an enrichment of the presented metamodels could avoid a lot heterogeneities. But there exist amounts of legacy systems with legacy models, that can not be easily adopted to cooperate with new adjusted metamodels.

So far our MTBE approach as presented in [11] has only dealt with *simple 1:1 mappings* on the concrete syntax, see Figure 3 *a*. In case of BP model transformations it is necessary to introduce new kinds of mapping operators. Based on the specialties and problems stated above we developed new operators. In the following we describe the semantics of these new operators to provide a notion of how they can be used. However, we still have to develop some formal specification for these operators.

The first new operator is the *compound mapping operator* (cf. Figure 3 *b*). This mapping operator allows for n:m mappings on the CS layer. Although we encountered only 1:n mappings so far, we want the user to have the feature of n:m mappings. With this mapping operator we intended to support the mapping of common work flow patterns, such as the one we show in the corresponding example for this operator. The basic idea of our compound mapping operator is however to support pattern matching on the object graphs.

Along with the compound mapping operator comes a *string manipulation operator*, that works in the context of compound mappings but is not restricted to them. A first notation approach is shown in Figure 3 *c* together with an example. Note, that this operator is used for *Attributes* specified in the metamodel, which are represented as labels in the model. This operator consists of two main components, i.e., a body and an expression part, each separated into left and right hand side. The two body parts consist of a list containing references to *Attributes*, that are going to be mapped. Furthermore each *Attribute* of the body manages a list containing the unidirectional mappings from itself to some other *Attributes*. During the transformation rule generation from one language to the other only one list of mappings is of interest. In the expression part one can use some simple string operations or regular expression. In the example given in the Figure above we apply a *toLowerCase* operation on the first mapping of the first *Attribute* on the right hand side.

For the *XOR operator* depicted in Figure 3 *d* there are two ways to use it, i.e., in an explicit way or in an implicit way. In Figure 3 we only illustrated the explicit use of this operator. In general an XOR mapping shall indicate that only one element should be created although one CS element in one language is mapped to more than one element in the other language. Omitting the XOR in the example in Figure 3 *d* would lead to the creation of an *Initial* and an *Activity Final Node* for every *Event* that is matched by the corresponding transformation rule. When using the XOR operator in an implicit way the whole issue is hidden from the user. Instead all XOR mappings are derived automatically in the metamodel as will be shown in Section 5. The drawback of this approach is that one looses the possibility of multiple object creations as mentioned before.

At last we introduce the *anchor operator*. The notation and an example are given in Figure 3 *e*. The anchor operator marks the element, which the transformation rule shall use as single source pattern element. It is thus always used in conjunction with the compound mapping operator, which usually leads to the creation of multiple source pattern elements in the rules.

5   MTBE with UML AD and EPC By-Example

Our MTBE approach for the domain of Business Process modeling can be best explained in a by-example manner. Therefore, we use the two BP languages EPC and UML AD described in Section 3. Although the example given in Figure 4 is rather simple, it still covers a lot of interesting aspects for MTBE.

For the case study we assume that on the concrete syntax layer in EPC's *Events* and *Basic Functions* to always occur pairwise connected through a *Control Flow* edge. Furthermore, in UML AD modeling it could be possible to omit a *Join* node and therefore model joins implicitly. However, in our first MTBE approach for BPM we do not jet cope with implicit joins or merges.

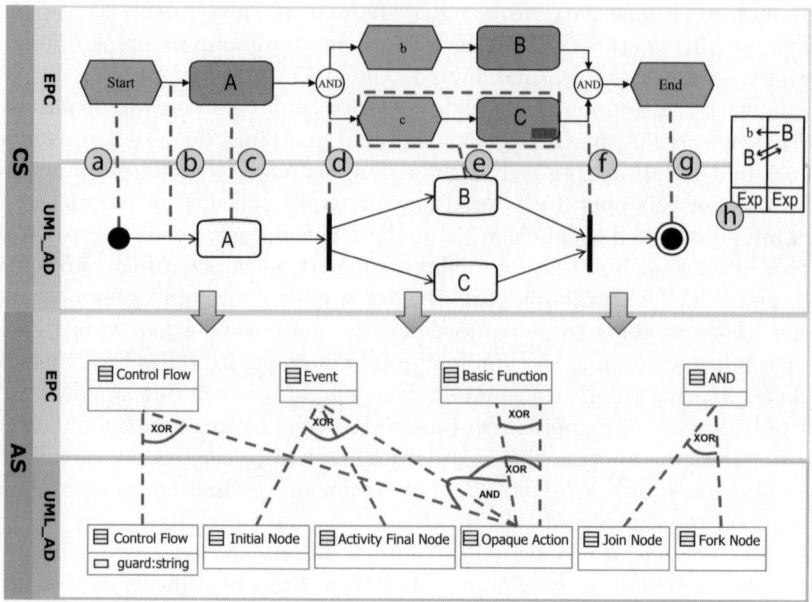

Fig. 4. Mapping EPC and UML Activity diagram - CS + AS perspectives

5.1   Model and Metamodel Mappings

As a first step one has to define manual mappings between two languages, which the transformation model shall be derived from. In the example in Figure 4 we specified eight mappings that capture all concepts being used in the two sample models. Mappings *a,b,c,d,f,* and *g* are of type simple mapping.

Mapping *e* is of type compound mapping with multiplicity 1:3. Consequently, whenever the pattern *Event, Control Flow, Basic Function* is matched this corresponds to a single *Opaque Action*. We also marked the *Basic Function C* in our compound mapping as anchor element, which has implications specific to transformation code generation. In our case the ATL code generator would use this *Basic Functions* metamodel element as single source pattern element instead of using multiple source pattern elements. During our implementation attempts we realized, that an anchor feature can be desirable in some transformation scenarios.

Mapping *h* in our example takes care of the labels used in *Events, Basic Functions* and *Opaque Actions*. To maintain usability this string manipulation operator is used in a separate modeling element and references the involved labels. To define string equivalences one can use only unidirectional mappings, which will be applied transforming from one set of labels to another. An optional expression allows us for example in mapping *h* to apply a *toLowerCase()* operation on the first mapping of the right hand side set of labels.

In EPC's there are no distinct metamodel elements nor distinct concrete syntax elements for start and end nodes, although these concepts are used in the modeling language implicitly. In UML AD we do have explicit concepts for start and end nodes both, in the model and the metamodel. If a transformation from EPC2UML_AD has to be performed the transformation model must know how to distinguish between start and end nodes even without having these concepts specified in EPC.

To keep our illustration in Figure 4 transparent and clear we omitted the mappings between CS and AS. Also these mappings are quite straightforward to define, as there are no constraints specified in the notation.

At last the mappings between the two metamodels can be derived from the user mappings and the notation. To highlight the existence of a compound mapping in the metamodel we marked the three involved mappings with an *and* operator. On the metamodel mapping level we now make use of our new XOR operator we introduced in Section 4. To keep the mapping task user-friendly the XOR between mappings can be reasoned automatically based on information in the metamodels. Whenever a meta class contains at least two outgoing mapping edges, an XOR relation can be set in an implicit way.

5.2   How to Make Mappings Executable

As the automatic generation of transformation rules is a difficult task, we do not claim to support fully automatic rule generation. Instead we believe in a semi-automatic approach. To face the new domain of business process models we implemented a paradigm, which can be best compared to Architecture-Centric MDSD. First of all we have implemented correct ATL transformation code, which acts as reference implementation. Thereby we have avoided imperative code sections and concentrate on coding in a declarative fashion.

In the next step we have developed the mapping operators described in Section 4. During this step we have turned our attention to the user-friendliness.

Next we have looked at the example models, the user mappings and the metamodels and tried to deduce the reference implementation. Code segments that could not be deduced automatically then lead to further refinement of the underlying heuristics. After refinement we tried again to deduce the reference implementation. This process can be seen as an iterative way to deduce heuristics on how to generate ATL transformation rules from a given set of models, metamodels and user mappings. The aim of this process is to optimize the relation between user-friendly mapping operators and the ability to generate executable transformation rules.

6   Related Work

To our best knowledge, there exists no approach for finding semantic correspondences between business process models so far. However, there exists general approaches that allow the definition of semantic correspondences between two (meta)models, which have been applied in the area of structural models. The first approach is a *model-based* approach from Varró, while the second approach from Fabro et al is *metamodel-based* which allows automatically finding correspondences directly between metamodels.

Model-based: Parallel to our MTBE approach [11] Varró proposed in [10] a similar approach. The overall aim of Varró's approach is comparable to ours, but the concrete realizations differ from each other. Our approach allows the definition of semantic correspondences on the concrete syntax, from which ATL rules can then be derived. In contrast, Varró's approach uses the abstract syntax to define the mappings between source and target models, only. The definition of the mapping is done with reference nodes leading to a mapping graph. To transform one model into the other, graph transformation formalisms [2] are used. Furthermore, there is no publication on applying Varro's approach on business process models which is the general aim of this paper.

Metamodel-based: Orthogonal to MTBE there exists the approach of using matching transformations combined with weaving models [3] in order to generate an ATL transformation model. Matching transformations are defined such that they use common matching algorithms or modifications, and then create a weaving model from the calculated similarity values. Afterwards these weaving models are taken as input for another transformation, called higher order transformation (HOT), to produce the desired ATL model describing the transformation rules between two metamodels. Because there will be always some mappings that can not be matched fully automatically, this approach is also to be considered semi-automatic. The model transformation generation process described in [3] currently focuses on using mappings between metamodels and is therefore based on the abstract syntax, while our approach aims at generating model transformation code from M1 mappings.

7   Conclusion and Future Work

In this work we have proposed an MTBE approach for the area of business process modeling languages. We introduced special mapping operators to give the user more expressivity for defining model mappings. Still, there remain a few heterogeneity issues we can not target with the introduction of new alignment operators. These heterogeneity issues will be on the one hand subject to special reasoning algorithms and on the other hand as seldom as possible to manual code adjustment done by the user.

Concerning future work, we particularly strive for first, the refinement of the proposed MTBE approach. We want to experiment with different versions of mapping models which possess different levels of granularity, size and modeling patterns. In particular we want to evaluate how much a model can be altered compared to the original model, which is mapped by-example, and can still be transformed properly. Therefore, we want to map two models A and B, generate the transformations, and then alter the models to A' and B' and test the generated transformations with the new versions of the models. Also, we plan to conduct student experiments to underpin the benefits coming along with MTBE in comparison to common model transformation approaches.

References

1. Akehurst, D.H., Bordbar, B., Evans, M.J., Howells, W.G.J., McDonald-Maier, K.D.: SiTra: Simple Transformations in Java. In: MoDELS/UML 2006. Proceedings of the ACM/IEEE 9th International Conference on Model Driven Engineering Languages and Systems, Genova, Italy, pp. 351–364 (October 2006)
2. Ehring, H., Engels, G., Kreowsky, H.-J., Rozenberg, G.: Handbook on Graph Grammars and Computing by Graph Transformation, vol. 2. World Scientific, Singapore (1999)
3. Fabro, M.D.D., Valduriez, P.: Semi-automatic Model Integration using Matching Transformations and Weaving Models. In: SAC. Proceedings of the 2007 ACM Symposium on Applied Computing, Seoul, Korea, pp. 963–970. ACM Press, New York (2007)
4. Jouault, F., Kurtev, I.: Transforming Models with ATL. In: Bruel, J.-M. (ed.) MoDELS 2005. LNCS, vol. 3844, pp. 128–138. Springer, Heidelberg (2006)
5. Kappel, G., Kapsammer, E., Kargl, H., Kramler, G., Reiter, T., Retschitzegger, W., Schwinger, W., Wimmer, M.: Lifting Metamodels to Ontologies - A Step to the Semantic Integration of Modeling Languages. In: MoDELS/UML 2006. Proceedings of the ACM/IEEE 9th International Conference on Model Driven Engineering Languages and Systems, Genova, Italy (2006)
6. Keller, G., Nüttgens, M., Scheer, A.-W.: Semantische Prozeßmodellierung auf der Grundlage "Ereignisgesteuerter Prozeßketten (EPK)". Technical report, Institut für Wirtschaftsinformatik Universität Saarbrücken
7. Murzek, M., Kramler, G.: Business Process Model Transformation Issues. In: Proceedings of the 9th International Conference on Enterprise Information Systems, Madeira, Portugal (2007)
8. OMG. QVT-Merge Group: Revised submission for MOF 2.1 Query/View/Transformation, version 2.0 formal/05-07-04 edition (2005)
9. OMG. UML 2.1 Superstructure Specification. Object Management Group (April 2006), http://www.omg.org/docs/ptc/06-04-02.pdf
10. Varró, D.: Model Transformation By Example. In: MoDELS/UML 2006. Proceedings of the ACM/IEEE 9th International Conference on Model Driven Engineering Languages and Systems, Genova, Italy (October 2006)
11. Wimmer, M., Strommer, M., Kargl, H., Kramler, G.: Towards Model Transformation Generation By-Example. In: HICSS-40 2007. Proceedings of the 40th Hawaii International International Conference on Systems Science CD-ROM / Abstracts Proceedings, Big Island, HI, USA, p. 285 (2007)

Extending OCL to Ensure Model Transformations

François Lagarde[1], François Terrier[1], Charles André[2], and Sébastien Gérard[1]

[1] CEA, LIST, Boîte 94, Gif-sur-Yvette, F-91191, France
{francois.lagarde, francois.terrier, sebastien.gerard}@cea.fr
[2] Université Nice-Sophia Antipolis/CNRS, I3S, 06903 Sophia Antipolis, France
charles.andre@unice.fr

Abstract. Effective use of model transformations in a MDA development relies on complex model generation scenarios. Their correct realizations require dedicated methods and constructs for specifying model transformations and assessing the resulting models. The aim of the work described here is to combine OCL language for expressing model constraints with a set of a dedicated high-level constructs that assess UML models involved in the transformation process. The proposed constructs are formalized in a UML profile that allows modelers to use intuitive constraint specification support. Our approach is illustrated by a model refinement operation within the context of real-time system design. The dedicated tool supporting our approach is also described.

1 Introduction

Automatic model generation is a key asset of Model Driven Architecture (MDA) approach. It covers a wide range of applications including: code generation, model refinement and metamodel transformation.

Many of the proposed approaches provide modelers with constructs for expressing transformations as well as with dedicated tools. Transformations are specified using metamodel information to define a function that evaluates available models and produce new ones. While we can confidently perform numerous transformations in a such way, others may be more difficult to write in one framework than in another, and their success thus depends heavily on the modeler's ability to determine the proper function. Obviously some transformations may not allow any straightforward implementation and none programming solution can be found. One such case is the explicit specification of a non-uniqueness of a model to produce, in transforming one source model element into at least two model elements in a target model.

Therefore effective use of model transformations in software development relies on the capacity to deal with heterogeneous transformation practices and involves dedicated procedures to assess their correct implementations and executions. This means devising criteria for assessment of transformation results independently from the underlying mechanisms.

Even if certain transformation mechanisms can also be used to check models, their effectiveness is limited by the fact that executing and checking do not share

J.-L. Hainaut et al. (Eds.): ER Workshops 2007, LNCS 4802, pp. 126–136, 2007.
© Springer-Verlag Berlin Heidelberg 2007

the same concerns. Transformation writing focuses on implementation concerns while checking is concerned with satisfying model properties. For instance, an expected result transformation is to ensure that each identified element in a source model matches another element in a target model. A more interesting result could ascertain that some elements present in one model are absent from another.

In this context, our work[1] introduces a set of high-level constructs with OCL constraint facilities for the purpose of constraining and verifying model transformations. The constructs are formalized in a UML profile. It allows modelers to use a single UML environment and provides an intuitive approach to specify constraints.

The paper is organized as follows: Section 2 briefly presents relevant work in the field of model constraints and model transformations. The main contribution of the article is Section 3, where the constructs are presented. Section 4 demonstrates its applicability to real-time system design. Section 5 presents our tooling support. Before concluding, Section 6 discusses the limitations of our approach and provides orientation for future work.

2   Related and Prior Work

2.1   Object Constraint Language

Object Constraint Language (OCL) [1] is a standard language for describing expressions in UML models. It is a twofold (query and constraint) language and relies on 4-layer metamodel architecture. It is thus a legitimate candidate for expressing constraints on model elements.

An OCL constraint has a contextual classifier and evaluation is done on each instance of its classifier. As we wish to constrain model elements belonging to the layer $M1$, which contains user model elements or model elements resulting from a transformation, we have to express constraints at the metamodel level. This can be done by creating a stereotype and using it as a context. Navigation to the element to which the stereotype is applied is possible with a one-way navigable association relationship from Stereotype to MetaClass [2, p.686].

For example, to express constraints on elements between packages we must create a stereotype extending the Model metaclass and use it as a context. The complexity of determining and maintaining paths to each of the elements depends on the relative position of an element from the root element and may be heavy to deal with.

To illustrate our comments, we examine how we may express constraints with a simplified real-time domain profile called SRT. This profile defines the Service

---

[1] This work has been performed in the context of the Usine Logicielle project (www.usine-logicielle.org) of the System@tic Paris Region Cluster. This project is partially funded by the "Direction Générale des Entreprises of the French administration", the "Conseil Régional d'Île de France", the "Conseil Général des Yvelines", the "Conseil Général de l'Essonne" and the "Conseil Général des Hauts de Seine".

concept, which extends the metaclass Action. It depicts any actions that the system must provide. We also identify the concept of Task which is logical entity to supporting execution of services.

We then need to make sure that: (a) each Task belonging to a package stereotyped «system» has exactly one operation stereotyped «entryPoint», (b) each of these tasks also matches a action stereotyped «service» with a same name in a package stereotyped «architecture».

The following OCL expression ensures these constraints:

```
package SRT
context Task inv:
 self.base_Class.ownedOperation->select(op |
  not op.extension_EntryPoint.isOclUndefined())->size=1
context RTModel inv:
 -- collect all the classes stereotyped task
 let Tasks :
  self.base_Model.nestedPackage->iterate(p, s :Set{} |
  ((not p.extension_System.isOclUndefined()
   and(s->union((p.packagedElement->iterate(c1, s1:Set{} |
   not c1.extension_Task.isOclUndefined()
      and (s1->including(c1))))))))) in
   -- collect all the actions stereotyped service
    let Services :
     self.base_Model.nestedPackage->iterate(p,
     s:Set{} | ((not p.extension_Architecture.isOclUndefined()
     and(s->union((p.packagedElement->iterate(c1,
      s1:Set{} | (not c1.extension_Service.isOclUndefined())
       and(s1->including(c1))))))))) in
      -- make them match
      Tasks->forAll(c1 |
       Resources->one(c2 | c1.name=c2.name))
endpackage
```

The complexity of this expression makes it hard to understand and its maintainability is only achieved at the cost of effort. Dominik Stein et al [3] have stressed that even a simple query quickly results in a complex query expression.

Another downside of OCL for checking is that the OCL specification does not provide the means for reporting information on constraint violations. Without execution traces, a simple report in which only the validity of a model transformations is given is useless if no additional helps is provided to locate the element which triggered the violation.

2.2 Model Transformation

This section examines how transformation mechanisms are limited for addressing constraints concerns.

Query View Transformation

OMG defines a Query/View/Transformation [4] (QVT) specification for querying and transforming models built-on the MOF. It includes a read-only mechanism to check models. This specification is built on a two-level declarative architecture.

The relationship language specifies a transformation using a set of declarations of relationships that must hold between model elements. This language is declarative and implicitly creates traces, i.e: the linkage between models established by a transformation execution.

The core language specifies relationship transformation by means of imperative expressions and declarative statements. Evaluation of variables and conditions allow expression of matching between model elements. The navigation across model elements is supported by OCL. Use of OCL to navigate through the model raises much the same difficulties as already mentioned in maintaining links between different model elements. Devon Simmonds et al [5] have shown that queries can also quickly produce complex expressions.

Moreover, QVT specification lacks suitable implementation tools, despite the existence of efficient transformation frameworks, such as the one proposed by Frédéric Jouault et al [6] who promotes ATL as a possible implementation.

Graph Transformation

Graphs are useful means for representing complex data structures such as models and for supporting model transformations through rewriting mechanisms. This approach benefits from a well-founded theory and there is abundant literature available for it [7, 8, 9].

Most of model transformations use triple graph grammar as a way to rewrite graphs. Rules describe how to substitute element occurrences. Source rules (*left hand side*) and target rules (*right hand side*) are coupled by use of shared variables (*host graph*).

This approach is declarative and has no explicit means for manipulating traces. Consequently, its applicability for constraint purposes has the same flaws as described above.

3   Definition of a Dedicated Language

This section presents our dedicated language for assessing models involved in a transformation process. The design choices made to identify elements of our language are governed by a simple intuition. Model transformations produce elements from a source model, hence we have to devise criteria for expressing existence quantification on produced model elements. The next parts define the identified elements.

3.1   Domain Model View

Our dedicated language specializes the Constraint concept existing in the UML Superstructure. The central element is SelectionConstraint (Fig. 1). It defines a

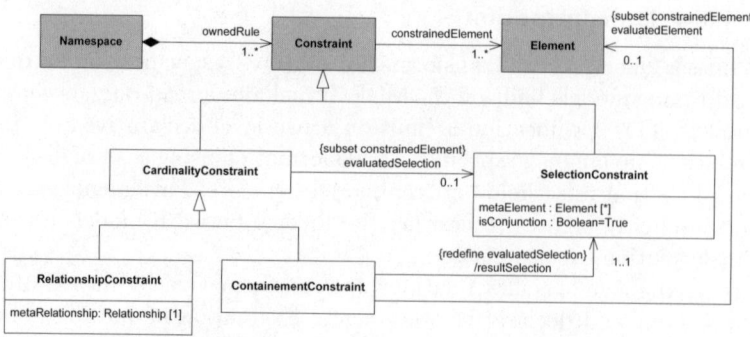

Fig. 1. Domain view

means for gathering model elements complying with meta-criteria. A criterion may be a type: Class, Action and consequently the selected model elements will be their instances. If a stereotype is given, then the selection is made up of elements to which the stereotype is applied.

In order to state whether the matched model elements meet transformation requirements, we provide the members of a set defined by a SelectionConstraint with a validity information. We introduce CardinalityConstraint to express an existence quantification condition among members of two SelectionConstraint. It comes an element is valid if the cardinality of valid elements in another set falls between a lower bound value and an upper bound value. The cardinality is the multiplicity of the association end property evaluatedSelection.

Constraint specification often entails use of recurrent expressions and led us to further specialize CardinalityConstraint in two additional constraints:

- ContainmentConstraint, which requires that one element respects a content-ment relationship (is a subset of the ownedMember) with another element. For example, we may restrict Operation by specifying its ownership. This construct is the main one for navigating through the UML hierarchy and populating sets,
- RelationshipConstraint, imposes a particular relationship between elements of two sets. For example, it can help to select classes belonging to a set having association relationships with another class belonging to another set.

3.2   Constraints Definition

This section further details the previous concepts by showing the arithmetic operation performed on sets with OCL-like expression. The mechanisms used to evaluate the overall constraints are discussed in section 5 which describes our tooling support.

CardinalityConstraint

First, we create two constraints named A and B stereotyped «constraintSelection».

To restrict the two sets, we create a constraint stereotyped «cardinalityConstraint» between them (Fig. 2). The evaluatedSelection is the selection on which the validation of members will be performed.

This constraint may enclose an OCL expression and refers to any member belonging to a ConstraintSelection with a same name.

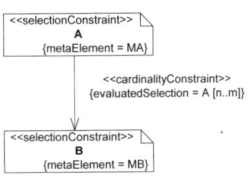

This constraint declares existence quantification in a similar way as a OCL invariant declaration. The condition must *always* be satisfied. We define this invariant using the following expression:

Fig. 2. Cardinality use

```
— evaluateOCL (String OCL): Boolean
— evaluate a OCL expression
A->forAll(a | (B.select(b |
    a.evaluateOCL(oclRule)))->size()>n)
  and (B.select(b |
    a.evaluateOCL(oclRule))->size()<m)
```

ContainmentConstraint

This constraint acts similarly to the CardinalityConstraint. In addition, it ensures that members respect a containment relationship.

RelationshipConstraint

This relationship condition looks for members of A having a relationship, in the example Fig. 3 a Dependency, with members of B.

This condition cannot be effortlessly written with an OCL expression. We have to distinguish scenarios according to the nature of their relationships. If the relationship is an association, we must fetch all associations in the models and make sure that association ends match the elements of the set. If the relationship is a directed one, such as Generalization relationship or a Dependency relationship, we have to investigate the corresponding source or target

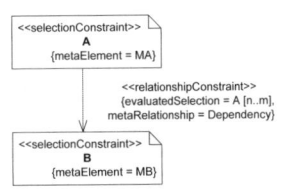

Fig. 3. Relationship use

roles present on the elements. We thus introduce an ad-hoc function called hasARelation to write an equivalent OCL:

```
— hasARelation (Relationship r, Element e)
— return true if there is a relation between elements.
A->forAll(a | (B.select(b |
    a.hasARelation(Dependency, b))->size()>n)
  and (B.select(b |
    a.hasARelation(Dependency, b))->size()<m))
```

4   Language Usage: A Case Study

To illustrate the applicability of our approach, we have enhanced a refinement operation for embedded real time system design with identified model constraints.

4.1   A Refinement Operation

Richard F. Paige et al [10] define refinement operation as the production of new models that necessarily enrich previously constructed models. To facilitate this process as well as to make it safer, refinement may be partially supported with model transformations. This usually involves simple, but efficient manipulation of models, e.g. renaming or typing operations. However, only few cases may be fully performed in this way. One of the main reasons is that a refinement operation must *enrich* a previous model and consequently cannot be reduced to a bijective function.

Our refinement study case falls into such a scenario. To support our example we reuse the previously introduced profile called SRT. We assume that our system benefits from an oversimplified distributed context and we complete this profile with the introduction of a Processor.

Modeling of a system encompasses two distinct steps and comes with a methodology. Design starts with a first abstract model called architectural model. This model is intended to identify services and processor entities. The second modeling stage provides a more concrete model of the system in which tasks are introduced. For the sake of simplicity, we consider a system providing only two services with two processors. The architecture of our system is represented in the figure (Fig. 4):

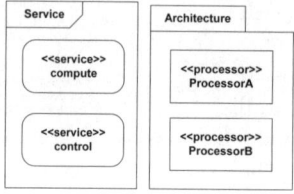   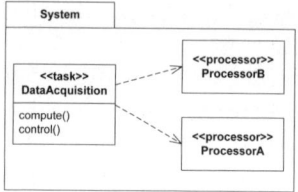

Fig. 4. Architecture   **Fig. 5.** System

We provide rules part of a hypothetical methodology to define the refinement operation: (a) an element stereotyped «processor» remains a processor entity, (b) a Service has to be allocated to a Task. To do so, a class stereotyped «task» has to own an operation with the same name as the service, (c) each Task has to maintain two dependency with two processors.

Application of these guidelines offers a partial support of the refinement operation with model transformations. The first rule can be fully specified with a simple function. The remaining rules cannot be automatically applied. All the rules make possible to infer a family of system models from the architectural model. For instance, one design choice could be to create two tasks, each in charge of one service, or to create one task to manage both.

We choose the first solution and consider the refined system shown in the above figure (Fig. 5).

4.2 Refinement Specification

The refinement specification (Fig. 6) ensures correct realization by determining the existence or non existence of expected model elements.

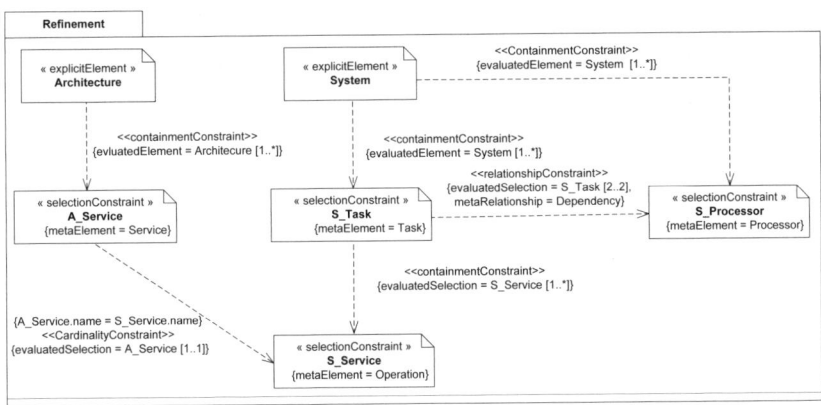

Fig. 6. Refinement specification

Its top elements are constraints stereotyped «explicitElement». They handle a mapping to the architectural model (i.e the package containing the services) and the package containing at least the expected refined model elements. They are the root context of the evaluation.

The three elements System, S_Task and S_Processor make sure that each Task maintains two Dependency members of S_Processor. To achieve this, we characterize members of the set S_Task as model elements on which the stereotype «task» is applied. S_Processor is defined in a similar way. Up to this stage, we have only stated the existence of such elements without imposing any relationship for task allocating to a processor. We must thus declare a Constraint stereotyped «relationshipConstraint» between S_Task and S_Processor. We therefore must specify the kind of relationship Dependency and pay attention to use a multiplicity value equal to 2..2.

To make sure that each service is allocated to a task, we simply add a CardinalityConstraint between A_Service and S_Service and give a proper OCL expression to respect the naming convention.

Assessment of Models

We have supported our approach with a tool (presented in sec. 5) that runs rules on models. The result is provided by means of a graph. Elliptical nodes are matched elements, square nodes give the existence quantification.

For purpose of illustration, we have deliberately evaluated an architectural model and system model that do not fulfill our requirements. We create another service called transmit in the input model while keeping the same refined model (Fig. 5). After running the refinement specification on both models, we obtain the results shown on the graph (Fig. 7).

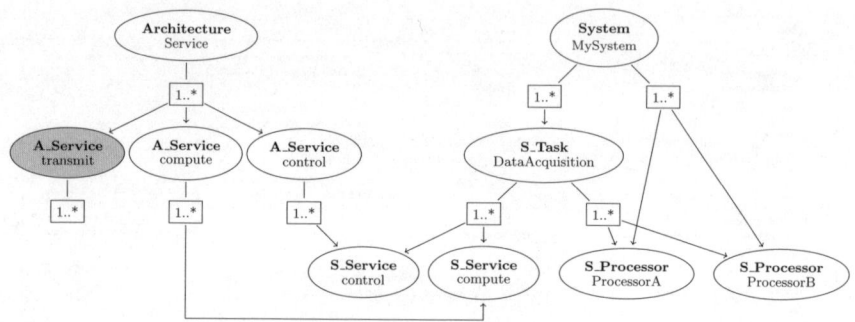

Fig. 7. Valid evaluation

The evaluation shows that the refined model does not satisfy the requirements due to the invalidity of the node transmit (filled with a gray color). With a little effort, graph helps for debugging models. Invalid elements are identified and it is easy to guess which rules and conditions are not satisfied.

5   Tooling Support

Our research has led us to develop a constraint assessment tool based on the Eclipse Modelling Model Development Tools (MDT) library.

The main issue is to jointly evaluate the constraints on sparse model elements. We rely on graph support to provide an efficient resolution mechanism. It is performed in three stages: (a) translate the model containing the rules into a directed graph, (b) do queries on user models to collect the elements, (c) evaluate the quantification conditions and the constraints on the previous graph.

6   Discussion and Future Direction

In some cases, it is sometimes difficult to specify model transformations by means of constraints. The modeler cannot provide exhaustive coverage of the constraints to be met by models part of the transformation process.

Constraint specifying may be made easier by braking down the specification into unitary constraints (or simpler sets of constraints). This means manually building a scenario for applying unitary constraints to obtain global constraints. We are currently planning support function to manage rule application sequence and to running scenarios.

7   Conclusion

Within the context of complex model transformations scenarios, we defined high-level constructs for the purpose of validating generated UML models involved in the model transformations process. These constructs include OCL facilities for specifying constraints and state relationships that must occur on the models. The constraints consist of rules that are made up of sets. They identify patterns of model elements and are supplemented by an existence condition for assessing models.

Models assessment is achieved using a graph and identifies the invalid element(s). This information is especially vital to designers for correcting their models.

Our approach is illustrated by a refinement operation for a simplified distributed real-time system. This refinement makes possible a partial transformations model but needs to be completed by hand. To make the transformation safer, we distinguish rules and constraints on the models. Finally the tool developed to assess models is introduced and limitations of our approach are discussed.

The proposed constructs appear general enough for application to a wide range of contexts. They can support a methodology in translating main modeling stages into rules and can also be used to verify well-formedness of models.

References

1. Object Management Group (OMG): Object Constraint Language formal/06-05-01 (May 2006)
2. Object Management Group (OMG): Unified Modeling Language: Superstructure ptc/06-04-02 (April 2006)
3. Stein, D., Hanenberg, S., Unland, R.: A Graphical Notation to Specify Model Queries for MDA Transformations on UML Models. In: Aßmann, U., Aksit, M., Rensink, A. (eds.) MDAFA 2003. LNCS, vol. 3599, pp. 77–92. Springer, Heidelberg (2005)
4. Object Management Group (OMG): MOF 2.0 Query/Views/Transformations ptc/05-11-01 (November 2005)
5. Simmonds, D., France, R., Ghosh, S.: Using Directives to Implement Model Transformations. In: From MDD to Experiments and Illustrations, Hardback (2006)
6. Jouault, F., Kurtev, I.: On the Architectural Alignment of ATL and QVT. In: SAC 2006. Proceedings of the 2006 ACM Symposium on Applied Computing, pp. 1188–1195. ACM Press, New York (2006) (chapter Model transformation (MT 2006)
7. Baresi, L., Heckel, R.: Tutorial introduction to graph transformation: A software engineering perspective. In: Corradini, A., Ehrig, H., Kreowski, H.-J., Rozenberg, G. (eds.) ICGT 2002. LNCS, vol. 2505, pp. 402–429. Springer, Heidelberg (2002)

8. Heckel, R.: Graph transformation in a nutshell. In: Bezivin, J., Heckel, R. (eds.) IBFI. Language Engineering for Model-Driven Software Development. Number 04101 in Dagstuhl Seminar Proceedings, Internationales Begegnungs- und Forschungszentrum fuer Informatik, Schloss Dagstuhl, Germany. Dagstuhl Seminar Proceedings (2005)
9. Kuster, J.M., Heckel, R., Engels, G.: Defining and validating transformations of UML models. In: IEEE Symposium on Human Centric Computing Languages and Environments, 2003, pp. 145–152. IEEE Computer Society Press, Los Alamitos (2003)
10. Paige, R.F., Kolovos, D.S., Polack, F.A.: Refinement via Consistency Checking in MDA. Electronic Notes in Theoretical Computer Science 137(2), 151–161 (2005)

A UML Profile for Modeling
Data Warehouse Usage*

Veronika Stefanov and Beate List

Women's Postgraduate College for Internet Technologies
Institute of Software Technology and Interactive Systems
Vienna University of Technology
{stefanov, list}@wit.tuwien.ac.at

Abstract. Data Warehouse (DWH) systems represent a single source
of information for analyzing the status, the development and the results
of an organization.

Today's DWH systems provide many different services to different
kinds of users. People involved in designing and managing DWH systems need to see the big picture of how the DWH is being used, to have
an overview of the current situation, and to be able to visualize future
scenarios. Currently, there is a lack of such general models in Data Warehousing.

We introduce the UML Profile for Modeling DWH Usage for modeling
the different kinds of DWH usage on a conceptual level. It uses features
of UML intended for the purpose of creating abstract, general models.
The profile distinguishes four perspectives of usage, and allows to model
details of the users. The UML Profile is applied to examples illustrating
some of the application scenarios.

1 Introduction

A Data Warehouse (DWH) system is more than just a big database. Defined
as "a subject-oriented, integrated, time-variant, nonvolatile collection of data in
support of management's decision-making process" [1], DWH systems represent
a single source of information for analyzing the status, the development and the
results of an organization [2]. Analysts and decision makers take measures such
as the number of transactions per customer or the increase of sales during a
promotion and use them to recognize trends or warning signs and to decide on
future investments.

Today's DWH systems provide many different services to different kinds of
users: Users retrieve summaries and reports relevant to them, or analyze data
with specialized visualization tools. The system may send them messages via
e-mail or sms, or provide a quick overview visualization in a dashboard or an

* This research has been funded by the Austrian Federal Ministry for Education,
Science, and Culture, and the European Social Fund (ESF) under grant 31.963/46-
VII/9/2002.

J.-L. Hainaut et al. (Eds.): ER Workshops 2007, LNCS 4802, pp. 137–147, 2007.

intranet portal. Users need access to data at different times, some need it occasionally, others more often, suddenly urgently, or regularly and predictably.

People involved in managing, designing or evolving today's DWH systems need to see the big picture of all these different ways the DWH is being used, in order to have an overview of the current situation, and to be able to visualize future scenarios. Overview diagrams are needed to facilitate communication with users and decision makers.

Surprisingly, today there are no existing models to describe the different aspects of DWH usage on a conceptual level. There is a lack of general models that provide a broader view over several aspects, eventhough there exist many detailed models of sub-areas. We identify a need for a model that shows on the conceptual level:

1. Who are the users and how are they grouped together?
2. Which part of the DWH system do they use? How do they use it?
3. How intensely are which parts of the DWH being used by which users?
4. When do users need to use which part, and how time critical is it?
5. How important is it?

To fill this gap, we use the UML extension mechanism to specify the *UML Profile for Modeling DWH Usage*. Our profile uses some of the lesser known features of UML, intended for the purpose of creating preliminary models with a "less precise but more general representation" [3]. We have grouped the features of the profile into four perspectives, which focus on the most common application scenarios of DWH usage modeling (Section 3). The *UML Profile for Modeling DWH Usage* (Section 4) offers the following contributions:

- It allows to model who uses the DWH, to group the users, and to model their organizational affiliation, skill level, and an approximate number of instances for each user role.
- Modelers can show how often users use something, and how time critical and how important a certain usage is, as well as active or passive usage types.
- The model allows the analysis of the implications of changing scenarios (e.g. adding a component, increasing numbers of users) on various levels of detail.
- It can be used to identify critical patterns (many important accesses, rapid growth) and to identify parts of the DWH that are not used or not used very often or importantly.
- DWH usage models can be used to support the design of user access controls or personalized user interfaces.
- In general, the models make the overall structure of DWH usage visible on the conceptual level, thus replacing the custom of creating *ad hoc* diagrams and drawings for the communication with users and decision makers.

DWH usage models are intended to provide an overview without aiming at a design process. Compared to requirements analysis in Data Warehousing, our approach to DWH usage is broader, and not necessarily focused on a future system to be built. The UML Profile allows to model the users in detail and does not explicitly include (design) goals of any kind. In MDA [4] terms, our approach is located in the CIM (Computation Independent) area, where models are not necessarily intended to be transformed into code.

2   Modeling Data Warehouses: Background

Our approach applies UML to the Data Warehousing domain. It is aimed at encompassing all the different ways that users may use a DWH. Our goal is to provide an overview over all aspects of DWH usage, not only focussing on the data model. Nevertheless, due to the special characteristics of DWH data, it is necessary to take the data model especially into account.

DWH applications involve complex queries on large amounts of data, which are difficult to manage for human analysts. In Data Warehousing, data is often organized according to the multidimensional paradigm, which allows data access in a way that comes more natural to human analysts. The data is located in n-dimensional space, with the dimensions representing the different ways the data can be viewed and sorted (e.g. according to time, store, customer, product, etc.).

A multidimensional model, also called star schema or fact schema, is basically a relational model in the shape of a star (see Fig. 1 for an example). At the center of the star there is the *fact* table. It contains data on the subject of analysis (e.g. sales, transactions, repairs, admissions, expenses, etc.). The attributes of the fact table (e.g. cost, revenue, amount, duration, etc.) are called *measures* or *fact attributes*. The spokes/points of the star represent the *dimensions* according to which the data will be analyzed (sorted/aggregated by data, by store). The dimensions can be organized in hierarchies that are useful for aggregating data (e.g. store, city, region, country). Stars can share dimensions, thus creating a web of interconnected schemas that makes drill-across operations possible.

There are many approaches to modeling the multidimensional data structures of data warehouses (see [6,7] for comparisons), some of which are object-oriented models or based on the Unified Modeling Language (UML) [7,5,8].

For modeling multidimensional data, we choose to use the UML Profile of Luján-Mora et al. as described in [5]. This Profile allows to model not only the core features of multidimensional models (facts, measures, and dimensions), but

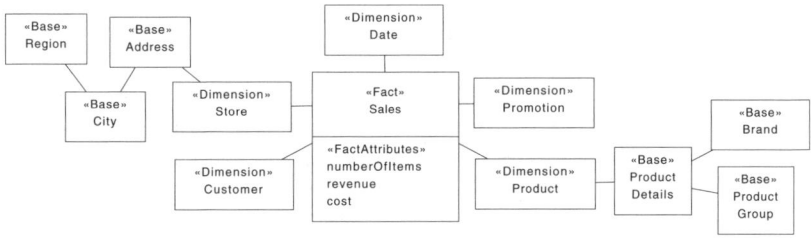

Fig. 1. A simple multidimensional model modeled in UML notation with stereotypes from [5]. Aggregation levels are only shown for the Product and the Store dimensions.

also many advanced features such as degenerate dimensions or nonstrict and complete dimensional hierarchies, and also provides three levels of detail.

3 Data Warehouse Usage: Perspectives and Application Scenarios

In order to provide models of DWH usage that are useful to different application scenarios, we need to define our notion of usage. Our goal is to achieve a broad view of usage, while maintaining concise models.

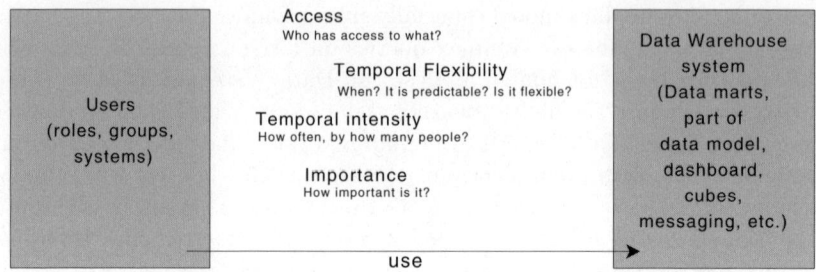

Fig. 2. Perspectives to consider when describing the usage of a data warehouse

Usage occurs between different kinds of *users* (i.e. roles of users, groups of users, external users) which use different *parts of the DWH system* (data marts, facts, overview dashboards) in different ways (only passively, very often, more resticted), as illustrated in Figure 2. For greater clarity, we have divided the general notion of DWH usage into four perspectives:

1. **Access control:** Who is allowed to use what?
2. **Temporal intensity:** How often do they use it?
3. **Temporal flexibility:** Do they have to use it at a certain time, or can it wait? Is it predictable when they need to use it?
4. **Importance:** How important is this usage?

We have identified a number of application scenarios of DWH usage models. They vary with the target user group and the perspectives to be modeled, and offer modeling solutions for typical every-day requirements in DWH management, maintenance and (re-)design.

To gain an overview of the current system, *DWH engineers* and *architects* as well as *managers* can employ usage models containing details from the *access control perspective* and an approximate number of instances for the user roles or groups. This answers general questions such as "who is using this?", "how many people would complain if we remove this?", etc.

A more detailed model using the *access types* from the access perspective can serve as input for specifying *access restriction* policies, and/or for setting *predefined views* and queries in data access tools.

Temporal intensity and flexibility considerations can be used (a) during the *planing phase* of a DWH design project, user requirements have to be matched to the available resources. DWH Usage models offer a way to capture a general overview of both aspects. Designers can then procede from the usage models to more detailed models later on.

Additionally, if (b) changes become necessary to an existing DWH, usage models can help to *identify critical patterns*. For instance, if due to mergers or reorganizations the number of users in a certain area rapidly increases or decreases, the intensity and flexibility perspectives can provide an overview of the implications.

As in any real-life setting, often not all that is desirable can be achieved. With the help of usage models with elements from the *importance perspective*, managers can decide how to resolve *resource conflicts*.

4   UML Profile for Modeling Data Warehouse Usage

This section introduces our UML Profile for Modeling DWH Usage. We use the extension mechanism of UML and import elements from a well-known UML Profile of the Data Warehousing domain, in order to achieve a conceptually sound model, with (a) tool support and (b) well-known notation elements.

Figure 3 gives an overview of our UML Profile and its stereotypes and enumerations, and also shows which classes are used as base classes of the stereotypes. We import the Profile of Luján-Mora et al. [5], and also some packages of the UML metamodel (for the convenience of not having to use fully qualified names). Table 1 describes the characteristics of the stereotypes not shown above.

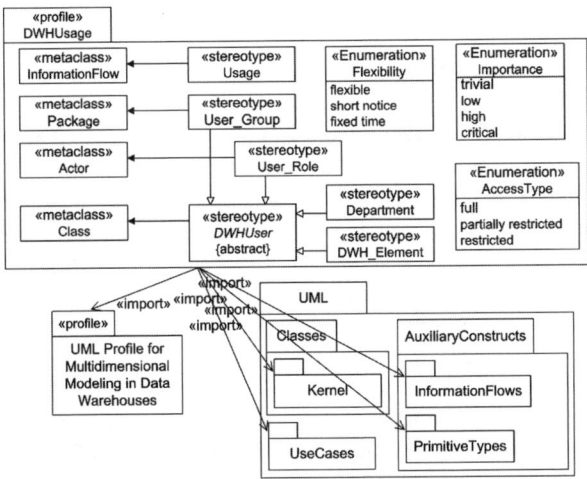

Fig. 3. The UML Profile for Modeling Data Warehouse Usage

Table 1. The UML Profile for Modeling Data Warehouse Usage

Name	*DWH User*
Description	An entity using the DWH. Abstract.
Tag Definition	**numberOfInstances**
	Type: Integer, Multiplicity: 1
	Description: The number of instances of this role. Visualized in the icon e.g. as a number in the "head" of an actor symbol.
	skillLevel
	Type: SkillLevel, Multiplicity: 1
	Description: The skill level of the user, i.e. whether able to write queries
Name	**User Role**
Description	A role that users/actors take when they access a DWH. One physical person (or external software system) may have several roles, and there may be several instances of one role.
Name	**User Group**
Description	Group of similar roles (orthogonal to Department)
Name	**Department**
Description	Organizational department (orthogonal to User Group)
Name	**DWH Element**
Description	An Element of the DWH system that users can access, and that can access other elements, e.g. a dashboard, a portal, etc.
Name	**Usage**
Description	A *Usage* indicates an "information channel" [3] between the DWH and its users. See text below for details.
Tag Definition	**accessType**, Type: AccessType, Multiplicity: 1
	Description: Indicates whether the access is (partially) restricted
	temporalFlexibility, Type: Flexibility, Multiplicity: n
	Default value: full
	Description: Indicates whether the usage is flexible in terms of time.
	temporalIntensity, Type: String, Multiplicity: 1
	Description: A textual description of the intensity of usage, e.g. number of instances per time interval, as scalar or as interval, probability range, etc.
	importance, Type: Importance, Multiplicity: 1
	Description: Indicates the level of importance attached to this usage.
Name	**SkillLevel**
Stereotype	Enumeration
Values	{basic, intermediate, expert}
Name	**AccessType**
Stereotype	Enumeration
Values	{full, partially restricted, restricted}
Name	**Flexibility**
Stereotype	Enumeration
Values	{flexible, short notice, fixed time}
Name	**Importance**
Stereotype	Enumeration
Values	{trivial, low, high, critical}
Additionally, all elements imported from the UML Profile for MD modeling in DWHs [5] (see Figure 3).	

Usage is defined as an InformationFlow, which is a type of directed relationship that specifies that information items circulate from sources to targets[1]. Information flows are defined in UML as a very general concept to be used in *"preliminary models, before having taken detailed modeling decisions on types or structures. One other purpose of information items and information flows is to abstract complex models by a less precise but more general representation of the information exchanged between entities of a system"* [3]. This makes information flows very suitable for our purpose, which is to provide models that capture an overview of the general structure of DWH usage.

The direction of the Usage arrow indicates whether the users actively initiate the access to the DWH or whether they wait to receive messages from the system, i.e. *push* or *pull* mode. A user analyzing OLAP data would be pull, whereas an e-mail alert is push.

---

[1] Sources and targets of an information flow may be: Actor, Node, UseCase, Artifact, Class, Component, Port, Property, Interface, Package, and InstanceSpecification [3].

5   Examples

In this section we illustrate the use of the UML Profile for Modeling Data Warehouse Usage with a number of examples. The examples each focus on a subset of the features of the profile and together provide an overview over the perspectives described in Section 3.

5.1   Users Accessing Data Warehouse Data

Figure 4 focuses on the questions "who needs which data?" and "who should be allowed to see what?". Diagrams of this type can be used in discussions with (future) users and in a later stage of the DWH design process may serve as rough input for specifying access restriction controls. Diagrams like this make it possible to identify preliminary groups of users, based on their data needs.

For each hospital admission it is recorded who was admitted (Patient dimension), what was the primary Diagnosis, which bed the patient was given (Placement) and which Insurance will cover the expenses. Health care professionals (nurses, doctors, therapists) need to access data on the patient, the diagnosis and the placement (the latter resticted to the ward they are working at), whereas the administration is interested in overall figures of how many patients were admitted where, but should not access patient details or diagnoses. For the billing clerks of the accounting department, all data for charging the hospital bills to the insurance companies has to be accessible. Finally, if the data is made available to medical researchers (e.g. research on the seasonal occurrence and duration of certain medical conditions), only aggregated patient data and diagnoses are relevant.

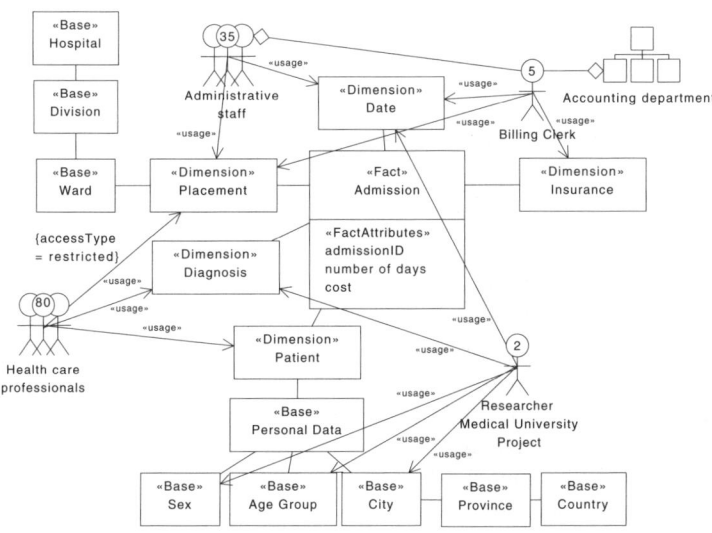

Fig. 4. Hospital admissions: Different roles and groups

5.2   Temporal Aspects of DWH Usage, Evaluating Scenarios

The example shown in Figure 5 illustrates the concepts of Intensity and Flexibility as described in Section 3.

Consider a Sales fact, the typical example of Data Warehousing (see Section 2), which contains data on items sold, to be sorted, aggregated and analyzed by time, product group, store, etc. In this example we take the Sales fact as a whole (a "StarPackage") and focus on the different users from various parts of the enterprise who all want to access this data.

Branch managers want to analyze the sales of their branch regularly once a week, and more or less predictably at the same time. The marketing department on the other hand will want access to sales data occasionally, but not necessarily at a given time or urgently. Product managers need to access the data with a varying intensity: If their new product is launched for example, they will watch the sales closely, but not at other times.

A new component that provides sales history data to sales agents is added to the system. Via this component, 150 individuals will want to access the sales data. The diagram provides an overview of the situation and supports discussions about this design decision.

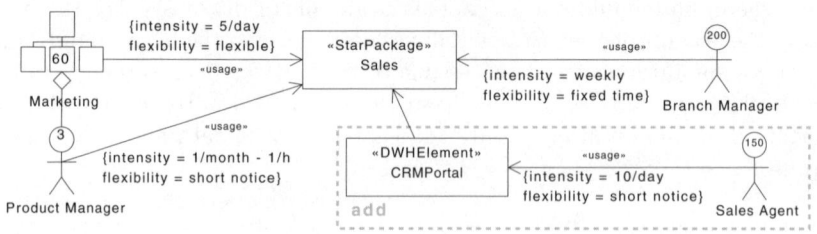

Fig. 5. Sales Example: Temporal aspects of DWH usage: Intensity and Flexibility

5.3   Importance in DWH Usage

Assessing the relative importance of features is crucial for design decisions. For our purpose, we restrict it to four levels, (critical, high, low, and trivial) and subsume economical as well as "political" importance (see [9] for examples of political issues in Data Warehousing) under one item, as shown in Figure 6. In this example the use of the attribute skillLevel and passive DWH usage (via an e-mail alert service) are also shown.

Sales and Marketing people need to access sales data for their everyday work, which is of high importance. Top managers occasionally browsing sales data should also be treated as important, which is an example of "political" and not so much economical importance. Aggregated sales data is also fed into the Intranet portal, but this considered a "nice to have" feature of trivial importance.

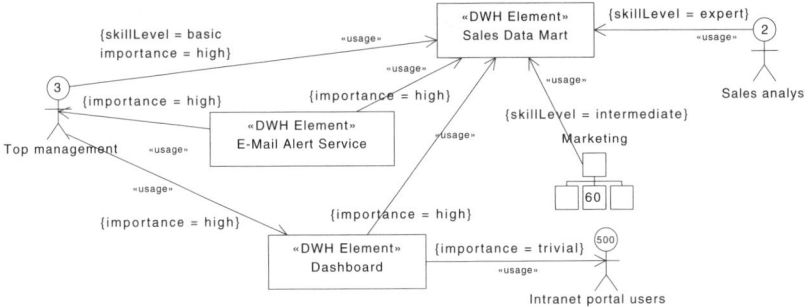

Fig. 6. Importance, passive usage (push instead of pull), and user skills

6   Related Work

Our approach to modeling DWH usage on the conceptual level as presented in this paper touches many different areas, with access control, temporal intensity, temporal flexibility and importance, active or passive usage, details of the users such as their skill level, number of instance or affiliation. To the best of our knowledge, there is no comparable work with the same focus. Nevertheless, regarding individual aspects, we can discuss our relationship to previous work.

Modeling the users who access different parts of data warehouse systems is also an issue for security and access control in Data Warehousing. In [10], the authors present a UML Profile for secure DWHs that includes user profiles and user roles contained in hierarchies. The users are granted privileges to access parts of a multidimensional data model. [11] define a different kind of authorization model integrated with MD modeling, also based on user roles. [12] or [13] are other example of approaches to modeling user access to parts of multidimensional models and/or OLAP operations.

As these approaches have a different aim in modeling users (i.e. grouping users with similar privileges for access control and security measures), they do not include the organizational affiliation of users, their skill level, or the importance, intensity or temporal flexibility of their access to the DWH.

Our approach can also be compared to Requirements Engineering. [14] describe an approach to model DWH requirements with UML use cases. [15] and [16] describe approaches for goal-oriented DWH requirements. [17] reconcile the available data sources with the requirements, and [18] present an MDA approach for building conceptual and logical data models from requirements models.

Our UML profile for modeling DWH usage is not intended as a means to create data models, but can be used to support the requirements analysis phase of a DWH project.

7   Conclusion

Today's DWH systems provide many different services to different kinds of users. In order to have a big picture of the current situation and to visualize future scenarios, people involved in designing and managing today's DWH systems need an overview of all these different ways the DWH is being used.

In this paper, we have introduced the UML Profile for Modeling DWH Usage for modeling the different kinds of DWH usage on a conceptual level. It distinguishes four perspective of usage (access control, temporal intensity, temporal flexibility and importance) as well as active or passive usage, and allows to model details of the users such as their skill level, number of instances, functional grouping or organizational affiliation. We base "usage" on UML information flows, which are intended for a general representation of information exchanges.

References

1. Inmon, W.H., Hackathorn, R.D.: Using the data warehouse. Wiley-QED Publishing, Somerset, NJ, USA (1994)
2. Kimball, R., Reeves, L., Ross, M., Thornthwaite, W.: The Data Warehouse Lifecycle Toolkit. John Wiley & Sons, Inc., England (1998)
3. Object Management Group, Inc.: UML 2.0 Superstructure (2005),
   http://www.omg.org/cgi-bin/apps/doc?formal/05-07-04.pdf
4. Object Management Group, Inc.: Model Driven Architecture (MDA) (2004),
   http://www.omg.org/cgi-bin/doc?formal/03-06-01
5. Luján-Mora, S., Trujillo, J., Song, I.Y.: A UML profile for multidimensional modeling in data warehouses. Data Knowl. Eng. 59(3), 725–769 (2006)
6. Vassiliadis, P., Sellis, T.K.: A Survey of Logical Models for OLAP Databases. SIGMOD Record 28(4), 64–69 (1999)
7. Abelló, A., Samos, J., Saltor, F.: YAM^2 (Yet Another Multidimensional Model): An Extension of UML. In: IDEAS 2002, pp. 172–181. IEEE Computer Society Press, Los Alamitos (2002)
8. Nguyen, T.B., Tjoa, A.M., Wagner, R.: An Object Oriented Multidimensional Data Model for OLAP. In: Lu, H., Zhou, A. (eds.) WAIM 2000. LNCS, vol. 1846, pp. 69–82. Springer, Heidelberg (2000)
9. Demarest, M.: The politics of data warehousing (1997),
   http://www.noumenal.com/marc/dwpoly.html
10. Ferández-Medina, E., Trujillo, J., Villarroel, R., Piattini, M.: Developing secure data warehouses with a uml extension. Inf. Syst. 32(6), 826–856 (2007)
11. Priebe, T., Pernul, G.: A pragmatic approach to conceptual modeling of olap security. In: Kunii, H.S., Jajodia, S., Sølvberg, A. (eds.) ER 2001. LNCS, vol. 2224, pp. 311–324. Springer, Heidelberg (2001)
12. Kirkgöze, R., Katic, N., Stolba, M., Tjoa, A.M.: A security concept for olap. In: DEXA 1997, Washington, DC, USA, p. 0619. IEEE Computer Society, Los Alamitos (1997)
13. Wang, L., Jajodia, S., Wijesekera, D.: Securing olap data cubes against privacy breaches. In: IEEE Symposium on Security and Privacy, p. 161. IEEE Computer Society Press, Los Alamitos (2004)
14. Bruckner, R., List, B., Schiefer, J.: Developing requirements for data warehouse systems with use cases. In: Proceedings AMCIS 2001, pp. 329–335 (2001)

15. Mazón, J.N., Pardillo, J., Trujillo, J.C.: A model-driven goal-oriented requirement engineering approach for data warehouses. In: Hainaut, J.-L., Rundensteiner, E.A., Kirchberg, M., Bertolotto, M., Brochhausen, M., Chen, Y.-P.P., Cherfi, S.S.-S., Doerr, M., Han, H., Hartmann, S., Parsons, J., Poels, G., Rolland, C., Trujillo, J., Yu, E., Zimbányi, E. (eds.) ER 2007. LNCS, vol. 4801, pp. 56–71. Springer, Heidelberg (2007)
16. Giorgini, P., Rizzi, S., Garzetti, M.: Goal-oriented requirement analysis for data warehouse design. In: Proceedings DOLAP 2005, pp. 47–56. ACM Press, New York (2005)
17. Mazón, J.N., Trujillo, J.C., Lechtenbörger, J.: Reconciling requirement-driven data warehouses with data sources via multidimensional normal forms. Data Knowl. Eng. (2007) (doi:10.1016/j.datak.2007.04.004)
18. Mazón, J.N., Trujillo, J.C.: An mda approach for the development of data warehouses. Decision Support Systems (2006) (doi:10.1016/j.dss.2006.12.003)

Preface to ONISW 2007

Mathias Brochhausen[1,2], Martin Doerr[3], and Hyoil Han[4]

[1] Institute for Formal Ontology and Medical Information Science, Saarland University, Saarbrücken, Germany
[2] European Centre for Ontological Research, Saarland University, Saarbrücken, Germany
[3] Institute of Computer Science, Foundation for Research and Technology - Hellas (FORTH), Greece
[4] Drexel University, USA

The emergence of the World Wide Web made massive amounts of data available. Data exist in many scattered electronic data sources (e-sources) over the Web. Even though some of the data are in well-organized data sources, interoperability and integration with data from other sources, semantic coordination and conflict resolution are required for its full exploitation. Semantic Web-enabled applications can potentially produce better results for semantic integration, interoperability and search. In particular, ontologies are widely regarded as the best solution to global information integration and semantic interoperability.

A crucial question is whether ontologies can replace information models. But whereas ontologies work quite well as virtual schemata in mediation systems, they may perform poorly as information models and on the user interface level. On the theoretical side, there is a lack of understanding of the effective relation and interplay of ontological and epistemological features in information models and systems. Furthermore there are still open questions concerning good scientific practice in developing ontologies. On the practical side, there is still a lack of good practice of how to integrate existing information systems into ontology-driven applications and few experiences at all with creating good new data structures from ontologies directly for interoperation in complex and diverse application environments.

A Method for Semi-automatic Creation of Ontologies Based on Texts

Luiz C.C. Carvalheira and Edson Satoshi Gomi

Escola Politecnica da Universidade de Sao Paulo
Av. Prof. Luciano Gualberto, 158 tv. 3. 05508-900 Sao Paulo, SP, Brazil
{luiz.carvalheira, edson.gomi}@poli.usp.br

Abstract. The recent developments related to knowledge management, the semantic web and the exchange of electronic information through the use of agents have increased the need for ontologies to describe in a formal way shared understanding of a given domain. For computers and people to work in cooperation it is necessary that information have well defined and shared definitions. Ontologies are enablers of that co-operation. However, ontology construction remains a very complex and costly process, which has hindered its use in a wider scale. This article presents a method for the semi-automatic construction of ontologies using texts of any domain for the extraction of concepts and relations. By comparing the relative frequency of terms in the text with their typical, expected use, the method identifies concepts and relations and specifies the corresponding ontology using OWL for use by other applications.

1 Introduction

Ontologies have been shown as very useful tools for managing, reusing and disseminating knowledge in areas where information integration and communication are of crucial importance. Systems inter-operability, communication among intelligent agents, shared services, information search based on metadata and concepts, are all examples of the application of ontologies that have been reported ([1], [2]). By providing an organization with a unified structure to reduce confusion about concepts and terminology, ontologies can promote knowledge sharing among people inside and outside the organization. Another use of ontologies is the establishment of integrated environments where different software tools and components can be used harmoniously in multi-agent architectures and corporate models that deal with activities, products, services and organizations. Ontologies can, not only serve as basis for the operation of applications and systems, but also support the design and development of those systems, facilitating the requirement identification process and the understanding of which components exist and how they relate to each other.

In spite of the vast applicability of ontologies, their creation and validation remain a significant obstacle to their widespread application, given the effort that is required. Given that context, the objective of this article is to propose a method for the semi-automatic creation of domain ontologies based on the

J.-L. Hainaut et al. (Eds.): ER Workshops 2007, LNCS 4802, pp. 150–159, 2007.
© Springer-Verlag Berlin Heidelberg 2007

analysis of non structured sources of information, such as texts, that contain and express the main concepts of that knowledge domain. Once collected, those texts can be processed with linguistic and statistical techniques to identify the main objects discussed and their exposed relations. What we suggest is that this method can support the work of a human specialist, by extracting and submitting to his or her evaluation the majority of the concepts and relations that the ontology should describe, thus reducing the cost and complexity associated with the ontology definition for the domain. Additionally, to allow for its use in other applications that involve the domain in study, the ontology is specified in OWL.

To evaluate it, the method was applied in the construction of an ontology for the domain of Project Management, using the texts of the Project Management Body of Knowledge (PMBOK) [3] as base corpus (here called "domain corpus").

This article is organized in 5 sections. Following this introduction, session 2 highlights some important notions related to the construction of ontologies. Session 3 offers a summary of the proposed method, while session 4 presents and discusses the results that were obtained. Finally, session 5 concludes and suggests evolution opportunities and future works.

2   Ontologies and Their Creation

Following [4], an ontology is a formal, explicit specification of a shared conceptualization. By conceptualization it is meant an abstract model of some phenomenon or fact about the world, a representation of which becomes explicit if all the concepts and restrictions for its utilization are clearly defined. Moreover, this conceptualization becomes shared when all the parts involved in its use agree and accept such representation. The formalism in the specification allows the reduction in communication ambiguity between human beings and makes its use by machines possible. Uschold and Gruninger [2] state that the ontology creation process involves the following basic activities: 1) the identification and description of the key concepts and relationships that exist in the domain; 2) the codification of the ontology with the use of a formal language; 3) the integration of the ontology with other ones through the reuse of concepts.

Although many tools have been developed in recent years to support these activities, the construction of ontologies continues to be a manual, labour intensive exercise carried out by specialists in the domain [5]. The high costs and long dedication required are difficulties that have seriously restricted an expansion in the use of ontologies. However, the scientific community has endeavored to elaborate new methods and tools to speed up that process. Aussenac-Gilles, Biebow and Szulman [6] have established a method based on the analysis of a corpus through the use of natural language processing techniques for the creation of ontologies. The method also includes the use of existing ontologies, as well as other resources, for this construction. Alfonseca and Manandhar [7] established an approach for extending an existing ontology which was based on the identification of contextual properties of words that co-occur with each group of concepts. These properties are used to cluster concepts in a new ontology or to add new concepts to one that exists already. Agirre et al. [8] designed a method for enriching the concepts of an ontology but control for the proliferation of meanings, a very common issue in more extensive ontologies such as the WordNet [9].

3   Method Overview

The ontology construction method presented here is founded on the recognition of concepts and relations that are present in the text through the use of a linguistic treatment and the identification of morphological and lexical patterns that are characteristic of the way concepts and relations are expressed.

The most important concepts of a given domain are typically represented in texts with the use of nouns or noun phrases. Considering the premise that the text or corpus on which the construction will be based effectively deals with the domain in focus, it is reasonable to assume that the most frequently used terms in the corpus are candidates for concepts in the ontology of that domain. To prevent the influence of common terms that are generally used in the language (such as *man, person, word*) it is necessary to analyse the frequency of each term in the corpus vis-à-vis its frequency in the common writings of the language.

To establish the expected frequency of a term in general American English, we used the Brown Corpus [10], a set of documents that contain a balanced sample of texts from a variety of sources and different genres, such as political, financial, cultural, economy-related and medicine-related. The basic idea here is that, if a term appears in the domain corpus with a relative frequency that is significantly higher than in the Brown Corpus, then we can safely assume that the term is a particularly important or specific concept for the domain. The term should then be considered for the purpose of ontology construction.

Another fundamental assumption of the method is that the domain corpus should make references to the relations that exist among the concepts. Amongst the interesting relations we are looking for, we can mention the "isA" relation that refers to the class-subclass subsumption relation. Examples of this relation extracted literally from the PMBOK are "a *project charter* **is a** *document* that formally recognizes the existence of a project", "the *project plan* **is a** formal, approved *document* used to manage and control project execution", "a *deliverable* **is a** tangible, verifiable *work product*". The natural language processing on the domain corpus was accomplished through the use of GATE (General Architecture for Text Engineering) developed in the University of Sheffield [11], one of most used frameworks for construction of linguistic engineering systems [12]. GATE implements an annotation system used to define and register the characteristics of portions of the text, such as sentences, words, noun phrases. GATE also allows the specification of transducers through the use of JAPE (Java Annotation Patterns Engine), a language for defining regular expressions that operate on the existing annotations and, in the sequence, create new annotations. The grammars written in JAPE are compiled into finite state machines. Figure 1 shows the main processes of the suggested method.

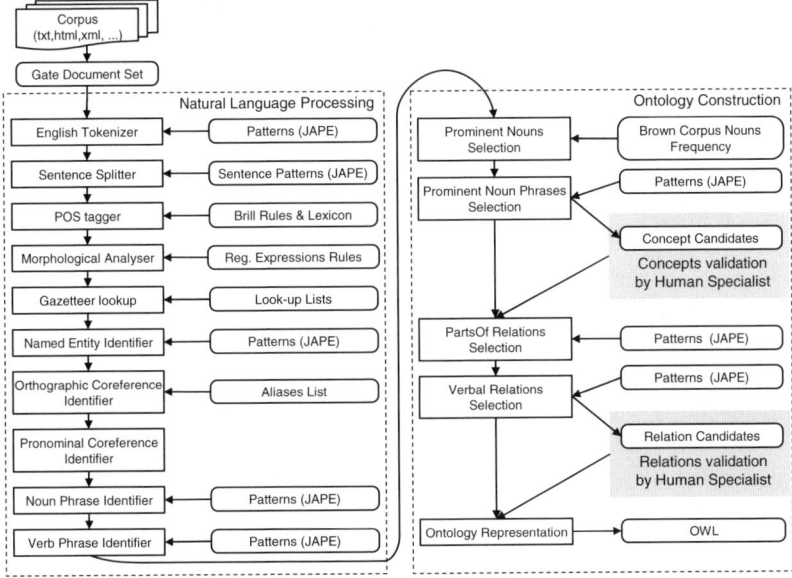

Fig. 1. Ontology Construction Method

3.1   Basic Linguistic Processing

To allow the execution of the concept and relation identification processes, the domain corpus is constituted and initially treated with the use of natural language processing techniques:

- the texts that will serve as base for the analyses are incorporated into a corpus and converted into the GATE document format;
- *tokens* (words and symbols) and sentences are identified and annotated;
- a lexical and morphologic analysis breaks the sentence into lexical items. A POS tagger based on Brill's algorithm [13] associates each lexical item to its grammatical category (article, noun, adjective, verb, etc.);
- the root (or lemma) of each word, as well as other morphological characteristics such as gender, number, etc, are recognized by a morphological analyzer;
- each word in the text is checked by a "gazetteer" process against a lists of cities, countries, etc., in order to incorporate semantics of known and common words. For example, *"dollar"* is annotated as representing a currency;
- a Named Entity recognizer is then used to identify and annotate people's name, types of construction (*Brooklyn Bridge*), places (*Mount Washington* or *New York City*), and other pre-established linguistic patterns;
- a module for anaphora resolution is executed to identify the term to which a pronoun refers to (as in "A project involves *risks* and *they* are associated with events.");

- based on a JAPE grammar and on the annotations established by previous components, a noun phrase identifier annotates constructions such as "project management" or "project communication plan", focusing on noun phrases composed exclusively of nouns;
- a transducer specified in JAPE analyzes the annotations created by the POS tagger and identifies verb phrases.

The annotations established during the basic processing are then used for the identification of concepts and relations and the construction of the ontology.

3.2   Concept Extraction

To analyze the importance of each noun in the text and to eliminate the influence of terms that are simply very common in the writings of the language (thus, not domain specific), it is necessary to compare the relative frequencies of each noun in the two corpora, the domain corpus and the Brown Corpus representing the typical, current use of the English language in texts.

Hofland and Johansson [14] proposed a test of statistical fit based on the non-parametric test of Chi-Square (χ^2) [15] for comparing the frequency of terms in two corpora. Assuming that the term is used with the same relative frequency in the texts of the domain and in general, current English, the test measures the probability of encountering the differences observed for every single noun. The Chi-Square statistic is calculated in the following manner:

$$\chi^2 = \sum_i \frac{(O_i - E_i)^2}{E_i} \tag{1}$$

where

$$E_i = \frac{N_i \sum_i O_i}{\sum_i N_i} \tag{2}$$

In this formula O_i represents the observed frequency of the word in corpus i, E_i is the expected frequency of this word and N_i is the total frequency of words in corpus i. The χ^2 statistic follows the known Gamma distribution, which allows us to calculate the probability of observing the value of that statistic for any given word. If the probability is very small we can reject the null hypothesis that states there is no difference in relative frequency of the word in the two corpora, or, putting differently, the word appears more or less frequently in the texts of the corpus domain than it does in regular, standard English writings.

In table 1 some of the terms found by our processes are shown with their corresponding statistics. Values of χ^2 larger than 10.82 indicate a confidence level larger than 99.9% that the frequencies are indeed different. We also suggest the use of a metric, which we call "*Prevalence in PMBOK*", to demonstrate how many times the relative frequency of the word in the domain corpus is higher than in the Brown Corpus, a simple ratio between those two relative frequencies. In the versions used in this work, the Brown Corpus contains 1,015,945 words, while the PMBOK 37,238 has words. A combination of large "*Prevalence in PMBOK*" and large χ^2 values indicates the most prominent nouns in the corpus.

Table 1. Frequencies and Statistics for some nouns found in the PMBOK

Term	Absolute Freq. in Brown Corpus	Absolute Freq. in Brown Corpus	Prevalence in PMBOK	χ^2
project	1192	93	349.6	30,031.4
scope	164	27	165.7	3,796.3
management	367	91	110.0	7,881.5
schedule	134	36	101.5	2,825.7
risk	182	54	91.9	3,747.3
estimates	73	24	82.9	1,463.1
processes	160	57	76.5	3,135.7
team	171	84	55.5	3,017.6
tools	63	34	50.5	1,072.7
documents	34	19	48.8	571.0

The nouns identified through this method are considered as very characteristic of the domain corpus. If one or more of those nouns appear as components of a frequent noun phrase identified previously then we consider that the noun phrase is a strong candidate for a concept of the domain. Those candidates are gathered and presented to the ontologist, together with the statistics that supported the selection. The noun phrases accepted by the ontologist, here called noun phrases of the domain, are then used in the next phases of the method.

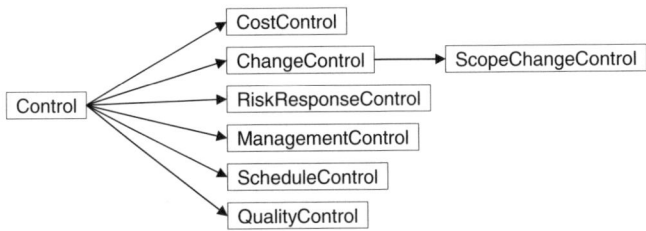

Fig. 2. Example of taxonomic relation construction. "*ScopeChangeControl*" is a subclass of "*ChangeControl*", which in its turn is a subclass of "*Control*".

The list of noun phrases of the domain is then structured in the form of taxonomy by a process that analyses noun phrases composed of more than one noun, for instance, "*scope change control*". The noun that confers more specificity to the noun phrase is extracted for the creation of taxonomic relations of class-subclass. This occurs as many times as required so to create more generic classes until only the head of the original noun phrase, i.e. the least specific noun, is left. Figure 2 demonstrates the result of the analysis in the example mentioned above, where the classes "*ScopeChangeControl*", then "*ChangeControl*" and finally "*Control*" are created and linked by a sublass-class relationship.

3.3   Relation Extraction

The basic assumption of this relation extraction method is the strong correlation
between the verbal syntactic structures and the semantic relations among the
entities that appear in a sentence [16]. Two distinct processes are used to extract
relations between concepts here.

The first process is based on the assumption that relations of the kind "is-
PartOf" or "has" can be expressed with linguistic constructions such as "departments of the organization" or "project's objectives". In those specific examples,
it is reasonable to think that a department *is part of* an organization (or that a
organization *has* departments), and, similarly, that a project *has* objectives.

The second process involves the analysis of verbal and noun phrases for the
identification of a subject (i.e. entity that performs an action), the action and
its objects (focus of the action performed). This process is implemented through
a series of grammars expressed in JAPE and which are based on the noun and
verbal phrase annotations previously discovered during the linguistic treatment.

For implementing that search for relations, verb phrases present in the text
are extracted together with the noun phrases that refer to the subject and the
object of the verbal action. If both noun phrases have previously been selected
in the concept extraction stage, then the verbal relation is considered relevant
and is marked as a candidate relation of the ontology.

As an example of the application of that process, in the sentence "*A change
control system is a collection of formal, documented procedures*" the relation
"*isACollectionOf*" is identified and associated to the concepts "*ChangeCon-
trolSystem*" and "*Procedure*". In another example that appears literally in the
PMBOK, the sentence "*The project plan contains the various baselines that will
be used to assess project performance.*" shows us evidences of the relation *contain*
between the concepts *ProjectPlan* and *Baseline*, as can be seen in figure 3.

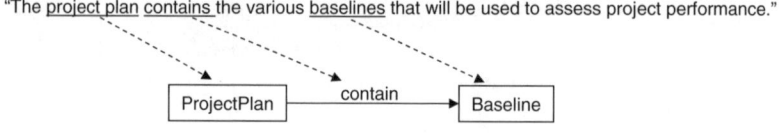

Fig. 3. Example of relation construction based on verbal structures

The specification of the patterns that identify these relations is made in the
GATE environment with the use of the JAPE language for detailing the appropriate grammars. Naturally, relations are selected at this stage only if they
connect concepts that have been previously identified and validated by the ontologist. Once selected by the process, the relations are presented to the ontologist
for evaluation. The concepts and relations that were accepted will form the ontology which is then represented in the language OWL, for further deployment
and use. A fragment of the generated code is shown below. As a reminder to
the ontologist, the passage that was the basis for recognizing the relation is
represented as comment (as can be seen in lines 11 to 14).

```
1    <owl:Class rdf:ID="Procedure"/>
2    <owl:Class rdf:ID="System"/>
3    <owl:Class rdf:ID="ChangeControlSystem">
4      <rdfs:subClassOf>
5        <owl:Class rdf:ID="System"/>
6      </rdfs:subClassOf>
7    </owl:Class>
8    <owl:Class rdf:ID="ChangeControlSystem">
9      <rdfs:subClassOf>
10       <owl:Restriction>
11         <rdfs:comment rdf:datatype="http://www.w3.org/2001/XMLSchema#string"
12         >A change control system is a collection of formal, documented procedures
13          that defines the steps by which official project documents may be
14          changed.</rdfs:comment>
15         <owl:onProperty>
16           <owl:ObjectProperty rdf:ID="isACollectionOf"/>
17         </owl:onProperty>
18         <owl:someValuesFrom rdf:resource="#Procedure"/>
19       </owl:Restriction>
20     </rdfs:subClassOf>
21   </owl:Class>
```

4   Evaluation of the Method

A clear consensus has not yet been reached regarding the best way to evaluate an ontology, specially when it is created by an automatic method. In this work we consider two main evaluation criteria: the degree of coverage that the created ontology has on the domain being studied and the support that it offers the ontologist during concept extraction.

As a gold standard reference for the set of important concepts that should appear in an ontology of Project Management, we used the glossary of definitions supplied at the end of the PMBOK by its authors (with a total of 239 definitions in the version we used). After careful analysis of the definitions in the glossary, some lexical items, such as "as-of date", "link", "node", "subnet" were removed from the gold standard as they were not deemed important enough for the domain to merit a definition. Likewise, some items such as "accountability matrix" or "contingency allowance" were removed as they were not mentioned in the texts of the PMBOK, in spite of being mentioned in the glossary.

For the evaluation of the ontology built and its coverage of the domain, we used a quantitative analysis of the concepts based on the traditional metric of recall from the field of information retrieval. In our context, recall can be understood as the number of concepts extracted from the domain corpus which are also present in the gold standard ($ExtractedConcepts_{Gold}$) as a proportion of the total amount of concepts in the gold standard ($Concepts_{Gold}$):

$$Recall = \frac{ExtractedConcepts_{Gold}}{Concepts_{Gold}} \tag{3}$$

Recall indicates how well the method managed to extract the concepts that were judged important by the authors (in this case). Similarly, precision, also commonly used in information retrieval, can be calculated in the context of extraction of concepts, as well. Here, precision is defined as the ratio between the amount of extracted concepts that also appear in the gold standard

($ExtractedConcepts_{Gold}$) and the total amount of extracted concepts ($ExtractedConcepts$):

$$Precision = \frac{ExtractedConcepts_{Gold}}{ExtractedConcepts} \tag{4}$$

Precision shows how the method succeeded in focusing exclusively on the gold standard concepts. Table 2 shows the results of the evaluation of the extracted ontology. It is important to notice that, in the case of ontology creation, precision does not demonstrate a possible benefit of the method, which is to offer the ontologist new concepts that were not part of the gold standard. For instance, many of the new concepts identified, such as *budget, assumption, change control, change request* were considered fundamental for the ontology. In fact, of all concepts that were extracted but were not mentioned in the gold standard ($NewConcepts$), 83% of them were judged pertinent to compose the ontology of the domain ($NewConcepts_{Pertinent}$).

Table 2. Concept Extraction Evaluation Metrics

$Concepts_{Gold}$	106
$ExtractedConcepts_{Gold}$	77
$ExtractedConcepts$	279
$Recall$	73%
$Precision$	28%
$NewConcepts$	211
$NewConcepts_{Pertinents}$	176
% $NewConcepts_{Pertinent}$	83%

Table 3. Qualitative evaluation of the relations found

Relevance	# of Relations	% of Relations
High	48	29,4%
Good	30	18,4%
Medium	30	18,4%
Little	29	17,8%
None	26	16,0%

Since there is no gold standard with a list of fundamental relations between concepts in the Project Management domain, the evaluation of relations extracted by the method was carried out with a qualitative analysis performed by a panel of three professionals, all experts in the domain. Each relation selected was judged by the panel and given one of five possible degrees to indicate relevance to the domain: high, good, medium, little or none. As shown in table 3, 57.8% of the relations were judged to be of high or good relevance to the domain.

5   Conclusion

This article presented a method for the semi-automatic creation of domain ontologies, based on the analysis of texts that express the main concepts of that domain of knowledge, as well as the relations that may exist between them. The objective is to support the work of a human specialist, the ontologist, decreasing the cost and effort related to ontology creation, thus enabling the dissemination of its use and application. As detailed in the previous session, this method presented good performance, with potential for expansion and refinement.

Ontologies are very complex specifications. Many different ontologies may be conceived and exist for the same domain. Still, all can be considered correct and appropriate to represent the knowledge it contains. That characteristic makes it non-trivial to compare ontologies and to analyse its quality or adequacy to a specific purpose, which constitutes one of the current challenges of the area. In spite of inherent difficulties of the theme, this method for ontology creation can establish a work basis for a specialist of virtually any domain, be it in the initial phases of the ontology creation or later when finding new concepts and relations to revise an existing ontology. This type of support is especially valuable when a glossary of definitions for the domain does not exist or is not readily available.

References

1. Uschold, M., Jasper, R.: A framework for understanding and classifying ontology applications (1999)
2. Uschold, M., Gruninger, M.: Ontologies: principles, methods and applications. Knowledge Engineering Review 11, 15–93 (1996)
3. Project Management Institute: A Guide To The Project Management Body Of Knowledge (PMBOK Guides). Project Management Institute (1996)
4. Studer, R., Benjamins, R., Fensel, D.: Knowledge engineering: Principles and methods. Data & Knowledge Engineering 25(1-2), 161–198 (1998)
5. Maedche, A., Staab, S.: Ontology learning for the semantic web. IEEE Intelligent Systems 16(2), 72–79 (2001)
6. Aussenac-Gilles, N., Biebow, B., Szulman, S.: Revisiting ontology design: A methodology based on corpus analysis. In: Proceedings of the 12th European Workshop on Knowledge Acquisition, Modeling and Management, pp. 172–188 (2000)
7. Alfonseca, E., Manandhar, S.: An unsupervised method for general named entity recognition and automated concept discovery (January 17, 2002)
8. Agirre, E., Ansa, O., Hovy, E., Martinez, D.: Enriching very large ontologies using the WWW (October 17, 2000)
9. Fellbaum, C. (ed.): WordNet: An Electronic Lexical Database. The MIT Press, Cambridge, Massachusetts (1989)
10. Francis, W.N., Kucera, H.: Computational analysis of present-day American English. Brown University Press, Providence (1967)
11. Cunningham, H., Gaizauskas, R.J., Wilks, Y.: A general architecture for language engineering (GATE) - a new approach to language engineering R&D (1996)
12. Maynard, D., Cunningham, H., Bontcheva, K., Catizone, R., Demetriou, G., Gaizauskas, R., Hamza, O., Hepple, M., et al.: A survey of uses of gate
13. Brill, E.: Transformation-based error-driven learning and natural language processing: A case study in part-of-speech tagging. Computational Linguistics (1995)
14. Hofland, K., Johansson, S.: Word frequencies in british and american english (1982)
15. Pearson, K.: On the theory of contingency and its relation to association and normal correlation. Biometric series no. 1 (1904)
16. Maedche, A., Staab, S.: Discovering conceptual relations from text. In: Proceedings of the 14th European Conference on Artificial Intelligence, 2000 (2000)

Enriching OWL with Instance Recognition Semantics for Automated Semantic Annotation

Yihong Ding[1], David W. Embley[1], and Stephen W. Liddle[2]

[1] Department of Computer Science
[2] Information Systems Department
Brigham Young Univeristy, Provo, Utah 84602, U.S.A.

Abstract. Although OWL provides a solid basis for many semantic-web applications, it lacks sufficient declarative semantics for instance recognition. This omission prevents OWL from being a satisfactory ontology language for automated semantic annotation. We can resolve this problem by adding to ontological OWL declarations epistemological instance recognition semantics that include external representations and context recognition information for atomic, lexical ontology concepts. Our implementation shows that the new automated annotation prototype system using OWL ontologies with rich instance-recognition semantics not only has high precision and recall, but also overcomes the post-processing problem of linking extracted data to semantic-web ontologies. Our study also shows that the use of instance recognition semantics in ontologies can lead to enhanced knowledge sharing and reuse through the Semantic Web.

1 Introduction

Semantic annotation research is fundamental for the Semantic Web. A *semantic annotation* process adds formal metadata to web pages. This metadata links data in a web page to defined concepts in an ontology. Because machine agents are capable of interpreting data with respect to an ontology, annotated content becomes machine-processable.

Automated semantic annotation is the primary means of adding machine-processable metadata to existing web pages. Several researchers have suggested various ways to automate semantic annotation (e.g., [1,3,9,11,14]). Each of these approaches has adapted a data-extraction engine to wrap and annotate existing web pages. None of the adapted data-extraction engines, however, was originally designed to produce annotations linking extracted data to an ontology [11,12]. To provide machine-processable semantics for annotated content, these approaches therefore need to do post-processing to map extracted data to an ontology. Some researchers identify this problem as the "main drawback" of current approaches to automating annotation and suggest the direct use of ontologies in the extraction process to help with semantic annotation [11].

Our ontology-based semantic annotation research shows that this suggested approach does indeed work [4]. There is, however, a hidden problem to address: any system that does not conform to Semantic Web standards will not

J.-L. Hainaut et al. (Eds.): ER Workshops 2007, LNCS 4802, pp. 160–169, 2007.

be interoperable and thus will not be generally accepted. The current standard (W3C-recommended) Semantic Web ontology language is OWL (Web Ontology Language) [15]. But OWL lacks sufficient declarative semantics for instance recognition, which are needed to extract data directly with respect to ontologies. Our ontology language [4] supports rich declarative instance-recognition semantics, but it is not a standard.

To resolve this issue, we propose an extension of OWL that contains enriched declarative instance-recognition semantics. Because this extension is specifically designed for automated annotation, we call it OWL-AA (OWL for Automated Annotation). Rather than just propose our OWL-AA syntax, the goal of this paper is to illustrate the essence of a sound solution to the problem.

This OWL-AA extension is epistemological in nature. While ontology (in our context) focuses on defining formal specifications of facts and the relations between facts, epistemology (in our context) focuses on how we know a presented fact is about a formal specification, i.e., the recognition of instances [10]. Ontological definitions are *independent* of the form of knowledge representation. In contrast, epistemological definitions are often *sensitive* to the form of knowledge representation [8].

The primary contribution of this work is that we have proposed OWL-AA as a way to extend OWL to provide for automated semantic annotation. With the use of OWL-AA, we present a new semantic annotation model that (1) embeds instance-recognition semantics declarations in ontologies and data-extraction tools and (2) provides enhanced knowledge sharing and reuse through the Semantic Web. Furthermore, as a significant consequence, OWL-AA separates the creation of domain knowledge from the implementation of a processor to use domain knowledge for the purpose of annotating web pages. With OWL-AA, domain experts need not know how to implement extraction and annotation programs, and system developers need not be domain experts.

To explain these contributions, Section 2 presents the details of our proposed instance-recognition semantics, and Section 3 explains their role within the automated semantic-annotation paradigm. Section 4 introduces OWL-AA—the motivation for our choices and its definition and usage. We discuss related work in Section 5 and make concluding remarks in Section 6.

2   Instance Recognition Semantics

Instance-recognition semantics, which we present as *instance semantics recognizers (ISR)*,[1] are formal specifications that identify instances of a concept in ordinary text. The text may be unstructured, semi-structured, or fully structured. For Semantic Web applications, the concept should be a lexical element of a formal ontology (e.g. concepts such as *date*, *time*, *place*, *location*, *name*, *telephone number*, *email address*, various weights and measures, etc.). Thus, an ISR of an ontology concept (e.g. *Telephone Number*) interprets an instance

---

[1] We avoid the acronym IRS to avoid association with the U.S. Internal Revenue Service.

in a text fragment (e.g. the contact number in "Call me at 222-1234.") to have the intensional meaning of the defined concept (e.g. *Telephone Number*).

We have used information-extraction (IE) ontologies that include ISRs to do data extraction [6] and semantic annotation [4]. In our IE ontologies, we use a *data frame* construct [5] that describes information about a concept—its external and internal representations, its contextual keywords or phrases that may indicate the presence of an instance of the concept, operations that convert between internal and external representations, and other manipulation operations that can apply to instances of the concept along with contextual keywords or phrases that indicate the applicability of an operation. Thus, a data frame contains ISR declarations together with other elements that are not the focus of this paper.

BedroomNr
    external representation: [1-9]|10
    left context phrase: \b
    right context phrase: .*r(oo)?ms?
    exception phrase: \s.*ba(th)?s?\b.*r(oo)?ms?
    context keywords: b(r|d)s?|bdrms?|bed(rooms?)?
    ...
end

Feature
    external representation: ApartmentFeature.lexicon
    ...
end

Fig. 1. ISR Declarations for `BedroomNr` and `Feature`

Figure 1 shows partial ISR declarations for two concepts: *BedroomNr* and *Feature*. We use Perl-style regular expressions to declare recognition patterns. Essentially, the ISR declaration of a concept contains two categories of information: self-recognition patterns and context patterns. The *external representation* clause defines self-recognition patterns, which are the typical signatures for any instantiations of a concept. In our example, a valid instantiation of *BedroomNr* must match the regular pattern that describes strings of digits from 1 to 10, and any valid instantiation of *Feature* must match one of the strings declared inside the lexicon file *ApartmentFeature.lexicon*.

Self-recognition patterns by themselves, however, are not enough to precisely recognize a valid instantiation. For example, an integer up to 10 can also be the number of bathrooms or the number of people who can share an apartment. To help resolve potential ambiguities, we use context information. There are two types of context declarations: context phrases and context keywords. A *context phrase* describes contextual information that is directly adjacent (either on the left or the right) to the self-recognition patterns. For example, a valid *BedroomNr* must have a word boundary for its direct left context, and its right context must contain the regular phrase "r(oo)?ms?" with possibly several other words in between. This pattern thus recognizes text content such as "3rms" or "1 large room". A *context keyword* is a word that typically appears close to a self-recognition pattern. In our example, the presence of keywords such as "bds", "bdrms", or "bedrooms" provides a strong hint that we should interpret a small number close to these keywords as a bedroom count.

Another part of an ISR declaration is the *exception* clause, which declares an exception from a previously declared pattern. An *exception* clause negates some external-representation or required-context clause by omitting candidates that match the exception pattern. In our example, the *BedroomNr* ISR declaration will recognize the number "1" in both "1 large room" and "1 bath room". But the exception phrase specifies that we should ignore bathrooms when attempting to recognize a bedroom count.

Although we only present one *external representation* and one set of context declarations for *BedroomNr* in our example, in general there could be many. Each concept can have zero or more self-recognition patterns and each self-recognition pattern can have zero or more context restrictions and exceptions.

3   Automated Semantic Annotation

We now show how we use declarative ISRs for automated semantic annotation. Our experiments have shown that by using ISR declarations, we can success-fully annotate many web pages automatically. The following text is a real-world example taken from *Sale Lake City Weekly*.[2]

> **CAPITOL HILL** Luxury 2 bdrm 2 bath, 2 grg, w/d, views, 1700 sq ft. $1250 mo. Call 533-0293

We illustrate the annotated results in Figure 2. Based on the ISR declarations in Figure 1, the annotation system can automatically recognize the first "2" to be a bedroom number, while the second and third "2" are not since neither of them satisfies the right-context requirement. Similarly, the system can automatically recognize the second "2" to be a bathroom number. The third "2" should be part of "2 grg"—a two-car garage. If we have an ISR regular-expression declaration for it in our *ApartmentFeature.lexicon*, it will be recognized along with "w/d"—washer and dryer—as an apartment feature. Based on the same ISR processing, we can annotate the remaining instances of the *aptrent* ontology—"1250" as a monthly rate, "533-0293" as a contact phone number.

Figure 3 shows a screen shot of annotated results from our ISR-based, auto-mated semantic annotation demo.[3] After annotating the page using ISR decla-rations, our demo lets a user move the mouse over annotated text, such as the "2" in the text, and see its detailed annotation information—its ontology name, concept name, record number, and multiple-value index. As Figure 3 shows, the

---

[2] http://www.slweekly.com/
[3] http://www.deg.byu.edu/

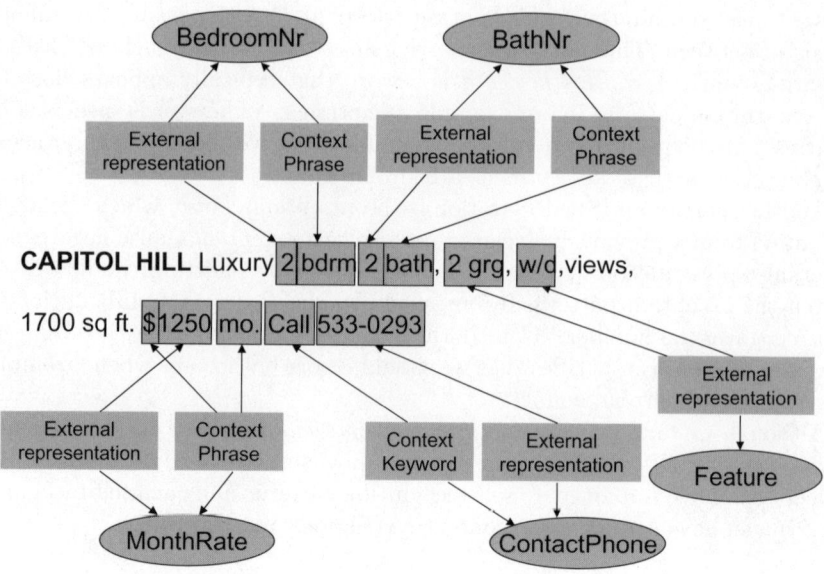

Fig. 2. Example of Apartment-Rental Annotation Results

highlighted number "2" is about a bedroom number that has been defined in the *aptrent* ontology. The system also automatically produces a table that contains all the annotated data with respect to every record entry in a web page. For some concepts (such as *Feature*) that may contain multiple values in one record, there is a clickable button. By clicking the button we can retrieve a sub-table that contains detailed information. In our example, when we click the "Show" button under *Feature*, "2 grg" and "w/d" appear as Figure 3 shows.

A significant feature of ISR-based annotation is the exclusion of document layout information from ISR declarations. Many current wrappers use web-page layout information, which allows the wrapper to do fast and accurate data recognition (when a target layout matches the defined layout) [12]. However, the primary purpose of layout is for displaying, not for describing data. More importantly, relying on layouts prevents an annotation system from working continually on a web page when its layout changes. A reality of the current web is that page layouts do change periodically, and rewriting or regenerating wrappers to accommodate these changes is costly. Hence, layout-independence, which is also called *resiliency* [12], is a preferred property of automated semantic annotation approaches. With the property of resiliency, an annotation system can automatically be applied to new web pages, independent of their specific layouts. Our premise is that the layout of information is not as important as its content. Our ontology-based, data-extraction engine is resilient, and it generally works well [6,12].

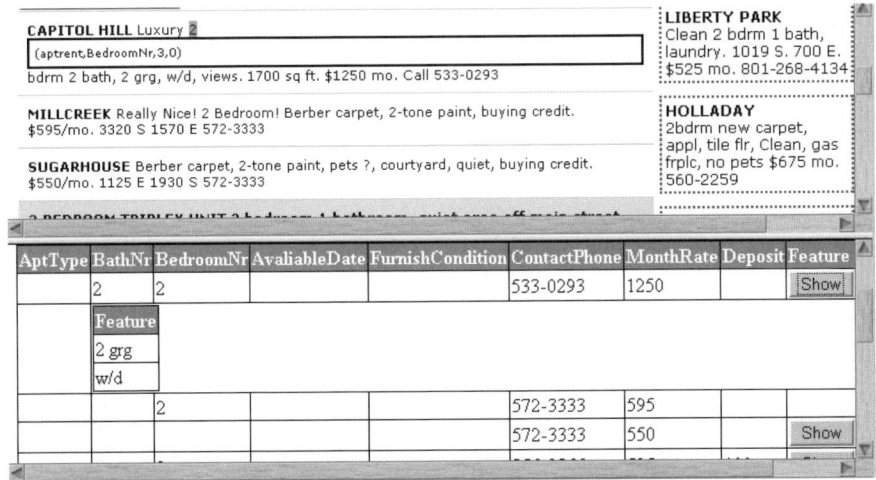

Fig. 3. Screen-shot of ISR-based Annotation Demo

4 OWL-AA: OWL for Automated Annotation

The importance of ISR declarations to semantic annotation, as well as to the Semantic Web in general, is still not widely recognized. Semantic Web researchers besides us (e.g, [2] and [13]) have also observed that the current Semantic Web ontology-language standard, OWL, is not sufficient to resolve all issues for the Semantic Web. We need various extensions of OWL for particular application scenarios. To leverage automated semantic annotation, we thus propose OWL-AA based on our experience with ISR declarations.

4.1 OWL-AA Description

Figure 4 shows the complete RDF schema of OWL-AA. The RDFS class *ISR* provides an abstraction mechanism for grouping different types of instance recognition semantics. Like RDF or OWL class declarations, every ISR declaration is associated with a set of individuals, each of which can have an *ISRvalue* that explicitly provides its content. Each *ISRvalue* has an XML string data type. An ISR individual has no conceptual meaning beyond the represented pattern itself. An individual obtains its conceptual meaning only after it is explicitly bound to an ontology concept. For example, the regular-expression pattern "[1-9]|10" does not have any conceptual meaning beyond the pattern itself. Only after it is bound to a concept such as *BedroomNr*, does it become conceptually meaningful. Two ISR individuals may have the same ISR value, but still mean something different by being assigned to different concepts.

There are three subclasses of *ISR*: *ExternalRepresentation*, *ContextualRepresentation*, and *Exception*. *ExternalRepresentation* declarations formalize the body of an ISR individual. *ContextualRepresentation* declarations formalize the

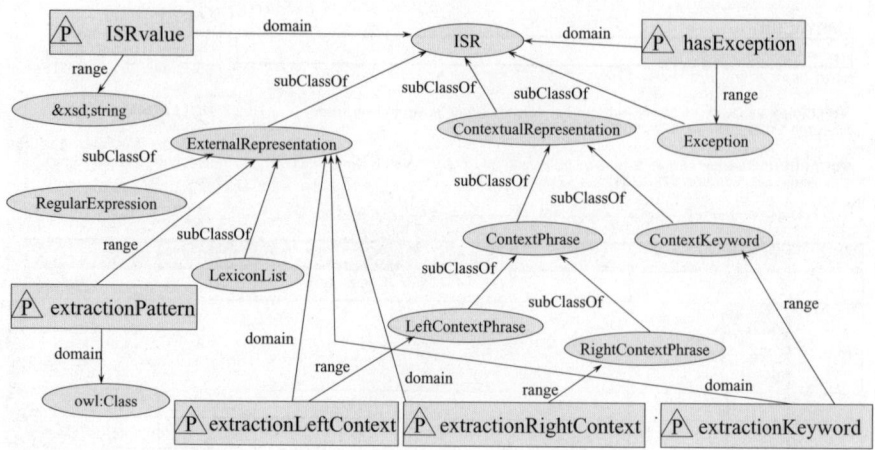

Fig. 4. RDFS Graph of OWL-AA

contextual constraints to the body of an ISR individual. *Exception* declarations formalize exceptions to ISR declarations.

An *ExternalRepresentation* could be either a *RegularExpression* declaration or a *LexiconList* declaration. The reason for two forms is simple convenience. Sometimes it is easier to enumerate matching strings (e.g. country names), and other times it is easier to write a descriptive formula (e.g. the integers between 1 and 10,000,000).

ContextualRepresentation also has two subclasses, *ContextPhrase* and *ContextKeyword* that declare context phrases and context keywords respectively. As a further subdivision, a *ContextPhrase* could be either a *LeftContextPhrase* or a *RightContextPhrase*.

Five properties apply when declaring an ISR. An *extractionPattern* binds an *ExternalRepresentation* to an *owl:Class*. An *extractionLeftContext*, an *extractionRightContext*, and an *extractionKeyword* respectively bind a *LeftContextPhrase*, a *RightContextPhrase*, and a *ContextKeyword* to an *ExternalRepresentation*. The *hasException* property binds an *Exception* to an *ISR*.

We now show how to declare our example in Figure 1 using OWL-AA. We start with a standard OWL ontology that describes the apartment rental domain. We then attach OWL-AA statements to respective concepts, such as *BedroomNr*.

At the beginning of the apartment rental OWL ontology, we first add a new *owlaa* namespace.

xmlns:owlaa="http://www.deg.byu.edu/OWL-AA#"

Within the OWL Class *BedroomNr*, we next add a new property *owlaa:extractionPattern* that associates an ISR declaration to *BedroomNr*.

```
<owl:onProperty rdf:resource="owlaa:extractionPattern" />
<owl:hasValue rdf:resource="BedroomNr-1" />
```

Then we can declare the ISR value and contextual restriction associated to this external representation as follows.

```
<owlaa:RegularExpression rdf:ID="BedroomNr-1"/>
    <owlaa:ISRvalue rdf:datatype="&xsd;#string">[1-9]|10</owlaa:ISRvalue>
    <owlaa:extractionLeftContext rdf:resource="#leftContext-1"/>
    <owlaa:extractionRightContext rdf:resource="#rightContext-1"/>
    <owlaa:extractionKeyword rdf:resource="#contextKeyword-1"/>
</owlaa:RegularExpression>
```

Finally, we add context phrases, context keywords, and exceptions as follows.

```
<owlaa:LeftContextPhrase rdf:ID="leftContext-1"/>
    <owlaa:ISRvalue rdf:datatype="&xsd;#string">\b</owlaa:ISRvalue>
</owlaa:LeftContextPhrase>
<owlaa:RightContextPhrase rdf:ID="rightContext-1"/>
    <owlaa:ISRvalue rdf:datatype="&xsd;#string">
        .*r(oo)?ms?.</owlaa:ISRvalue>
    <owlaa:hasException rdf:resource="#exception-1"/>
</owlaa:RightContextPhrase>
<owlaa:ContextKeyword rdf:ID="contextKeyword-1"/>
    <owlaa:ISRvalue rdf:datatype="&xsd;#string">
        b(r|d)s?|bdrms?|bed(rooms?)?</owlaa:ISRvalue>
</owlaa:ContextKeyword>
<owlaa:Exception rdf:ID="exception-1"/>
    <owlaa:ISRvalue rdf:datatype="&xsd;#string">
        \s.*ba(th)?s?\b.*r(oo)?ms?</owlaa:ISRvalue>
</owlaa:Exception>
```

4.2 Discussion

Syntactically, OWL-AA is attachment-independent with respect to OWL. We add OWL-AA statements without modifying a single line of an existing OWL ontology. We can obtain the original OWL ontology by removing these attached OWL-AA statements.

Semantically, OWL-AA is also attachment-independent with respect to OWL. The addition of ISR declarations to ontology concepts only enriches their details without changing the meaning at the conceptual level. An OWL ontology enriched by OWL-AA can, for example, be used for reasoning as usual. A regular reasoning engine can simply skip attached OWL-AA statements since they are syntactically attachment-independent.

In order to process OWL-AA for automated annotation, we have implemented a syntax convertor that can convert an OWL-AA ontology to our IE ontology. We can then apply our ontology-based annotation tool using the converted ontology. In our implementation, we have used Jena, a standard Semantic Web framework, to help with this ontology conversion. Since we have captured all of an IE-ontology ISR declaration in OWL-AA, OWL-AA ontologies can work as well as the original data-extraction ontologies for automated annotation tasks.

5   Related Work

The idea of ISR declarations is not new. Many researchers have coded ISR declarations into their procedures, such as wrappers. IBM's UIMA (Unstructured Information Management Architecture)[4] is a typical approach that has addressed the issues of sharing and reusing ISRs in procedures. The text analysis engines (TAEs) in the UIMA project are typical examples of small reusable ISR processors. Each TAE does a particular data analysis on source documents, for example, pulling out chemical names and their interactions [7].

Although OWL is the current W3C-recommended web-ontology language, researchers have pointed out that OWL is not sufficient for all Semantic Web work. There have been several proposed extensions of OWL. Among them, two extensions are closer to our work than others: C-OWL and OWL-Eu. Context OWL (C-OWL) [2] provides an extension for ontology mapping that localizes ontology content to allow for limited and totally controlled forms of global visibility. Similar to ours, C-OWL declarations can be expressed independent of ordinary OWL ontologies. OWL-Eu [13] enriches OWL with customized datatypes. The authors point out that "many potential users will not adopt OWL unless [the datatype-support problem] is overcome." In our experience, support for customized datatypes could also improve the performance of data extraction. Unlike our extension, OWL-Eu modifies existing OWL constructs to support customized data types. OWL-AA, on the other hand, is an attachment-independent extension. As a result, OWL-AA appears to be fully compatible with OWL-Eu too, so we could use them together if desired. With enriched customized data types, we expect that we could declare richer ISR representations by attaching our OWL-AA extension to OWL-Eu ontologies.

6   Concluding Remarks

Automated semantic annotation is an important and fundamental problem for the Semantic Web. The key to automated annotation is the ISR declaration. Our epistemological OWL-AA extension augments OWL to formalize ISR declarations. OWL-AA is fully compatible with ordinary OWL, and fully attachable and detachable from standard OWL ontologies. OWL-AA does not introduce new complexity and decidability issues into OWL. Our prototype implementation demonstrates that the OWL-AA extensions work well for our automated semantic annotation system. Embedding epistemological declarations within ontologies is an interesting approach to realizing the Semantic Web, and this approach merits further study.

References

1. Arlotta, L., Crescenzi, V., Mecca, G., Merialdo, P.: Automatic annotation of data extracted from large web sites. In: WebDB 2003. Proc. Sixth International Workshop on the Web and Databases, San Diego, California, pp. 7–12 (June 2003)

---

[4] http://www.research.ibm.com/UIMA/

2. Bouquet, P., Giunchiglia, F., van Harmelen, F., Serafini, L., Stuckenschmidt, H.: Contextualizing ontologies. Journal of Web Semantics 1(4), 325–343 (2004)
3. Dill, S., Eiron, N., Gibson, D., Gruhl, D., Guha, R., Jhingran, A., Kanungo, T., McCurley, K.S., Rajagopalan, S., Tomkins, A., Tomlin, J.A., Zien, J.Y.: A case for automated large scale semantic annotations. Journal of Web Semantics 1(1), 115–132 (2003)
4. Ding, Y., Embley, D.W., Liddle, S.W.: Automatic creation and simplified querying of Semantic Web content: An approach based on information-extraction ontologies. In: Mizoguchi, R., Shi, Z., Giunchiglia, F. (eds.) ASWC 2006. LNCS, vol. 4185, pp. 400–414. Springer, Heidelberg (2006)
5. Embley, D.W.: Programming with data frames for everyday data items. In: Proc. 1980 National Computer Conference, Anaheim, California, pp. 301–305 (May 1980)
6. Embley, D.W., Campbell, D.M., Jiang, Y.S., Liddle, S.W., Lonsdale, D.W., Ng, Y.-K., Smith, R.D.: Conceptual-model-based data extraction from multiple-record web pages. Data & Knowledge Engineering 31(3), 227–251 (1999)
7. Ferrucci, D., Lally, A.: Building an example application with the Unstructured Information Management Architecture. IBM Systems Journal 43(3), 455–475 (2004)
8. Guarino, N.: Formal ontology in conceptual analysis and knowledge representation. International Journal of Human-Computer Studies 43(5/6), 625–640 (1995)
9. Handschuh, S., Staab, S., Ciravegna, F.: S-CREAM Semi-automatic CREAtion of Metadata. In: Gómez-Pérez, A., Benjamins, V.R. (eds.) EKAW 2002. LNCS (LNAI), vol. 2473, pp. 358–372. Springer, Heidelberg (2002)
10. Johnson, P.: Epistemology and ontology: helping the folks at the Web Ontology Workshop Group, XML.com, mantic Web, http://www.xml.com/cs/user/view/cs_msg/635
11. Kiryakov, A., Popov, B., Terziev, I., Manov, D., Ognyanoff, D.: Semantic annotation, indexing, and retrieval. Journal of Web Semantics 2(1), 49–79 (2004)
12. Laender, A.H.F., Ribeiro-Neto, B.A., da Silva, A.S., Teixeira, J.S.: A brief survey of web data extraction tools. SIGMOD Record 31(2), 84–93 (2002)
13. Pan, J.Z., Horrocks, I.: OWL-Eu: Adding customised datatypes into OWL. Journal of Web Semantics 4(1), 29–39 (2006)
14. Vargas-Vera, M., Motta, E., Domingue, J., Lanzoni, M., Stutt, A., Ciravegna, F.: MnM: Ontology driven tool for semantic markup. In: SAAKM 2002. Proc. Workshop Semantic Authoring, Annotation & Knowledge Markup, Lyon, France, pp. 43–47 (July 2002)
15. W3C (World Wide Web Consortium) OWL web ontology language reference, http://www.w3.org/TR/owl-ref/

Making Web Users' Domain Models Explicit by Applying Ontologies

Tarmo Robal[1], Hele-Mai Haav[2], and Ahto Kalja[1,2]

[1] Dept. of Computer Engineering, Tallinn University of Technology
Raja 15, 12618 Tallinn, Estonia
[2] Institute of Cybernetics at Tallinn University of Technology
Akadeemia 21, 12618 Tallinn, Estonia
tarmo@pld.ttu.ee, helemai@cs.ioc.ee, ahto@cs.ioc.ee

Abstract. When searching the web, users have an implicit conceptual model of the domain in their mind. This model is based on their knowledge of the domain and as a rule does not entirely match to the given web site topology. In this paper, we provide a solution to this mismatching problem by making web users' domain models explicit by using ontologies created on the basis of the user profile mining on the web. The provided method and system enable to improve web ontologies and existing web site topologies to become closer to the user preferences as well as derive new and specific ones.

Keywords: Domain ontology modelling, Semantic Web, Web mining.

1 Introduction

How users are browsing and searching the web is heavily dependent on their own conceptual model of the subject area rather than the given web site topology. There exists a well-known problem of possible mismatch between web users domain model and web site topology. Several solutions are provided to solve the problem in the fields of web users profile mining, personalisation and adaptive web [1,2,3,4,5] but there is still lack of works devoted to making users domain models explicit. Research on semantic portals is dealing with this issue by supporting web sites with domain ontologies, which should be commonly shared conceptualisations of a domain [6,7]. Nevertheless, users typically have their own personal views to the domain. A general solution to this problem is to provide a kind of personalisation service.

We claim that implicit domain models of web users, if made explicit, can be valuable for improving web site topologies in order to bring them closer to the user preferences and making the search of the web efficient as well as helping to solve above-stated mismatch problem.

In this paper, we provide an approach to construct web users' domain models by using ontologies created on the basis of user profile mining on the web. Fully automated log system is used for collecting data about preferences and behaviour of users. Based on original log data mining procedure, user preferences are found and concepts corresponding to user profile ontology are automatically extracted from the mining result and added to the ontology description. Using ontology reasoning

J.-L. Hainaut et al. (Eds.): ER Workshops 2007, LNCS 4802, pp. 170–179, 2007.
© Springer-Verlag Berlin Heidelberg 2007

services, extracted user profile concepts are classified under predefined user profile concepts. This allows specifying rather general user profile concepts according to mined user preferences giving as a result a definition of domain ontology of users.

Our contribution can be seen as a new method for learning users' domain models in order to provide personalisation services. Comparing to semantic portals, our approach makes users' domain models (ontologies) explicit instead of concentrating to domain ontologies of content made explicit by domain experts.

Novel ideas of our proposal lie in giving conceptual meaning to web usage mining results by using ontologies and automatic classification of concepts to ontologies via ontology reasoning as a part of the system.

For empirical studies, a real website access data, collected from the usage of the website of the Department of Computer Engineering (DCE) at Tallinn University of Technology (TUT), was used. Examples provided in this paper are based on that data. The Web Ontology Language (OWL) [8] and the description logic reasoner RacerPro [9] are used for ontology construction. Predefined ontologies for web site information and general user profiles are created using the Protégé 3.2 ontology editor [10].

In [11], we have explored the possibilities of mining recommendations for web site tactic adaptation that are partially used in this paper.

The paper is organized as follows. In section 2, we discuss the web mining for users' profile extraction based on locality model. Section 3 provides our method of constructing users domain ontology in a semiautomatic way. In Section 4, we consider the system architecture and implementation, followed by related works discussed in Section 5.

2   Users Profile Mining

Collecting users' interaction data during their web sessions and analysing it in further is inevitable for understanding their behaviour and developing assistance systems. To reason about the collected information, techniques of web mining have to be applied. In general, web mining is divided into three categories: (1) web content mining, (2) structure mining and (3) usage mining [12]. However, some authors also highlight fourth category called user profile mining [13]. Hereby, we will concentrate on the usage and structure mining for site modification and usage characterization through profile mining. The process of user profile mining, we developed, consists of (1) access data capturing, (2) data preparation for locality profiles extraction, and (3) user preferences profile extraction with domain ontology construction.

2.1   The Log System

To be able to reason about users' actions and behaviour on web pages, we applied implicit data collection methods, i.e., a special log system to capture users' interactions with system. The need for special log system emerged as web server logs contain all the accesses to objects on that server and are developed for general purposes. Obviously, these logs are not designed for specific analyses. Moreover, as proven in [14], HTTP traffic logs appear to be flawed and also there exist some major difficulties due to data incompleteness [1,2,15]. One of the many problems with web server log files is that they do not allow identifying visitor sessions [16].

The log system was developed in 2003 being based on a preliminary system introduced in 2002, which allowed to store only some basic properties of actions users performed and was only aimed on general usage statistics. The major improvement towards the new log system was the ability to capture distinct and recurring user sessions, which is also the basis of users' profiles construction. Presently, the log system is an extension to the web systems kernel developed at TUT, enabling to capture data for analyses discussed in [15] and also provides sufficient basis for studies proposed here. From the data captured by the log system [15], we used the following: (1) requested page, (2) client identifier (session based ID), (3) operations performed during a session.

The log system stores user sessions as series of page requests. If a previous session cookie is found, a reference to the prior visit is made. The aim of the log system is to capture a precise snapshot of the web system at the moment it was accessed. The log produced by the system contains raw access data, which is then further processed using the log analyser. Currently the log is about 189.5 MB in raw data size, consisting of more than 1 022 984 records (269 782 sessions) captured from the DCE website, which was chosen for data analyses demonstration as it has been used for the longest period of time. An overview of the system is provided on Fig. 4.

2.2   Data Preparation and Profiles Extraction Based on the Locality Model

The web log contains raw access data, which had to be cleaned from noise and filtered before it could be used for the purpose of our study. Preliminary data cleaning and classification was already performed by the log analyser (Fig. 4), during which website crawler robot records were identified and marked; user sessions were collected and annotated. This was followed by the mining of user navigational paths from session data. We derived from the log 269 782 navigational paths as series of unique numerical identifiers set for each page in the web site management system. These paths were further filtered by removing duplicate neighbouring items, and paths with only one operation; leaving after these two steps a total of 87 953 paths to be dealt with. The process of users profile extraction is shown on Fig. 1.

We believe that if a large number of users frequently access a set of pages, then these pages must be related. This belief is also the basis of our locality model. We define the locality as the nearest activity history of a user within a site as a sliding window of past w actions. Thereby, during the web session, users are moving from one locality to another, which can be represented by the w latest operations (ordered accesses to pages) they made. A user session is defined as a sequence of accessed pages $s=<p_i, p_{i+1},...p_n>$, where $p_i \in P$ and P is the set of all pages. Thus, we were searching for localities L of size w, such as $L=p_j, p_{j+1}, ...p_m$, where $p_j \neq p_{j+1} \neq ... \neq p_m$, p_j is a visited page ID. For each $s_t \in S$ we apply a function $L=CalculateLocality(s_t, w)$.

In order to properly apply the locality model onto the navigational paths, the appropriate size of the window $w \in W$ had to be determined. As w is not a fixed value and depends mainly on the absolute menu depth of a particular web site, an empirical study to discover the best value for the sliding window w was performed. The absolute menu depth for the site used herein is 3. For the evaluation the following attributes were observed: (1) cover percentage for the number of combinations

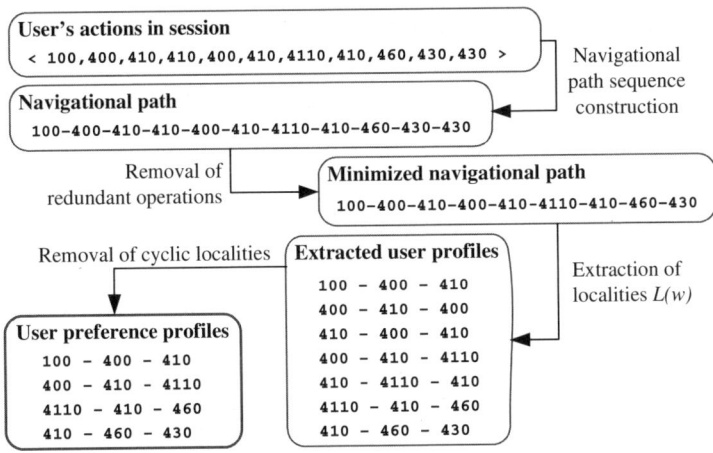

Fig. 1. The process of user preference profiles extraction from navigational paths

computed from the paths, (2) average frequency of finding these combinations in paths, (3) average number of possible localities in path, and (4) the availability of next item for each locality. The attribute "next item" is found as p_{j+1}, if available. The experiments were performed with W={2,3,4,5} (Table 1).

The locality with window size w=3 performed the best in respect to observed parameters (higher values provide better performance for w), and considering the fact that w=2 would be too short for describing recent actions in sense of user profiling.

Table 1. Results of the empirical study for window size w

Properties observed	Studied window size w			
	2	3	4	5
(1) Combination coverage [%]	31.2	35.5	20.7	12.6
(2) Combination frequency	1.1	1.0	1.0	1.0
(3) No of localities in path	6.3	6.6	6.5	5.9
(4) Availability of next item [%]	76.6	77.4	74.1	76.3

Knowing the appropriate value for the window size w of locality L, now allows to extract user locality profiles consisting of w pages accessed as a series of actions $s=<p_i, p_{i+1}, p_{i+2}>$. As cyclic localities do not carry any value of progress (during the locality users end up at the same point they started), further filtering may be needed, to remove such $L= p_1 p_2..p_w$, where $p_1=p_w$. The next step in profiles extraction is the elimination of infrequent localities that are probably the result of random operations. Thus, we are only interested in the top frequent localities, as they represent the user preference profiles. At this point, we have finally extracted user preference profiles from the access logs. However, these profiles do not contain any semantic information about the content represented on the website. Therefore, the extracted user profiles are annotated with metadata (Fig. 4), allowing to automatically add those profiles into ontology description.

3 Constructing Web Users' Domain Models

Our approach to creating web users domain models is starting by manual construction of two initial ontologies: web ontology and predefined user profiles ontology. As the ontology description language we use the Web Ontology Language (OWL) [8], which is the standard language for the Semantic Web [17]. In particular, we use OWL DL (Description Logic variance of OWL) in order to support automatic reasoning. Both predefined ontologies could be created by using, for example, ontology editor Protégé 3.2 [10] as we did for our work. In principle, there is no limitation to editors used.

In addition to predefined concepts (classes), user profile ontology contains also a class for extracted profiles. As described in previous section, the user preferences are found by original log data mining procedure. Based on semantic annotations of web ontology, OWL definitions of corresponding concepts are generated automatically and added to the user profile ontology OWL description file as definitions of subclasses of extracted profiles. Using ontology reasoning, extracted user profile concepts are classified under predefined user profile concepts. This gives as a result ontology definition for users' domain model.

In the following, we present details of modelling of both ontologies as well as the reasoning process.

3.1 Web Ontology

Web ontology should be built up based on the informational needs as concepts from the point of view of potential users. Semantic portals typically rely on domain ontologies to structure knowledge provided on the web site.

We argue that it is not possible to create in advance a web ontology, which entirely matches conceptual interests of potential users but it is possible to learn from web usage how users are browsing the web and then make necessary improvements to the web ontology as well as to the web site topology.

According to our view, web ontology defines a set of concepts captured by web site and their relationships. A domain expert may create this ontology. For instance, a simple university department web site ontology we use in this paper (Fig. 2.) as an example may have classes for department information, for students' information, for news. Each of those classes can have subclasses, e.g., students' information can be divided to subclasses for computer resource information, for lecture schedule, etc.

3.2 User Profiles Ontology

In addition to web ontology, we need to create user profile ontology, where we define rather general user profiles as concepts that capture typical preferences of some well known types of possible web site users as shown on Fig. 2. We call them predefined user profiles. User profiles, which are extracted from mining results, are called extracted profiles. User profiles ontology is defined as superclass of extracted and predefined profiles classes.

The classes in the user profile ontology link the user profile ontology concepts to the web ontology concepts via the hasItem property. The hasItem property represents relationship between user profiles and web ontology concepts. Using this property,

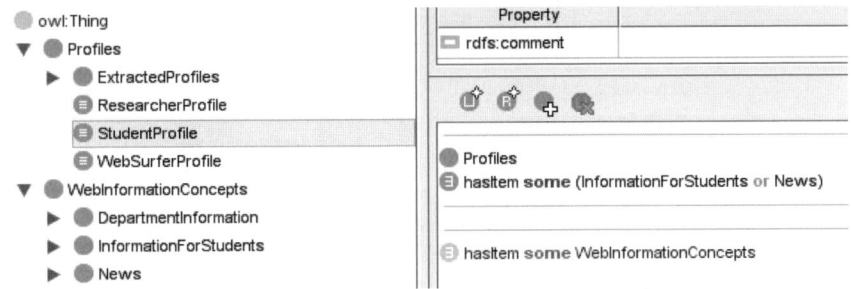

Fig. 2. Definitions of predefined user profile classes (Screenshot from the Protégé ontology editor)

one can define user profile classes that have items from some classes from the web ontology. The hasItem property is transitive and its domain is the class Profiles and its range is the class WebInformationConcepts.

We now use the hasItem property for defining predefined user profiles. In order to do that, we define certain property restrictions. For example, to represent predefined student profile, we add restrictions to the class StudentProfiles in order to specify that the class has at least one item that is from the class InformationForStudents or from the class News (see Fig. 2). This restriction is defined as necessary and sufficient condition in order to make the definition complete for using automatic classification of extracted profiles under predefined user profiles. The restriction in this example is very simple but in general the restriction to predefined profiles can be defined according to the needs.

For each extracted user profile the corresponding OWL class definition is automatically generated. It is added to the predefined user profile ontology OWL file. The generated classes are defined as subclasses of the extracted profile class from profiles ontology.

For class definitions of extracted profiles, we specify closure axiom on the property hasItem. This closure axiom consists of universal restriction with filler that is union of all three components of a given extracted profile. For example, in the case of extracted profile P_410_4110_490, the closure axiom is as follows ∀hasItem (Subject OR SubjectDetails OR InformationForStudents). Closure axioms guarantee that the classification of extracted profiles under predefined profiles will be correct.

Automatic generation of the above descriptions is made possible due to the fact that extracted profiles are semantically annotated on the basis of web ontology (see Section 4).

3.3 Automatic Classification of Extracted User Profiles

The OWL representation of web ontology and profiles ontology described above permits automatic classification of extracted user profiles. In order to provide this reasoning service, we used description logic inference engine RacerPro [9]. New user profile ontology can be derived from the logical definitions in the original asserted ontology. As a result we get an inferred ontology, which shows how the extracted user profiles are classified as subclasses of predefined user profile classes.

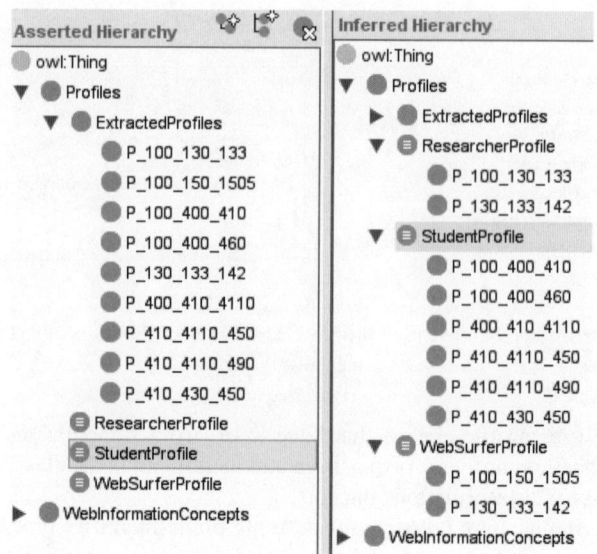

Fig. 3. Inferring domain ontology: (left side) extracted profiles, (right side) result of reasoning. Screenshot from the Protégé ontology editor.

The result of the classification task is the specification of definitions of predefined user profile classes that can be considered as web users' explicit model of the domain. We may see, what concepts constitute users' model of domain, when looking for subclasses of some of predefined profiles. For example, the predefined user profile class StudentProfile was defined in rather general way as seen in Fig. 2 above. After classification, as seen from inferred ontology (Fig. 3), it is specialized by subclasses and we may recognize, what concepts from web ontology really were interesting for users.

It might happen that not all extracted profiles fall under some of predefined user profile classes. This indicates the need for introducing a new profile.

The result of our approach is users' domain model in the form of ontology. When comparing asserted predefined user profiles and inferred ones, we may learn from differences. This can be used for creating web ontologies and based on them web topologies that are closer to users preferences. Obviously, users' preferences are changing over time. Thus, the processes of topology design and learning of users' domain models are iterative.

Our objective is not only to construct users' domain model ontology but also to go further and to find collection of all web ontology concepts different groups of users were interested in. For that we need to derive the concepts of web ontology linked by the user domain ontology via the hasItem property. This will be done in our future work.

4   System Architecture and Implementation

We already have described the important parts of the system in terms of this work. Let us now consider the implementation and architecture of the system in general.

To be able to construct the users' domain ontology, the original web ontology composed by a domain expert and an activity log of users' operations on the site are needed. From the log the modalities of the site access can be mined and processed for users locality profiles construction as discussed in Section 2.2. The statistically relevant locality profiles are then manually annotated with references to the concepts of web ontology to construct user preference profiles. These user preference profiles are then automatically placed into OWL file as subclasses of the predefined class called ExtractedProfiles (Fig. 3). The metadata database is a collection of data, where pages represented in the web site management system are referred to the concepts of web ontology by their URI references. Thus, we have generated a new OWL-file, which comprises extracted user profiles, predefined user profiles and web ontology.

As can be seen (Fig. 4), we have planned to go further with our research and discover web ontology concepts for predefined user profiles with the aim to provide new or refined web site topology for that user by applying topology construction.

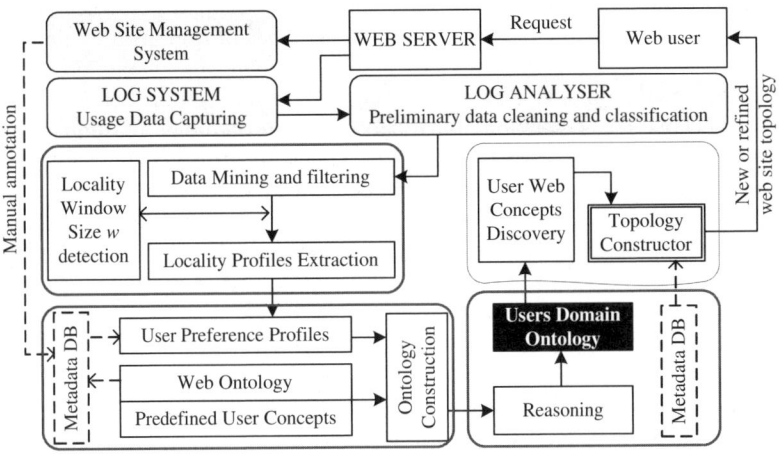

Fig. 4. System architecture and implementation: from page request to topology construction

5   Related Works

The works related to this study fall into three categories: web mining, adaptive web sites and recommender systems, and web personalization. The majority of research is tightly connected with web mining and its techniques, being explored by many researchers [1,3,18]. Web personalization based on usage mining and conceptual relationships between web documents has been explored in [4], data processing and mining for such processes in [19]. Researchers of the domain have also focused on applying several methods for improving web mining techniques by using ontologies. Lim and Sun [20] have proposed to define web pages as concept instances and apply ontology-based structure mining to derive linkage patterns among concepts from web pages for the site's design improvement. However, they analyse user historical data, clustering previously visited pages based on content similarities in order to recommend users similar pages. In our approach, the key factor is the users locality model, which follows the paths users are taking while surfing the web and through which also the user profiles are modelled.

The use of ontologies in terms of topology refinement has also been studied by Mikroyannidis and Theodoulidis [21]. Also, several recommender systems based on web mining with ontological approach to help users to find information easily, have been developed [5,13]. However, in none of the works the approach of our locality model is present. Though, clustering of actions and weighting algorithms are being generally used.

In contrast to semantic portals [6,7], we also use web ontology, but we learn user preferences for making users' domain model explicit for web ontology improvement as well as for providing personalisation services.

6 Conclusion

In this paper, we provided the approach and the system to construct web users' domain models by using ontologies created on the basis of user profile mining on the web. Fully automated log system is used for collecting data about preferences of users. Based on the mined user preferences OWL descriptions of extracted user profiles concepts are created automatically. Using ontology reasoning services, extracted user profile concepts are classified under predefined user profile concepts giving as a result a definition of domain ontology of users.

Our main contribution is in providing a method for learning user preferences for making users' domain model explicit in order to provide personalisation services, improving existing web site domain ontologies and corresponding web site topologies. Our future work will constitute web site topology discovery for a particular set of users applying the users' domain models developed in this work.

Acknowledgments. We appreciate the support of Estonian Information Technology Foundation, Doctoral School in ICT of Measure 1.1 of the Estonian NDP, and the Estonian Scientific Foundation grant no. 5766.

References

1. Srivastava, J., Cooley, R., Deshpande, M., Tan, P.N.: Web Usage Mining: Discovery and Applications of Usage Patterns from Web Data. SIGKDD Explorations 1(2), 12–23 (2000)
2. Mobasher, B., Cooley, R., Srivastava, J.: Automatic Personalization Based on Web Usage Mining. Communications of the ACM 43(8), 142–151 (2000)
3. Kolari, P., Joshi, A.: Web Mining - Research and Practice. IEEE Computing in Science and Engineering - Web Engineering Special Issue 6(4), 49–53 (2004)
4. Eirinaki, M., Lampos, C., Paulakis, S., Vazirgiannis, M.: Web Personalization Integrating Content semantics and Navigational Patterns. In: 6th ACM International Workshop on Web Information and Data Management, Washington DC, USA, pp. 72–79. ACM Press, New York (2004)
5. Middleton, S., De Roure, D., Shadbolt, N.: Capturing Knowledge of User Preferences: Ontologies in Recommender Systems. In: 1st Int. Conference on Knowledge Capture, pp. 100–107. ACM Press, New York (2001)

6. Hartmann, J., Sure, Y.: An Infrastructure for Scalable, Reliable Semantic Portals. IEEE Intelligent Systems 19(3), 58–65 (2004)
7. Jin, Y., Decker, S., Wiederhold, G.: OntoWebber: Model-Driven Ontology-Based Web Site Management. In: SWWS 2001. 1st Int. Semantic Web Working Symp., pp. 529–547 (2001)
8. OWL Web Ontology Language, http://www.w3.org/TR/owl-features/
9. RacerPro, http://www.racer-systems.com/
10. The Protégé Ontology Editor and Knowledge Acquisition System, http://protege.stanford.edu/
11. Robal, T., Kalja, A.: Applying User Profile Ontology for Mining Web Site Adaptation Recommendations. In: Ioannidis, Y., Novikov, B., Rachev, B. (eds.) ADBIS 2007. LNCS, vol. 4690, Springer, Heidelberg (2007)
12. Kosala, R., Blockeel, H.: Web Mining Research: A Survey. ACM SIGKDD Explorations 2(1), 1–15 (2000)
13. Li, Y., Zhong, N.: Mining Ontology for Automatically Acquiring Web User Information Needs. IEEE Transactions on Knowledge and Data Engineering 18(4), 554–568 (2006)
14. Davison, B.: Web Traffic Logs: An Imperfect Resource for Evaluation. In: INET 1999. Proceedings of Ninth Annual Conference of the Internet Society, San Jose, CA (1999)
15. Robal, T., Kalja, A., Põld, J.: Analysing the Web Log to Determine the Efficiency of Web Systems. In: DB&IS'2006. Proc. of the 7th International Baltic Conference on Databases and Information Systems, Technika, Lithuania, pp. 264–275 (2006)
16. Kimball, R., Margy, R.: The data warehouse toolkit: the complete guide to dimensional modelling. John Wiley & Sons, England (2002)
17. Semantic Web, http://www.w3.org/2001/sw/
18. Berendt, B., Hotho, A., Mladenic, D., Someren, M., Spiliopoulou, M., Stumme, G.: A Roadmap for Web Mining: From Web to Semantic Web. In: Berendt, B., Hotho, A., Mladenić, D., van Someren, M., Spiliopoulou, M., Stumme, G. (eds.) EWMF 2003. LNCS (LNAI), vol. 3209, pp. 1–22. Springer, Heidelberg (2004)
19. Baglioni, M., Ferrara, U., Romei, A., Ruggieri, S., Turini, F.: Preprocessing and Mining Web Log Data for Web personalization. In: Cappelli, A., Turini, F. (eds.) AI*IA 2003. LNCS, vol. 2829, pp. 237–249. Springer, Heidelberg (2003)
20. Lim, E-P., Sun, A.: Web Mining – the Ontology Approach. In: Int. Advanced Digital Library Conference (IADLC'2005), Nagoya University, Nagoya, Japan (2005), http://iadlc.nul.nagoya-u.ac.jp/archives/IADLC2005/Ee-Peng.pdf
21. Mikroyannidis, A., Theodoulidis, B.: Web usage Driven Adaptation of the Semantic Web. In: Proceedings of UserSWeb: Workshop on End User Aspects of the Semantic Web, Heraklion, Crete, pp. 137–147 (2005)

Provability-Based Semantic Interoperability
Via Translation Graphs

Joshua Taylor, Andrew Shilliday, and Selmer Bringsjord

Rensselaer AI & Reasoning (RAIR) Lab
Departments of Cognitive and Computer Science
Rensselaer Polytechnic Institute (RPI), Troy NY 12180, USA
{tayloj,shilla,selmer}@rpi.edu

Abstract. *Provability-based semantic interoperability* (PBSI) is a kind
of interoperability that transcends mere syntactic translation to allow
for robust, meaningful information exchange across systems employing
ontologies for which mappings or matchings may not exist, and which
can be evaluated by provability-based (PB) queries. We introduce a system of *translation graphs* to formalize the relationships between diverse
ontologies and knowledge representation and reasoning systems, and to
automatically generate the translation axioms governing PB information exchange and inter-system reasoning. We demonstrate the use of
translation graphs on a small number of simple systems to achieve interoperability.

Keywords: translation graphs, provability-based semantic interoperability.

1   What is Semantic Interoperability?

The proliferation of knowledge-rich systems has led to the creation of myriad
intelligent systems possessing diverse reasoning capabilities. Unfortunately, cooperative efforts among these systems are hindered by lack of a common representation scheme or effective general methods for information exchange. Ideally,
these systems could reason about their peers' representation schemes and work
out a way to exchange information automatically—a capability well beyond the
abilities of current systems.

Many systems today achieve various levels of interoperability and information
exchange using ontology mapping [1] and schema matching [2]. These techniques
are useful and have achieved high levels of information sharing, but cannot capture all the relationships that semantic interoperability requires.

In the general tradition of logicist AI and cognitive science [3,4,5], and specifically in the tradition of logic-based semantic interoperability [6,7], we maintain
that semantic interoperability can be evaluated only with respect to provability-
based queries. This stems from the fact that ontology mapping and schema
matching cannot always capture asymmetry of translation [8], nor can information from a source ontology always be translated into a corresponding form in a

J.-L. Hainaut et al. (Eds.): ER Workshops 2007, LNCS 4802, pp. 180–189, 2007.

target ontology, even if the information has semantic consequences in the target ontology.

Ontologies contain complex relationships among their own terms, and any approach to semantic interoperability must be able to capture not only these, but also the relationships between between multiple ontologies. A system which does not use a sufficiently expressive formalism or language to describe these relationships is inherently specialized and cannot be used for general applications.

Furthermore, consumers of the products of semantic interoperability should have access to the *justifications* that bring about those products. Consumers should have, then, in a schema-mapping approach, access to the mapping itself, in an axiomatic approach, access to the axioms, and in a *provability*-based approach, access to the *proofs*. Ideally, the proofs would be couched in a format that is readily understood by non-specialists; e.g., proofs in natural language are far superior to resolution based proofs. Herein we describe a new brand of PBSI that meets the desiderata just enumerated.

2   The Need for Semantic Interoperability

In a Truly Useful Semantic Web. To achieve a useful Semantic Web, information on the web must (1) be structured in a meaningful way and (2) information from different systems must be able to be combined easily and *meaningfully*. (1) is being addressed as more and more information is stored in databases, and by the adoption of regular markup languages such as XML and XHTML. (2) is only happening partially. Service-oriented architectures *are* sharing information meaningfully, but need complete knowledge of the ontologies employed by the systems involved. When systems with web presence can share information without having to have extensive knowledge of their peers' ontologies or schemata, web-based agents can be built that are capable of deep reasoning and planning [9].

In the Defense and Intelligence Communities. The defense and intelligence communities have, for some time, recognized the need for semantic interoperability, and have sponsored research in tools and languages to address this need. Some results of this research are the DARPA Agent Markup Language [10], DAML, and DAML+OIL, for the markup of information and for the description of ontologies. Though languages such as KIF [11] and Common Logic [12] have been developed for describing the *relationships* between ontologies and for the exchange of information between ontologies, in 2005 the Disruptive Technology Office sponsored the Interoperable Knowledge Representation for Intelligence Support (IKRIS) workshop [13], which resulted in the IKRIS Knowledge Language (IKL), an extension to Common Logic that addresses specific needs of the intelligence community.

3   Relevant Past Approaches

We review a number of past approaches to the problem of semantic interoperability. We note how these approaches fare with respect to the aforementioned desiderata; in particular, whether: an approach is logically based, asymmetry of translation is preserved, information in a foreign ontology can influence a query in a native ontology even when the foreign information cannot be directly translated, and the quality of available justifications is sufficiently high.

When the subject domains and vocabularies of the ontologies to be related are similar and the information represented within them is not too complex, *schema matching* can be effective in translating information from one ontology to another. With schema matching, corresponding terms from the ontologies are selected, and information from one is recast in another. There are automated tools that aid in schema matching [2]. Evaluating whether a schema matching is correct can be difficult, particularly if the matching has been generated (even partially) automatically. A schema matching can be provided as primitive justification for results. It seems difficult, with schema matchings, to capture semantic influence when information translation is not possible.

The use of *schema morphisms* to map the sentences of one ontology to sentences of another allows for more complex transformations between ontologies. This approach can be used when ontologies are treated as *institutions* [14]. Within this framework, it is possible to determine whether a schema morphism is correct [15], and to impose constraints that capture some of the asymmetry of translation and semantic influence. Morphisms are not trivial to construct, but can capture relationships between ontologies using different logics [16]. Signature morphisms are expressed with a different formalism and notation than the ontologies themselves, however, and so the justification for a particular translation requires human intervention or specialized reasoning outside of the ontologies themselves.

Simple syntactic manipulation of sentences does not afford the meaningful translations that are desired. In fact, to answer queries expressed in a query ontology using information from various source ontologies often requires making use of information from many source ontologies. Unfortunately, sentences in a source ontology that have semantic consequences in a target ontology cannot always be translated into the target ontology [8]. Semantic interoperability is still attainable, however, by relating the ontologies logically, and evaluating queries with respect to provability. Ontologies can be related axiomatically using *lifting axioms* [6] or by merging the ontologies to be related and expressing *bridging axioms* in the new merged ontology [7].

The techniques reviewed above have been used in real applications and have successfully enabled varying levels of interoperability. No individual system, however, possesses all of the necessary qualities for top-notch semantic interoperability. Building on these excellent foundations, we believe that our system of translation graphs takes a step closer to the ideal.

4   Formal Preliminaries

The work herein described is, clearly, logic-based, and partakes of the paradigms of logic-based AI [3], cognitive science [4], and computational cognitive modeling [5].

We treat ontologies as pairs of the form $\langle \Sigma, \Phi \rangle$ where Σ is a signature in a many-sorted logic, and Φ is a set of sentences in Σ. While many-sorted logic is not employed by all ontology designers, it is appropriate for describing many ontological constructs including modalities, has a impressive history within computer science and mathematics, and is reducible to standard first-order logic [17].

A *sort* is a domain, a universe, or a set of objects. There is a global set of sorts, S^*. Generally, every signature will contain a sort corresponding to truth values. In traditional logics, this sort is the set $\{\mathbf{true}, \mathbf{false}\}$, but this needn't be the case. Many-valued logics, for instance, will use a different sort for truth values. A *functor* f is a function $s_0 \times \ldots \times s_{n-1} \to s_n$ where $s_0, \ldots, s_n$ are elements of S^*. $\langle [s_0, \ldots, s_{n-1}], s_n \rangle$ is the *rank* of f and denoted $Rank(f)$.

A *signature* Σ is a tuple $\langle \sigma, \phi \rangle$ where σ is a subset of S^*, called the sorts of Σ and ϕ is a partial injective function from string-rank pairs to functors of the same rank. The range of ϕ is the set of functors of Σ. There is a restriction on ϕ that for every functor f among Σ's functors, each sort in f's rank is one of Σ's sorts.

A well-formed term of Σ has a particular *interpretation* which denotes the application of corresponding functors to their arguments. E.g, if man(Socrates) is sentence of Σ_1 and human(Sokrates) is a sentence of Σ_2, but both Σ_1 and Σ_2 map man and human, respectively, to the same functor f of rank $\langle [s_1], s_0 \rangle$, and Socrates and Sokrates to the same functor g of rank $\langle [], s_1 \rangle$, then the two sentences have the same interpretation.

5   Ontology Modifications

A number of operations can be defined on signatures which correspond to incremental modifications that might be performed on the signatures of ontologies. Four primitive operations on signatures are defined by the following equations

$$AddSort(s, \langle \sigma, \phi \rangle) = \langle \sigma \cup \{s\}, \phi \rangle \tag{1}$$

$$RemoveSort(s, \langle \sigma, \phi \rangle) = \langle \sigma - \{s\}, \phi \rangle \tag{2}$$

$$AddFunctor(w, f, \langle \sigma, \phi \rangle) = \langle \sigma, \phi \cup \{\langle \langle w, Rank(f) \rangle, f \rangle\} \rangle \tag{3}$$

$$RemoveFunctor(w, r, \langle \sigma, \phi \cup \{\langle \langle w, r \rangle, f \rangle\} \rangle) = \langle \sigma, \phi \rangle \tag{4}$$

subject to several restrictions. $RemoveSort(s, \langle \sigma, \phi \rangle)$ is undefined if any of the functors of $\langle \sigma, \phi \rangle$ use s. $AddFunctor(w, f, \langle \sigma, \phi \rangle)$ is undefined if $\langle w, arity(f) \rangle$ is already mapped to some functor. $RemoveFunctor(w, r, \langle \sigma, \phi \rangle)$ is undefined if ϕ does not map $\langle w, r \rangle$ to any functor.

With the primitive methods, simple ontologies can be constructed that specify only the vocabulary of a language. However, ontology consists not only in vocabulary, but also in the *meaning* of the vocabulary and the relationships among these terms. As a result, many knowledge representation languages include forms analogous to Athena's [18] `define-symbol` for defining symbols *axiomatically*. For instance, MatGrandmotherOf(x), denoting the maternal grandmother of x can be defined in KIF using MotherOf(x) by (`deffunction MatGrandmotherOf (x) := (MotherOf (MotherOf x)))`.

Both classical mathematicians and logicians along with modern knowledge representation language designers have devoted a great amount of time to the subject of the forms that can be used in axiomatic definitions. Some definitions may be implemented as macro-like substitutions, while in other cases, the entire axiom must remain available for subsequent reasoning [11, Ch. 11].

6   Translation Graphs

We implemented a prototype of the structures and modifications described in the previous section, thereby providing a framework in which to perform natural ontology-related activities, such as ontology construction and mapping. Ontology construction becomes easy: Starting from an empty signature (i.e., a signature with no sorts or functors), existing ontologies can be recreated by adding the ontology's sorts, and then relations and function symbols. These reconstructed ontologies can then be related by adding the functors of one ontology to another with axiomatic definitions. Displaying the process graphically inspired translation graphs.

After initial experiments demonstrated the feasibility of this approach, we realized that the process could be used to describe the interoperability in the IKRIS workshop and experiments in interoperability between robust software systems, such as Oculus' GeoTime [19,20], SUNY Albany's HITIQA [21], Attempto Controlled English [22,23], and the RAIR lab's own Slate [24] and Solomon [25].

A *translation graph* is a directed graph whose vertices are signatures, and whose edges denote axiomatic relationships between the signatures of the graph. If signatures Σ_i and Σ_j are vertices of some translation graph and the edge $\langle \Sigma_i, \Sigma_j \rangle$ is in the graph, there is information associated with it that describes how information represented in an ontology employing Σ_i can be used in an ontology employing Σ_j. This property is transitive, and so a Σ_u, Σ_v *path* contains information for using information under Σ_u in Σ_v.

7   An Example

We present an example to show that translation graphs can be used to enable interoperability between ontologies whose subject domains intersect but are not identical, that queries can be answered with information from multiple ontologies, and that the information used to answer the query is not representable in all of the ontologies presented. (For the sake of readability and conciseness, we will ignore issues such as namespaces and the use of fundamental datatypes such as strings and numbers.)

We consider four separate software systems operating with four distinct ontologies amongst which information will be shared.

The first two systems are social networking programs which represent information about phone calls. The first system, $\mathcal{A}$, keeps records of the form $\mathsf{Called}(x, y)$ to denote that x called y, where x and y are names of individuals. The second system, $\mathcal{B}$, uses $\mathsf{CalledBy}(x, y)$ to denote that x was called by y, where x and

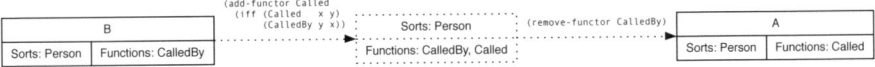

Fig. 1. Ontologies $\mathcal{A}$ and $\mathcal{B}$ are related

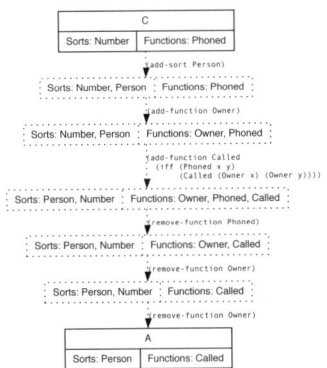

Fig. 2. Phone company $\mathcal{C}$ is related to $\mathcal{A}$

y are names of individuals. $\mathcal{A}$ and $\mathcal{B}$ can be related with the primitive operations described earlier; the result is shown in Figure 1. The function Called is added to $\mathcal{B}$ with an axiomatic definition, yielding an intermediate signature. CalledBy is removed from the intermediate signature, resulting in $\mathcal{A}$. Tracing the path between the ontologies and collecting axioms along the way gives all the information needed to use information from one ontology in the other.

The axiomatic definition between $\mathcal{A}$ and $\mathcal{B}$ is a biconditional and could be optimized as a rewriting rule. That is, assertions in one ontology could be *rewritten* in terms of the other's vocabulary. The translation here is symmetric, and could be handled by schema matching tools.

Next, we introduce a cellular phone company database $\mathcal{C}$ which has information about phone calls made on the cellular network, and keeps records of the form Phoned(n_1, n_2) where n_1 and n_2 are phone numbers between which calls have been placed. Figure 2 illustrates the relationship between $\mathcal{C}$ and $\mathcal{A}$.

While no individual link in Figure 2 is particularly complicated, the addition of the axiomatically defined Owner deserves special note. Owner(x) denotes the person who owns a phone number x. Owner is present in neither $\mathcal{A}$ nor $\mathcal{C}$, but its use in relating them does seem clear: Owner functions as a sort of semantic placeholder. Without an interpretation of Owner, information exchange would not be possible; there would be information missing. However, the use of translation graphs has allowed us to capture *what* is needed to exchange information meaningfully.

Another possibility is that Owner may stand for a non-logical function. For instance, in the process of exchanging information, occurrences of Owner(x) might be replaced with the results of a database lookup or some procedural

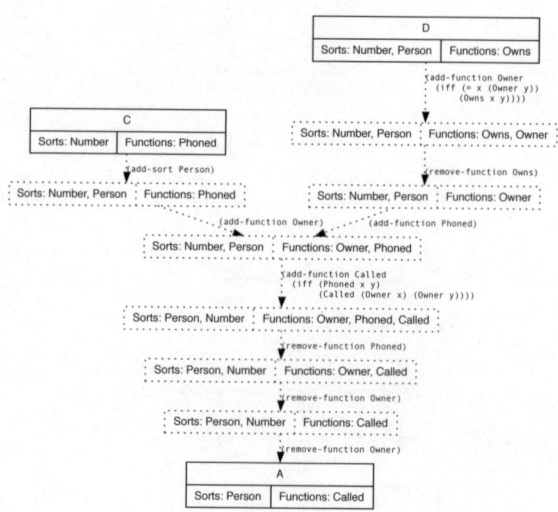

Fig. 3. The information in $\mathcal{D}$ is made available to $\mathcal{A}$ and $\mathcal{C}$

transformation (e.g., if phone numbers were a function of the characters comprising a person's name).

In this example, however, we integrate the database of a reverse phone number lookup system, $\mathcal{D}$. In this case, the information that $\mathcal{D}$ provides is not phone records, but pairs of phone numbers and their owners' names. $\mathcal{D}$ records that $\mathsf{Owns}(x, y)$ when x, a person, owns the phone number y. The integration, shown in Figure 3, is straightforward.

Having connected $\mathcal{A}$ with $\mathcal{B}$, and then $\mathcal{A}$, $\mathcal{C}$, and $\mathcal{D}$, enough work has been done to yield the translation graph shown in Figure 4. The graph can be used to describe the relationships between the ontologies, and the axiomatic relationships needed to answer queries about the contents of the four knowledge bases can be automatically extracted from it.

Remarks. In such a small example, the overall *structure* of the translation was not given much thought. In real systems, however, engineers must consider the implications of their translation structures. For example, in some situations, an interlingua and intertheory may be preferred, or in some cases it may not be appropriate or feasible [26,27]. However, we present translation graphs without expressing preference among these possible architectures; translation graphs general enough to be applied in an architecture-agnostic manner.

With the translation graph as given, it would be possible to run automated reasoners directly on the union of the knowledge bases and all the axioms extracted from the edges of the graph. Of course, intractability and undecidability make this a tricky technique, but there is an interesting parallel to Green's method. Green's method extracts plans that achieve particular goals from proofs that such plans exist [28]; with the naive method above, interoperability and translation are achieved as a *side effect* of automated theorem proving.

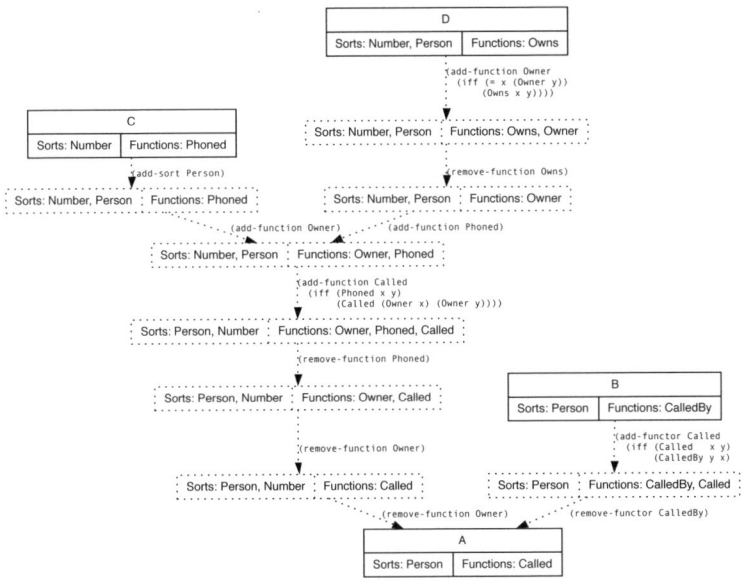

Fig. 4. The final translation graph of the relationships between the systems

8   Conclusion and Future Work

Automaticity. The ultimate dream of this sort of R&D is full automaticity. Following a divide and conquer approach, translation graphs allow for the automatic production of bridging axioms. So, if translation graphs could be automatically produced, the dream would be reality. We are investigating the application of automatic programming [29] toward this goal. More immediately, some of the approaches in automated schema matching could be applied.

Sophisticated Ontology Representation. We built translation graphs with the signatures of many-sorted logic as nodes, for flexibility and convenience of expression, though such graphs lack some desirable features such as subsorting, sort hierarchies, and a standard language for describing the signatures themselves. There has been a great deal of research in what kind of reasoning [30] must be performed over ontologies [31], and there are many languages, such as RDF, DAML, and OWL, designed for the purpose of ontology description. Building translation graphs from ontologies represented in these languages would allow us to work with many ontologies already constructed and in use today.

Categorizing Axiomatic Definitions. From certain types of axiomatic definitions we can extract rewriting rules (inline translations); indeed, to make the translation graph approach scale well, optimizations such as inline translations are almost certainly necessary. We believe more sophisticated rewriting rules and other types of procedures can be developed by examining *paths* in a translation graph, and will be pursuing this line of work.

Acknowledgements. With much gratitude, we acknowledge the financial support provided by the Disruptive Technology Office (DTO), contract # N61339-06-C-0128 to RPI, A-SpaceX program.

References

1. Choi, N., Song, I.Y., Han, H.: A survey on ontology mapping. SIGMOD Rec. 35(3), 34–41 (2006)
2. Wang, G., Goguen, J.A., Nam, Y.K., Lin, K.: Critical points for interactive schema matching. In: Yu, J.X., Lin, X., Lu, H., Zhang, Y. (eds.) Advanced Technologies and Applications: Proceedings of Sitch Asia Pacific Web Conference, Hangzhou, China, pp. 654–664. Springer, Heidelberg (2004)
3. Bringsjord, S., Ferrucci, D.: Logic and artificial intelligence: Divorced, still married, separated..? Minds and Machines 8, 273–308 (1998)
4. Bringsjord, S., Yang, Y.: Representations using formal logics. In: Nadel, L. (ed.) Encyclopedia of Cognitive Science, vol. 3, pp. 940–950. Nature Publishing Group, London, UK (2003)
5. Bringsjord, S.: Declarative/logic-based computational cognitive modeling. In: Sun, R. (ed.) The Handbook of Computational Cognitive Modeling, Cambridge University Press, Cambridge (forthcoming)
6. Buvač, S., Fikes, R.: Proceedings of the fourth international conference on Information and knowledge management. In: CIKM 1995. Proceedings of the fourth international conference on Information and knowledge management, pp. 340–347. ACM Press, New York, NY, USA (1995)
7. Dou, D., McDermott, D., Qi, P.: Ontology Translation by Ontology Merging and Automated Reasoning. In: Ontology Translation by Ontology Merging and Automated Reasoning, Birkhäuser Basel, pp. 73–94 (2005)
8. Dou, D., McDermott, D.: Deriving axioms across ontologies. In: AAMAS 2006. Proceedings of the fifth international joint conference on Autonomous agents and multiagent systems, pp. 952–954. ACM Press, New York, NY, USA (2006)
9. Hendler, J.: Agents and the Semantic Web. IEEE Intelligent Systems, 30–37 (2001)
10. DARPA Agent Markup Program: DARPA Agent Markup Language Homepage (2003), http://www.daml.org/
11. Genesereth, M.R., Fikes, R.E.: Knowledge Interchange Format Version 3 Reference Manual (1997)
12. Common Logic Working Group: Common Logic Standard (2007), http://cl.tamu.edu/
13. MITRE: IKRIS (2007), workshop site http://nrrc.mitre.org/NRRC/ikris.htm
14. Goguen, J.A., Burstall, R.M.: Introducing institutions. In: Clarke, E., Kozen, D. (eds.) Logics of Programs. LNCS, vol. 164, pp. 221–256. Springer, Heidelberg (1984)
15. Goguen, J.A.: Information integration in institutions. In: Moss, L. (ed.) Paper for Jon Barwise memorial volume Moss (2004)
16. Goguen, J.A.: Data, Schema, Ontology and Logic Integration. Logic Journal IGPL 13(6), 685–715 (2005)
17. Manzano, M.: Extensions of First Order Logic. Cambridge University Press, Cambridge (1996)

18. Arkoudas, K.: Athena (2005), http://www.cag.csail.mit.edu/~kostas/dpls/athena
19. Kapler, T., Harper, R., Wright, W.: Correlating Events with Tracked Movements in Time and Space: A GeoTime Case Study. In: IA 2005. Proceedings of the 2005 International Conference on Intelligence Analysis, McLean, VA, USA (2005)
20. Chappell, A., Bringsjord, S., Shilliday, A., Taylor, J., Wright, W.: Integration Experiment with GeoTime, Slate, and VIKRS. ARIVA Principle Investigator Meeting Handout (2007)
21. Strzalkowski, T., Small, S., Hardy, H., Yamrom, B., Liu, T., Kantor, P., Ng, K.B., Wacholder, N.: HITIQA: A Question Answering Analytical Tool. In: Conference Proceedings of the 2005 International Conference on Intelligence Analysis, McLean, VA, USA, MITRE, Sponsored by the Office of the Assistant Director of Central Intelligence for Analysis and Production (2005)
22. Fuchs, N.E., Kaljurand, K.: Attempto Controlled English: Language, tools and applications. Lecture / Presentation (2006)
23. Bringsjord, S., Arkoudas, K., Clark, M., Shilliday, A., Taylor, J., Schimanski, B., Yang, Y.: Reporting on some logic-based machine reading research. In: Proceedings of the 2007 AAAI Spring Symposium on Machine Reading (2007)
24. Bringsjord, S., Shilliday, A., Taylor, J.: Slate (2007), http://www.cogsci.rpi.edu/slate/
25. Bringsjord, S., Clark, M., Shilliday, A., Taylor, J.: Solomon (2007), http://www.cogsci.rpi.edu/solomon/
26. Barker-Plummer, D., Greaves, M.: Architectures for Heterogeneous Reasoning on Interlinguae (1998)
27. Menzel, C.: Common Logic Standard. In: Santa Fe, Metatdata Forum Symposium on Ontologies (2003) Presentation of the case for Common Logic
28. Genesereth, M.R., Nilsson, N.J.: Logical foundations of artificial intelligence. Morgan Kaufmann Publishers Inc., San Francisco, CA, USA (1987)
29. Rich, C., Waters, R.C.: Automatic programming: Myths and prospects. Computer 21(8), 40–51 (1988)
30. Antoniou, G., Franconi, E., van Harmelen, F.: Introduction to semantic web ontology languages. In: Eisinger, N., Małuszyński, J. (eds.) Reasoning Web. LNCS, vol. 3564, Springer, Heidelberg (2005)
31. Smith, B.: The basic tools of formal ontology. In: Guarino, N. (ed.) Formal Ontology in Information Systems, pp. 19–28. IOS Press, Amsterdam (1998)

Preface to QoIS 2007

Samira Si-Saïd Cherfi[1] and Geert Poels[2]

[1] CEDRIC-CNAM, France
[2] Ghent University, Belgium

Welcome to the proceedings of the 3rd International Workshop on Quality of Information Systems, which was held in conjunction with the 26th International Conference on Conceptual Modeling (ER2007) at Auckland, New Zealand.

Nowadays, rapid technological advances imply a greater use of information systems in organizations. Furthermore, strategic and vital decisions rely heavily on the quality of information provided by these systems. As a consequence, quality is emerging as a key issue for information systems researchers and practitioners. Information system quality aims at the evaluation of its main components, i.e., system quality, data quality, information quality as well as model quality and method quality. Ongoing research encompasses theoretical aspects including quality definition and/or quality models. These theoretical contributions lead to methods, approaches and tools for quality measurement and/or improvement. Most approaches focus on specific application domains, such as model-driven systems engineering, Web application engineering, data warehousing, decision support, ontology, etc.

The workshop that we organize is the premium forum for the presentation and discussion of the most recent research related to theoretical, empirical and pragmatic aspects of information system quality. Special attention is paid to research on the relationship between conceptual modeling and quality assurance.

This year's workshop attracted 11 papers, 9 of which resulted in full-paper submissions. After a rigorous peer review process, the Program Committee finally selected only three papers (acceptance rate 33%). The accepted papers were further improved by their authors based on the reviewer comments, before being included in the ER2007 workshop proceedings.

We would like to thank all the authors who submitted papers to our workshop. We would also like to thank the Program Committee members and all the referees who gave up their valuable time to review the papers and helped us in preparing the workshop program. We would also like to thank J.C. Barrez for maintaining the workshop Web site since 2005. Finally, our thanks go out to the Organizing Committee of the ER2007 conference for recognizing the relevance of this workshop.

We hope this workshop will be an opportunity for stimulating exchange between researchers and practitioners about the role of conceptual modeling in information system quality assurance.

PQM vs. BPQM: Studying the Tailoring of a General Quality Model to a Specific Domain

Coral Calero[1], Cristina Cachero[2], Julio Córdoba[2,3], and Mª Ángeles Moraga[1]

[1] Grupo ALARCOS. University of Castilla-La Mancha (Spain)
[2] Departamento de Lenguajes y Sistemas Informáticos. University of Alicante (Spain)
[3] Centro de Investigacón en Informática Aplicada. Universidad Latina de Costa Rica
{Coral.Calero, MariaAngeles.Moraga}@uclm.es,
{ccachero, jcordoba}@dlsi.ua.es

Abstract. In this paper a comparative study carried out between a generic model for portal quality (PQM) and a specific model for eBanking portal quality (BPQM) is presented with the aim to identify the differences and to justify why these have appeared. The goal is to detect gaps in BPQM in order to think about including them in future versions.

Keywords: Web portals, e-Banking portals, quality model.

1 Introduction

Web Portals are emerging Internet-based applications, enabling access to different sources (providers) through a single interface (Mahdavi et al., 2004). They provide personalization, single sign on and content aggregation from different sources, as well as hosting the presentation layer of Information Systems (Java Community Process, 2003). Moreover, they can help users to find the desired information, service or product from among a (large) number of providers effectively, without having to navigate through these one by one (Mahdavi et al., 2004).

The primary objective for the development of portals may vary from one organization to another (Hazra, 2002) although, in general, it can be formulated as 'creating a working environment where users can easily navigate to find the information they specifically need to perform their operational or strategic functions quickly and to make decisions' (Collins, 2001). In order to assure the fulfilment of this objective, certain software quality criteria should be met.

The term *"software quality"* is used to assess both processes and products. In our case, a product-based view is adopted based on the opinion that the clearer we are about what to achieve in terms of product quality, the easier it will be to tune the process accordingly. This view is also backed by ISO (2001), who stated that evaluating a product can provide feedback for the improvement of a process.

Software product quality is evaluated in accordance with a *"quality model"*. Quality models should consider criteria that satisfy the needs of the developers, maintainers, buyers and end users (ISO, 2001). As for their objectives, these models should be used for building better products and to assess and ensure their quality.

J.-L. Hainaut et al. (Eds.): ER Workshops 2007, LNCS 4802, pp. 192–201, 2007.

Building a quality model is a complex undertaking. Normally, the software product quality is broken down hierarchically into characteristics and sub-characteristics which can be used as a checklist. However, trying to define a quality model that fits any software product regardless of the differences between it and other types is not sufficient. Faced with this fact, two approaches are possible: either to define a new quality model from scratch, or to tailor any of the existing models (e.g. the ISO 9126 quality model) to fit specific domains (ISO, 2001).

Whatever the case, if developing a quality-assured Web portal is important for those organizations aiming to provide services through the Web, this fact is especially relevant for eBanking Web portals. The reason is threefold:

- Firstly, as a banking channel, the Internet has certainly experienced strong growth in customer uptake in recent years, with predicted European compound annual growth rates of approximately 9%; that may be up to 15% in countries like France [Reuters Report]. This banking channel competes in importance with actual physical branches and call centres. The importance of this means may grow if we focus on simple transactions (e.g. making payments, transferring funds and checking balances). In fact, according to a recent survey (Córdoba el al., 2007) the frequency of use of eBanking solutions among bank clients oscillates between daily and weekly, a rate that is similar to that of Automatic Teller Machines (ATMs). Indeed, banks are the driving force in the economic development of countries, as they are the support for the importation and exportation of most products and services.

- Secondly, banks acquired a competitive advantage by being the first to market with Internet banking services before there was widespread availability of these. This fact served as a useful marketing tool to attract more technologically-aware consumers. Now, however, when Internet banking services are virtually omnipresent, a bank's promise of access to these services is no longer enough to woo consumers. Instead, banks are forced to give their services some features that make them different, by using quality "hooks", such as enhanced functionalities and effective Web design. If they don't do this, there is a risk that technologically-aware consumers who chose a bank in the first place because of its eBanking solution, will now switch to another because they have found an even better product.

- Thirdly, certain bank quality characteristics are bound up with the core of the business itself. For example, the conditions under which some kinds of operations are performed are regulated by tight requirements and governmental laws, so the developers must be sure that they are enforced in the application.

BPQM is our attempt to meet all these needs. This model used expert consultation to extract which of the general portal quality characteristics, gathered mainly from PQM (a portal quality model), were relevant for the eBanking domain. Next, the model was completed with a set of new characteristics/sub-characteristics, not present in any of the sources. Finally, the quality model was validated through a survey performed on 168 eBanking users. In this paper, we present the comparative study carried out between PQM and BPQM models to identify the differences and to justify why these have appeared. The goal is to detect gaps in the model, in order to think about including them in future versions of BPQM.

Sections two and three provide an introduction to the quality models, PQM and BPQM. Section 4 presents the comparative study and the conclusions obtained from those. Finally, the overall conclusions and proposed future work can be found in the last section.

2 PQM: The Portal Quality Model

In (Moraga et al., 2006), a generic quality model for portals, namely PQM (Portal Quality Model) was put forward and compared to other portal quality model proposals. After the comparison, PQM stood out as being the most complete and this is the reason why we used it as basis for the creation of BPQM. This model was made using the SERVQUAL model proposed by (Parasuraman et al., 1998) as the basis, along with the GQM (Goal Question Metric) method (Basili et al., 1994). The characteristics identified for the PQM model are:

- **Tangible:** characteristic of the portal that indicates whether it contains all the software and hardware infrastructures needed, according to its functionality. The sub-characteristics are:
  - *Adaptability*. Ability of the portal to be adapted to different devices (for instance PDA's, PCs, mobile phones, etc).
  - *Transparent access*. Ability of the portal to provide access to the resources, while at the same time isolating the user from their complexity.
- **Reliability:** ability of the portal to perform its functionality accurately. In addition, this characteristic will be affected by:
  - *Fault tolerance*. Capability of the portal to maintain a specified level of performance in the event of software faults (for example, a fault during the sending of information or the execution of a job).
  - *Resource utilization*. Capability of the portal to offer its resources to the user according to his profile or particular role or privileges.
  - *Availability*. Capability of the portal to be always operative, so that users may be able to access it and use it anywhere, anytime.
  - *Search Quality*. Appropriateness of the results that the portal provides when undertaking a search/request made by the user.
- **Responsiveness:** willingness of the portal to help and provide its functionality in an immediate form to the users. In this characteristic, the sub-characteristics are:
  - *Scalability*. Ability of the portal to adapt smoothly to increasing workloads which come about as the result of additional users, an increase in traffic volume, or the execution of more complex transactions.
  - *Speed*. Ability of the portal to remain within the response time boundaries tolerated by portal users.
- **Empathy:** ability of the portal to provide caring and individual attention. This dimension has the following sub-characteristics:
  - *Navigation*: Simplicity and intuitiveness of the navigation paths provided by the portal.
  - *Presentation*: Clarity and uniformity of the interface.

- *Integration*: Degree of global portal coherence achieved after the inclusion of the components that make up the portal. All the components of the portal must be integrated in a coherent form.
- *Personalization*: The portal's capability to adapt to the user's priorities.
- **Data quality (DQ):** This characteristic is defined as the quality of the data contained in the portal. It has four sub-characteristics:
  - *Intrinsic DQ.* Degree of care taken in the creation and preparation of information.
  - *Representation DQ.* Degree of care taken in the presentation and organization of information for users.
  - *Accessibility DQ.* Degree of freedom that users have to use data, define and/or refine the manner in which information is input, processed or presented to them.
  - *Contextual DQ.* Degree to which the information provided meets the needs of the users.
- **Security:** capability of the portal to prevent, reduce, and respond to malicious attacks adequately. Its sub-characteristics are:
  - *Access control.* Capability of the portal to allow access to its resources only to authorized people. Thus, the portal must be able to identify, authenticate, and authorize its users.
  - *Security control.* Capability of the portal to carry out auditing of security and to detect attacks. The auditing of security shows the degree to which security personnel are enabled to audit the status and use of security mechanisms by analyzing security-related events. In addition, attack detection seeks to detect, record and notify attempted attacks as well as successful attacks.
  - *Confidentiality*: Ability to guard the privacy of the users.
  - *Integrity.* Capability of the portal to protect components (of data, hardware and software) from intentional or unauthorized modifications.

3 BPQM: The eBanking Portal Quality Model

The purpose of creating the BPQM has been the integration of relevant characteristics for eBanking that are present in PQM, completing them by filling in the gaps detected in a survey performed among domain experts.

These gaps were filled in through the validation of the model, performed using an "eBanking User Satisfaction Survey" that was carried out during the months of February and April 2007 with the aim of evaluating satisfaction levels based on Internet banking user experience. In the production of the survey, the suggestions proposed in (Pfleeger and Kitchenham, 2006) were taken into account. The survey was printed and physically distributed to 200 people randomly selected from the eBanking user community of Costa Rica. From the total number of individual surveys given out, 22 were left in blank, 10 were partially completed, and 168 were fully answered, which gives a result of an 84% response index. For the validation of the quality model, these last 168 results were considered.

This work generated the following list of characteristics:

- **Empathy:** capacity of the portal to provide personalised help and attention. Within this characteristic we would highlight:
  - *Navigability*: capacity of the portal to offer users an intuitive and easy-to-use navigation system.
  - *Personalisation*: capacity of the portal to allow users to customise their services so that user effort is reduced and client satisfaction is increased.
- **Usability:** capacity of a software product to be understood, learnt and used, as well as to be attractive, always bearing in mind regulations and usability guides. Within this characteristic the following sub-characteristics are found:
  - *Degree of attractiveness*. The portal must be able to satisfy the wishes of users, not only in terms of visual appearance, but also through its services and the degree of satisfaction achieved in its deployment.
  - *Accessibility*. This refers to the freedom of users when using data, by defining and/or refining the manner by which the information is entered, processed or presented to them.
  - *Learning capacity*. It focuses on the amount of effort needed by the user to learn how to use services and functionalities that are available in the portal.
- **Efficiency:** capacity of a software product to provide an appropriate performance which takes into account the amount of resources used and which is adaptable to specific conditions. This characteristic includes:
  - *Response time*: It focuses on the response times after a user request. There are various important response times for a banking portal: the time that elapses between when the solution indicates to the user that it is processing a request (state bar or progress bar) and the time taken by the solution in delivering a final response, whether positive or negative, the latter through an error message).
- **Security:** It represents the "capacity of the software product to avoid unauthorised access, either accidental or deliberate, to programs and data". For this characteristic we point out:
  - *Integrity*: The portal must protect the data and information in such a way that no deliberate or unauthorised modifications take place.
  - *Confidentiality*: The main focus is on preserving the privacy of users.
  - *Fault tolerance*: It refers to the capacity of offering a clear response to the user when faults, errors, or attacks occur. The portal must, at all times, recover in the shortest period of time and affect user services as little as possible.
- **Functionality:** It refers to the "capacity of a software product to satisfy the established functional requirements and the implicit needs of users". It includes:
  - *Maintainability*: The portal must be built in such a way that it facilitates the easy and rapid solution of any type of error, as well as the maintenance and updating of information.
  - *Interoperability*: It focuses on the ability of the portal to interact with other systems and services.
  - *Available services*: The portal must have the maximum amount of services, which should be available 24 hours a day.

4 PQM vs. BPQM

Having given an overview of both models, this section presents the set of similarities and differences that have been detected between them. Possible reasons have been found to explain why, with respect to some of the discrepancies, those characteristics have not been identified either by experts or by the eBanking users we interviewed as being relevant for eBanking portals. For other differences, however, no explanation could be found, which in turn suggests a need for further empirical research to decide about their incorporation into the BPQM model.

Table 1 shows the comparison made between both models. Black cells mean equality in the sub characteristic in both models. Dark grey cells mean some kind of similarity (partial coverage of a sub characteristic, same meaning but different name, etc) while rows or columns in grey imply that there is not any kind of coincidence. In the next sub-sections, the results of the study are explained.

Table 1. PQM sub-characteristics vs. BPQM sub-characteristics

PQM \ BPQM		Empathy		Usability			Efficiency	Security			Functionality		
		Navigability	Personalization	Attractiveness	Accessibility	Learnability	Response time	Integrity	Confidentiality	Fault tolerance	Maintainability	Interoperability	Availability
Tangibility	Adaptability						▨						
	Transparent access												
Reliability	Fault tolerance									▨			
	Availability												■
	Search Quality												
	Resource utilization												
Responsiveness	Scalability						▨		▨				
	Speed												
Empathy	Navigation	■											
	Presentation			▨									
	Integration												
	Personalization	■											
Data Quality	Intrinsic DQ												
	Representation DQ												
	Accessibility DQ				■								
	Contextual DQ												
Security	Access control								■				
	Security control								■				
	Confidentiality								■				
	Integrity							■					

4.1 Sub-characteristics in PQM and in BPQM

As shown in Table 1 above, there are nine sub-characteristics that are common to both models: (1) Fault tolerance, (2) Availability, (3) Navigation, (4) Personalization, (5) Confidentiality, (6) Integrity, (7) Accessibility DQ, (8) Access control and (9) Speed.

Most of these characteristics are mainly related to the comfort of the user when using the portal (2, 3, 4, 7, 9) and to security aspects, which are very relevant when the portal is used as part of the user activities (1, 5 6, 8). These aspects coincide with the definition of a portal given by Collins (2001): "a working environment where users can easily navigate to find the information they specifically need to make decisions and to perform their operational or strategic functions quickly ". We would like to point out how our survey showed that, for 68% of the interviewees, a 24/7 service availability is absolutely essential if user objectives are to be fulfilled. Whether the weight of this characteristic is similarly important in other kinds of portals remains an open issue and this will be investigated.

It must be noted that BPQM considers access control as an integral part of confidentiality and that personalization includes both static and dynamic techniques.

4.2 Sub-characteristics in PQM and Partially in BPQM

There are four sub-characteristics of PQM that are partially supported in BPQM: (1) Presentation (with attractiveness), (2) Integration (with attractiveness), (3) Transparent access (with learnability) and (4) Scalability (with response time and fault tolerance).

Learnability (3) is a characteristic that, although not included in PQM, plays a dominant role in eBanking applications. This is because the Internet does not make personal interaction with a bank employee possible and to input information online can be difficult. For this reason, banks consider the Internet to be a poor channel for advice and sales. In fact, their branches continue to generate more business for banks than other means and the number of branches has therefore not decreased significantly, despite their being the most expensive to maintain (Reuters, 2004). Improving help mechanisms and guiding the customer throughout online interaction is a necessary step in increasing clients' confidence in Internet so that they see it as a convenient point of sale (POS) for them to access high-value, complex products. These kind of products are those where customers still require the reassurance of speaking to a bank advisor.

Given that (1) bank customers are not Internet experts and that (2) eBanking applications are a service channel, there is a need to educate and train clients in the use of the Internet as a means or channel of access to banking services. The survey demonstrated that the incorporation of services such as online help, tutorials or simulations increases user satisfaction. According to Hudson (2007) and Centeno (2003) training is needed to improve the penetration, acceptance, and satisfaction of banking portal users. We believe that learnability (BQPM) is related to transparent access (PQM) since both aim to reduce application complexity for the user.

BPQM considers that attractiveness involves, though is not limited to, presentation and integration of the application (1, 2).

Lastly, BPQM points out that scalability (4) is covered by response time and fault tolerance. The reason is that an eBanking customer is not interested in how many accesses the application can support at the same time, but rather in whether his requests are carried out quickly and without errors.

4.3 Sub-characteristics in PQM and Not in BPQM

There are seven characteristics in PQM that have been not considered in BPQM: (1) Adaptability, (2) Resource utilization, (3) Intrinsic DQ, (4) Representational DQ, (5) Contextual DQ, (6) Search quality and (7) Security control.

PQM includes a whole section devoted to quality of data (3, 4, 5) which, in the case of BPQM, is not covered completely. The reason is that most data are very specific and/or directly come from databases (e.g. account states) or external sources (e.g. stock prices) and so authoring of content was not considered to be a significant activity during eBanking portal development. Nevertheless, the survey showed that data-intensive help mechanisms are necessary and would be a candidate for its inclusion in the BPQM model.

As far as adaptability and resource utilization (1, 2) is concerned, so far we have not found interest on this topic among the users interviewed. The lack of concerns about adaptability (support for different devices) might be due to the fact that some channels, such as mobile devices, have not been considered important. We think that this may change in the future, however, so that banks should start getting ready to meet this need.

From the user point of view, the background to the e-banking solution is not something she knows or cares about; the user just knows that the application is very slow or that it fails. In any case, it would be necessary to work more on these important aspects in order to determine their inclusion in BPQM.

It seems logical that users have not said that search quality (6) is an important sub-characteristic of BPQM, since searching is not a prime functionality of this kind of portals. Basically, the user expects to find in the portal the same kind of services as are available in a branch. It is not normal for users to look for information, much less information requiring search tools. However, this aspect must be studied in depth in order to decide whether it must be added to the model or not, due to the increasing diversification of services that banks are starting to offer.

Finally, in this section, the security control (7) is the most suspicious result from the ones obtained in this section. eBanking users do not seem to be interested in how the bank monitors attacks and audits the status and use of security mechanisms, as long as their data remain confidential and show integrity. We do not expect such a situation to change in the near future, so we have dismissed this characteristic as being relevant for BPQM.

4.4 Sub-characteristics Not in PQM and in BPQM

Finally, there are two sub-characteristics included in BPQM but not considered in PQM: (1) Maintainability and (2) Interoperability.

Users expect eBanking applications to have a high maintainability. This means that they expect the bank to maintain the accuracy of the data provided and to respond rapidly to any problems encountered to any lack of functionality detected during use (1). In addition, users require the eBanking portals to provide the possibility of interconnecting with different applications. Since a bank is a financial operator, the interoperability (2) with third parties is vital for users. In the survey, 90% of eBanking users said they were not satisfied with the number of third parties that could interoperate with their bank.

4.5 Conclusions of the Study

As result of the study, we obtained that the BPQM includes 45% of PQM sub-characteristics completely and that 20% of the PQM sub-characteristics are also considered, although not completely. The remaining sub-characteristics are not considered in the BPQM for the time being. We believe that all of the characteristics, with the exception of security control, should be considered for inclusion in BPQM, as they could become quality indicators to evaluate how well the eBanking application has foreseen changes in the way bank service types and user access devices may evolve in the future. Finally, by means of the survey, two sub-characteristics excluded in PQM have been identified as important for BPQM.

Conversely, we need to further give consideration to whether scalability and security control should be regarded as characteristics of PQM. On one hand, our survey has shown that from a user perspective, scalability is related to fault tolerance and speed. Both of these characteristics are also included in PQM. Regarding security control, it seems that users are not interested in how security personnel audit information as long as the control is confidential and shows integrity (these characteristics are also present in PQM).

5 Conclusions and Future Work

Nowadays, eBanking solutions are strengthening their position as the main channels for banking operations. This makes us conclude that the quality of these channels will be fundamental for the survival of banks in the global economy. Furthermore, banks have seen the need to transform their eBanking solutions from simple transactions to systems that provide integrated services similar to those available in an actual bank office (with customer service and support).

In an initial work on this (Córdoba et al., 2007), and basing ourselves on PQM, we tried to define a quality model for bank portals. This work included a 168-user survey, designed to determine the characteristics and sub-characteristics of the new model. However, since the creation process of a model is not a trivial issue (even when it may be based on a generic model), the objective was to know whether the model defined was correct or rather inaccurate. It might have left something out (due to possible subjectivity of users in the survey participation).

To this end, we have carried out a comparison between both models, concluding that BPQM includes 45% of PQM sub-characteristics completely, while 20% of the PQM sub-characteristics are also considered, although not completely. The remaining sub-characteristics are not considered in BPQM for now. Although it is logical for some of these not to be included in the model, others need to be studied in depth in order to determine if they must be components of the final model or not.

Lastly, two sub-characteristics not included in PQM have been identified in the survey as being important for BPQM.

For future work, it would be important to validate the current characteristics, with new users and the new characteristics. Once the model is finalised, and with the purpose of making it applicable in practice, we would aim to define measures for each sub-characteristic of the model so that it can be applied to different eBanking solutions by different people, attempting to generate a quality comparison (ranking) between the banks within a specific region (i.e.: Costa Rica, Spain, etc.).

Acknowledgements

This paper has been supported by the MEC Spanish Ministry projects CALIPSO (TIN20005-24055-E), MEC-FEDER (TIN2004-03145), ESFINGE (TIN2006-15175-C05-05), METASIGN (TIN2004-00779) and DSDM (TIN2005-25866-E)). Also, it is part of the DADASMECA project (GV05/220), financed by the Valencia Government and DIMENSIONS (PBC-05-012-1) financed by the Castilla-La Mancha Government.

References

Basili, V.R., Caldiera, C., et al.: Goal Question Metric Paradigm. Encyclopedia of Software Engineering, vol. 1. John Wiley & Sons, England (1994)

Centeno, C.: Adoption of Internet Services in the Enlarged European Union. Lessons from the Internet banking case. Institute for Prospective Technological Studies, European Commission, Joint Research Centre, Report EUR 20822 EN (June 2003)

Collins, H.: Corporate Portals, New York, Amacom (2001)

Córdoba, J., Cachero, C., Calero, C., Genero, M., Marhuenda, y.: Modelo de calidad para portales bancarios. Sent to CLEI 2007 (in Spanish, 2007)

Hazra, T.K.: Building Enterprise portals: principles to practice. In: 24 International Conference on Software Engineering, pp. 623–633 (2002)

Hudson, W.: The Lost World of E-Banking. SIGCHI Bulletin September/October 2002, Syntagm Ltd USA. [Abr. 14, 2007], available on the Internet http://www.syntagm.co.uk/design/articles/ebanking.htm

ISO/IEC TR 9126-1. Software engineering — Product quality — Part 1: Quality model

Mahdavi, M., Shepherd, J., Benatallah, B.: A Collaborative Approach for Caching Dynamic Data in Portal Applications. In: Proceedings of the fifteenth conference on Australian database, vol. 27, pp. 181–188 (2004)

Moraga, M., Calero, C., Piattini, M.: Comparing different quality models. Online Information Review Journal (2006)

Parasuraman, A., Zeithami, V.A., et al.: SERVQUAL: a multi-item scale for measuring consumer perceptions of service quality. Journal of Retailing 67(4), 420–450 (1998)

Pfleeger, S., Kitchenham, B.y.: Principles of Survey Research, Part1-Part6. Software Engineering Notes 26(6), 27(1), 27(2), 27(3), 27(5), 28(2). ACM SIFSOFT (2001–2003)

Reuters (2004). The European eBanking and ePayments market outlook. Reuters-Business insights. MBA Group Limited (August 2004)

An Ontological Approach for the Quality Assessment of Computer Science Conferences*

Maria Aparecida M. Souto, Mariusa Warpechowski, and José Palazzo M. de Oliveira

Instituto de Informática – Universidade Federal do Rio Grande do Sul (UFRGS)
Caixa Postal 15.064 – 91.501-970 – Porto Alegre – RS – Brasil
{souto, mariusa, palazzo}@inf.ufrgs.br

Abstract. Today the proliferation of the availability of the information of scientific events on the Web has created the necessity to offer a quickly access to up-to-date information about the quality of these events. This requirement demands for (semi) automatic tools to speedily provide this information. The human-performed activity of the information quality evaluation is extremely time consuming and easily leads to failures. The application OntoQualis here described was motivated to support the quality evaluation of Scientific Conferences, in the Computer Science area, based on the graduated programs evaluation protocol of the Brazilian agency CAPES. The evaluation mechanism is specified in the QUALIS document specifically designed to assess journals and conferences ranking. This paper presents a brief vision of the ongoing process of domain analysis and ontology prototyping aiming to classify Scientific Conferences: the OntoQualis project. Some results of OntoQualis preliminary evaluation have shown a satisfactory classification level in comparison with CAPES-QUALIS ranking.

1 Introduction

The quality evaluation of scientific productions is mandatory to allocate the constrained resources to an increasing research founding demand. The researchers, research grants demands and graduated programs evaluation work is a real burden to the founding agencies where a high percentage of the total resources is employed in this supporting activity. In some research areas as Biology, Physics and Computer Science the amount of available data in the Web allows an automated quality evaluation process, or at least a semi-automatic process. A single numerical index as the h-index [1] applied to quality assessment may be a dangerous approach if adopted without precautions. In one hand, it is easy to the decision-makers to take a simple measure to support the evaluation process in order to eliminate the complexity of the decision process and the individual decision. On the other hand, it is not fair the use of only one numeric indicator to represent the quality of a researcher's production as a single index measuring only one characteristic of the production. As example we have the

---

* This work was partially supported by the projects Pronex FAPERGS, grant 0408933; PerXML CNPq. Grant 475.743/2004-0 and CTInfo CNPq, grant 550.845/2005-4. The second and the last authors are partially supported by CNPq.

J.-L. Hainaut et al. (Eds.): ER Workshops 2007, LNCS 4802, pp. 202–212, 2007.

endless discussion about quality versus popularity of a paper. The quality is a multidimensional aggregate variable and for a more complex task as the quality assessment of a scientific conference or a graduate program a full multidimensional evaluation criterion is needed.

The main motivation to the present work on conferences evaluation is the national graduated program evaluation performed triennially by CAPES[1]. The following data was obtained from a public speech of Jorge Guimarães, president of CAPES, Dec 2006: "The Brazilian National System for Graduate Studies encompasses 2.313 programs responsible for 3.624 courses being 2.386 masters and 1.238 doctorates programs. The evaluation of graduate programs is responsibility of the federal agency CAPES and is based in a triennial evaluation performed by academic peers with the participation of 46 evaluation committees and it involves more than 800 evaluators. The triennial evaluation is composed by four processing stages: 1. Data gathering from graduate programs and transmission to CAPES data base; 2. QUALIS Classification of publication media; 3. Definition of evaluation criterion for each area and provide reports for analysis; 4. Data base check and the generation of the evaluation report".

"… The evaluated courses receive grades according to their performance and the following scale: grades 1 and 2 disapprove the program; grade 3 means a regular performance, meeting the minimum demanded; grade 4 is considered a good performance for Masters courses; grade 5 is the maximum grade given to Ms courses; grades 6 and 7 indicate high standard performance with an international level of the course. All the results are public (approved courses and programs, grades, reports) and the evaluation has legal effect. Only the titles and courses approved by the evaluation process have national validity".

The gathering and maintenance of all this data is a huge human effort. The second processing stage the QUALIS generation is a very time-consuming task. QUALIS is a classification of the Journals and Conferences were Brazilian researchers have published papers [2] and is an important referential not only to the graduate programs evaluation but also for the individual researchers evaluation and for grants attribution by the National Research Council – CNPq and other research founding agencies. Presently this work is human-developed and consumes some months to be achieved.

The purpose of the work described in this paper was to develop a classification model to support the (semi) automatic evaluation of Computer Science Conferences (CSC) based on the QUALIS document. This model has generated an ontology prototype and the inference rules to perform the classification. In a first validation test twelve CSCs were evaluated, showing a satisfactory classification level in comparison with CAPES-QUALIS document. We are working in a web-based system that will not only classify a CSC but also presents a dashboard with a multidimensional set of quality indexes.

The paper is organized as follows: Section 2 discusses related work. Section 3 presents the Classification Model for CS Conferences. Section 4 describes the OntoQualis ontology and the classification criteria. Section 5 presents OntoQualis preliminary evaluation, and finally the section 6 presents the conclusion and future works.

---

[1] CAPES is a Brazilian federal agency charged to promote the human qualification to support the University education.

2 Related Work

For an individual researcher evaluation a single numeric index was recently proposed by J.E. Hirsh, the h-index [1]. This is an interesting numerical evaluator but has some pitfalls. In our opinion the most relevant is the confusion caused by the association of popularity with scientific value. This is a central point in all the quality evaluation processes as the quality of a work must not be considered as the popularity of the work but as the result of a multifaceted evaluation. A superficial paper may be more mediatic than a deep scientific one. This impact-index ranking suffers also a drawback from positive feedback. It is a well-known fact that users access only the first page of a search service answer, this happens in more than 85% of the cases; as a consequence the most cited authors – these appearing in the first page of Scholar Google, as example – are even more cited.

Another bias was recently discovered [3]: the scientific papers being published by bigger and bigger teams, the authors show clearly that the increase in the number of citations is strongly correlated with the number of authors. Exists also a tendency of papers citations of the production of potential reviewers or from colleagues from the same country and others similar bias [4]. These authors have shown in engineering an increasing of 3.72 times in the citations of a paper with more than 5 authors in comparison to a paper of a single author this rate increases to 13.01 times in social sciences for papers published in 2000. Apart these difficulties, the use of an aggregated and easily computable index as the h-index, give an estimate of the importance of the cumulative scientists' production.

For a more complex evaluation, as a scientific conference grading, or the classification of a graduate CS program a wider and multidimensional model of quality must be developed. One alternative for Journals evaluation is the experts' opinion collection and ranking. A recent study [5] presented an extensive analysis of IS Journals. In this survey the perception and impact of IS journals where assessed by an impact factor. An extensive investigation of global community perceptions of IS journals, [6], was employed as reference. One conclusion was: "The data suggests there is consistency in several of the top-rated journals, but wide variety in others". Another analysis from this work is: "Whether perceptions meet with the reality of research dissemination is a question that is open for debate". This analysis supposes that the impact factor is the correct indicator of the dissemination power of a paper; if this assertion is not completely true associated with the above mentioned uncertainties will open a wide research schedule on quality evaluation.

The more extensively employed index, the impact index, alone is clearly subject to some well-founded criticism. We are facing a clear interpretation problem; a consistent experimental protocol must be developed to evaluate the real importance of the index and we need a multi-faceted quality evaluation for the scientific production. This is our research point, how to develop and implement multidimensional metrics to evaluate the scientific production.

3 A Classification Model for CS Conferences

To be able to mange the complexity of the CS Graduated Program assessment we decided to consider initially only the task of classifying CSC. In this work we focused on the (semi) automation of this task based on the indexes and rules set specified in the CAPES Computer Science area QUALIS document [7] for Conferences evaluation.

According to QUALIS document a CSC could be classified as level "A", "B", and "C". The information is temporally dynamic meaning that the classified CSC could vary annually. The index set defined by the CAPES Committee to classify a CSC is composed by the following indexes: *Impact Index, Edition, Sponsorship, Program Committee, Accepted Paper Type, Associated Conference, Publication* and *Scope*. These indexes are a limited subset of the possible criteria. One of the not considered, in this work, is the acceptance rate, the main criterion was that this index is not widely available and, more important; it is extremely dependent of the conference quality. A poor quality conference attracts a large number of weak papers; in the other hand a top-level conference eliminate the non-competitive ones before the submission. It is clear that some relevant internal data on the reviewing process as the actual number of reviewers per paper is not publicly available for all the conferences. Other interesting issue is an h-index composed from the PC member indexes; we are evaluating some statistical properties of this index. The future inclusion of these indexes and of others will be considered, our expectation is to find the minimal set of indexes allowing a good classification.

The selected indexes were interpreted as follows.

Impact Index – corresponds to that captured from CiteSeer[2] impact index, at the present we are working with evaluation of the h-index[3].

Edition - corresponds to the CSC edition number, this index can influence the classification when combined with others. A well established conference has a higher rating than a first edition one.

Sponsorship – the main CSCs are sponsored by recognized Scientific Institutions. As recognized Institutions we can mention: ACM, IEEE, SIAM, IFIP, W3C and others.

Program Committee – a good CSC has a program committee whose researcher group is composed by recognized researchers assuring serious papers evaluation.

Accepted Paper Type – submitted papers may be classified as: full, short and poster. A full paper has 6 or more pages; short paper are the papers from 3 to 5 pages; and posters the ones having 2 or one pages.

Associated Conference – a Conference may be "Principal" or an "Associated" one. A Workshop is considered as an Associated Conference when it happens in conjunction with Principal Conference. For example, when a workshop is associated to the principal Conference which was classified as level "A", the workshop will be classified as, at last, level "C".

Publication – proceedings published by recognized publisher institutions as ACM, IEEE, SIAM, IFIP, and W3C, as examples.

Scope – the scope index of a Conference is: "Regional", "National", or "International".

The dynamic of these eight criteria according QUALIS document are summarized in Table 1. As we can see in Table 1, conferences level "A" and "B" have two alternatives of classification and conferences level "C" has four. Each alternative corresponds to different sort of combination of the indexes considered. Currently, the *impact index* is considered predominant information to classify a CSC. The alternative to classify a

---

[2] CiteSeer, http://citeseer.ist.psu.edu/
[3] Publish or Perish software from harzing.com

conference as level "A" take into consideration four indexes: *edition, sponsored, program committee*, and *accepted paper type*, i.e., the conference *edition* must be greater or equal 4, and the conference must be *sponsored* by recognized institutions in Computer Science area, and it must have a qualified *program committee*, and it have *accepted paper* type "full". *Edition* rule (i.e., >=4) means a traditional conference. Presently, about *program committee*, we are just considering whether the conference has a *program committee*. Nevertheless, colleagues of our team project are working towards a *program committee* quality model.

Another criterion is related to associate conferences. We considered associated conference the workshops which occur together with principal conference. The classification of associated conference depends of classification of the principal conference, i.e., if the principal conference is level "A", the classification of the workshop would be at least level "C".

Table 1. CSC classification criteria summary

	Values	Level "A"		Level "B"		Level "C"	
1. Impact Index	<= 40	x					
	>= 41 and <= 79			x			
	>= 80					x	
2. Edition	<= 3			x			
	>= 4		x				
3. Sponsorship	ACM, IEEE, IFIP, SIAM, W3C, …	x		x			x
4. Program Committee	Yes	x		x		x	
5. Accepted Paper Type	Full	x		x		x	x
6. Associated Conference	Conference level "A" ➔ associated event level "C"					x	
7. Publication	ACM, IEEE, IFIP, SIAM, W3C, …					x	
8. Scope	Regional						x

4   The OntoQualis Prototype

The OntoQualis prototype was developed base on the works of [8, 9]. It implements an ontology based on the classification model for CS Conferences as is shown in Figure 1. The ontology was developed using the plug-in OWL [10] of the Protégé environment [11].

The ontology represents the indexes employed to the conferences classification. The classification task is usual in the context of Information Systems when is required the specification of the pertinence relation within a domain. The use of ontologies in Information Systems generally is oriented to specify and communicate domain knowledge in a generic way and to structure and define the concepts meaning. Besides this, in our work, we also explore the use of the reasoning services aiming to classify CS Conferences in a pre-defined category level.

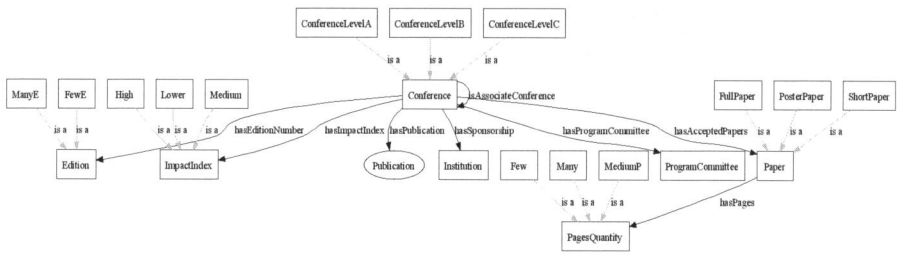

Fig. 1. The OntoQualis Ontology

To accomplish the classification task it is essential to specify which indexes will be measured and how the measurement will be performed. In the following we present the OntoQualis indexes description.

Criterion 1 - ImpactIndex
In the ontological model the Impact Index is represented as the class ImpactIndex and its respective subclasses High, Medium and Lower. The subclass High corresponds to the individuals whose Impact Index value is less or equal 40%; the subclass Medium corresponds to the individuals whose Impact Index value is greater or equal 41% and less or equal 79%; and finally, the subclass Lower corresponds to the individuals whose Impact Index value is greater or equal 80%. In the OntoQualis we have:

ImpactIndex ≡ High ⊓ Medium ⊓ Lower

Criterion 2 – Edition
In the ontological model this index was represented as the class Edition and its respective subclasses FewE and ManyE. The subclass ManyE corresponds to all individual of the domain for which the Edition number value is less or equal 3; and FewE corresponds to that individuals for which this value is greater or equal 4. Thus, in the OntoQualis we have:

Edition ≡ ManyE ⊓ FewE

Criterion 3 – Sponsorship
In the ontological model it represents a property hasSponsorship that relates the Conference class and Institution class. The Institution class can represent three different types of Institutions: Scientific Society, University and Enterprise, which are modeled as subclasses of Institution.

Institution ≡ ScientificInstitutionSociety ⊓ UniversityInstitution ⊓
               EnterpriseInstitution

Criterion 4 - Program Committee
In the ontological model it represents the ProgramCommittee class. A Program Committee is composed with chair and members. This is modeled in the ontology through the specification of the properties hasChair and hasMember, which have ProgramCommittee class as domain and Person class as range.

ProgramCommittee ≡ hasChair Person ⊓ hasMember Person

Criterion 5 - Accepted Paper Type

This index was represented in the ontological model as the Papers class and its respective subclasses FullPapers, ShortPapers and PosterPapers. In order to be classified an individual of the domain as pertaining into one of these subclasses it must satisfy some criterion associated with the number of pages of the accepted papers. Then, was created the class PagesQuantity and the respective subclasses Many, MediumP, and Few. The sub-class Many represents the number of pages that is greater or equal 6; the MediumP the number of pages that is less or equal 3 and greater or equal 5; and the Few the number of pages that is less or equal 2.

Papers ≡ FullPapers ⊓ ShortPapers ⊓ PosterPapers

PagesQuantity ≡ Many ⊓ MediumP ⊓ Few

In the Ontology, the papers' types were modeled as the property "hasPages" having the Paper class as Domain and "PagesQuantity" as range, as showed below.

FullPapers ≡ Papers ⊓ hasPages Many

ShortPaper ≡ Papers ⊓ hasPages MediumP

PosterPapers ≡ Papers ⊓ hasPages Few

Criterion 6 - Associated Conference

In the Ontology, all Conferences pertain to the Conference class. In order to distinguish between "Principal" and "Associated" it was defined the "isAssociateConference" property which has Conference class as domain and range.

Criterion 7 – Publication

This index was model as value criterion. To model it in the Ontology it was defined the Publication class and the respective subclasses: BookPublication, ProceedingsPublication and ScientificSocietyPublication, where:

Publication ≡ BookPublication ⊓ ProceedingsPublication ⊓
            ScientificSocietyPublication

To represent a recognized Scientific Publisher the "isPrint" property was defined having ScientificSocietyPublication class as domain and the individuals: ACM, IEEE, IFIP, SIAM, W3C, among others, of the ScientificInstitutionSociety class as range. Then we have:

ScientificSocietyPublication ≡ Publication ⊓ isPrint
            ScientificInstitutionSociety {ACM IEEE IFIP SIAM W3C}

Criterion 8 – Scope

This index was modeled as value criterion. In the ontology it was represented as the Scope class with the individuals "Regional", "National" and "International":

Scope ≡ {Regional National International}

Conference Classification

As we have mentioned before, our objective is to classify CSC as level "A" or "B" or "C". These levels were modeled in the ontology as subclasses of the class Conference: ConferencelevelA, ConferencelevelB, and ConferencelevelC:

Conference ≡ ConferenceLevelA ⊓ ConferenceLevelB ⊓ ConferenceLevelC

The Conference level classification task involves the evaluation of the eight criteria; each individual of the Conference class is classified as being "A", "B" or "C" considering the evaluation of some combination of these eight criteria. A Conference is classified as level "A" when it has ImpactIndex pertaining to the subclass "High" or when it is sponsored by a recognized Scientific Society and has a Program Committee and has Edition pertaining to the subclass "ManyE" and has Paper pertaining to the subclass "FullPapers". Then, we have:

ConferenceLevelA ≡ (∀ hasImpactIndex High)

ConferenceLevelA ≡ ((∃ hasSponsorship {ACM IEEE IFIP SIAM W3C}) ⊓

(∀ hasProgramCommittee ProgramCommittee) ⊓ (∀ hasEditionNumber ManyE) ⊓
(∃ hasAcceptedPapers FullPapers))

To classify a Conference as level "B" the reasoner must verify if the respective ImpactIndex pertains to the subclass "Medium" or if it is sponsored by a recognized Scientific Society and has a Program Committee and has Edition pertaining to the subclass "FewE" and has Paper pertaining to the subclass "FullPapers".

ConferenceLevelB ≡ (∀ hasImpactIndex Medium)

ConferenceLevelB ≡ ((∃ hasSponsorship {ACM IEEE IFIP SIAM W3C}) ⊓

(∀ hasProgramCommittee ProgramCommittee) ⊓ (∀ hasEditionNumber FewE) ⊓
(∃ hasAcceptedPapers FullPapers))

At last, to classify a Conference as level "C" the reasoner must verify if the respective ImpactIndex pertains to the subclass "Lower" or if the Conference has Paper pertaining to the subclass "FullPapers" and it is associated to the principal Conference previously classified as level "A" or if it has publication pertaining to the subclass "ScientificSocietyPublication" and it has a Program Committee and Paper pertaining to the subclass "FullPapers" or if it is sponsored by a recognized Scientific Society and its scope is "Regional". In order to model this, we have:

ConferenceLevelC ≡ (∀ hasImpactIndex Lower)

ConferenceLevelC ≡ (∃ hasAcceptedPapers FullPapers) ⊓
(∀ isAssociateEvent ConferenceLevelA)

ConferenceLevelC ≡ (∃ hasPublication ScientificSocietyPublication) ⊓

(∀ hasProgramCommittee ProgramCommittee) ⊓ (∃ hasAcceptedPapers FullPapers)

ConferenceLevelC ≡ (∃ hasSponsorship {ACM IEEE IFIP SIAM W3C}) ⊓
(hasScope {Regional})

5   OntoQualis Preliminary Evaluation

The evaluation of the ontological classification model was accomplished in two steps: (i) Conferences data collection and instantiation; (ii) Conferences classification. Our goal was make the comparison between CAPES and OntoQualis CSC classification ranking.

Conferences data collection and instantiation: we chose 12 conferences of Computer Science area, totalizing 154 papers, 840 researchers, 346 Universities, 26 Scientific Societies, 49 enterprises, 12 program committee, and 12 publication titles. Conferences data were collected through web search engines. The information was collected manually by students of a graduated course on Ontology. Presently we are integrating Web collecting tools in an automated environment to populate the Protégé ontology. The collected data corresponds to: *Edition, Program Committee* (chair and members), *Publication, Sponsorship Institutions, Accepted Papers* (title, authors and quantity of pages), *Scope,* and *Impact Index.* The conferences selected were: AH 2006, CAISE 2005, DAMON 2006 (associated with SIGMOD 2006), ER 2005, ICDM 2005, MOBIDE 2006 (associated with SIGMOD 2006), SIGIR 2006, SIGKDD 2005, SIGMOD 2006, UM 2005, VLDB 2005 and WISME 2005 (a standalone conference named workshop).

Conferences classification: In order to perform the conferences classification task, the Racer OWL Reasoner [12] was selected. Through Racer tool, the taxonomy and the ontology consistence were checked. Next, the Racer proceeds with the inference of CSCs level and with the CSCs instantiation into the fitting class (ConferenceLevelA, ConferenceLevelB, ConferenceLelveC).

Table 2. Comparison between the classification results

Scientific Conference	OntoQualis		Qualis-Capes
AH	B	C	A
CAiSE	A	C	A
DAMON	B	C	-
ER	A	C	A
ICDM	B		B
MOBIDE	B		-
SIGIR	A	C	-
SIGKDD	A	B	-
SIGMOD	A	C	-
UM	B	C	A
VLDB	A	C	A
WISME	C		C

It is possible that a conference should satisfy more than one criterion; in this case the reasoner instantiates the conference level in more than one class. As defined in the CAPES document a conference will receive the higher classification level. Table 2 shows the results of the classification generated by OntoQualis and the QUALIS-CAPES official classification which has ranked only the conferences with papers published by Brazilian researchers in the last tree years. The official classification is available on-line[4].

Among these 12 CSCs considered, five were classified as level "A" fulfilling criterion 1 and one was classified as level "A" fulfilling criterion 2. Four CSCs were classified as level "B" fulfilling criterion 3 and one was classified as level "B" fulfilling

---

[4] http://www.capes.gov.br/avaliacao/webqualis.html
http://qualis.ic.unicamp.br/conferencias/consulta_congressos

criterion 4. Only one CSC was classified as level "C" fulfilling criterion 7. Analyzing these results we can observe that 10 CSCs were classified according official Qualis-Capes rank and only two were not: the AH and UM conferences. We have interpreted these misclassifications as a consequence of the impact index information from Cite-Seer that currently has not been updated.

6 Conclusions and Future Works

The quality evaluation of the scientific work is a complex task. In contrast to other quality evaluation areas as engineering, where some well-defined quantitative measures may be employed, the intellectual production is a multidimensional and subjective task. One of the important quality evaluations of scientific work is associated with Conferences. In this paper we have described our concern towards the development of a (semi) automatic tool taking into consideration several criteria to evaluate the quality of a Conference and, in a near future, alleviate the manual work of the academic community in obtaining this important information.

The first experiments demonstrate that it is possible to reach an (semi) automatic evaluation, compatible with the human-developed classification available from CAPES analysis. In order to obtain an expressive data volume to extensively validate the OntoQualis model, we are working and developing data extraction tools on the Web from the Conferences' homepage.

Up to now, we have worked with the implementation of the QUALIS-CAPES classification model. Nevertheless, we expected to refine the considered indexes and incorporate other ones such as the *program committee* quality index aiming to enrich the present model.

References

1. Hirsh, J.E.: An index to quantify an individual's scientific research output. In: Proceedings of the National Academy of Sciences of the United States of America, November 15, 2005, vol. 102(46), pp. 16569–16572 (2005)
2. CAPES 2007 (March 2007), http://servicos.capes.gov.br/webqualis/
3. Wuchty, S., Jones, B.F., Uzzi, B.: The Increasing Dominance of Teams in Production of Knowledge. Science (published online April 12, 2007), doi:10.1126/science.1136099
4. Leimu, H., Koricheva, J.: What determines the citation frequency of ecological papers? TRENDS in Ecology and Evolution 20(1), 28–32 (2005)
5. Barnes, S.J.: Assessing the value of IS journals. Communications of the ACM 48(1), 110–112 (2005), http://doi.acm.org/10.1145/1039539.1039573
6. Mylonopoulos, V.T.: On site: global perceptions of IS journals. Communications of the ACM 44(9), 29–33 (2001), http://doi.acm.org/10.1145/383694.383701
7. CAPES 2005, http://qualis.ic.unicamp.br/documentos/doc_area_cc-2005-ago05.pdf
8. Noy, N.F., McGuinness, D.L.: Ontology development 101: a guide to creating your first ontology. Technical report KSL-01-05, Stanford Knowledge Systems Laboratory
9. Guarino, N.: Formal ontology and Information Systems. In: FOIS 1998. International conference on Formal Ontologies in Information Systems, Trento, Italy, pp. 3–15 (1998)

10. Horridge, M., Knublauch, H., Rector, A., Stevens, R., Wroe, C.: A Practical Guide to Building OWL Ontologies Using The Protégé-OWL Plugin and CO-ODE Tools (2004), http://www.co-ode.org/resources/tutorials/ProtegeOWLTutorial.pdf
11. Protegé, julho (2006), http://protege.standford.edu
12. Racer OWL Reasoner, julho (2006), http://www.sts.tu-harburg.de/ r.f.moeller/racer/

Using Practitioners for Assessing the Understandability of UML Statechart Diagrams with Composite States

José A. Cruz-Lemus[1], Marcela Genero[1], Sandro Morasca[2], and Mario Piattini[1]

[1] ALARCOS Research Group
Department of Information Technologies and Systems
Indra-UCLM Research and Development Institute
University of Castilla-La Mancha
Paseo de la Universidad, 4 – 13071 Ciudad Real, Spain
{JoseAntonio.Cruz, Marcela.Genero, Mario.Piattini}@uclm.es
[2] Dipartimento di Scienze Della Cultura, Politiche e dell'Informazione
Università degli Studi dell'Insubria. Como, Italy
Sandro.Morasca@uninsubria.it

Abstract. We have carried out a family of empirical studies to investigate whether the use of composite states improves the understandability of UML statechart diagrams. Our hypothesis derived from conventional wisdom, which says that hierarchical modeling mechanisms are helpful to master a system's complexity. We carried out three studies that have gradually evolved in the size of the UML statechart models, the type of subjects (students vs. professionals), the familiarity of the subjects with the domains of the diagrams, and other factors. In this work we briefly review the first and second studies and present the third one, performed with practitioners as experimental subjects. Surprisingly, our results do not seem to show that the use of composite states improves the understandability of UML statechart diagrams.

1 Introduction

Models help us understand a complex problem and its potential solutions through abstraction. This is why software systems can benefit from using models and modeling techniques [21]. The Model-Driven Development (MDD) paradigm [1] focuses the effort of development on the design of models, rather than coding. Correspondingly, the focus of software quality assurance is shifting from system implementation towards system modeling. So, to be useful and effective, an engineering model must possess the following five key quality characteristics to a sufficient degree [21]: abstraction, understandability, accuracy, predictiveness and inexpensiveness.

In this paper, we focus on understandability because it is recognized as one of the main factors influencing maintainability, and it is well-recognized that a large part of the effort invested in the development of any software product is devoted to maintenance [19]. More specifically, we focus on the understandability of UML statechart diagrams, since UML has now become the de facto standard for modeling software systems and UML statechart diagrams have become an important technique for describing the dynamic aspects of a software system [7].

J.-L. Hainaut et al. (Eds.): ER Workshops 2007, LNCS 4802, pp. 213–222, 2007.

UML statechart diagrams are also considered as one of the most important UML diagrams [3] and should be used by practitioners as a starting point to train UML newcomers. Although the use of this kind of diagram is not critical for many systems, they are very useful for their intended purpose, have low redundancy and provide new helpful information in some systems [8], specially in real-time systems [9].

We have previously studied [5] the effect that some UML constructs may have on the understandability of UML statechart diagrams, including composite states, which are an important construct of the UML statechart diagrams metamodel [17] and a fundamental modeling abstraction mechanism. UML statechart diagrams with composite states extend finite state machines to facilitate the description of highly complex behaviors [13]. A composite state is a state that contains other states within it. Composite states are believed to be useful to model the behavior of complex classes, as we can group the simple states that are part of a larger common one, and so they help structure a system and master a system's complexity by dividing the system in smaller, less complex parts. However, it is necessary to provide empirical support to this belief to show if it is actually true and, if so, under what conditions.

Thus, based on the common use of hierarchical structures in modeling techniques for making easier to reason about a system, both for those who write the design of a system and those who read it [11, 20, 23], we hypothesized that grouping simple states that are highly related into a composite state could help improve the understandability of a diagram. We have performed a family of experiments to test this hypothesis, and in this paper we present the third empirical study corresponding to this family. We have gradually modified and improved the design and execution of these experiments, so as to alleviate some threats to the validity of the different studies that are part of it.

In Section 2 of this paper, we provide a roadmap of the family of experiments that we have performed. Section 3 explains the main features and conclusions of the two first members of the family of experiments that has been performed. Section 4 comments in detail the experimental process performed in order to carry out the third experiment of the family. Finally, section 5 summarizes the main lessons learned from this family of experiments and the future work planned to perform.

2   The Family of Experiments

An experiment can be viewed as part of common families of studies, rather than being an isolated event [2]. Common families of studies can contribute to devising important and relevant hypotheses that may not be suggested by individual experiments.

Fig. 1 shows the chronology of the family of experiments we have carried out in our study on the understandability of UML statechart diagrams.

In the initial studies (E1, R1, E2, and R2), we used students as experimental subjects. The tasks to be performed did not require high levels of industrial experience, so we thought that this experiment could be considered appropriate [2, 12]. In addition, working with students implies a set of advantages, such as the fact that the prior knowledge of the students is rather homogeneous, the availability of a large number of subjects [26], and the possibility of testing experimental design and initial hypotheses [22].

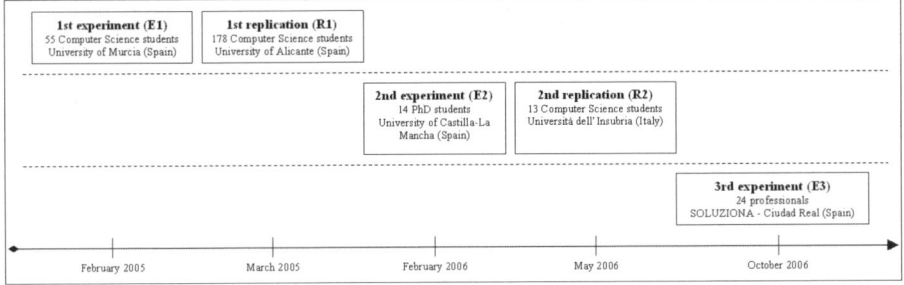

Fig. 1. Chronology of the family of experiments

The main difference between these studies and the third experiment (E3) presented in this paper, lies both in having professionals as experimental subjects, and also having renewed and improved materials and tasks.

In studies E1 and R1, we used the understandability effectiveness variable, defined as the ability to understand the presented material correctly. In studies E2, R2 and E3, we added two new variables related to the CTML, retention and transfer. Retention is defined as the comprehension of material being presented. Transfer is the ability to use knowledge gained from the material to solve related problems not directly answerable from it.

To measure all these variables, we have used three separate tests based on questionnaires. Each understandability characteristic measure is computed as the proportion of correct answers provided in each test. In this work, we will call these values UEffec for understandability effectiveness, UTrans for transfer, and UReten for retention, respectively.

We carried out the experimental processes based on well-known guidelines for empirical studies [14, 15, 28] and we used SPSS [24] to perform all the statistical analyses. In all of our studies, we set a statistical significance threshold $\alpha = 0.05$.

3 First and Second Empirical Studies

In this section, we briefly outline the main features (see Table 1 and Table 2) and conclusions reached after the performance of the two first members of the family of experiments. The complete details about them can be found in [4] and [6].

Table 1. E1 And R1 main features

Location	Universities of Murcia, Spain (E1) and Alicante, Spain (R1)
Date	February (E1) and March (R1) 2005
Subjects	55 (E1) and 178 (R1) Computer Science students
Dependent Variable	Understandability of UML statechart diagrams, measured through the measure of *UEffec*.
Independent Variables	The use or not of composite states (CS) in the diagrams.
Alternative-Hypotheses	H_{1a}: using composite states improves the *UEffect* in subjects when trying to understand an UML statechart diagram.

The materials and tasks to be performed in the first experiment and its replication (E1 and R1) were quite simple and the background knowledge of the subjects was not advanced.

The results obtained in these experiments indicated that using composite states did not improve the understandability of UML statechart diagrams. This assertion was completely against our original hypothesis, so we considered the study as preliminary and decided to perform further empirical studies for accepting or definitively reject our hypothesis.

Table 2. E2 and R2 main features

Location	Universities of Castilla-La Mancha, Spain (E2) and Insubria, Italy (R2)
Date	February (E2) and May (R2) 2006
Subjects	14 PhD students (E2) and 13 Computer Science students (R2)
Dependent Variable	Understandability of UML statechart diagrams, measured through the measures of UEffect, UTrans, and UReten.
Independent Variables	The use or not of composite states (CS) in the diagrams.
Alternative-Hypotheses	H1a: using composite states improves the UEffect in subjects when trying to understand an UML statechart diagram. H1b: using composite states improves the UTrans in subjects when trying to understand an UML statechart diagram. H1c: using composite states improves the UReten in subjects when trying to understand an UML statechart diagram.

With respect to the second experiment and its replication (E2 and R2), the materials and tasks were improved, especially with the use of the Cognitive Theory for Multimedia Learning (CTML) [16] for assessing the complete set of variables of the experimental design. More about this theory can be found in [6, 10, 16].

The results obtained with the UEffec variable agreed with the ones obtained in the E1 and R1, i.e., those UML statechart diagrams modeled without using composite states were easier to understand. Besides, the values obtained for the new variables introduced in this study, retention and transfer, were in almost all cases middle values, neither especially high nor low. This way, no strong assertions could be made about them.

So, we decided revising and improving the materials and tasks to perform in a new experiment, trying to throw light to these uncertain conclusions.

4 Third Experiment

In this section, we explain the process we followed when we carried out the third member of the family of experiments (E3).

The experiment was carried out in the facilities of Indra Company, located in Ciudad Real, Spain. Indra currently holds a top position in the market of professional services in software, with a sales volume which is close to 800 million euros and, after a long period of expansion, the company has spread to over 28 countries in 4 different continents. The company has recently reached maturity level 3, according to the CMMi model, and the achievement of level 4 is planned for the current year 2007.

4.1 Design

In this study, we based on the design of a digital watch [27], through a model which has a size and complexity representative of a real-life case. The working hypotheses and part of the procedure to follow were similar to E2 and R2. Table 3 outlines the main features of E3.

Table 3. E3 main features

Location	Ciudad Real (Spain)
Date	October 2006
Subjects	24 professionals (12 in each group)
Dependent Variable	Understandability of UML statechart diagrams, measured through the measures of *UEffect, UTrans,* and *UReten*.
Independent Variables	The use or not of composite states (CS) in the diagrams.
Alternative-Hypotheses	H1a: using composite states improves the *UEffect* in subjects when trying to understand an UML statechart diagram. H1b: using composite states improves the *UTrans* in subjects when trying to understand an UML statechart diagram. H1c: using composite states improves the *UReten* in subjects when trying to understand an UML statechart diagram.

4.2 E3 Procedure

The experiment was divided into two sessions, in the afternoon and morning of two consecutive days.

In order to have a homogenous knowledge background in the subjects, the first session began with a seminar about "Dynamic Modeling with UML." Twenty-five professionals attended the first session, and they were provided with a summary of the main concepts of dynamic aspects in modeling in general and in UML in particular. The last part of the seminar focused on the UML statechart diagrams, although there was not an explicit mention which made the subjects guess the relation between the seminar and the subsequent experiment.

After the seminar, the instructor explained several examples related with the test that the subjects were performing in that session (Test 0). These examples consisted of questions about navigation through several statechart diagrams. Then, the subjects performed Test 0. This test was used for balancing the subjects into groups depending on their knowledge and performance.

This first session lasted approximately two hours. After it, all of Test 0 questionnaires were corrected and the subjects were grouped into two groups depending on their results.

The subjects were ordered according to the number of correct answers and the time spent on the questionnaire [18]. After that, those subjects who occupied an odd rank were assigned to Group A, and the others to Group B. Thus, we obtained two balanced groups as Table 4 shows.

Table 4. Means and standard deviations for groups after Test 0

Group	N	Correct Answers		Time	
		Mean	Std. Dev.	Mean	Std. Dev.
A	13	5.3846	1.3409	628.62	122.88
B	12	5.4583	1.0967	620.67	135.18

One of the subjects who had been assigned to Group A did not show up for Session 2. We decided to not rearrange the groups, which would now have the same number of subjects.

Session 2 was composed of three phases. In phase 1, the subjects were informed that they had been grouped depending on their performance in the test collected the day before. After that, they received one UML statechart diagram and a copy of Test 1. To avoid possible learning effects, we adopted a balanced between-subjects and blocked design, i.e. each subject was assigned only one diagram. The subjects of Group A received a diagram modelled using composite states and those in Group B received exactly the same system but modelled without using composite states.

As in E2 and R2, this test was used for measuring the UEffec of the model. The questions in Test 1 covered all the different parts of the diagram so that we could make sure that all parts of the diagram had been covered by the subjects before we removed them. This phase lasted for 25 minutes.

Then, all diagrams and tests were collected and Tests 2 and 3 were handed out. Test 2 was used for measuring the UReten variable and consisted of a fill-in-the-blanks text with 10 gaps that the subjects had to complete in order to build the text with the specifications of the system. The subjects had 15 minutes for this phase.

Test 3 was used for measuring the UTrans variable and consisted of a list of 6 tasks to perform based on the information taken from the diagram. The subjects had 35 minutes to solve them.

All the questions and tasks to perform in the different tests were exactly the same for both groups.

After the end of Test 3, all the materials were collected and the subjects were handed out a final questionnaire, to collect the subjects' impressions about the difficulty of the tests and the main positive and negative points that they had found during the experiment.

Fig. 2 graphically describes all this process.

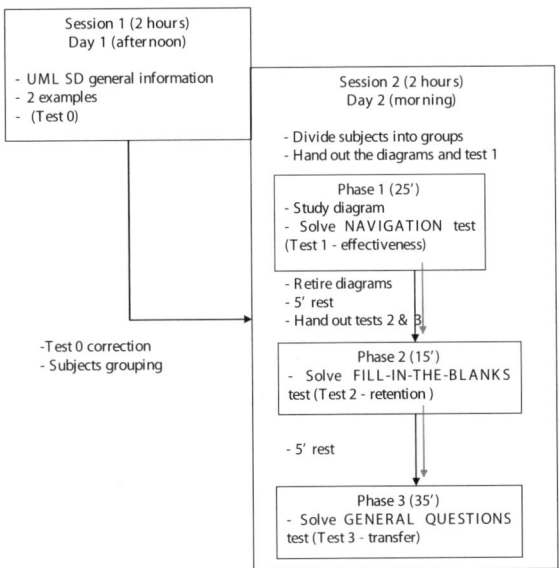

Fig. 2. E3 procedure schema

All the experimental material and data of the complete family of experiments can be found at the website http://alarcos.inf-cr.uclm.es/CSExperiments/

4.3 E3 Data Analysis and Interpretation

As in the previous studies, we carried out an analysis of the descriptive statistics of the data. Table 5 presents the descriptive statistics for the measures of the dependent variables studied in E3.

We can observe how, in this case, the results obtained for the understandability effectiveness and transfer variables are higher when the subjects worked with the diagram modelled with composite states, while the retention variable was higher in the diagram modelled without using composite states.

Table 5. Descriptive statistics for E3

CS	With (n=12)				Without (n=12)			
Stat.	Mean	Std. Dev.	Min.	Max.	Mean	Std. Dev.	Min.	Max.
UEffec	0.750000	0.1446	0.5	0.9	0.641667	0.1564	0.4	0.9
UTrans	0.369048	0.2079	0.071	0.786	0.261906	0.1677	0.036	0.571
UReten	0.775000	0.1390	0.5	1.0	0.891667	0.0515	0.8	1.0

After removing the outlier values, those which were too high and low and had fallen out of the range of typical values, we performed a t-test for testing the three hypotheses shown in Table 3 (see Table 6).

Table 6. E3 t-test results

Domain	UEffec		UTrans		UReten	
	t	sig.	t	sig.	t	sig.
Digital Watch	1.7617	0.092	1.3895	0.179	-3.6093	0.002

The results indicate that not using composite states positively affects the retention of the subjects while using composite states (weakly) improves UEffec, since the result is actually not statistically significant with our 0.05 threshold, but would be significant with a 0.1 threshold.

Regardless of the use of composite states, the subjects obtained a high level on the retention test and a low value in the transfer one. The CTML suggests that in this case, there has been only fragmented understandability but not memorization. In our opinion, the increased difficulty of the diagrams and the tasks might have produced this effect.

5 Lessons Learned and Future Work

In this work, we have presented a controlled experiment, last member of a family of empirical studies, for assessing how the use of composite states in an UML statechart diagram affects the understandability of the diagram.

We have followed the steps suggested in the Empirical Software Engineering, beginning our study with students for testing the original designs [25] and gradually improving the materials used and the experience of the subjects until performing the last study with a group of professionals.

After reviewing the obtained results in the family of experiments, we reach the following conclusions:

1. Our first idea, and the one most commonly accepted in the Software Engineering field, was that using composite states helps make a UML statechart be more understandable. The results of our empirical study do not support this idea. In most of the cases, the understandability effectiveness of the experimental subjects was better when working with systems modelled without using composite states. Nevertheless, in our last experiment, in which the subjects were professionals and hence more experienced, and the model was quite complex, the results for the understandability effectiveness were somewhat better for the subjects that used a model with composite states.
2. There is not a clear effect either in using or in not using composite states on the concepts of transfer and retention introduced by the CTML [16]. The increasing difficulty of the diagrams used and the task proposed to the subjects might have affected them, by making the diagrams more difficult to be understood.

Unexpectedly, despite performing a family of empirical studies, the evidence we collected does not support our initial hypothesis as clearly supported as we thought, and composite states do not seem to be highly useful for UML statechart diagrams understandability.

We have focused on understandability and we are planning to study some other quality aspects, such as modifiability, that could be affected by the presence of composite states. Tasks related to modifiability will probably be more complex and we will try to check whether the use of composite states makes them easier to perform.

Acknowledgements

This research is part of the MECENAS project (PBI06-0024) financed by "Consejería de Ciencia y Tecnología de la Junta de Comunidades de Castilla-La Mancha" and the ESFINGE project (TIN2006-15175-C05-05) and the CALIPSO Network (TIN20005-24055-E) supported by the "Ministerio de Educación y Ciencia" (Spain). The research presented in this paper has been partially funded by the IST project "QualiPSo," sponsored by the EU in the 6th FP (IST-034763), the FIRB project "ARTDECO," sponsored by the Italian Ministry of Education and University, and the project "La qualità nello sviluppo software," sponsored by the Università degli Studi dell'Insubria.

The authors would like to thank the staff in SOLUZIONA – Ciudad Real for their time and understanding during the preparation and performance of this experiment.

References

1. Atkinson, C., Kühne, T.: Model Driven Development: a Metamodeling Foundation. IEEE Transactions on Software Engineering 20, 36–41 (2003)
2. Basili, V., Shull, F., Lanubile, F.: Building Knowledge through Families of Experiments. IEEE Transactions on Software Engineering 25, 456–473 (1999)
3. Bolloju, N., Leung, F.S.K.: Assisting Novice Analysts in Developing Quality Conceptual Models with UML. Communications of the ACM 49(7), 108–112 (2006)
4. Cruz-Lemus, J.A., Genero, M., Manso, M.E., Piattini, M.: Evaluating the Effect of Composite States on the Understandability of UML Statechart Diagrams. In: Briand, L.C., Williams, C. (eds.) MoDELS 2005. LNCS, vol. 3713, pp. 113–125. Springer, Heidelberg (2005)
5. Cruz-Lemus, J.A., Genero, M., Piattini, M.: Metrics for UML Statechart Diagrams. In: Genero, M., Piattini, M., Calero, C. (eds.) Metrics for Software Conceptual Models, Imperial College Press, UK (2005)
6. Cruz-Lemus, J.A., Genero, M., Piattini, M., Morasca, S.: Improving the Experimentation for Evaluating the Effect of Composite States on the Understandability of UML Statechart Diagrams. In: Proceedings of 5th ACM-IEEE International Symposium on Empirical Software Engineering, Rio de Janeiro, Brazil, pp. 9–11 (2006)
7. Denger, C., Ciolkowski, M.: High Quality Statecharts through Tailored, Perspective-Based Inspections. In: Proceedings of 29th EUROMICRO Conference New Waves in System Architecture, Belek, Turkey, pp. 316–325 (2003)
8. Dobing, B., Parsons, J.: How UML is Used. Communications of the ACM 49(5), 109–113 (2006)
9. Erickson, J., Siau, K.: Can UML Be Simplified? Practitioner Use of UML in Separate Domains. In: EMMSAD 2007. Proceedings of Twelfth International Workshop on Exploring Modeling Methods in Systems Analysis and Design, Trondheim, Norway, pp. 89–98 (2007)

10. Gemino, A., Wand, Y.: Evaluating Modeling Techniques based on Models of Learning. Communications of the ACM 46(10), 79–84 (2003)
11. Ghezzi, C., Jazayeri, M., Mandrioli, D.: Fundamentals of Software Engineering. Prentice Hall, Englewood Cliffs (1992)
12. Höst, M., Regnell, B., Wohlin, C.: Using Students as Subjects - a Comparative Study of Students & Proffesionals in Lead-Time Impact Assessment. In: EASE 2000. Proceedings of 4th Conference on Empirical Assessment & Evaluation in Software Engineering, Keele, UK, pp. 201–214 (2000)
13. Hu, Z., Shatz, S.M.: Explicit Modeling of Semantics Associated with Composite States in UML Statecharts. Automated Software Engineering 13(4), 423–467 (2006)
14. Juristo, N., Moreno, A.: Basics of Software Engineering Experimentation. Kluwer Academic Publishers, Boston (2001)
15. Kitchenham, B., Pfleeger, S., Pickard, L., Jones, P., Hoaglin, D., El-Emam, K., Rosenberg, J.: Preliminary Guidelines for Empirical Research in Software Engineering. IEEE Transactions on Software Engineering 28(8), 721–734 (2002)
16. Mayer, R.E.: Multimedia Learning. Cambridge University Press, Cambridge (2001)
17. OMG, UML 2.0 - 2nd Revised Submission, Object Management Group (2003)
18. Otero, M.C., Dolado, J.J.: Evaluation of the Comprehension of the Dynamic Modeling in UML. Information and Software Technology 46(1), 35–53 (2004)
19. Pigoski, T.: Practical Software Maintenance. Wiley Computer Publishing, New York, USA (1997)
20. Pressman, R.: Ingeniería del Software. McGraw-Hill, New York (1993)
21. Selic, B.: The Pragmatics of Model-Driven Development. IEEE Software 20(5), 19–25 (2003)
22. Sjoberg, D.I.K., Hannay, J.E., Hansen, O., Kampenes, V., Karahasanovic, A., Liborg, N.K., Rekdal, A.C.: A Survey of Controlled Experiments in Software Engineering. IEEE Transactions on Software Engineering 31(9), 733–753 (2005)
23. Sommerville, I.: Software Engineering. Addison Wesley, Reading (1995)
24. SPSS, SPSS 12.0, Syntax Reference Guide, SPSS Inc.: Chicago, USA (2003)
25. Tichy, W.F.: Hints for Reviewing Empirical Work in Software Engineering. Empirical Software Engineering 5, 309–312 (2000)
26. Verelst, J.: The Influence of the Level of Abstraction on the Evovability of Conceptual Models of Information Systems. In: ISESE 2004. Proceedings of 3rd International Syposium on Empirical Software Engineering, Redondo Beach, USA, pp. 17–26 (2004)
27. Webb, K.: Xholon Digital Watch Project (2006)
28. Wohlin, C., Runeson, P., Hast, M., Ohlsson, M.C., Regnell, B., Wesslen, A.: Experimentation in Software Engineering: an Introduction. Kluwer Academic Publishers, Boston (2000)

Preface to RIGiM 2007

Colette Rolland[1] and Eric Yu[2]

[1] Université Paris1 Panthéon Sorbonne, France
[2] University of Toronto, Canada

RIGiM aims to provide a forum for discussing the interplay between requirements engineering and conceptual modeling, and in particular, to investigate how goal- and intention-driven approaches help in conceptualizing purposeful systems. What are the fundamental objectives and premises of requirements engineering and conceptual modeling, respectively, and how can they complement each other? What are the demands on conceptual modeling from the standpoint of requirements engineering? What conceptual modeling techniques can be further taken advantage of in requirements engineering? What are the upcoming modeling challenges and issues in GORE? What are the unresolved open questions? What lessons are there to be learnt from industrial experiences? What empirical data are there to support the cost–benefit analysis when adopting GORE methods? Are there applications domains or types of project settings for which goals and intentional approaches are particularly suitable or not suitable? What degree of formalization and automation or interactivity is feasible and appropriate for what types of participants during requirements engineering? e.g., business domain stakeholders, requirements modelers, ontology engineers, etc.

An Ontology for Requirements

John Mylopoulos[1], Ivan J. Jureta[2], and Stéphane Faulkner[2]

[1] Department of Computer Science, University of Toronto, Ontario, Canada
jm@cs.toronto.edu
[2] Information Management Research Unit (IMRU), University of Namur, Belgium
iju@info.fundp.ac.be, stephane.faulkner@fundp.ac.be

Abstract. In the good old days, the world of Requirements Engineering (RE) was simple: there were functional requirements to be modelled, somehow, and non-functional ones that usually consisted of a product quality wish list. Solving a particular requirements problem amounted to (loosely-understood) accommodation of functional requirements and doing one's best with non-functional ones. This world changed dramatically with the advent of Goal-Oriented Requirements Engineering. The primitive concepts in terms of which requirements are now conceived are no longer functions, states and things. Instead, the brave new world is populated with goals, stakeholder intentions and social settings.

We review, contrast and compare some of the new and old concepts, including goal, intention, function, preference, priority, softgoal, quality, criterion, and non-functional requirement. In addition, we attempt to organize them into a new ontology for requirements. We also present first results on a theory of requirements where, given a requirements problem, we define precisely what is a solution and what is an optimal solution.

J.-L. Hainaut et al. (Eds.): ER Workshops 2007, LNCS 4802, p. 224, 2007.
© Springer-Verlag Berlin Heidelberg 2007

GOORE : Goal-Oriented and Ontology Driven Requirements Elicitation Method

Masayuki Shibaoka[1], Haruhiko Kaiya[2], and Motoshi Saeki[1]

[1] Dept. of Computer Science, Tokyo Institute of Technology
Ookayama 2-12-1, Meguro-ku, Tokyo 152, Japan
[2] Dept. of Computer Science, Shinshu University
Wakasato 4-17-1, Nagano 380-8553, Japan
saeki@se.cs.titech.ac.jp, kaiya@cs.shinshu-u.ac.jp

Abstract. Goal oriented modeling methods are one of the promising approaches to elicit requirements. However, the difficulties in goal decomposition during requirements elicitation processes and a lack of the technique to utilize domain knowledge are obstacles to make them widely used in industry community. This paper proposes a method call GOORE where a domain ontology is utilized to support goal decomposition as domain knowledge to support goal decomposition. Although the experimental case study of feed readers suggested some points necessary to be improved, our approach can make goal oriented approach more powerful and more widely used.

1 Introduction

Goal-oriented modeling is one of the promising methodologies for requirements elicitation. In this methodology, customers' needs are modeled as goals, and the goals are decomposed and refined into a set of more concrete sub-goals. After finishing goal-oriented requirements analysis, the analyst obtains an acyclic (cycle-free) directed graph called goal graph. I There are several excellent studies on goal oriented methods such as I* [7], Tropos [2], KAOS [9], AGORA [4] etc. and their results are being put into practice. However, almost of them focused on languages including graphical languages and supporting tools. One of the obstacles to be spread to industries is a lack of powerful support of goal decomposition and refinement processes, i.e. supports for creating sub-goals of high quality. In these processes, stakeholders performing goal oriented analysis need high level of knowledge. For example, in early requirements elicitation phases for modeling business processes whether they are as-is or to-be [10], knowledge of a problem domain such as banking business is necessary, while the analysts should have much knowledge on the domain of software technology and an application such as Web computing and cryptography algorithms. Utilizing this kind of knowledge effectively allows us to perform goal decomposition and refinement processes of high quality.

This paper proposes two points; one is the technique to represent knowledge of a specific domain (domain knowledge) as an ontology and another is to embed the ontological system with goal-oriented analysis processes. We consider that an ontology consists of a set of concepts represented with a thesaurus and inference rules on them. By using inference rules, the sub-goals are deduced and suggested to the analysts. The

J.-L. Hainaut et al. (Eds.): ER Workshops 2007, LNCS 4802, pp. 225–234, 2007.

rest of the paper is organized as follows. The next section presents the basic idea and the structure of our ontologies. In sections 3, we show our method GOORE (Goal Oriented and Ontology driven Requirements Elicitation) and its supporting tool. Section 4 presents a case study of developing a feed reader to assess our approach.

2   Basic Idea

2.1   Goal Graph

Goal oriented methods are for supporting hierarchical decomposition and refinement of customers' requirements as the goals to be achieved. Basically, the resulting artifact after performing this method is an AND-OR graph whose nodes are elicited goals. Each goal is decomposed and refined into more concrete goals, sub-goals during the process. We have two types of goal decomposition; one is AND decomposition and another is OR. In AND decomposition, if all of the sub-goals are achieved, their parent goal can be achieved or satisfied. On the other hand, in OR decomposition, the achievement of at least one sub-goal leads to the achievement of its parent goal. Figure 1 illustrates a part of the goal graph of a software music player which can play the sound files of MP3, WPA etc, like Windows Media Player. The edges attached with an arc outgoing from a parent node show an AND decomposition, and for example, all of three goals "Control Playing", "Have Visualization Effects" and "Lightweight Processing" should be achieved in order to achieve their parent goal "Music Player" in the figure. On the other hand, either "Use Skins" or "Visualize Music", or both are necessary for the achievement of "Have Visualization Effects".

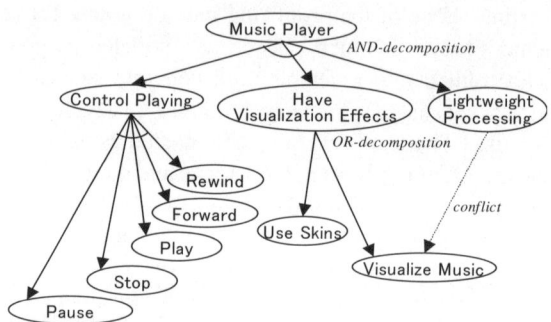

Fig. 1. An Example of a Goal Graph for Music Players

2.2   Using a Domain Ontology

As discussed in [6], we consider an ontology as a thesaurus of words and inference rules on it, where the words in the thesaurus represent concepts and the inference rules operate on the relationships on the words. Each concept of a domain ontology can be considered as a semantic atomic element that anyone can have the unique meaning in a problem domain. The content of each goal is written in natural language and we use semantic processing of the goal content using a domain ontology.

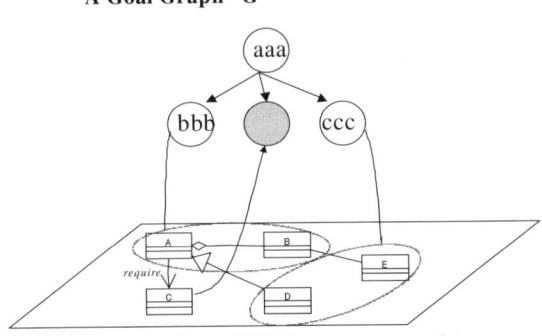

Fig. 2. Mapping from a Goal Graph to an Ontology

The words appearing in a goal description are mapped into words of the thesaurus part of the ontology, as shown in Figure 2. In the figure, the ontology is written in the form of class diagrams. Suppose that the goal description "bbb" contains the concepts A and B or their synonyms. The goal of "bbb" is mapped into the words A and B of the ontology, as shown in the figure. The logical inference on the ontology suggests to the analyst how to evolve his or her goal graph G. In the figure, although the graph G includes the goal node "bbb" mapped into the concept A, it has none of the goals mapped into the concept C, which is required by A. The inference resulted from "C is required by A" and "A is included" suggests to the analyst that a goal having C should be added to the graph G, say as a sub-goal shown in a grayed goal of the figure. In our technique, it is important what kind of relationship like "required by" should be included in a domain ontology for inference, and we will discuss this issue in the next sub section.

2.3 Domain Ontology

Figure 3 (a) shows the overview of a meta model of the thesaurus part of our ontologies. Thesauruses consist of concepts and relationships among the concepts, and they have varies of subclasses of "concept" class and "relationship". In the figure, "object" is a sub class of a concept class and a relationship "apply" can connect two concepts. Concepts and relationships in Figure 3 are introduced so as to easily represent the semantics in software systems. Intuitively speaking, the concepts "object", "function", "environment" and their subclasses are used to represent functional requirements. On the other hand, the concepts "constraint" and "quality" are used to represent non-functional requirements. The concept "constraint" is useful to represent numerical ranges, e.g., speed, distance, time expiration, weight and so on. The figure also includes a part of the inference rules. For example, the first and second rules express the reflectivity of synonym relationship and the inheritance of require relationship to a sub concept, respectively. Figure 3 (b) illustrates a part of the thesaurus for the domain of music players.

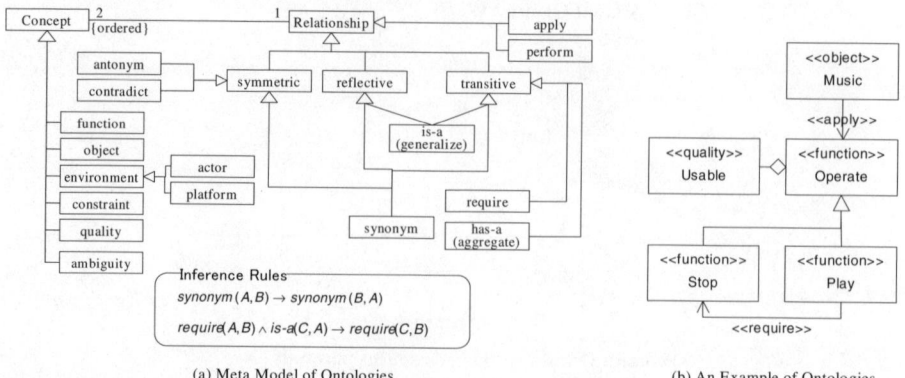

(a) Meta Model of Ontologies   (b) An Example of Ontologies

Fig. 3. Ontology Meta model

3   GOORE Method and Supporting Tool

3.1   Elicitation Process

We call our method GOORE (Goal-Oriented and Ontology Driven Requirements Elicitation) and Figure 4 sketches the process for eliciting requirements following GOORE. The tasks to be preformed during the GOORE process can be divided into two cate-

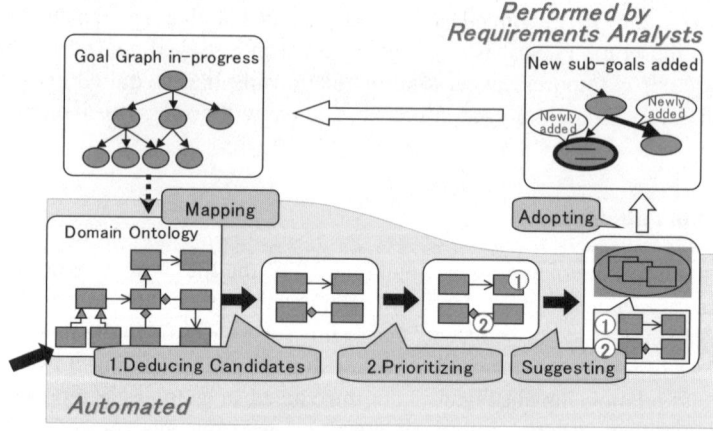

Fig. 4. GOORE Process

gories; one is a set of the activities that human requirements analysts should perform and another is the task that can be automated. The lower part, gray-shaded area of the figure expresses the automated tasks. The inputs of the automated task are a domain ontology and a goal graph in-progress that will be developed and evolved further. In the first step of the automated task, the natural language sentences of a goal description are semantically analyzed by using a morphological analyzer[1]. The morphological analyzer also tells the part of speech of a morpheme (lexical categories such as nouns, verbs, adjectives, etc.), several morphemes that are not so significant can be removed from the sentences. For example, articles such as "a" or "the" are removed in general. After filtering out morphemes and identifying words and their parts of speech, the tool finds corresponding ontological concepts to each morpheme using a synonym dictionary. Thus mapping from the sentences into thesaurus part of the ontology is semi-automatically constructed. In the example of Figure 5 (a), the sentence of a goal includes the word "play" and it can be mapped into the ontological concept "play".

After constructing mappings, the inference mechanism on the ontology deduces and detects a set of the ontological concepts that should be added to the goal graph. The inference rules are described as a Prolog program. In Figure 5, the current version of the goal graph does not include a goal related to "stop", even though there is a require-relationship between Play and Stop in the ontology. Thus, the goal presenting Stop operation should be added to the goal graph shown in the left part of the figure. In addition, since the generalized concept Operate of Play has a quality concept Usable, the goal presenting the quality "Usable" can be suggested to add to the goal graph. Our inference rules are used for tracing the ontological relationships between the concepts and for detecting ones missing out of the goal descriptions. On account of space, the details of the rules are omitted and for the readers who have an interest to them, refer to [3].

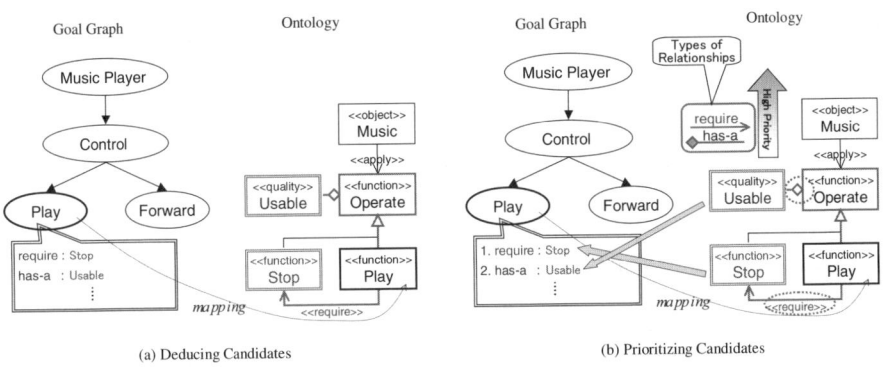

(a) Deducing Candidates   (b) Prioritizing Candidates

Fig. 5. Deducing and Prioritizing Candidates

Until the above steps, we can get several candidates to be added to the goal graph, e.g. Stop and Usable. Generally, there can be much more candidates suggested to the requirements analysts, as the goal graph becomes larger. The next step is prioritization. To avoid the explosion of candidates suggested, we attach priority to the deduced candidates and the candidates having high priority are selectively suggested. The priority can be calculated from numerical degrees attached to ontological elements, the structure of the goal graph being constructed, the inference process and records of the

---

[1] Since our supporting tool is currently Japanese version, we use Japanese morphological analyzer called SEN (https://sen.dev.java.net) written in Java.

analyst's selection activities. This calculation mechanism is embedded into the Prolog program and during the inference on the ontology, a candidate and its priority are simultaneously calculated. In the example of Figure 5 (b), consider that two candidates "Stop" and "Usable" are deduced from "Play". We attach to relationship type "require" higher priority degree rather than "has (aggregation)". Furthermore the relationship "is (generalization)" is used to deduce the "Usable". That is to say, to reach "Usable", two relationships "is" and "has" should be traced, while only one relationship is used for getting "Stop". The case where the fewer relationships are used for getting deduction results has the higher priority. Thus, the candidate "Play" is listed with higher priority than "Usable", as shown in Figure 5 (b). According to the ontological relationships and concepts, we provide constant numerical values as priority degrees, e.g. "require" has 5 and "has" has 3 etc. And the priority degree is also attached to each inference rule, and whenever the rule is used for deducing a candidate, its priority degree is numerically added. The number of inference steps is a factor to calculate the priority and the fewer step gives the higher priority value. The priority value is dynamically calculated and changed. Suppose that the analyst does not select the highest priority "Stop" in Figure 5 (b). At the next calculation, the priority degree of the candidate "Stop" is reduced. One of the benefits of using Prolog is flexibility of calculation rules of priority degrees.

3.2   Using a Supporting Tool

We have developed a prototype of the supporting tool for GOORE method. It architecturally consists of two parts; an editor of goal graphs and an inference engine to deduce and prioritize candidates. The former part was implemented as Java program, while in the latter Prolog was used. Thus, our ontologies are defined as facts of a Prolog program. In the example of Figure 3, the concepts Play and Stop and their require-relationship can be represented as a Prolog fact `require(function(Play), function(Stop))`. Each ontological concept is defined using a Prolog functor labeled with its type and the concept, e.g. `function(Play)` and `quality(Usable)`, while a relationship corresponds to a binary predicate like `require(x,y)` and `has(z,w)`.

The tool has the functions; 1) inputting and editing goal graphs in graphical form, i.e. a graphical editor for goal graphs, 2) getting an ontology described in Prolog, 3) deducing candidates and prioritizing them as background tasks of Prolog execution, and 4) showing a list of the candidates following their priorities to an analyst as a menu so as to make him or her select suitable ones out of them. The tasks of deducing, prioritizing and showing the candidates are done in real-time during the analyst's activities of inputting and editing the goal graph.

Figure 6 illustrates a screenshot of the tool. In the figure, an analyst is constructing a goal graph of a feed reader (news aggregator), which is a client software to retrieve frequently updated Web contents of certain topics from blogs and mass media web sites etc. She uses the ontology of Feed Reader and is refining the goal "User Registration" (for the feed reader). The candidates of newly added goals related to "User Registration" are shown in a small window. The mark ⊞ attached to a goal shows the existence of the deduced candidates, and she can get the candidate list if she clicks this mark. Figure 7 shows the process of selecting a candidate as a sub-goal of "User Registration".

As shown in the left screen, the analyst selects as a sub-goal the candidate "Setting a Personal Profile", which has the second priority in the candidate list, and gets a new goal graph shown in the right screen of the figure.

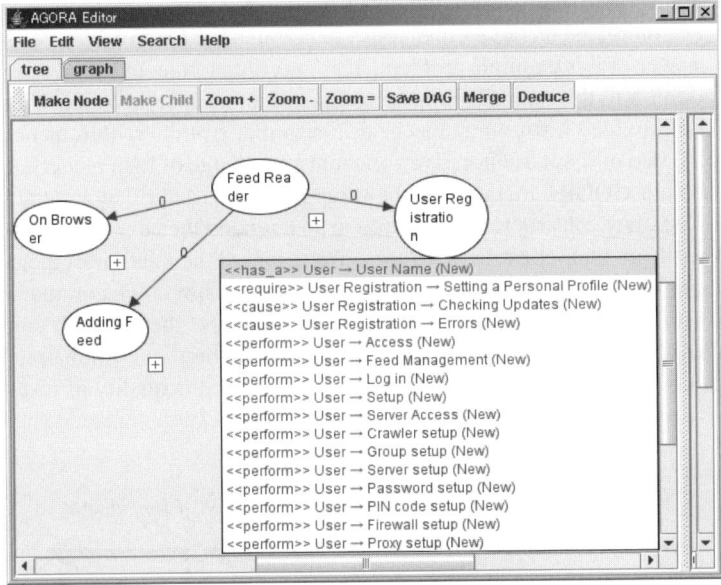

Fig. 6. Screen of the Supporting Tool

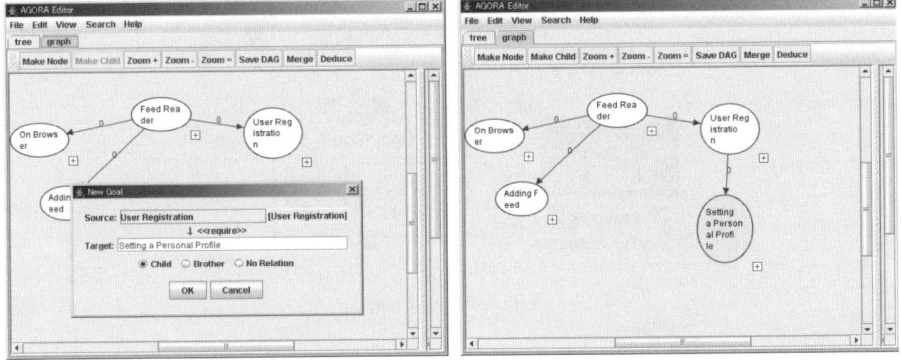

Fig. 7. Selecting a Candidate on the Supporting Tool

4   Case Study

We had a small experimental case study to investigate the usefulness of our approach. We used the example of Feed Reader, which was mentioned in the previous section, and asked our subjects to construct its goal graph. Before starting this experiment, we developed a domain ontology of Feed Reader. It consists of 593 concepts and 1023 relationships as a thesaurus part, and has 8 types of inference rules and 10 types of calculation rules for prioritizing candidates. Our supporting tool is set to suggest 16 candidates at a time for each goal.

Figure 8 illustrates the process of our experimental case study. One of us, the expert of Feed Reader, completed its requirements specification in the form of itemized sentences as a right answer of this example problem. This specification included 93 requirements items. In addition to this specification, he composed an initial requirements list having 10 items and provided it for our subjects as a customer's initial requirements. We had four subjects, two of them did not have sufficient knowledge of feed readers, while two of them used our GOORE method and the supporting tool. After they got the initial requirements list, they constructed goal graphs in an hour and then extracted requirements specifications from their constructed graphs. We recorded their activities including tool manipulations with a video camera and a screen recorder. They could communicate with one of us, who played a role of customers, in order to resolve their unclear points. After they finished these construction tasks, we had interviews and questionnaires to evaluate usability of our method and tool. In addition, to check the quality of resulting goal graphs, we compared them with the right answer that the expert developed before.

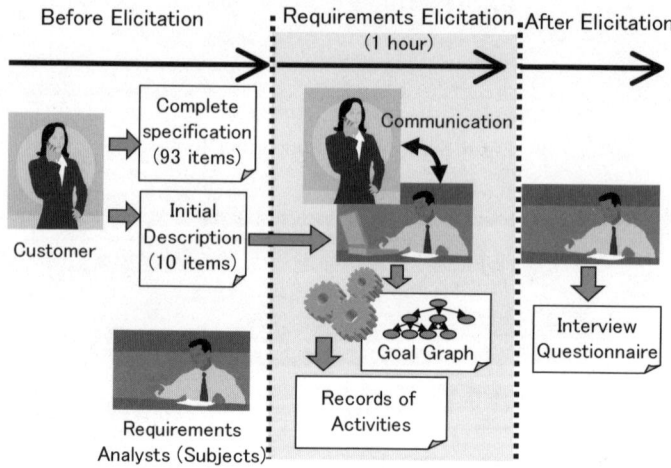

Fig. 8. Experiment

Table 1 summarizes the experimental result. Each column from the second expresses the result of a subject. In the example of the third column, the subject (No, Yes) means that the subject had no knowledge of feed readers and used our supporting tool. He got 43 goals, 10 of which resulted from the candidates suggested by the tool. He selected the candidate of rank average 5.9 out of 16 suggested candidates, and it shows that our adopted priority calculation would be reasonable. Furthermore 40 requirements items could be extracted from the 43 goals, and 34 of them could be considered as right answers because these 34 were also included in the complete specification. 9 of the 10

Table 1. Experimental Result

Evaluation Items	Result of a Subject[†]			
	(No, No)	(No, Yes)	(Yes, No)	(Yes, Yes)
#Goals	42	43	63	48
#Goals obtained by Suggesting Candidates	—	10	—	13
#Goal Relationships by Suggesting Candidates	—	0	—	0
Average of Ranks of Adopted Goals (in 16 items)	—	5.9	—	2.4
# Requirements Items included in the Goal Graph[‡]	38	40 (10)	71	60 (11)
# Requirements Items included in the Goal Graph and the Completed Artifacts[‡]	32	34 (9)	56	55 (10)
Expiration Time for Constructing a Goal	70sec.	76sec.	55sec.	75sec.
Longest Time for Deducing Candidates	—	23sec.	—	68sec.

† (having domain knowledge or not, using a GOORE method & a tool or not)
‡ the number in ( ) stands for the requirements items included in the adopted candidates

adopted candidates were also included in the completed specification, and it shows that the suggested candidates had high quality. In the case of another subject using the tool, i.e. (Yes, Yes), the reasonability of prioritization (rank 2.4) and the high quality of the suggested candidates were clearer.

Negative results are that 1) the tool users did not find goals much more than non-users and 2) the tool users spent longer time for constructing a goal.

5   Conclusion and Future Work

In this paper, we presented an extended version of a goal oriented modeling method which was combined with an ontological technique to utilize domain knowledge. Furthermore we developed a supporting tool following the proposed method and had a experimental case study. From the results of the case study, although our approach has still some deficiencies, we found that we should keep our approach, GOORE. The future research agenda can be listed as follows.

1. Combining GOORE with other methods, in particular scenario based approach
   Since our approach uses natural-language description, it would be more effective in the cases when more natural-language sentences are described such as scenario oriented methods [8].
2. Improving how to suggest deduced candidates to analysts
   The current version of the tool shows the candidates in the form of words and phases only. In our case study, the subjects could not understand the correct meanings of the suggested words and phases, and were sometime confused to hold precise intents of the suggestions. This shortcoming caused longer time consumption for the tool users to construct their goal graphs in our experiment. Ontologies can include this kind of information.

3. Constructing ontologies of high quality
   Constructing an ontology by hand is time-consuming. Currently, we consider the application of text mining techniques to extract ontological elements from natural-language documents in a domain [1,5].

References

1. KAON Tool Suite, http://kaon.semanticweb.org/
2. Castro, J., Kolp, M., Mylopoulos, J.: A Requirements-Driven Development Methodology. In: Dittrich, K.R., Geppert, A., Norrie, M.C. (eds.) CAiSE 2001. LNCS, vol. 2068, pp. 108–123. Springer, Heidelberg (2001)
3. Kaiya, H., Saeki, M.: Using domain ontology as domain knowledge for requirements elicitation. In: RE 2006. Proc. of 14th IEEE International Requirements Engineering Conference, pp. 189–198. IEEE Computer Society Press, Los Alamitos (2006)
4. Kaiya, H., Horai, H., Saeki, M.: AGORA: Attributed Goal-Oriented Requirements Analysis Method. In: RE 2002. IEEE Joint International Requirements Engineering Conference, pp. 13–22. IEEE Computer Society Press, Los Alamitos (2002)
5. Kitamura, M., Hasegawa, R., Kaiya, H., Saeki, M.: An Integrated Tool for Supporting Ontology Driven Requirements Elicitation. In: ICSOFT 2007. Proc. of 2nd International Conference on Software and Data Technologies (2007)
6. Maedche, A.: Ontology Learning for the Semantic Web. Kluwer Academic Publishers, Boston (2002)
7. Mylopoulos, J., Chung, L., Yu, E.: From Object-Oriented to Goal-Oriented Requirements Analysis. Communications of the ACM 42(1), 31–37 (1999)
8. Rolland, C., Souveyet, C., Achour, C.B.: Guiding Goal Modeling Using Scenarios. IEEE Transaction on Software Engineering 24(12), 1055–1071 (1998)
9. van Lamsweerde, A.: Goal-Oriented Requirements Engineering: A Guided Tour. In: RE 2001, pp. 249–263 (August 2001)
10. Yu, E.: Towards Modeling and Reasoning Support for Early-Phase Requirements Engineering. In: RE 1997. Proc. of 3rd IEEE International Symposium on Requirements Engineering, pp. 226–235. IEEE Computer Society Press, Los Alamitos (1997)

Early Prioritisation of Goals

Sarah Hatton

The University of Western Australia,
School of Computer Science and Software Engineering,
35 Stirling Highway
Crawley, Western Australia, 6009
sarah@csse.uwa.edu.au

Abstract. Prioritisation is important in keeping a software project on track and ensuring that the stakeholder's needs are being fulfilled. In this paper four methods for goal prioritisation are examined. Using a case study as its basis, each method is evaluated on its ease of use for a stakeholder, the time it takes to complete and the amount of confidence they have that the method accurately reflects their priorities. The type of information yielded by each method is also described. The results show the techniques give a consistent priority. Grouping and ranking techniques are easier for stakeholders to perform and generally take less time than techniques showing magnitude of preference. There is also a strong correlation between the stakeholder's confidence in the technique and how easy they feel it was to perform.

Keywords: Goal Prioritisation, Agile, Analytical Hierarchy Process (AHP), MoSCoW, Hundred Dollar Method, Simple Ranking, Software Engineering.

1 Introduction

The process of defining and prioritising a clear and precise set of requirements is inherently difficult. However it is a task which must be done in order to avoid costly project failures [1]. In many cases the phase in which a projects' requirements are elicited and defined, is not allocated the time and resources required to ensure customer satisfaction [2]. In reality this is unlikely to change and rather than pushing for increasing the time and money allocated to this phase, there is a need to find a method, or combination of methods, which are relatively quick and simple to perform and thus minimise the cost associated with these tasks.

Performing prioritisation early on in the software development life-cycle has a number of important advantages, including minimising the risk of overlooking important functionality [3] and helping the development team to allocate project resources to the most valuable requirements [4]. It was shown by Ryan and Karlsson [5] that in an example software project 94% of the value of the project could have been delivered for about 78% of the possible maximum cost, thus highlighting the benefit of prioritisation for conserving resources.

While the benefits of prioritisation are well known [1-5], it is usually not performed at the very start of a project, but rather once the majority of the systems

J.-L. Hainaut et al. (Eds.): ER Workshops 2007, LNCS 4802, pp. 235–244, 2007.

requirements are discovered. This is where looking at a system in terms of goals can provide the basis for performing prioritisation early on, and as such minimise the risk of wasting project resources on areas not directly related to the main system goals. Even from the early stages of a project, when requirements are rapidly changing and evolving, the goals remain relatively stable [6], thus making them an ideal starting point for gauging priorities.

This paper describes a study undertaken to examine the client's confidence, time taken and ease of use for four prioritisation methods. It compares their resulting priorities to look at consistency and makes some suggestions about the use of the methods. It further suggests how using a combination of methods during different stages of a software project is likely to yield the most useful results in terms of minimising risk, gauging client value and perspectives and providing traceability between requirements and overall goals.

1.1 Benefits of Using Goals

Goals provide the framework for the desired end product; they can relate to the functional, non-functional and business objectives of a system [6]. As goals are initially at a high level of granularity, they provide an excellent starting point to examine priorities, look for interactions and define the scope of a system. Once goals are refined into more specific goals or individual system requirements, having a record of where a requirement originated from provides traceability.

1.2 Performing Prioritisation

A software development team may overlook the fact that stakeholders in the majority of projects are often not from an Information Technology background. One of the implications of this is that they may not understand, or be comfortable with, some of the tasks they are asked to perform. Somé [7] highlights this by stating that a gap exists between the software development process and the client which results in requirements induced software failures. He emphasises the importance of keeping the stakeholders involved and sourcing and representing information in a way that is understandable for the stakeholder. If a stakeholder does not understand what is being asked of them, there is a strong possibility that they will inadvertently provide misleading information, which could result in the failure of the project.

2 Prioritisation Methods

Many requirements prioritisation techniques were developed for domains other than software engineering but have applicability to this domain as well. The various techniques yield different degrees of information about the stakeholder's preferences. In this section four different requirements prioritisation techniques are briefly described.

2.1 Selection of Methods

Previous studies have been carried out which aim to compare prioritisation methods on various criteria [8-10]. These works were used to aid in the selection of methods for this study. Karlsson *et al* [8], in 1998, conducted a study focusing on comparing pair-wise prioritisation methods. They selected six techniques and compared them on ease of use, time taken and reliability of results. They noted that AHP was difficult and time consuming, but still found it to be the best technique [8].

Other studies have compared the Extreme Programming Planning Game technique with a pair-wise method [9] showing the Planning Game method to be quicker. The scales of some prioritisation methods including the Hundred Dollar Method and numeral assignment (Simple Ranking) have also been examined and results indicated there was a trade off in the time taken compared to the richness of information [10].

This study selected four methods representing the range of the different approaches mentioned above. AHP was chosen as Karlsson *et al* [8] identified it as being the most promising pair-wise method. The other three techniques, Simple Ranking, MoSCoW and the Hundred Dollar Method, were commonly used and gave a range of information to the development team.   It was thought that they each possessed valuable characteristics which could be investigated.

2.2 Simple Ranking

Ranking elements into a single list has long been useful for prioritisation. It is fairly intuitive for most people as writing a list is something that is done in every day life. For *n* requirements the list is simply ranked *1...n*, where the most important requirement is ranked one and the least *n*. The use of this ordinal scale gives the order of preference but no information on the magnitude of the preference.

When there are a fairly small number of goals, ranking via this method is likely to be easy for people to do. There is well established research in the field of psychology, that people have difficulty remembering more than seven (plus or minus two) elements [11]. It would therefore be difficult and probably inaccurate to have a client try to rank a list of 15 or more elements. This is where grouping of goals may become important.

2.3 MoSCoW

MoSCoW is an acronym commonly used to represent four different priorities for requirements [12 – 14].

*"**M**UST have"*: Requirements are not negotiable; the failure to deliver these requirements would result in the failure of the entire project.

*"**S**HOULD have"*: Features that would be nice to have if at all possible

*"**C**OULD have"* Features that would be nice to have if at all possible but slightly less advantageous than the "S"

*"**W**ON'T have"* (also known as wish list). These requirements are not unimportant, but they will definitely not be implemented in the current software project. They may, at a later stage, be created.

MoSCoW is currently incorporated into the Dynamic Systems Development Method (DSDM) [14]. One problem with MoSCoW is that it doesn't give the software developer as much information as the other three methods presented in this paper. Although MoSCoW uses an ordinal scale it groups the requirements into hierarchal categories. As such all requirements grouped into one category are represented as having equal priority and no information on the magnitude of preference is given.

2.4 Hundred Dollar Method

The Hundred Dollar Method uses the concept of giving the user $100 to spend on getting goals implemented. It is a simple and quick method for getting a magnitude ranking which is suitable for use on an individual, or a group of stakeholders [15]. There are basically three steps that make up the Hundred Dollar method.

1. *The user is asked to assume that they have $100 to spend*
2. *They are asked to allocate the $100 across the options that they have to evaluate.*
3. *The distribution of money is re-evaluated. There are usually a few options which stand out as being higher priority. If the money is spread too evenly then another hundred dollars is given but it must be spent as one lot of $50 and two lots of $25.*

The hundred dollar method makes use of peoples' familiarity with dealing with money. Having money distributed over the requirements means it effectively uses a ratio scale giving detailed information about the magnitude of preferences.

There are a number of problems with this technique which have been identified by authors such as Davis [15] in that people may simply place all $100 on one requirement. Other problems include groups of people who will guess what options others will think are important. If an option which is important to them is also important to someone else then they may 'bet' that the other person will spend their money on it and then will spend their money on another they would also like [15].

2.5 The Analytical Hierarchy Process (AHP)

The Analytic Hierarchy Process is a well known method for aiding in the decision making process. It is applicable to many different problem situations where comparative judgment is required such as: selecting between competing alternatives in a multi-objective environment, allocating scarce resources and forecasting [16, 17].

Within the domain of Requirements Engineering, AHP provides a powerful tool for comparing alternative design concepts and also the relative value of components in a system. It is recognised that the full power of AHP is demonstrated when multi-objective decisions need to be made [18]. However, AHP is also commonly used when only one decision factor is being examined.

AHP uses pair-wise comparisons to examine a list of objectives or alternatives [17]. It requires components to be ranked [19], using an ordinal scale, to show their relative importance to one another. The scale of relative importance uses the numbers one to nine to signify the importance of one requirement over the other [15].

Each requirement is compared to each other one (a pair-wise comparison). Pair-wise comparisons are meant to be easy for a user to perform. This is because the user is only required to compare two options at any given time rather than examining a long list of alternatives. AHP is also non repetitive meaning that the user will only have to compare two goals once i.e. if they have compared α to β there is no need to compare β to α as the reciprocal can be assumed.

Once a value has been assigned for each pair-wise comparison, and the matrix of the relative importance of the goals has been completed, the relative value for each option is calculated [20]. The AHP also enables a consistency index to be calculated which indicates the accuracy of the results of the pair-wise comparisons. The AHP effectively transforms the information of the original ordinal scale into a ratio scale through a series of mathematical operations.

AHP has been described as being tedious and only realistic for small sets of elements [21]. However, in many projects, there tends to be smaller numbers of high level goals which means that AHP could work quite well.

3 Methodology

This research was conducted using a case study comprising two sections. The first section consisted of a scenario about a traffic monitoring system. A narrative description was provided and a list of seven goals to prioritise was given. Each prioritisation method was then walked through from the perspective of a fictitious traffic officer known as Sam and the results were filled out 'as Sam wanted them'. The participant was then asked to fill in each method to reflect their viewpoint. The first scenario was an introduction to prioritisation and no information was recorded.

The second scenario was about a mobile phone software system. It provided the user with a context and a list of twelve goals. The number of goals was intentionally kept small as literature indicates that this is the upper limit of the amount of information that can be processed [11]. Each participant was required to fill in their priorities using each method and record information about the time taken, rate the method's difficulty, and also to rate their confidence in it.

Time information was recorded by getting the participant to record the time they started and completed each page without interruption. Difficulty of the method (Ease of Use) was measured on an ordinal scale of 1-10 scale where 1 represents 'very easy' and 10 'very difficult'. Confidence was measured on a similar ordinal scale where 1 represented 'not confident' and 10 'very confident'.

A total of 31 studies were completed and have been used in this research.

3.1 Validity

Possible threats to the validity of this study were identified early on and measures taken to try and prevent them. These threats included:-

Participant selection – The use of a group of university students was deliberately avoided as they would have similar backgrounds which could bias results. Instead a wide variety of people of different ages, genders, levels of education and occupations were selected. It was felt that due to the diverse nature of the participating group, the results were more applicable to the wider population than results would have been from a larger group consisting solely of university students.

Study timing – It was thought that each study would take between 30 and 60 minutes to complete. Participants were allowed to complete it over the course of a day so they would not become bored or fatigued but would still retain information.

Order of methods – The order of the sheets within the study was varied to prevent bias caused by people getting more familiar with prioritisation as they progressed.

Explanation of methods – It was important to be consistent in the way that the participants were introduced to each method. To ensure this a 'learning case study' was provided which walked them through each method and allowed them to practice doing it. Actual results were only recorded about the second section of the study.

Explanation of goals – the meaning of the specific goals was not explained to participants. This may mean that different participants had a disparate interpretation of what each goal actually meant. As the actual preference for mobile phone features was not being examined, and each participant's interpretation of a goal would be consistent throughout the study, the validity of results should not be affected.

4 Results

Time taken (to the nearest minute) was calculated from the start and end times recorded on each sheet. There was a considerable amount of variation in time between the 31 participants (see Table 1). For the AHP completion times ranged from 7 to 22 minutes. There was less variation with the other methods being 1-4 minutes for Simple Ranking, 1-5 minutes for MoSCoW and 1-8 minutes for the Hundred Dollar Method. Looking at the mean times taken it is apparent that the AHP method took considerably longer to perform than any of the other methods. AHP is also the most difficult and the stakeholders have the least confidence in it. MoSCoW and Simple Ranking took a similar amount of time and had a similar degree of confidence and difficulty.

Table 1. Times Taken (Minutes), Median Confidence and Difficulty (1-10 Scale)

	Minimum Time	Maximum Time	Mean Time	Standard Deviation	Median Confidence	Median Difficulty
MoSCoW	1	5	1.78	1.083	8	2
Simple Ranking	1	4	1.5	0.73	8	3
$100	1	8	3.6	2.42	7	4
AHP	7	22	14.03	4.4	2	9

The relationship between Confidence and Difficulty was graphed showing a fairly clear relationship; when there was a high degree of difficulty the confidence was quite low as shown in Figure 1. The Spearman's Rank Correlation Coefficient [22] was calculated to be -0.873 (*p<0.001 on 118 df*) showing a very strong negative relationship. The relationships between the time and difficulty and time and confidence were also statistically investigated. The Spearman's Rank Correlation Coefficient for the time and difficulty was 0.648 (*p<0.001 on 118 df*) showing a strong positive relationship. For time and confidence the Spearman's Rank Correlation Coefficient showed that there was a highly significant negative relationship with a coefficient of -0.923 (*p<0.001 on 118 df*).

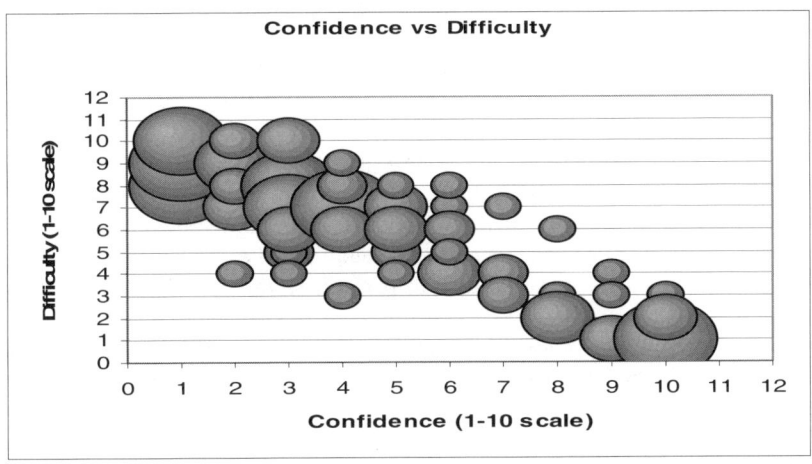

Fig. 1. Confidence and Difficulty for Prioritisation Methods

Since the consistency of the resulting priorities from the various methods is an important factor this was also examined. For each person their priorities for each method were compared using the Spearman's Rank Correlation Coefficient. The coefficient was then averaged; Table 2 shows the mean and minimum correlations between the priorities.

Table 2. Mean and Minimum Correlations between Prioritisation Methods

	AHP	$100	Simple Ranking	MoSCoW
AHP				
$100	0.86 mean 0.71 minimum			
Simple	0.80 mean 0.57 minimum	0.83 mean 0.59 minimum		
MoSCoW	0.86 mean 0.71 minimum	0.88 mean 0.79 minimum	0.91 mean 0.82 minimum	

5 Discussion

The results show that for each method there was substantial variation in the time taken by participants. This would be affected by the participants' previous experience with the methodologies or similar techniques. As the type of participants ranged from University lecturers in IT to business people who have no mathematical or scientific background, a fairly large variation was to be expected. The fastest time was recorded by a participant who had a background in software engineering and had used some of the methods before. The longest times were consistently recorded by a person who was not familiar with the concept of goal prioritisation.

There were considerable differences in the average time taken for the methodologies. This is best explained by the different complexities of each technique. For example AHP requires the participant to perform 66 individual comparisons compared to the ranking of 1-12 performed in the simple ranking method. This is further confirmed by the strong correlation (0.648) for the relationship between complexity and time.

The strong relationship between complexity and confidence is logical as it is generally accepted that simple methods of performing tasks are preferred; this has been investigated in domains such as marketing and psychology [23]. The strong relationship between time and confidence would also be related to this, the more difficult a task the longer it would take to perform so the confidence in the method would decrease.

While these simpler methods may be friendlier to the stakeholder, the requirements engineer must still be able to extract sufficient information out of the technique. Clearly techniques such as AHP and the Hundred Dollar Method give more information as a ratio scale is used. The Simple Ranking gives slightly less information as it uses an ordinal scale, giving order but not magnitude. The least information is provided by the MoSCoW method which hierarchically groups requirements with each member of the group having equal value.

The Hundred Dollar Method and AHP are both intended to provide ratio scale information. Participants had considerably more confidence in the accuracy of the priorities achieved using the Hundred Dollar Method, although the strong correlation between the priorities achieved, showed that they were giving the same results. If the same degree of information can be achieved, then it would be in the best interests of the software engineer to use the simpler technique. Possibly using the less confusing technique will reduce the amount of 'accidental' misleading information provided by the stakeholder.

Table 3. Properties of Prioritisation Methods

	Simple Ranking	MoSCoW	$100	AHP
Ratio Scale Information			√	√
High Confidence From User	√	√	√	
Consistent	√	√	√	√
Low Difficulty	√	√	√	
Low Effort	√	√	√	
Able to handle large numbers of alternatives		√		

This study showed that all methods were consistent in indicating the client's priorities. The high correlation is reassuring in that it indicates that whatever method is used, the client's priorities are likely to be accurately reflected. Table 3 summarises the main properties of each method and highlights MoSCoW and the Hundred Dollar Method as having many qualities which would make them suitable for the early prioritisation of goals.

6 Conclusion

Requirement Analysis is a critical but often error riddled phase of the software development process. The gap in understanding between the client and the software engineer is a reason for some of the errors. It is very important to ensure goal analysis has been systematically performed and that there is a clear and unambiguous list of goals which convey the stakeholder's views on importance to the developer.

The earlier prioritisation is performed, the quicker the project resources can be directed to where they will be of the highest value. Prioritising goals early on can thus be incredibly beneficial. When deciding on what prioritisation method to use it is important to consider the relative merits of the various methods. This study has shown that the MoSCoW method is one which client's have confidence in and find quick and simple to perform and from the engineers perspective it provides accurate priorities, and identifies groups which have higher value than others, thus making it ideally suited for the initial, early prioritisation of goals. When more detailed goals have been discovered and possibly more information on the "Must Have" and "Should Have" goals is needed, techniques such as the Hundred Dollar Method would be appropriate. While they are more time consuming, they provide detailed information about the relative value of these goals.

Software engineers need to take note that working closely with a client and using methods which are understandable to them, may be one of the main factors that determines a software projects success. Using simpler methods to get information from a client can mean that the client better understands what is being asked of them and thus provides the software team with more meaningful and useful information.

Acknowledgements. Thanks to Mark Reynolds and Terry Woodings for all their assistance and to the survey participants who made this research possible.

References

1. Standish, Group: "The High Cost of Chaos" from Few IS Projects Come in on Time, on Budget. Computerworld, 1994. p. 20 (December 12, 2004)
2. Davis, A.: Software Requirements: Analysis and Specification, 2nd edn. Prentice-Hall, Englewood Cliffs (1993)
3. Sommerville, I., Sawyer, P.: Requirements Engineering: A Good Practice Guide. John Wiley & Sons, England (1997)
4. van Genuchten, M.: Why is Software Late? An Empirical Study of Reasons for Delay in Software Development. IEEE Transactions of Software Engineering 17(6), 582–590 (1991)
5. Karlsson, J., Ryan, K.: Supporting the Selection of Software Requirements. In: Proceedings of the 8th International Workshop on Software Specification and Design, pp. 146–149 (1996)
6. Anton, A., Potts, C.: The Use of Goals to Surface Requirements for Evolving Systems. In: ICSE 1998. Proceedings of 20th international Conference on Software Engineering, Kyoto, Japan (1998)
7. Somé, S.: Supporting use case based requirements engineering. Information and Software Technology Amsterdam 48(1), 43 (2006)

8. Karlsson, J., Wohlin, C., Regnell, B.: An evaluation of methods for prioritizing software requirements. Information and Software Technology 39, 939–947 (1998)
9. Karlsson, L., Berander, P., Regnell, B., Wohlin, C.: Simple Is Better? - An Experiment on Requirements Prioritisation. In: SERPS 2003. proc. Third Conference on Software Engineering Research and Practice in Sweden, Lund, Sweden (2003)
10. Karlsson, L., Höst, M., Regnell, B.: Evaluating the practical use of different measurement scales in requirements prioritisation. In: ISESE 2006. Proceedings of the 2006 ACM/IEEE international Symposium on Empirical Software Engineering, pp. 326–335. ACM Press, New York, NY (2006)
11. Miller, G.A.: The Magical Number Seven, Plus or Minus Two: Some Limits on Our Capacity for Processing Information. The Psychological Review 63, 81–97 (1956)
12. Barritt, D.: IEC 61131 and DSDM in real-time process control applications. Computer and Control Engineering Journal, 94–100 (April 2002)
13. Tudor, D., Walter, G.A.: Using an Agile Approach in a Large, Traditional Organisation. In: AGILE 2006. Proceedings of AGILE 2006 Conference, pp. 367–373 (2006)
14. The DSDM Consortium: DSDM public version 4.2 online; last accessed (January 22, 2007), available www.dsdm.org
15. Davis, A.: Just Enough Requirement Management, Where Software Development Meets Marketing. Dorset House Publishing Company Incorporated New York, USA (2005)
16. Forman, E.H., Gass, S.I.: The Analytic Hierarchy Process - and Exposition. Operations Research 49(4), 469–486 (2001)
17. Mustafa, M.A, Al-Bahar, J.F.: Project Risk Assessment Using the Analytic Hierarchy Process. IEEE Transactions on Engineering Management 38(1), 46–52 (1991)
18. Saaty, T.L.: The Analytic Hierarchy Process, Planning, Priority Setting, Resource Allocation. McGraw-Hill, USA (1980)
19. Triantaphyllou, E., Mann, S.H.: Using the Analytic Hierarchy Process for Decision Making in Engineering Applications: Some Challenges. International Journal of Industrial Engineering: Applications and Practice 2(1), 35–44 (1995)
20. Bao, Y., Wu, T., He, Y., Ge, X.: An Improved AHP method in Performance Assessment. In: Fifth World Congress on Intelligent Control and Automation, pp. 177–180 (2004)
21. Ryan, K., Karlsson, J.: Prioritizing Software Requirements in an Industrial Setting. In: Proceedings of the 19th International Conference on Software Engineering, pp. 564–565 (1997)
22. Sidak, Z., Sen, P.K., Hajek, J.: Theory of Rank Tests (Probability and Mathematical Statistics), 2nd edn. Academic Press, San Diego, USA (1999)
23. Macintosh, G., Gentry, J.W.: Decision making in personal selling: Testing the "K.I.S.S. Principle". Psychology and Marketing. Hoboken 16(5), 393–399 (1999)

Goal-Aligned Requirements Generation

Kousik Sankar Ramasubramaniam[1] and Raman Venkatachar[2]

[1] Philips Electronics India Limited, MFAR Manyata Tech Park, Manyata Nagar, Nagavara,
Bangalore 560045. India
[2] Magnum Semiconductor, Salarpuria Infinity, No. 5, Bannerghatta Road, Byrasandra Village,
Bangalore 560029. India
Kousik.Sankar@philips.com, Raman.Venkatachar@magnumsemi.com

Abstract. During the process of eliciting requirements, it is important to be aware and constantly focused on the organizational goal like mission, vision, brand promise etc. It is important to understand purpose, goals and intentions as necessary conditions for successful design and development of products (for product companies). Losing focus on the organizational goal can lead to a disastrous situation especially where the customer completely relies on the supplier for full product development. This paper discusses a goal-oriented model that uses brand requirements to underpin the incoming requirements at every stage of development i.e. conceptualization, eliciting requirements, execution, and testing. The end result is a win-win relationship with the internal customer as well as the external customer.

Keywords: Organizational goal, requirements control, functional requirements, emotional requirements, product development, sense and simplicity.

1 Introduction

Traditional requirements management involves initial discussions with the customer, gathering requirements, executing them, rigorously testing them and agreement on a few planned releases. However, this traditional process is satisfactory for very low complexity product development. In today's world of consumerism and increasing influx of new technology applications in existing products, the customer would like to constantly align with the end-user requirements as they keep changing constantly. This constant change throws the traditional requirements process out of gear and even worse; this is mostly detected after 50% of the project timeline. By this time, it is too late to incorporate any design changes without compromising on the stability and the cost.

This paper focuses on how to achieve tighter requirements control driven by the original goal during each of the important milestones in product development viz. conceptualization and requirements, execution, testing (internal / field) and post-release maintenance [1].

J.-L. Hainaut et al. (Eds.): ER Workshops 2007, LNCS 4802, pp. 245–254, 2007.

2 Organizational Goal

All companies have an organizational goal. The difference lies in the goal depth and the *immediately apparent* relation to the product team's development life cycle process. Not taking the organizational goal into account results in a functional product that falls outside the boundary set of the requirements of the organizational goal.

It is known that any product development life cycle consists of six broad steps as depicted in Figure 1. The bubble boxes in Figure 1 indicate the important inputs that are required to effectively perform the desired step.

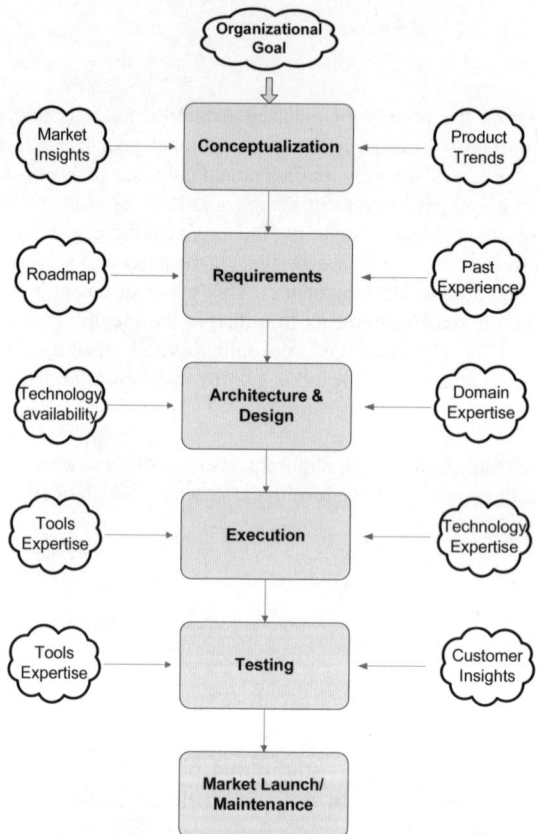

Fig. 1. External dependencies on product development life cycle

In this paper, we shall focus and delve deeper into the topic of requirements and associate the organizational goal as a means to cross-validate and verify the output of the requirements step [11]. We shall take a Consumer Electronics product sold worldwide. The product in discussion is a HDD DVD recorder with a 250 GB HDD and DVD Recordable drive with IEEE1394/cable inputs and HDMI output. The product was quite a complex one with lots of concurrent scenarios. The customer was a very high-end customer known for their professional expertise in audio and video. We shall arrive at a model in which most of the requirements [9] are underpinned before the commencement of the Architecture and Design phase [6], [8].

While it is idealistic to underpin 100% of the requirements during the requirements phase, it is not a realistic model. For the above-mentioned product development, we used a model whereby the requirements were progressively frozen [10] and aligned with the organizational goals at the end of each of the development life cycle phases (refer Table 1).

Table 1. Percentage of Requirements Frozen

Software Development Phase	Requirement Gathering	Architecture and Design	Execution	Testing	Launch / Maintenance
% of Frozen Requirements	70	10	10	8-10	0-2

In this paper, we shall focus on the requirements phase in order to ensure maximum benefit and alignment to the organizational goal. Following the same model for the other phases of the product development cycle ensures continuous alignment between the requirements and the organizational goals.

3   Goal ←→Requirements Link

The requirements phase is further sub-divided into the following activities [2] (see Figure 2):-

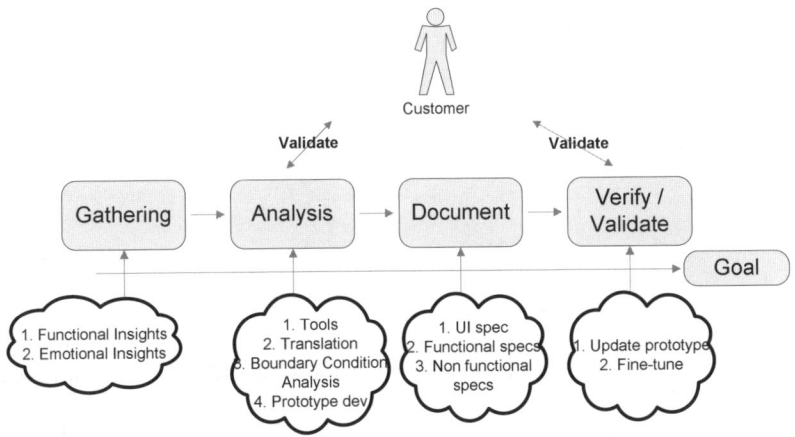

Fig. 2. Phases in requirements

3.1 Requirements Gathering

The main inputs for this phase are the consumer/market insights. These insights are basically of 2 types *Functional and Emotional*. For example, "I would like to view all my photos on my TV" is a pure functional requirement; compared to "I want to be able to see my family anytime anywhere" is an emotional requirement. Catering to the emotional requirement and delighting the customer will ensure a high degree of brand loyalty. Eliciting the emotional requirement is important during all stages of product development especially during design and execution.

Finding a strong link between the emotional requirement and the organizational goal (example, for Philips, the organizational goal is *Sense and Simplicity* as depicted in *Figure 3*), we realized that to get the right perspective of the organizational goal, we needed to strengthen the emotional requirement surrounding the organizational goal. This had to be done, keeping in mind the intricacies of the functional requirements, which in turn would reinforce the organizational goal [7].

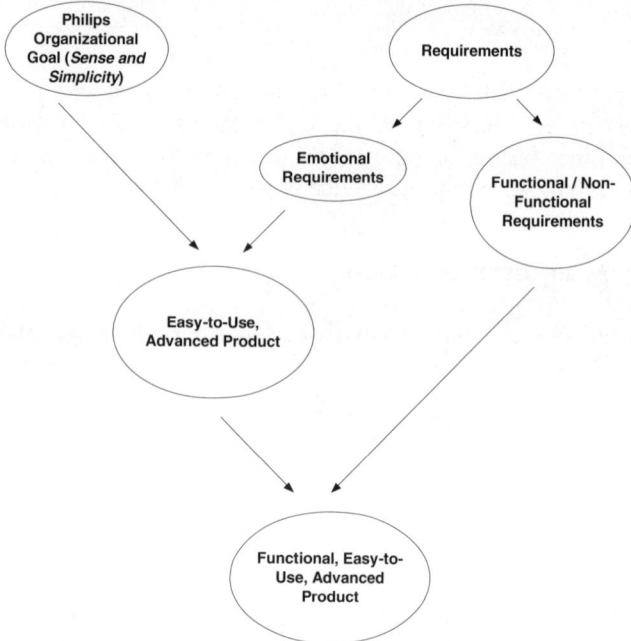

Fig. 3. Goal - Requirements Link

During the design and execution phases, we encountered a lot of situations where there were 2 or more possible options to choose from. Without burdening the customer with our queries, we were able to put on the customer's thinking hat and arrive at a recommended option out of the many possibilities. One of the main driving factors behind arriving at a recommended option was to ascertain the degree of alignment between each of the possible options and the customer's organizational goal. We then confirmed with the customer that this is indeed the right choice.

It is not so straightforward to elicit the emotional requirement. Eliciting the emotional requirement involves multiple facets of social and cultural maturity and fairly high degrees of emotional quotients (EQ)[4]. It is two-dimensional with both the customer and the supplier involved. It is important to understand the social and cultural nuances when interacting with diverse customers ranging from the Far East to western countries like US / Europe. For example, in some cultures like the European culture, it is perfectly acceptable to interrupt and ask questions if required. However, in some cultures in the Far East, this will be regarded as a social *faux pas*. Even a simple thing as addressing the customer in mails/conversations differs widely between various cultures. Some cultures are open to correcting others, whereas some others make pre-conceived impressions and stick to them. There is a possibility that such pre-conceived impressions can eventually lead to a downturn in the relationship. Hence there is a strong need to emotionally align with the customer and work towards perceiving requirements from the customer's perspective. One way of achieving this is via regular face-to-face meetings/workshops involving all the stakeholders. In such meetings, the architects/project managers on both sides (customer and supplier) have to be completely aligned, both functionally and emotionally.

The key point here is that understanding customer (both internal and external), their cultural nuances, emotional alignment, etc. helps building *rewarding relationships* which helps *free flow of information* and this contributes quite effectively to clarity of requirements. Due to this, any flux on the requirements would seem trivial to handle (for example, during the end of the execution phase, we got a new requirement to add a top-level directory menu structure with links to all existing menus/functionalities. This was a major change to the design and in spite of the increase in effort / cost; there were no bad feelings about such a new requirement creeping so late into the project. One of the main reasons for the lack of negative feelings was the alignment in viewing the new requirement from the customer's perspective). We had many sessions with the customer and gathered useful insights into the end consumer usage. All these needs were functional, except one that was emotional. This emotional need (The end-consumer of the product is less tolerable towards a 'wrong reaction' than a 'no reaction') played a very strong role in influencing decisions in later phases of the project.

3.2 Requirements Analysis

During the analysis phase, we involved the customer and reinforced their requirements in our framework via prototype demos that gave the customer a tangible feeling regarding the end product. We derived a lot of useful information from the demo reviews, especially reviewing the alignment with the organizational goal. The customer also got to know our perspective of his/her requirements and made some major changes to the requirements to suit our architecture.

Functional/non functional/emotional requirements as discussed earlier are the *positive scenarios* that satisfy the customer needs/wants. It is equally important to understand *negative scenarios* [13] (where and how the product can fail, and boundary conditions). Understanding negative conditions gives us opportunities to make the product design more robust before the design is implemented in the real world [6], [13]. Figure 4 depicts the contribution of positive/negative conditions in making a quality product.

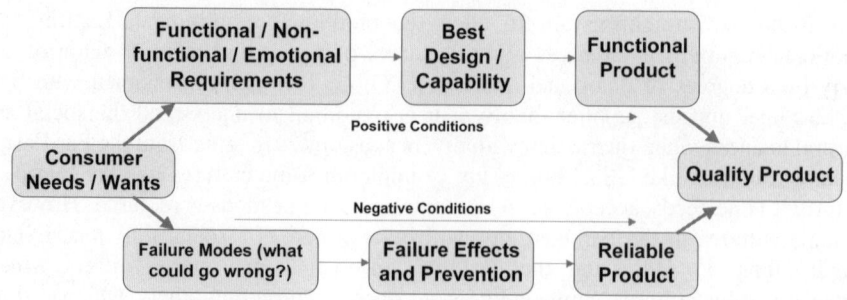

Fig. 4. Positive and Negative Conditions in Requirements

Following are the concept techniques to find the potential failures:-

Failure Mode Effect Analysis (FMEA): The purpose of FMEA is to identify beforehand possible failure modes of the system components, evaluate their influences on system behavior and propose proper countermeasures to suppress these effects. FMEA [3] is also a typical technical risk-assessment; the product architect should own and facilitate the proceedings. FMEA triggers members to seek answer for questions like:

- what could go wrong – *Failure mode*
- how badly might it go wrong – *Failure effect*
- what is reason for going wrong – *Failure cause*
- what can be done to prevent failures – *Preventive action*

For example, Figure 5 depicts one possible failure mode for adding USB functionality in the software.

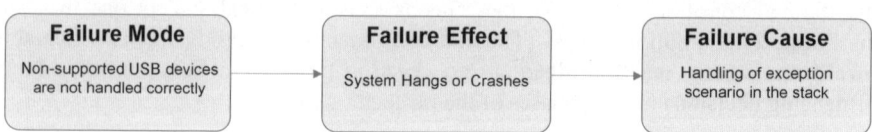

Fig. 5. Failure Mode analysis for USB

This is a many-many relationship i.e. a failure mode could have many effects which in turn could have many causes. Correspondingly, a single cause could result in multiple failure modes. For example, a memory leak may lead to hangs, crashes, intermittent instabilities etc. FMEA being an assessment of the technical risks for a project/product/feature/module etc there cannot be one universal team for doing FMEA. The success of FMEA greatly depends on the Product/domain knowledge, and hence there should be some domain experts in the team doing the FMEA.

Mistake proofing: This is a technique to use ingenuity to make potential errors impossible to occur. Identifying potential opportunities for errors during the development phase of a new product provides the possibility for easily implementing design changes to make these errors impossible to occur [5]. For example, we realized that sometimes excessive complexity in the user interface can cause mistakes (and dissatisfaction) during product usage. During the alignment with the organizational goal (viz. *Sense and Simplicity*), we decided to *simplify* the design. Once the design has been simplified, it is important to identify potential causes for mistakes and devise improvements that will prevent mistakes from occurring.

There are numerous "mistake proofing" opportunities in software development. For example, a menu that will not allow wrong choices. *Or* an application that won't allow the user to move to the next screen until all of the critical information is entered into the current screen.

Mistake-proofing can be achieved at different levels of effectiveness (see Figure 6), depending on the feasibility/cost/effort.

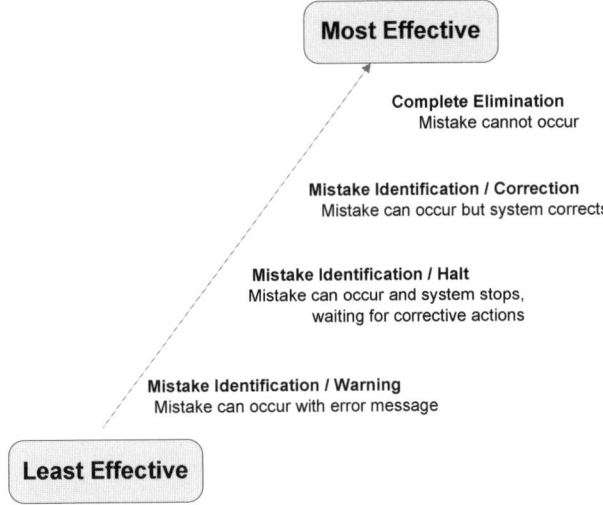

Fig. 6. Levels of effectiveness in mistake-proofing

Boundary condition analysis

In simple terms, boundary condition analysis means the maximum parameters to which the development entities can be stretched without causing a breakdown of the product. In our case, the boundary condition analysis was started from the requirement stage onwards. Specific boundary conditions were identified for the product as follows:-

Known boundary conditions arising out of the normal usage of the product. This is typically very straightforward to determine. For example, we encountered lots of scenarios where applying the organizational goal as a filter greatly simplified our choice.

Unknown boundary conditions arising out of the normal usage of the product. This is highly dependent on the geographical characteristics where the product is used, people behavior, presence/absence of technology assumed in that geographical section. For example, specific characteristics of the tuner signal in the geographical location.

Product Environment boundary conditions: This happens either during normal or abnormal usage of the product. For example: Operational temperatures of 45 degrees Celsius, product falls from a height of 1 meter.

We shared the anticipated boundary condition specifications based on our analysis and this helped to further fine-tune the requirements. We found the FMEA technique [3] extremely useful to capture all the above boundary conditions.

3.3 Requirements Documentation

We clearly demarcated the product behavior into the following 3 categories. We documented the requirements based on the categorization mentioned below:

User Interface specifications: consisting of the UI screen characteristics (like bitmaps, fonts, user interaction dialogs etc), the interaction model, the country specific features to be enabled / disabled.

Functional specifications: consisting of the product behavior model [11] excluding the UI. We set the upper and lower limits for each of the attributes for A/V. These were set in such a way that the customer (and hence the user) will accept these limits. For example, the features like the HDD recording, archiving from HDD to DVD etc together with the quantified upper and lower A/V specifications would constitute the functional behavior.

Non-functional specifications: consisting of all those behaviors that are experienced by the end-user but not directly highlighted as an explicit product function. For example, product performance parameters like the time taken to read a DVD disc, the time taken to identify an unknown recordable medium.

Such a demarcation allowed the development team and the customer to de-link and focus on improving the quality of the specs.

3.4 Requirements Verification and Validation

Verification vs validation is popularly distinguished as 'doing the thing right' vs 'doing the right thing'. After documenting the requirements, we verified the requirements with the internal customer (within Philips). The prototypes were updated and validated with the external customer. This proved useful in underpinning the finer details of the requirements and the overall alignment with the external customer's organizational goal.

For example, when switching to camera mode for DV recording, the time shift buffer needs to be flushed and a user dialog (UD1 – asking the user if the time shift buffer can be flushed or not) is shown. When HDD is full, the user dialog (UD2 – informing the user that the HDD is full and time shift buffer functionality is not possible) needs to be shown. However, when switching to camera mode in HDD full scenario, the time shift buffer is not used and the user dialog UD1 should not be shown.

4 Results and Conclusion

By understanding and translating the organizational goal into explicit functional and emotional requirements, the product development was on track throughout the development life cycle and the end results were very encouraging (a customer satisfaction rating of 9.2/10, whereas the previous average rating was about 8/10). The internal customers (within Philips) as well as the external customer were impressed with the mechanisms in place. We strongly believe that this can lay the foundation for mutually rewarding relationships and future co-operations to take place.

In reality, it is always difficult to the development team to be in the ideal situation where 100% of the requirements are aligned with the organizational goals. However, it is desirable to get as closely aligned as possible mainly to avoid major surprises during later phases of the development.

In this paper, we have showcased our experience of focusing on the organizational goal during the requirements phase. Extending this focus into the other development phases will ensure that the organizational goal is met at the end. Understanding both positive & negative conditions during requirements phase and aligning with the organizational goal during development will ensure product robustness and high chances of 'first time right'.

Acknowledgments. The authors wish to thank **Dr. Shyam Vasudevarao** (Philips Consumer Electronics Technology Office) for his valuable insights and technical breadth and Mr. **Dheeraj Baijal** (Philips Consumer Electronics Entertainment Solutions) for his project management insights.

References

[1] Software Engineering: A Practitioner's Approach, 6th edn., Roger Pressman, McGraw Hill, (2004)

[2] Software Requirements 2: Practical techniques for gathering and managing requirements throughout the product development cycle, 2nd edn., Wiegers, Karl Eugene, Tandem (2003)

[3] Failure Mode and Effect Analysis, FMEA: From Theory to Execution, 2nd edn., D.H. Stamatis, ASQ Quality Press (2003)

[4] Working with Emotional Intelligence, Daniel Goleman, Bantam Publishers (2000)

[5] Make No Mistake - An Outcome-Based Approach to Mistake Proofing, Dr. C. Martin Hinckley, Productivity Press (2001)

[6] Damas, C., Lambeau, B., van Lamsweerde, A.: Scenarios, Goals and State Machines: a Win-Win Partnership for Model Synthesis. In: Intl. ACM Symposium on the Foundations of Software Engineering, ACM Press, New York (2006)

[7] van Lamsweerde, A., Letier, E.: From Object Orientation to Goal Orientation: A Paradigm Shift for Requirements Engineering, Radical Innovations of Software and Systems Engineering. In: Post-Workshop Proceedings of the Monterey-02 Workshop. LNCS, Springer, Heidelberg (2003)

[8] De Landtsheer, R., Letier, E., van Lamsweerde, A.: Deriving Tabular Event-Based Specifications from Goal-Oriented Requirements Models. In: 11th IEEE Joint Intl. RE Conference, IEEE Computer Society Press, Los Alamitos (2003)

[9] van Lamsweerde, A.: Goal-Oriented Requirements Engineering: A Roundtrip from Research to Practice. In: 12th IEEE Joint Intl. RE Conference, IEEE Computer Society Press, Los Alamitos (2004)

[10] van Lamsweerde, A.: From System Goals to Software Architecture. In: Bernardo, M., Inverardi, P. (eds.) SFM 2003. LNCS, vol. 2804, pp. 25–43. Springer, Heidelberg (2003)

[11] Letier, E., van Lamsweerde, A.: Deriving Operational Software Specifications from System Goals. In: 10th ACM SIGSOFT Symp. on the Foundations of Software Engineering, ACM Press, New York (2002)

[12] van Lamsweerde, A.: Goal-Oriented Requirements Engineering: A Guided Tour. In: 5th IEEE International Symposium on Requirements Engineering, pp. 249–263. IEEE Computer Society Press, Los Alamitos (2001)

[13] van Lamsweerde, A., Letier, E.: Handling Obstacles in Goal-Oriented Requirements Engineering. IEEE Transactions on Software Engineering, Special Issue on Exception Handling 26(10), 978–1005 (2000)

A Model-Driven Goal-Oriented Requirement Engineering Approach for Data Warehouses*

Jose-Norberto Mazón, Jesús Pardillo, and Juan Trujillo

Dept. of Software and Computing Systems
University of Alicante, Spain
{jnmazon,jesuspv,jtrujillo}@dlsi.ua.es

Abstract. The development of a data warehouse has been traditionally guided by an in-depth analysis of the underlying operational data sources, thus overlooking an explicit development phase in which information requirements of decision makers are addressed. This scenario has prompted that the deployed data warehouse often fails in delivering the expected support of the decision making process. To overcome this problem, we propose to use the i^* modeling framework and the *model driven architecture* (MDA) in order to describe (i) how to model goals and information requirements for data warehouses, and (ii) how to derive a conceptual multidimensional model that provides the required information to support the decision making process.

1 Introduction

Data warehouse (DW) systems provide decision makers with information related to a business process. This information is useful for decision makers to fulfil their goals in order to improve the business process. Both practitioners and researchers agree that the development of these systems must be based on a conceptual *multidimensional* (MD) model [1] that allows designers to easily structure information into facts (which contain interesting measures of a business process) and dimensions (which represent the context for analyzing the measures of a business process). Since the DW integrates several operational data sources, the development of conceptual MD models has been traditionally guided by their detailed analysis [2]. However, several studies [3,4] have pointed out that most of these conceptual MD models fail in addressing the required information as a result of a poor communication between DW developers and decision makers. Actually, information needs cannot be understood by only analyzing the operational data sources, and a requirement analysis stage is needed in order to model the information requirements of decision makers and derive a suitable conceptual MD model [5,6]. Furthermore, this stage should be based on a *goal-oriented*

---

* This work has been partially supported by the METASIGN (TIN2004-00779) project and by the FPU grants AP2005-1360 and AP2006-00332 from the Spanish Ministry of Education and Science, and by the DADS (PBC-05-012-2) project from the Castilla-La Mancha Ministry of Education and Science (Spain).

J.-L. Hainaut et al. (Eds.): ER Workshops 2007, LNCS 4802, pp. 255–264, 2007.

requirement engineering (GORE) framework since (i) the DW aims at providing adequate information to support the decision making process, thus helping to fulfil goals of an organization [1], and (ii) information requirements for DWs are difficult to specify from scratch [4], since decision makers often only express general expectations about which goals the DW should support.

Therefore, in this paper, we propose a GORE approach for modeling organizational goals that the DW supports and relating them to information requirements. To this aim, we have used the profiling mechanism of the *unified modeling language* (UML) [7] to adapt the *i** modeling framework [8,9] to requirement analysis in DWs according to our previous work [10]. On the other hand, the success of a DW project highly depends on the acceptance of the DW as a valuable resource for the organization [11,12]. To assure this acceptance, the derived conceptual MD model should represent the MD elements that satisfy information requirements and goals, in such a way that the decision makers understand the purpose of the DW. To accomplish this, our GORE approach is integrated into a *model driven architecture* (MDA) [13] framework for the development of DWs that has been described in [14]. This framework is based on defining a *computation independent model* (CIM) which addresses goals and information requirements, a *platform independent model* (PIM) to specify MD properties at the conceptual level, and a *platform specific model* (PSM) tailored to a specific database technology. Following these considerations, our GORE approach for DWs is used to define a CIM, while a PIM for MD modeling is derived by establishing a formal transformation between these models via the *query/view/transformation* (QVT) language [15]. The main advantage is that the conceptual MD model, represented in a PIM, meets every goal and information requirement defined in the CIM, since links between elements of both models are implicitly created with the execution of each QVT transformation. Therefore, QVT provides mechanisms to assure traceability in such a way that changes into a CIM are propagated to a PIM by re-executing the transformation.

The remainder of this paper is structured as follows: related work is described in the next section. Our approach for requirement analysis for DWs is presented in section 3. How to specify a conceptual MD model for DWs is defined in section 4. Section 5 describes an example of applying our approach. Finally, in section 6, we present our conclusions and sketch some future work.

2   Related Work

Only few approaches have considered requirement analysis as a crucial task in early stages of the DW development. In [4], a method is proposed in order to both determine information requirements of DW users and match these requirements with the available data sources. The work in [16] presents the *data warehouse requirements definition* (DWARF) approach that adapts traditional requirements engineering process for requirements definition and management of DWs. The approach described in [5] focuses on describing a requirement elicitation process for DWs by identifying goals of the decision makers and the required information that supports the decision making process. Finally, in [17], the authors present a goal-oriented framework to model requirements for DWs, thus obtaining a conceptual MD model from them by using a set of guidelines.

Unfortunately, these approaches present one main drawback, since they do not provide formal mechanisms to assure the traceability between the requirement model and the conceptual MD model, which is a desirable property just as stated in [18]. At most, these approaches only provide a set of informal guidelines to derive a conceptual MD model from requirements. To overcome this problem, we propose to use a GORE approach within our MDA framework for DWs.

3   Requirement Analysis for Data Warehouses

A requirement analysis stage for DWs aims at obtaining informational requirements of decision makers [5], which are related to interesting measures of business processes and the context for analyzing these measures [19]. However, decision makers often ignore how to suitably describe information requirements, since they are rather concerned about goals which the DW helps to fulfil. Therefore, a requirement analysis phase for DWs should start discovering goals of decision makers. Afterwards, the information requirements will be easier discovered from these goals. Finally, information requirements will be related to the required MD concepts, i.e. the measures of the business process or the context for analyzing these measures. Within our MDA approach for the development of DWs, both goals and information requirements have to be modeled in a CIM by using a UML profile for the i^* modeling framework[1]. From the defined CIM, a conceptual MD model which contains the necessary elements to achieve information requirements can be derived in a PIM.

3.1   Goals and Information Requirements for Data Warehouses

GORE is concerned about modeling goals, thus obtaining user requirements by following a refinement process [20]. To ease the task of discovering and eliciting goals and requirements for DWs, we have defined a classification of the different kind of goals that decision makers expect to fulfil with the envisaged DW. We consider three kind of goals, depending on their level of abstraction:

- **Strategic goals** represent the highest level of abstraction. They are main objectives of the business process. They are thought as changes from a current situation into a better one. For example: "increase sales", "increase number of customers", "decrease cost", etc. Their fulfilment causes an immediate benefit for the organization.
- **Decision goals** represent the medium level of abstraction. They try to answer the question: "how can a strategic goal be achieved?", and they aim to take the appropriate actions to fulfil a strategic goal. For example

---

[1] The i^* framework is used because the DW implies many organizational units and it can often involve dependencies among many kind of users with different goals.

"determine some kind of promotion" or "open new stores". Their fulfilment
only causes a benefit for the organization if it helps to reach strategic goals,
since decision goals only take place within the context of a strategic goal.

- **Information goals** represent the lowest level of abstraction. They try to
answer the question: "how can decision goals be achieved in terms of information required?", and they are related to the information required by a
decision goal to be achieved. For example "analyze customer purchases" or
"examine stocks". Their fulfilment helps to achieve decision goals and they
only happen within the context of a decision goal.

These goals form a hierarchy that can de defined via two well-known strategies [20]: discovering goals by refinement or discovering goals by abstraction. The
former consists on asking "how" questions over goals already identified ("how can
this goal be satisfied?"), while the latter is driven by "why" questions ("why is this
goal useful?"). Once this hierarchy of goals is defined, information requirements
can be directly obtained from the information goals. Later, the different MD elements, such as *facts* or *dimensions* will be discovered from these information requirements in order to specify the corresponding conceptual MD model of the DW.

3.2 Designing a CIM for Data Warehouses

A CIM is a view of a system from a computation independent viewpoint [13], thus
taking into account the business environment of the system in order to improve
the communication between decision makers and DW developers. Goals and information requirements for the DW are modeled in a CIM by using an adaptation
of the i^* modeling framework [8,9]. This adaptation is based on two extensions of
UML [7] (see Fig. 1): (i) a profile for i^*, in order to integrate it within our MDA
framework; and (ii) a profile which adapts i^* to the DW domain. Due to space
constraints, we only show an overview of the designed UML profiles.

The i^* modeling framework [8,9] provides mechanisms to represent actors,
their dependencies, and structuring the business goals that the organization

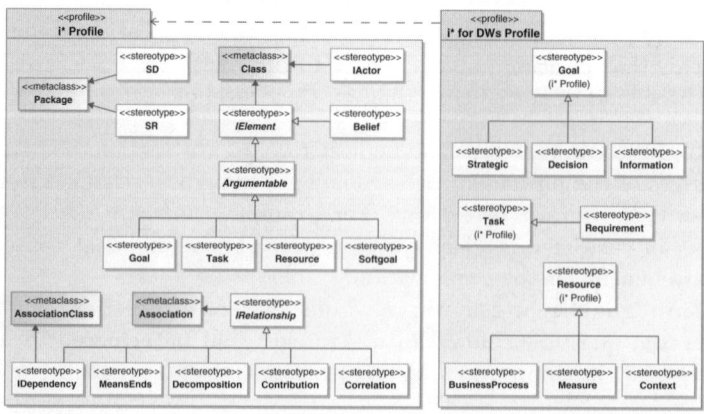

Fig. 1. Overview of the profiles for i^* modeling in the DW domain

pretends to achieve. This framework establishes two models: the *strategic depen-dency* (SD) model for describing the dependency relationships among various actors in an organizational context, and the *strategic rationale* (SR) model, used to describe actor interests and concerns, and how they might be addressed. From now on, we focus on describing the SR models to specify goals and information requirements of decision makers[2].

The SR model (modeled with the *SR* stereotype and represented as) pro-vides a detailed way of modeling internal intentional elements and relationships of each actor (*IActor*, ○). In our CIM, we use intentional elements such as goals (*Goal*, ○), tasks (*Task*, ◇), resources (*Resource*, □); and intentional relationships such as means-end (*MeansEnd*, ⊳) representing alternative ways for fulfilling goals, or task-decomposition (*Decomposition*, +) representing the necessary elements for a task to perform. Due to space constraints, we refer reader to [8,9] for a further explanation of *i**.

In order to define SR models for DWs, goals, tasks, and resources are rep-resented as intentional elements for each decision maker. In a CIM, goals of decision makers are defined by using the *Strategic, Decision*, and *Information* stereotypes by specializing the previously defined *Goal* stereotype; and inten-tional means-end relationships (*MeansEnds*, ⊳) between them. From informa-tion goals, information requirements (*Requirement*) are derived and represented as tasks. All of the described modeling elements are designed in the *i* Profile* (sketched in Fig. 1).

Furthermore, the requirement analysis for DWs needs some MD concepts to be added (in the sense of [17]). Therefore, the following concepts are added as resources in the CIM: business processes related to the goals of decision makers (*BusinessProcess* stereotype), relevant measures related to information require-ments of decision makers (*Measure*), and contexts needed for analyzing these measures (*Context*). Additionally, foreseen relations between context of analysis are modeled. For instance, the city and the country contexts are related be-cause cities can be aggregated in countries. For modeling these relationships, we use the (shared) aggregation relationship of UML (*Association* UML metaclass, represented as ─◇).

In summary, several steps must be followed to properly define a CIM: (i) discovering the intentional actors (i.e. decision makers), thus defining SR models for each one, (ii) discovering their goals according to the classification described in Sect. 3.1, (iii) deriving information requirements from information goals, and (iv) obtaining the MD concepts related to the information requirements.

4   Obtaining a Conceptual MD Model

Once goals and information requirements are specified in a CIM, a conceptual MD model that supports them must be derived in a PIM. Within our MDA

---

[2] Dependencies among different decision makers would be defined in a SD model, but for the sake of clarity this issue is out of the scope of this paper.

approach we apply several QVT transformation rules to perform this task, thus assuring traceability between information requirements and the necessary MD elements related to them.

4.1   A PIM for Data Warehouses

A PIM describes a system hiding the necessary details related to a particular technology [13]. This point of view corresponds to the representation of MD elements at the conceptual level, independently from the platform in which the DW will be implemented. Our PIM is based on our UML profile for MD modeling presented in [21]. Although in this section we focus on describing a subset of this UML profile, an overview of the whole profile (showing stereotypes and extended metaclasses) is given in Fig. 2.

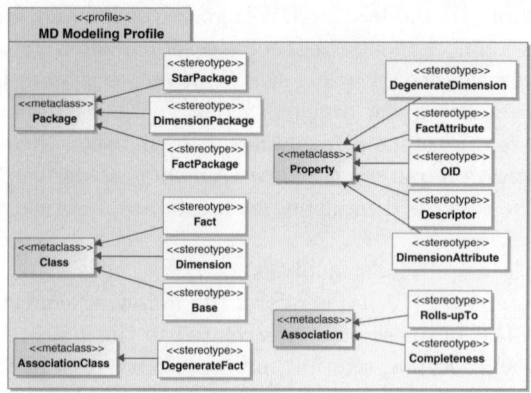

Fig. 2. Overview of the UML profile for MD modeling of DWs

This profile contains the necessary stereotypes in order to elegantly represent main MD properties at the conceptual level by means of a UML class diagram in which the information is clearly organized into facts and dimensions. These facts and dimensions are modeled by *Fact* (represented as ▦) and *Dimension* (⬀) stereotypes, respectively. Facts and dimensions are related by shared aggregation relationships (the *Association* UML metaclass). While a fact is composed of measures or fact attributes (*FactAttribute*, **FA**), with respect to dimensions, each aggregation level of a hierarchy is specified by classes stereotyped as *Base* (**B**). Every *Base* class can contain several dimension attributes (*DimensionAttribute*, **DA**) and must also contain a *Descriptor* attribute (*Descriptor*, **D**). An association stereotyped as *Rolls-UpTo* between *Base* classes specifies the relationship between two levels of a classification hierarchy. Within it, role *R* represents the direction in which the hierarchy rolls up, whereas role *D* represents the direction in which the hierarchy drills down.

4.2 From CIM to PIM

A QVT transformation has been developed to derive the corresponding PIM from the CIM. A QVT transformation is composed of several relations that must hold between model elements of a set of candidate models (source and target models). Each of these relations is defined by using the elements described in Fig. 3.

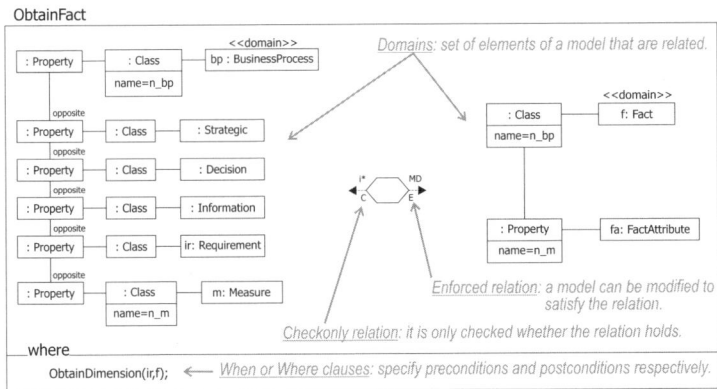

Fig. 3. QVT relation to obtain facts and their attributes

Due to space constraints, we only show one of the designed QVT relations (*ObtainFact*, Fig. 3). In this relation, the source domain is a set of elements of the CIM that represents the hierarchy of strategic, decision, and information goals of decision makers. This hierarchy is build on *Property* and *Class* (stereotyped as *Strategic*, *Decision*, or *Information*) UML metaclasses. A information requirement is also represented by using the *Requirement* stereotype. Finally, a strategic goal is related to a business process (*BusinessProcess*), while the information requirement is related to measures (*Measure*). The *ObtainFact* relation enforces the following set of elements in the PIM: a fact (*Fact* stereotype on *Class* instances) containing one fact attribute (*FactAttribute* on *Property* ones). Once this relation holds, the *ObtainDimension* relation must be carried out (according to the *where* clause) in order to obtain every required dimension and their hierarchies. After applying the QVT transformation, we obtain a PIM that provides the required information to support the decision making process.

5 Sample Application of Our Approach

In this section, we provide an example of our approach inspired from the case study presented in [21]. In this case study, a company sells automobiles across several countries. Therefore, we focus on the automobile sales business process which is related to one main actor, the sales manager, via the strategic goal *"increase automobile sales"*. From this strategic goal, three different decision goals are derived: *"decrease sale price"*, *"determine promotion according to a country"*, and *"give incentive to salespersons"*. From each of these decision goals the following information goals have been obtained: *"analyze automobile price"*, *"analyze automobile sales"*, and *"study salesperson sales"*. The derived information requirements are as follows: *"analyze automobile sale price"*, *"analyze the amount of sold automobile by customer age"*, *"analyze total amount by customer city and country"*, and *"analyze total amount by salesperson and date"*.

Each of these elements are defined in a SR model that represents a CIM according to our UML profile for *i** (see Fig. 4). In this CIM, strategic, decision, and information goals are represented as goals, and information requirements correspond to tasks. Furthermore, the necessary measures and contexts of analysis are associated to the information requirements as resources. Specifically, the measures are *"quantity"*, *"price"*, and *"total"*, and the elements that represent the context of analysis are *"salesperson"*, *"date"*, and *"automobile"*. *"Customer"*, *"age"*, *"city"*, and *"country"* also represent the context of analysis, and they are related each other, since they are useful for aggregating the *"customer"* data.

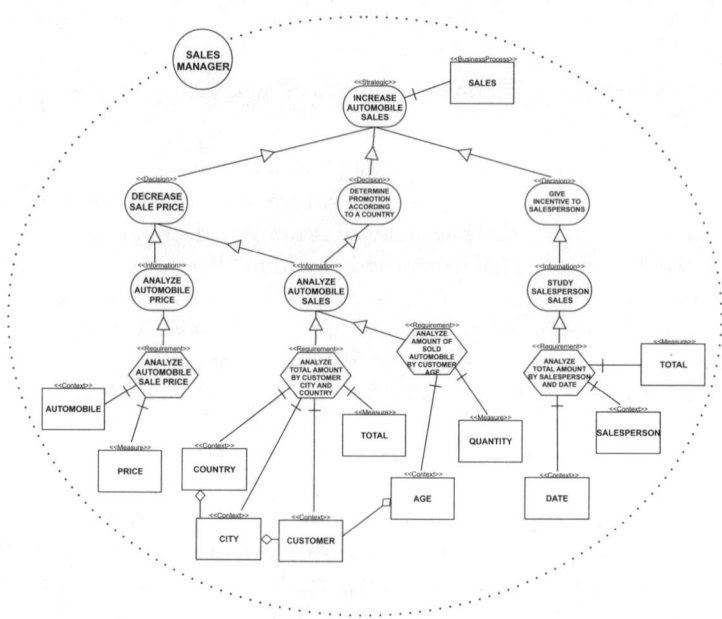

Fig. 4. Defined CIM for our example

From the defined CIM, the QVT transformation is applied to obtain a PIM (i.e. the conceptual MD model, Fig. 5) that delivers enough information in a suitable way to accomplish the information requirements and goals of decision makers. Table 1 overviews which MD elements are created in the PIM and their relations with information requirements. For instance, when the *ObtainFact* relation (Fig. 3) is executed, it takes the previously defined CIM as an input to create certain MD elements which conform the PIM. This relation would create a *Sales* fact related with the *"sales"* business process, and following the hierarchy of goals *"increase automobile sales"*– *"determine promotion according to a country"*– *"analyze automobile sales"*– *"analyze total amount by customer city and country"*, the relation would find the *"total"* measure and would create the corresponding fact attribute associated with the *Sales* fact. Additionally, the *where* clause of the relation would be executed, then dimensions and their hierarchies would be also created from the related contexts and aggregations modeled in the CIM. The resulting PIM is shown in Fig. 5.

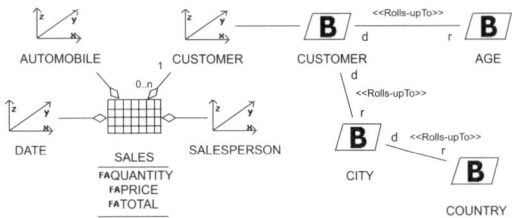

Fig. 5. Obtained PIM for our example

Table 1. Information requirements and their corresponding MD elements

Information requirement	Fact attribute	Base or dimension
Analyze automobile sale price	Price	Automobile
Analyze the amount of sold automobile by customer age	Quantity	Age
Analyze total amount by customer city and country	Total	Customer, city, country
Analyze total amount by salesperson and date	Total	Date, salesperson

6   Conclusions and Future Work

Modeling goals and requirements in DW projects is crucial for deploying a successful DW that properly supports the decision making process. In this paper, we have proposed a GORE approach within our MDA framework for the development of DWs. First, a CIM is specified by using the i^* modeling framework in order to model goals and information requirements for DWs. Then, the conceptual MD model of the DW is derived from this CIM into a PIM by using some QVT transformation rules. This PIM supports the goals that the decision makers plan to achieve with the DW. Our immediate future work comprises several tasks, such as adding quality measures to i^* models to analyze the understandability of i^* diagrams for DW developers, or adding softgoals in order to gather non-functional requirements apart from information requirements.

References

1. Kimball, R., Ross, M.: The Data Warehouse Toolkit. Wiley & Sons, Chichester (2002)
2. Inmon, W.: Building the Data Warehouse. Wiley & Sons, Chichester (2002)
3. Schiefer, J., List, B., Bruckner, R.: A holistic approach for managing requirements of data warehouse systems. In: Americas Conf. on Information Systems, pp. 77–87 (2002)
4. Winter, R., Strauch, B.: A method for demand-driven information requirements analysis in data warehousing projects. In: HICSS (2003)
5. Prakash, N., Singh, Y., Gosain, A.: Informational scenarios for data warehouse requirements elicitation. In: Atzeni, P., Chu, W., Lu, H., Zhou, S., Ling, T.-W. (eds.) ER 2004. LNCS, vol. 3288, pp. 205–216. Springer, Heidelberg (2004)
6. Rizzi, S., Abelló, A., Lechtenbörger, J., Trujillo, J.: Research in data warehouse modeling and design: dead or alive? In: DOLAP, pp. 3–10 (2006)
7. OMG: Unified Modeling Language Specification 2.0, http://www.omg.org/cgi-bin/doc?formal/05-07-04
8. Yu, E.: Modelling Strategic Relationships for Process Reenginering. PhD thesis, University of Toronto, Canada (1995)
9. Yu, E.: Towards modeling and reasoning support for early-phase requirements engineering. In: RE 1997, pp. 226–235 (1997)
10. Mazón, J.N., Trujillo, J., Serrano, M., Piattini, M.: Designing data warehouses: from business requirement analysis to multidimensional modeling. In: REBNITA (2005)
11. Chenoweth, T., Corral, K., Demirkan, H.: Seven key interventions for data warehouse success. Commun. ACM 49(1), 114–119 (2006)
12. Gam, I., Salinesi, C.: A requirement-driven approach for designing data warehouses. In: REFSQ (2006)
13. OMG: MDA Guide 1.0.1, http://www.omg.org/cgi-bin/doc?omg/03-06-01
14. Mazón, J.N., Trujillo, J.: An MDA approach for the development of data warehouses. Decision Support Systems doi:10.1016/j.dss.2006.12.003
15. OMG: MOF 2.0 Query/View/Transformation, http://www.omg.org/cgi-bin/doc?ptc/2005-11-01
16. Paim, F.R.S., Castro, J.: DWARF: An approach for requirements definition and management of data warehouse systems. In: RE 2003, pp. 75–84 (2003)
17. Giorgini, P., Rizzi, S., Garzetti, M.: Goal-oriented requirement analysis for data warehouse design. In: DOLAP 2005, pp. 47–56 (2005)
18. Stefanov, V., List, B.: Business metadata for the datawarehouse - weaving enterprise goals and multidimensional models. In: EDOC 2006 Workshops (2006)
19. Romero, O., Abelló, A.: Multidimensional design by examples. In: Tjoa, A.M., Trujillo, J. (eds.) DaWaK 2006. LNCS, vol. 4081, pp. 85–94. Springer, Heidelberg (2006)
20. van Lamsweerde, A.: Goal-oriented requirements enginering: A roundtrip from research to practice. In: RE 2004, pp. 4–7 (2004)
21. Luján-Mora, S., Trujillo, J., Song, I.Y.: A UML profile for multidimensional modeling in data warehouses. Data Knowl. Eng. 59(3), 725–769 (2006)

Visually Effective Goal Models Using KAOS

Raimundas Matulevičius and Patrick Heymans

PReCISE Research Center, Computer Science Department, University of Namur,
rue Grandgagnage 21,5000 Namur, Belgium
{rma, phe}@info.fundp.ac.be

Abstract. Goal modelling languages are visual modelling languages. To communicate ideas effectively with a visual modelling language, one should follow some basic principles. One is modularity, i.e. organising diagrams in manageable modules to avoid confusing the reader with overly complex diagrams. Another is emphasis, i.e. visually drawing the attention to the most important pieces of information. In this paper, we evaluate how the goal modelling language KAOS and its supporting tool, Objectiv*er*, help modellers respect nine visual modelling principles. From our observations, we formulate recommendations for modellers, language designers and tool developers.

1 Introduction

Goal modelling languages (GMLs) are visual modelling languages used primarily during the early stages – a.k.a. Requirements Engineering (RE) – of information system (IS) development. RE seeks to identify the requirements that the future IS has to fulfil and the constraints under which it has to operate. Failing to understand the stakeholders' requirements is still the primary reason of project failure.

In this context, GMLs appear to be useful means to facilitate the identification, structuring and validation of requirements. GMLs introduce new abstractions, most notably the notion of goal, to supplement other more traditional abstractions used in data and process modelling. Goal models make the purpose and rationale of the new IS explicit, thereby preventing the development team to waste time on detailed technical descriptions if the goals are not yet clear, and allowing them to justify each detailed requirement by reference to goals.

Over time, several GMLs have been proposed, together with their supporting methods and tools. Those languages include *i** [25], TROPOS [5], NFR [6], KAOS [12, 24], GBRAM [2], and Lightswitch [22]. Various applications [1, 14] and evaluations [3, 10, 11, 16, 23] of GMLs were also reported. Following [13], we situate qualitative assessments of GMLs along three dimensions: syntax, semantics and pragmatics (see Fig. 1).

In this paper, as in [3] and [8], we examine the link between the *syntax* of GMLs – which constructs they propose and how these can be combined – and their *semantics*– how models are understood by their audience. However, whereas [3] and [8] deal with the *abstract* syntax (metamodel) of GMLs, we concentrate on their *concrete*, or surface, syntax. More precisely, we investigate the relationship between the visual display of goal models and how humans understand them.

J.-L. Hainaut et al. (Eds.): ER Workshops 2007, LNCS 4802, pp. 265–275, 2007.

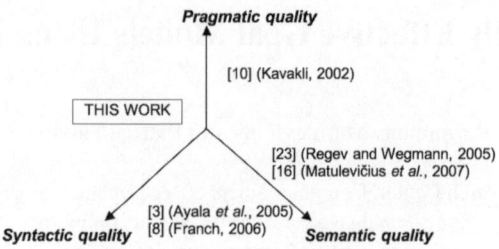

Fig. 1. Classification of GML research

Good practices for the definition and usage of visual modelling languages have been a subject of research for several decades now (see e.g. [4, 7, 20, 21]). In this work, we use a set of nine basic principles recently formulated by Moody in [18]. Those principles consolidate a large body of knowledge originating from disciplines such as human-computer interaction [20] and cognitive psychology [4]. One of these principles is modularity, i.e. organising diagrams in manageable modules to avoid confusing the reader with overly complex diagrams. Another is emphasis, i.e. visually drawing the attention to the most important pieces of information.

This paper evaluates how easy it is to follow the principles using the current version of the KAOS (Knowledge Acquisition in autOmated Specification) language [12, 24] and its associated tool, Objectiver (release 2.0.0 professional edition). The resulting critical analysis of KAOS' and Objectiver's visual abilities aims to formulate recommendations (*i*) for KAOS modellers, (*ii*) for language engineers, and (*iii*) for tool developers. Although this paper is not a comparative study, it is a first step towards such a systematic comparison of different GMLs.

Section 2 describes KAOS. Section 3 recalls the nine principles [18]. Section 4 analyses how the principles are addressed in KAOS and Objectiver. Section 5 formulates the recommendations. Section 6 discusses conclusions and future work.

2   KAOS

In this work we use KAOS as defined in [12, 24]. The KAOS approach consists of a modelling *language*, a *method*, and a *software environment*. The main purpose of KAOS is to ensure that high-level goals are identified and progressively refined into precise operational statements. These are then assigned to agents of the *software-to-be* and its environment, both forming the *system-to-be*. Along this process, various alternative goal assignments and refinements are considered until the most satisfactory solution is chosen.

A KAOS model consists of four kinds of diagrams: goal, object, agent and operation. KAOS constructs have a textual (Fig. 2) and a graphical (Fig. 3) syntax. Selected constructs (e.g., goal, operation) can be further defined using the KAOS real-time temporal logics facilitating reasoning (Fig. 2, FormalDef clause). The major construct in KAOS is goal, which is a prescriptive assertion that captures an objective that the system-to-be should meet. A goal can be refined through G-refinement, which relates it to subgoals whose conjunction contributes to the satisfaction of the goal. A

> **Goal** Achieve [Meeting held]
> **Def** Each requested meeting is eventually being held with the presence of all intended participants.
> **FormalDef**
>   ∀m : Meeting: m.Requested
>   ⇒◊ m.Holds ∧ (∀ p: Participant): Intended (p, m) → Participates (p, m)

Fig. 2. Textual goal syntax. A goal has a name (Meeting held), a natural language definition (Def), and optional attributes like pattern (e.g. Achieve) and formal definition (FormalDef) [12].

goal can have alternative G-refinements, which result in different software designs. Goals are refined until they are assigned to individual agents. A goal effectively assigned to a software agent is called a requirement. If a goal is operationalised and has a responsible agent, the latter performs the operations.

In this work we focus on the KAOS graphical syntax. Fig. 3 represents an excerpt of the meeting scheduler model [12]. The goal Meeting held is refined into two subgoals: Participant informed and Participant info known. The latter is further refined into Participant agenda is up to date and Participant info known from agenda. The agent Scheduler is responsible for Participant informed to become true. The agent performs operations Inform about the time and Inform about the place to fulfil the requirement.

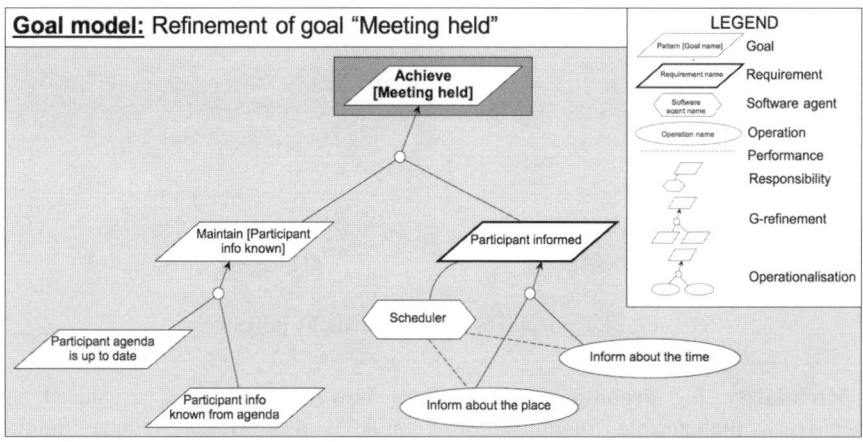

Fig. 3. KAOS diagram with drawing tools (MS Word's graphical editor)[1]

3 Principles of Effective Diagrams

The principles for the effective diagrams [18] introduce how a diagram should be prepared manipulating eight visual variables (vertical position, horizontal position, shape, colour, size, value, orientation, and texture) in order to communicate effectively *wrt* a "model of human graphical information processing, which reflects current research in human cognition and visual perception" [18]. We start with the

---

[1] Figures of higher resolution are available in [15].

principle of **discriminability**. There are two types of discriminability. *Absolute* discriminability is reader's ability to separate diagram elements from the background (see Fig. 4). It depends on element size (a), element proximity (b), and diagram contrast (c and d). *Relative* discriminability is the reader's ability to differentiate between different element types. It relies on the use of shapes, lines and visual variables. There are five basic geometric signs [9] – square (e.g., diamonds and rectangles are square variations), triangle, circle, cross and arrows – which are not likely to be confused. For example, KAOS uses three basic geometric signs: squares (e.g., goal and agent), circles (e.g., operation) and arrows (e.g., G-refinement, operationalisation, and assignment). Shape, colour, orientation, thickness and colour of borderline also play a very important part. For instance, in Fig. 3 constructs goal and requirement are discriminated only by borderline thickness (see also Table 2).

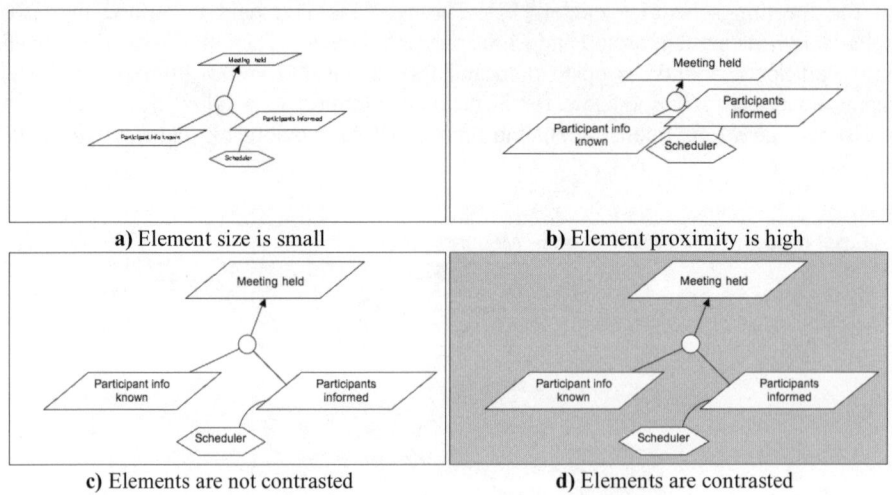

a) Element size is small   b) Element proximity is high

c) Elements are not contrasted   d) Elements are contrasted

Fig. 4. Examples of absolute proximity principle

Modularity (or **decomposition**), defines how a diagram is organised into cognitively manageable modules, or "chunks", that would reduce diagram complexity. In order to avoid cognitive overload, diagrams should be limited to seven plus/minus two elements [18]. In our example (Fig. 3) we respect these boundaries; however, we note that goal models can quickly become complex, and the modularity is highly dependent on the modeller's skills.

Structure organises diagram elements into distinct perceptual groups. Elements can be structured by proximity, similarity, or common region. In Fig. 5 operations Inform about the time and Informed about the place are structured by *similarity* (because of labels and shape), *proximity* (because they are physically close to each other) and *common region* (because they belong to the region Participants informed). Structuring is "an alternative and a complement to decomposition" [17]: instead of dividing a diagram into manageable modules, elements can be organised into groups.

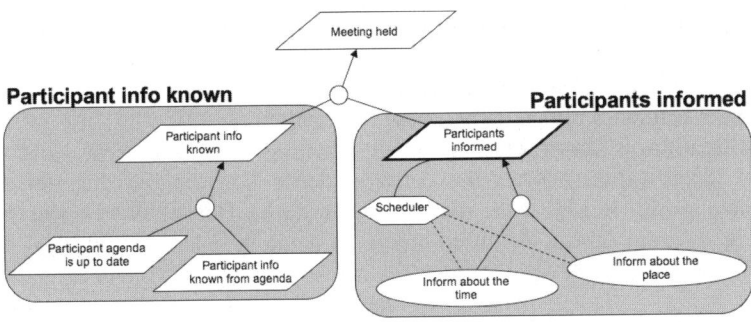

Fig. 5. Structuring the KAOS diagram

Cognitive integration deals with understanding the overall information covered by the whole set of diagrams. A model usually consists of multiple diagrams. Cognitive integration describes how different pieces of information are integrated from various diagrams. *Summarisation* is the process of creating more abstract representations of information (Fig. 6a). A *navigation map* is a representation of the entire system of diagrams and the navigation paths between them (Fig. 6b). *Signposting* includes navigation clues to show diagram transitions, helping user orientation in the system of diagrams (Fig. 6c). Other cognitive integration techniques are discussed in [17].

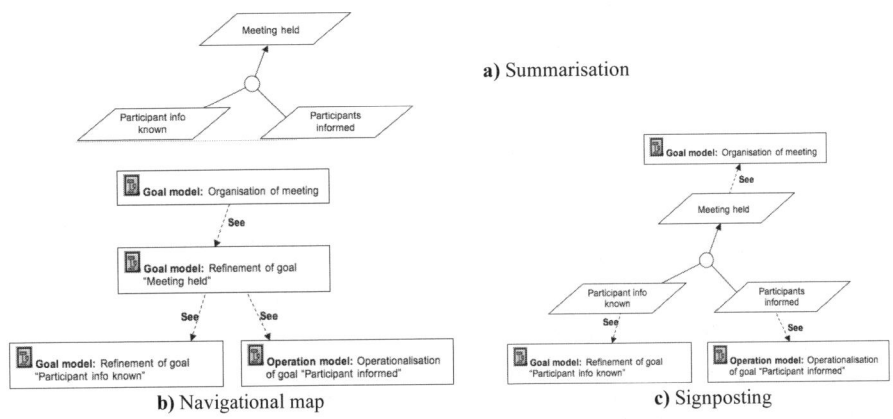

Fig. 6. Cognitive integration techniques supported in Objectiv*er*

Emphasis is about drawing attention to the most important information presented in a diagram. Emphasis is attained by visual variables, like shading, size, and colour of the element, font size, value and colour of the label, colour of the diagram background. Fig. 3 illustrates emphasis using font size of the label and by shading the element background (goal "Meeting held").

Perceptual directness describes the use of representations that have direct interpretation. In case of very abstract concepts, perceptually direct representations are difficult to find. Hence, arbitrary representation conventions are made, and a legend often facilitates remembering those conventions (Fig. 3).

Identification is about clear diagram labelling with title, type, and legend. *External* identification defines the correspondence between the diagram and the represented world. In Fig. 3 the diagram has a *name*: Refinement of goal "Meeting held". The diagram *type* is identified before its name in bold, **Goal model**. *Internal* identification defines the correspondence between graphical conventions and their meaning. In Fig. 3 all element types used in the diagram are indicated in its *legend*.

Visual expressiveness refers to the visual variables used to encode information. Visual variables may increase perceptual representation, accuracy and draw attention and interest. However the modellers need to be careful not to violate other principles (e.g., discrimination, emphasis, and structuring). All variables should be held constant or normalised. This helps avoiding undesirable and unintended messages of the diagram. In Fig. 3 all relative elements are normalised by size.

Graphical simplicity is about minimising the number of different conventions used. The span of absolute judgment (the ability to discriminate between perceptually distinct alternatives) is around 7 plus or minus two. In Fig. 3 we use eight KAOS constructs. In addition to graphical notations, textual information using attributes (not appearing in diagram, see Fig. 2) could be defined.

4 KAOS/Objectiv*er* Evaluation

Following the principles for effective diagrams, in Fig. 3 we present a KAOS goal model created using *drawing tools* (MS Word's graphical editor). However, this was time consuming. Thus in this section we investigate how the principles for effective diagrams can be fulfilled with Objectiv*er*. The resulting diagram is shown in Fig. 7.

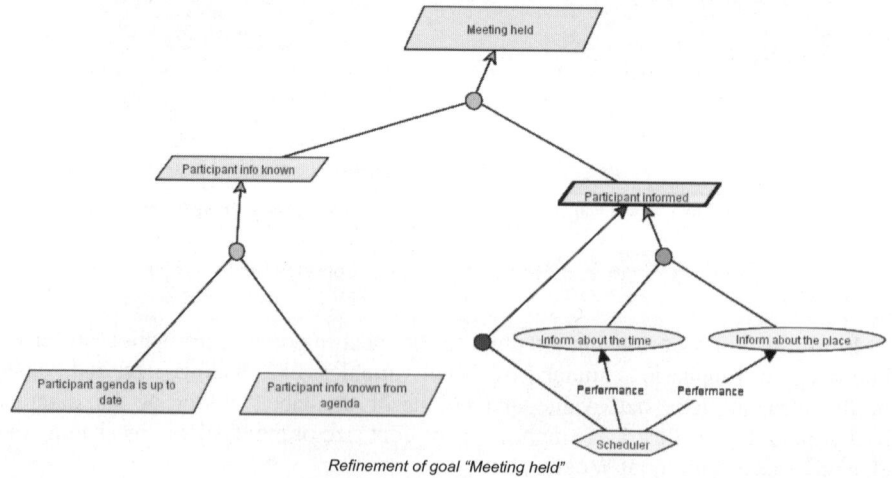

Refinement of goal "Meeting held"

Fig. 7. KAOS diagram with Objectiv*er*

KAOS includes constructs from graph and iconic classes [7], which are manipulated using visual variables (Tables 1 and 2). KAOS uses parallelograms (that we classify in squares), ellipses (circles) and arrows (Table 1). Visual variables (Table 2) are important for construct *discrimination* when modelling; but some of them (e.g. colour) play no role when a diagram is printed using a black and white printer (Fig. 7). Objectiv*er* does not provide means for contrasting elements versus background. The modeller can manually use element proximity and size; however, he has to be careful not to neglect other principles.

Table 1. Construct variation

Construct variation	KAOS/Objectiv*er*		
Square	Goal	Agent	
Circle	Operation		
Arrow	G-refinement	Operationalisation	Responsibility
	←—Output ——	——Input —▶	

Table 2. Relative discriminability of the KAOS/Objectiv*er* goal-related constructs

Construct	Shape	Background colour	Border line colour	Border line thickness	Orientation
Goal	Parallelogram	Light blue	Black	Thin	Right
Softgoal	Parallelogram	Light blue	Blue	Thin	Right
Requirement	Parallelogram	Light blue	Black	Thick	Right
Expectation	Parallelogram	Yellow	Black	Thick	Right
Obstacle	Parallelogram	Orange	Black	Thin	Left

To integrate *cognitively* different information from different diagrams, Objectiv*er* supports navigation map and signposting techniques (see Fig. 6 b, c). However, model *modularisation* and *structuring* depends entirely on the modeller's skills. Objectiv*er* also has no means to set boundaries for structured elements. Also there are no means to *emphasise* elements in a diagram.

When working with Objectiv*er* the diagram is *identified* by its name and type in the title bar of the active window. When printing the diagram, its name (but not type) is included at the bottom of the diagram (Fig. 7). Objectiv*er* includes no legend on the printed diagrams.

Visual expressiveness is limited in Objectiv*er*: very few visual variables (e.g., size) are used. But the tool can automatically normalise element *size*, *spacing* and *alignment*. KAOS turns out to be quite complex *wrt graphical simplicity*; it has 18 graphical conventions. In addition to graphical constructs, modellers also have to provide values for mandatory (e.g. Def in Fig. 2) and, if wished, optional (e.g. FormalDef in Fig. 2) attributes.

5 Recommendations

Based on the analysis of KAOS we formulate recommendations (Table 3) for modellers, language engineers and tool developers. Due to the page limit for this paper we can list the recommendations only on a very general level. More specific suggestions appear in the technical report [15].

Modeller. The only Objectiv*er* function that might help to deal with the element discriminability is proximity control (*M.1*) The modeller can apply clan or tree layout strategies. Although modularisation (*M.2*) is helpful for large diagrams [17], the modeller should not overestimate it, viz. it might be inefficient for small diagrams [21]. For structuring (*M.3*), modellers have to use proximity appropriately, and structure elements based on semantics but not on syntax [21]. To ease cognitive integration, the modeller must name each diagram, as well as specify its type (*M.6*).

Objectiv*er* does not provide legends automatically, nor does it allow attaching more "direct" icons to diagram elements (*M.5*); instead the modeller has to do it manually (e.g., with text-editing and drawing tools). This is a labour intensive activity, but might result in better model understanding, especially for unskilled diagram readers.

Table 3. Recommendations for modeller, language engineer and tool developer

Modeller	Language engineer	Tool developer
M.1: Use proximity normalisation functions to discriminate elements.	*E.1*: Design a concrete syntax that would allow discriminate language constructs relatively.	**Tool should:** *T.1*: have means for absolute discriminability.
M.2: Divide model into manageable modules.	*E.2*: Define clues supporting different cognitive integration techniques.	*T.2*: provide guidelines for decomposing the model into modules.
M.3: Group elements in the diagram according to the semantic relevance.	*E.3*: Develop icons and relationships to help remembering their meaning.	*T.3*: have means to structure elements in a diagram. *T.4*: provide cognitive integration techniques [16].
M.4: Define visual cues that might ease information integration from different diagrams.	*E.4*: Language constructs should be equipped with attributes for defining additional information.	*T.5*: have means for emphasis of diagram elements. *T.6*: be explained in its tutorials. *T.7*: guide creation of the diagram using scenarios.
M.5: Create legends for the constructs used in the diagrams.	*E.5*: Define simple language graphical conventions.	*T.8*: have means to define legend in a diagram. *T.9*: include diagram name and type on the printed diagram.
M.6: Name and specify type for each diagram in the model.	*E.6*: In documentation, explain each language element (relations and icons).	*T.10*: have element normalisation means. *T.11*: have means for controlling visual variables for every element in the diagram (model).
M.7: Learn the language and tool principles from documentations.		*T.12*: print comments, conceptual cues, textual and formal information provided in the element properties.
M.8: Define contextual information in the diagram.		*T.13*: have discussion means for modellers and diagram readers.

Language engineer. These recommendations include guidelines on how to improve KAOS, or support the development of specific domain languages based on KAOS. For example the concrete suggestion for *E.1* is to discriminate constructs not only by colours, but also include different shapes and icons (e.g., as they are suggested in [12] or [24]). Other concrete suggestions are provided in [15].

Tool developer. The modelling tool should provide different support for unskilled modellers (novices) and for experts. Petre notices that, on one hand, novices are distracted by "syntax and surface features" [21]. On the other hand, experts "handle information at a different level" [21]. Thus, the concrete suggestions for the tool improvement include the differentiation handled between the novices and experts – for example, including a legend (*T.8*) for novices and not including it for experts; guiding novices with scenarios (*T.7*) and switching off the scenario guidance for experts. Other existing tool functionality and suggested tool improvements are provided in the technical report [15].

6 Discussion

In this paper we have considered some basic principles [18] for producing the effective diagrams and analysed how they are fulfilled for KAOS and its supporting tool Objectiv*er*. The study was performed by one researcher, the first author of this paper. Therefore, the research is subjective *wrt* understanding and interpreting the principles. The study was carried out during one week, executing small examples. Hence, it might be that some properties were not observed. The observations would be more accurate if the principles were applied on large-scale models. But our study resulted in an arguably more fine-grained analysis in comparison to [1, 3, 14].

Applying the principles relies on the individual skills and insight of modellers and diagram readers. For example, normalisation can affect intentional use of visual variables for emphasis, grouping and discriminability [18]. Furthermore KAOS, like other GMLs, deals with abstractions, like goal, that have no physical representation in the real world. This makes perceptual directness difficult to achieve. The graphical conventions proposed in KAOS (as well as other GMLs) are thus arbitrary and have to be learnt before starting modelling.

The analysis indicates that KAOS can be substantially improved *wrt* most principles. The observations are in line with similar studies for other visual modelling languages (e.g. UML [19]). Thus, we have suggested recommendations for KAOS and Objectiv*er* users to produce models that communicate more effectively. Recommendations for language engineers and tool developers suggest (*i*) how to maintain and improve both KAOS and Objectiv*er*, and (*ii*) how to devise new domain specific languages and tools based on KAOS. Our future work includes validating the results and suggested recommendations, analysing other GMLs, and investigating the principles for effective diagrams in large-scale (industrial) goal models.

Acknowledgement. We wish to thank R. Darimont for the very useful comments, P.-Y. Schobbens and G. Saval for their help and proofread of the paper.

References

1. Al-Subaie, H.S.F., Maibaum, T.S.E.: Evaluating the Effectiveness of a Goal-oriented Requirements Engineering Method. In: CERE 2006. Proc. of the 4th Int. workshop on Comparative Evaluation in Requirements Engineering, pp. 8–19 (2006)
2. Anton, A.I.: Goal-based Requirements Analysis. In: ICRE 1996. Proceedings of the 2nd Int. Conference on Requirements Engineering, pp. 136–144. IEEE Computer Society, Los Alamitos (1996)

3. Ayala, C.P., Cares, C., Carvallo, J.P., Grau, G., Haya, M., Salazar, G., Franch, X., Mayol, E., Quer, C.: A Comparative Analysis of i*-based Agent-oriented Modelling Languages. In: SEKE 2005. Proc. of the 17th Int. Conf. China, pp. 43–50 (2005)

4. Blackwell, A., Green, T.: Notational Systems – the Cognitive Dimensions of Notations Framework. In: Carrol, J.M. (ed.) HCI Models, Theories and Frameworks: Towards a Multidisciplinary Science, Morgan Kaufmann, San Francisco (2003)

5. Bresciani, P., Perini, A., Giorgini, P., Giunchiglia, F., Mylopoulos, J.: Tropos: An Agent-oriented Software Development Methodology. Autonomous Agents and Multi-Agent Systems 8(3), 203–236 (2004)

6. Chung, K.L., Nixon, B., Mylopoulos, J., Yu, E.: Non-Functional Requirements in Software Engineering. Kluwer Academic Publishers, Boston (2000)

7. Costagliola, G., Delucia, A., Orefice, S., Polese, G.: A Classification Framework to Support the Design of Visual Languages. Journal of Visual Languages and Computing 13, 573–600 (2002)

8. Franch, X.: On the Quantitative Analysis of Agent-oriented Methods. In: Dubois, E., Pohl, K. (eds.) CAiSE 2006. LNCS, vol. 4001, pp. 495–509. Springer, Heidelberg (2006)

9. Frutiger, A., Bluhm, A.: Signs and Symbols: Their Design and Meaning. Watson-Guptill Publications, New York (1998)

10. Kavakli, E.: Goal-oriented Requirements Engineering: a Unifying Framework. Requirements Engineering Journal 6(4), 237–251 (2002)

11. Kavakli, E., Loucopoulos, P.: Goal Modeling in Requirements Engineering: Analysis and Critique of Current Methods. In: Krogstie, J., Halpin, T., Siau, K. (eds.) Information Modeling Methods and Methodologies, pp. 102–124. IDEA Group Publishing, USA (2005)

12. Letier, E.: Reasoning about Agents in Goal-Oriented Requirements Engineering. PhD thesis, Universite Catholique de Louvain (2001)

13. Lindland, O.I., Sindre, G., Sølvberg, A.: Understanding quality in conceptual modelling. IEEE Software 11(2), 42–49 (1994)

14. Matulevičius, R., Heymans, P.: Comparison of Goal Languages: an Experiment. In: REFSQ 2007. Proceedings of the Working Conference on Requirements Engineering: Foundation for Software Quality, pp. 18–32. Springer, Heidelberg (2007)

15. Matulevičius, R., Heymans, P.: Visually Effective Goal Models using KAOS. Technical report, University of Namur (laccessed 06.07.2007) (2007),
    http://www.info.fundp.ac.be/~rma/cigmol/deliverables/Deliverable-D2-1-b-kaosvisual.pdf

16. Matulevičius, R., Heymans, P., Opdahl, A.L.: Comparing GRL and KAOS using the UEML Approach. In: Concalves, R.J., Muller, J.P., Mertins, K., Zelm, M. (eds.) Enterprise Interoperability II. New Challenges and Approaches, pp. 77–88. Springer, Heidelberg (2007)

17. Moody, D.: Dealing with Map Shock: A Systematic Approach for Managing Complexity in Requirements Modelling. In: Proceeding of the REFSQ 2006, Luxembourg (2006)

18. Moody, D.: What Makes a Good Diagram? Improving the Cognitive Effectiveness of Diagrams in IS Development. In: ISD 2006. Proc. of the 15th Int. Conf. in Information Systems Development (2006)

19. Morris, S., Spanoudakis, G.: UML: An Evaluation of the Visual Syntax of the Language. In: Proc. 34th Annual Hawaii International Conference on System Sciences (HICSS-34), IEEE Computer Society, Los Alamitos (2001)

20. Narayanan, N.H., Hubscher, R.: Visual Language Theory: Towards a Human-Computer Interaction Theory. In: Marriott, K., Meyer, B. (eds.) Visual language theory, pp. 87–128. Springer, Heidelberg (1998)

21. Petre, M.: Why looking isn't Always Seeing: Readership Skills and Graphical Programming. Communications of the ACM 38(6), 33–44 (1995)
22. Regev, G.: A Systemic Paradigm for Early IT System Requirements Based on Regulation Principles: The Lightswitch Approach. PhD thesis, Swiss Federal Institute of Technology (EPFL) (2003)
23. Regev, G., Wegmann, A.: Where do Goals Come From: the Underlying Principles of Goal-oriented Requirements Engineering. In: RE 2005. Proc. of the 13th IEEE Int. Conf. on Requirements Engineering, IEEE Computer Society Press, Los Alamitos (2005)
24. van Lamsweerde, A.: The KAOS Meta-model: Ten Years After. Technical report, Universite Catholique de Louvain (2003)
25. Yu, E.: Towards Modeling and Reasoning Support for Early-phase Requirements Engineering. In: RE 1997. Proc. of the 3rd IEEE Int. symposium on Requirements Engineering, pp. 226–235. IEEE Computer Society Press, Los Alamitos (1997)

Agent Based Executable Conceptual Models Using i* and CASO

Aniruddha Dasgupta, Aneesh Krishna, and Aditya K. Ghose

Decision System Laboratory
School of Computer Science and Software Engineering
University of Wollongong, Wollongong, Australia
{ad844, aneesh, aditya}@uow.edu.au

Abstract. Agent-Oriented Conceptual Modelling (AOCM) is a novel approach to conceptual modelling that has gained considerable credence within the research community over the last decade. The key innovation in AOCM is the use of an agent, together with associated concepts such as goals, plans, commitments etc. as modelling constructs. This has been inspired, in part, by the growing popularity of agent-oriented approaches to the building of intelligent systems, within the artificial intelligence and related research communities. CASO is a formal agent programming language for process specification and agent programming. It supports the formal specification of complex multiagent systems and provides a tool for process simulation. On the other hand it lacks features for modeling the rationale behind design choices. This work allows us to benefit from the complementary representational capabilities of the two frameworks.

1 Introduction

Requirements Engineering is the process which identifies the purpose of a software system and documents it in a form which is responsive to detailed examination, communication followed by implementation [15]. The criticality and centrality of requirements engineering is now widely acknowledged, given that the majority of software errors can be traced back to errors in the requirements engineering phase, and given the difficulty of recovering from these errors in subsequent phases (i.e., design specification, coding, testing etc.) of the software life cycle. AOCM notations offer abstractions that are sufficiently high-level to be particularly suitable to early-phase requirements modelling. As problems grow in size and complexity, the need for formal analysis during the early RE phase may increase. Hence it is often desirable to include mechanisms which allow simulations of different scenarios which would be useful in analyzing its properties and could serve as a decision-support tool.

Our approach uses the technique of BDI (belief-desire-intention) [13] agent programming with CASO [2], which incorporates constraints into a reactive BDI agent programming language. This extends and builds upon our earlier work on co-evolution of i* model and CASO agent programs [2]. Unlike most of the other approaches we particularly focus on modeling non-functional requirements

J.-L. Hainaut et al. (Eds.): ER Workshops 2007, LNCS 4802, pp. 276–285, 2007.

also known as softgoals in the context of i* framework without making any modifications to the existing i* model. We use CASO to incorporate preferences or soft constraints to map to the softgoals in i* model. In this work a methodology for the combined use of i* and CASO is proposed and applied to a harbor supply chain specification. The goal of our approach is to devise a method for the analysis and validation of requirements models represented in i* with CASO agent programs without ignoring aspects of the i* model in the translation and analysis process.

The remainder of this article is organized as follows. Section 2 and gives some background on i* and CASO. Sections 3 describes how softgoals can be modeled in i* and section 4 discusses the combined use of i* and CASO. Section 5 uses a harbor supply chain system to discuss the methodology and shows using examples, how one can benefit from the combined us the two frameworks. Finally, related work and concluding remarks are presented in the last sections.

2   Background

We briefly provide an overview of the i* and CASO in this section. Considerable detail has been omitted in this section due to space limitations but examples and full versions of the text can be found in [10] and [2].

The i* framework

The i* framework [10] for agent-oriented conceptual modeling was designed primarily for early phase requirements engineering. An i* model consists of two main modeling components: the Strategic Dependency (SD) Model and the Strategic Rationale (SR) Model. The SD diagram consists of a set of nodes and links. Each node represents an "actor", and each link between the two actors indicates that one actor depends on the other for something in order that the former may attain some goal. The depending actor is known as depender, while the actor depended upon is known as the dependee. The SD diagram represents the goals, task, resource, and softgoal dependencies between actors. There are four types of dependencies in i* - goal dependency, task dependency, resource dependency and softgoal dependency. The notion of softgoals (quality goals) is related to the notion of non-functional requirements [1]. Softgoals are the goals that do not have a clear-cut satisfaction condition. Each contribution link [11] towards a softgoal is characterized by a label that specifies the contribution type and strength. The positive contribution types for softgoals are HELP (positive but not by itself sufficient to meet the higher goal), MAKE (positive & sufficient), SOME+ (partial positive contribution), and '?' (unknown). The dual negative types are HURT, BREAK, and SOME- respectively.

CASO

CASO(Constraint AgentSpeak with Objectives) [2] is a programming language based on the popular BDI (belief-desire-intention) language AgentSpeak [9]. It incorporates constraints and objectives into the symbolic approach of BDI model. CASO is based on a logical formalism and is very expressive. It can be well adapted to the early-design stages of system development, when detailed alternative process designs have to be specified and need to be compared. CASO incorporates Constraint Solving and Optimization (CSOP) techniques where the optimization is based on the objective function (softgoal). An agent program in CASO consists of a set of beliefs B, a set of constraints C, an objective function O, a set of events E, a set of intention I, a plan library P, a constraint store CS, an objective store OS and three selection functions S_E, S_P, S_I to select an event, a plan and an intention respectively to process and n_p and n_i are parameters which denote the number of steps to look-ahead for *plan* and *intention* selection respectively. Transition of agent program to process events depends on the event triggers. An event trigger, t, can be addition(+) or removal(-) of an achievement goal($\pm!g_i$) or a belief($\pm b_i$). The CASO interpreter manages set of events, a constraint store, an objective store and a set of intentions with three selection functions. The usefulness of using preferences comes when a CASO agent has to perform an option selection S_O operation in choosing a particular plan to pursue.

3   Modelling Softgoals as Preferences in i*

In order to decide among different courses of action, it is important to rank them according to some measure of utility. This decision involves the categorization of the softgoals according to the importance to the system. We extend the approach put forward in [6] for goal analysis in which simple preference scores are given to softgoals based on the contributions. A value of 1 is added for each positive contribution to a preference, 0 to each unmarked preference, and -1 for each negative contribution to a preference. Note that these numbers are arbitrary and depends on the modeler who may wish to give any other weights depending on the domain requirements. It is to be noted that although softgoal decomposition is possible, we are only concerned about the lowest level of softgoal decompositions here. Using this notion we observe that in Figure 1 which describes the *Meeting Initiator* agent, using the task UseMeetingScheduler is going to have a positive contribution (value of 1) towards the softgoal MinimizeEffort and the task ScheduleManually is going to have a negative contribution (value of -1) towards the same softgoal. If there are many softgoals in the system say p1, p2 p3 and tasks T1, T2, T3 and T4 then i* could be extended to incorporate a preference value associated with each softgoal. We annotate each softgoal with a weight

Table 1. Task-softgoal contribution

Task	p1	p2	p3
T1	3	0	-1
T2	2	1	1
T3	-1	2	0
T4	1	1	1

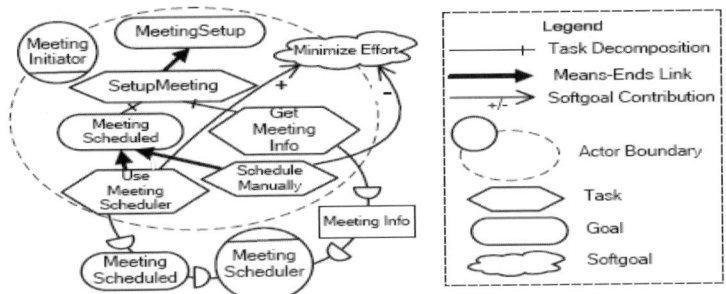

Fig. 1. SR diagram for Meeting Initiator

(say between 0 and 1) such that p1=0.5, p2=0.7 and p3=0.7. Let us further assume that the tasks have contribution towards each softgoal as given in Table 1. Now if we further assume that each of the tasks T1, T2, T3 and T4 are alternate ways of achieving the same goal, then using simple weighted sum for calculating the task contribution values ($ConVal$) towards all the softgoals, we get the following values for each task: $ConVal(T1) = 3*0.5 + 0*0.7 + (-1)*0.7 = 0.8$; $ConVal(T2) = 2*0.5 + 1*0.7 + 1*0.7 = 3.1$; $ConVal(T3) = -1*0.5 + 2*0.7 + 0*0.7 = 0.9$; $ConVal(T4) = 1*0.5 + 1*0.7 + 1*0.7 = 1.9$.

4   Combined Use of the i* and CASO Frameworks

Our methodology for the combined use of i* and CASO frameworks includes six steps described below.

1. Building the strategic Dependency Model(SD) for the System. The analyst develops a SD model that specifies the agents, roles, positions, and the intentional dependency relationships between them.

2. Building the Strategic Rationale Model(SR) for the System. The analyst further analyzes the requirements of the system based on the developed SD model, focusing on identifying the goals, softgoals, and tasks to be accomplished inside agents. It also specifies the decompositions of the tasks/goals, and the contributions to softgoals. The dependency relationships will be specified between nodes inside the related agents.

3. Developing the Initial CASO Model. The analyst maps elements in the SR model into entities in the CASO model using the defined mapping rules and builds the initial CASO model by specifying the actions, tasks, softgoals, plans, the initial beliefs and constraints, and the behavior of the agents in the system.

4. Validating the CASO Model by Simulation. The analyst evaluates the CASO model through simulation. Given a specification of an initial state for the system, the developed CASO model will be simulated using the interpreter and the results are used to check the correctness of the model. Then, we identify the shortcomings and refine the CASO model based on the results of the evaluation.

This step will help the analyst find those mistakes and revise the annotated SR model in the next step.

5. Refining i* and CASO Models Based on Validations Results. Whenever the analyst observes that the i* model or CASO models has to be modified based on the results of the validation step, he will refine both the CASO model and the corresponding part of the i* model. Also by communicating with the client about the current i* and CASO models, the analyst can obtain the feedback from the client and revise the i* model and the corresponding parts of the CASO model. This brings out new specification of the system of interest. Another case is when the analyst needs to add new features into the designed system after he finds some missing requirements have to be modeled, such as loops, exogenous actions, etc. He must modify the i* model and the corresponding part of the CASO model to ensure consistency between these two models.

6. Producing the Requirements Analysis Document. The models and specifications are collected in a document with appropriate explanations and discussion. The results of simulation and verification are also described.

5   Case Study: Harbor Supply Chain Management

In this section we take the example of supply chain management in the context of a harbor as described in [14]. In a harbor, cargo is loaded onto the ship, unloaded from the ship, and stowed on the pier where the receipt and delivery of freight happens. Harbor management system consists of ship operation system, cargo moving system, storage systems, receipt and delivery systems, gate operation systems, and management and operation information system. Ships arrive at the port, receive the service at the berth, and then de-berth after completing the activities of unloading and loading of the containers. Containers are unloaded by the gantry cranes from the ships, and then transported by yard trailer to the yard. At the yard, the transfer crane load/unload the container to the yard and the tractor moves the container to the gate for transferring outside the port. For the export containers, the reverse process applies. In addition, the operational policies of container terminal consists of berth allocation, yard planning, stowage planning, and logistics planning.

5.1   Building i* Models for the Harbor Supply Chain

From the case study described above, we can formulate the following logical entities in the harbor supply chain which control material and information flow: *Customer Relationship Agent* which is responsible for acquiring, modifying and cancelling orders and negotiating with customers with regards to due dates, prices etc. *Logistics Agent* which is responsible for coordinating the supplier, harbor, and distributor. *Transportation Agent* which is responsible for the assignment and scheduling of transportation resources to satisfy inter-harbor movement requests from the logistic manager. *Distributor Agent* which is responsible for delivery of goods to the customer. Other physical entities are ship and harbor. Figure 3 shows the Strategic Rationale (SR) model of the actor

Customer Relationship Agent. We will use this actor to describe our combined methodology of using i* and CASO. The main goal of the agent is to *make deals* with the customer and fulfil the order. To achieve this goal, it has to perform the task of *acquiring orders* which in turn calls for *handling requests* from the customer and *negotiating* with the customer in terms of price and date. It needs to also check with the logistics agent to see whether an order can be fulfilled (*order feasibility*). A soft goal of the agent is to *process orders fast*. Also the negotiation process could be carried out by an *automated software* or *manually* - the former contributing positively towards the soft goal and the later contributing negatively. The agent is also responsible for generating invoice to the customer and supply the *order information* to the logistics agent.

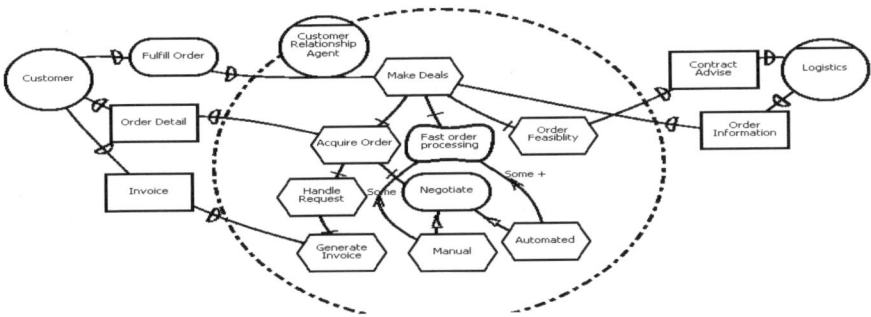

Fig. 2. SR diagram for Customer Relationship Agent in Harbor SCM

5.2   Developing the Initial CASO Model

A first step in developing the CASO model is to define a mapping from i* to CASO. For sake of continuity, we provide the results from the earlier work [3] where this mapping was initially defined. A multi-agent system (MAS) is defined as a pair {Agents, ESA} where Agents= $a_1 \cdots a_n$, each a_i is a CASO agent and *ESA* is a specially designated Environment Simulator Agent implemented in CASO. *ESA* holds the knowledge about the actions that might be performed by actors in SD model and the possible environment transformation after the executions of those actions. The environment agent can verify fulfillment properties such as creation conditions, invariant conditions, and fulfillment conditions of those actions associated with each agent. While *ESA* is a CASO agent, it must be provided with necessary beliefs as well as the plans. The context of the plans determines the constraints that must hold and actions in the body are how to react to the situation. The agents in the MAS are *Customer Relationship Agent, Logistics Agent, Transportation Agent,* and *Distributor Agent.* We map the edges and nodes for each agent from the SR diagrams for each actor which defines the goal, task and resource dependencies into CASO plans. It is to be noted here that besides the four agents, the ESA is also supplied by the analyst of the system (not shown here) which monitors all of the actions/tasks

---

Plans

(p1) +task(Make Deals):True ← !task(Acquire Order), !task(Order Feasibility), Supply(Order Information).

(p2) +task(Acquire Order):resource(Order Detail) ← !task(Handle Request), !goal(Negotiate).

(p3) +task(Handle Request):True ← !task(Generate Invoice).

(p4) +task(Generate Invoice)):True ←Supply(Invoice).

(p5) +goal(Negotiate):True ← !task(Automated).

(p6) +goal(Negotiate):True ← !task(Manual).

(p7) +task(Manual): ConVal(Manual , >=, 0) ← .

(p8) +task(Automated): ConVal(Automated , >=, 0) ← .

(p9) +task(Order Feasibility):resource(Contract Advice)← .

Fig. 3. CASO Customer Relationship Agent

performed by each agent, all of the messages exchanged and all of the beliefs communicated by individual agents for consistency and constraint violations. In CASO we use decision-theoretic planning which investigates alternatives with the aim of finding the best solution in terms of some measure of utility. For this purpose a softgoal is given a numeric value, which can then be incorporated into the utility computation. The mapping of softgoals in i* to the soft constraints extended CASO model as described earlier is quite straightforward. The option selection function S_o determines the particular plan to follow in case there are alternatives by calculating the task contribution towards each softgoal.

The result of applying the mapping rules generate the four CASO agents as mentioned above. We show the *Customer Relationship Agent (CRA)* CASO agent in Figure 3. Note that some of the plans that do not have any body do not exist in the actual programs. However, we show them in this figure to avoid the confusion and improve the clarity of the paper. There are 9 plans - p1, p2, p3, p4 p5, p6, p7, p8 and p9 as shown in the figure. There are 2 *beliefs* in the belief base of the ESA which quantifies the contribution of two tasks towards the softgoals as described in section 5. *ConVal(Manual , FastOrderProcessing =, -1)* signifies that the task *Manual* has a contribution of -1 towards the existing softgoal. Similarly *ConVal(Automated, FastOrderProcessing, =, 1)* implies that the task *Automated* has a contribution of +1. Plans p4 and p5 which describe the alternates that *CRA* can choose from has constraints as the *context* of the plans. In plan p7, the context *ConVal(Manual , >=, 0)* means that this plan will be pursued if and only if the task contribution of *Manual* is >= 0. Here the belief base has the contribution as -1 and hence this plan will not be selected. Thus plan p5, whose context clearly satisfies the constraint, will be selected as the alternate. It is to be noted here that it is purely up to the analyst to decide the context constraint of the plans when dealing with softgoals. Thus any plan context of the form *ConVal(T, cond, val)* could be used by the analyst where T refers to the task, *cond* could be any of $>, =>=, <, <=$ and *val* refers to a number. Thus CASO captures the softgoal contribution of each task into its plans as shown here.

5.3   Validating and Refining the CASO Model by Simulation

The analyst can evaluate the CASO model through simulation. First, we specify an instance of the system and then we run a simulation. By checking and comparing the results of simulation on different system instances or initial states, we can see whether the system behaves as expected. The analyst first creates an instance of *ESA*, and each of the other agents in the system. As an example, the Harbor Supply Chain may not have capabilities to handle more that 20 shipments in a day. This constraint can be supplied to the *CRA* but it cannot be depicted in the i* model. Thus this constraint is given to the *ESA* by the user and thus this MAS can detect in its execution cycle whether the constraint is violated. Based on the order fulfillment perspective, simulation model identifies not only the product and information flow, but also the delay of delivery, information, order decision, and order filling. The input parameters of the experiment are the number of entities, their association, properties and the coordination strategy, which determines the information propagation depth upstream or downstream.

In the initial SR model, the possibility of failing to achieve a goal is not clearly shown, but in the CASO model, this is handled. Moreover the CASO model gives hints about how to achieve the softgoals using quantitative modeling. The six steps as described in section 4 will need to be repeated if errors are found or if aspects of the i* model and CASO model do not satisfy the client's needs. Based on the CASO model and simulation experiments, if the i* model lacks some part of the desired requirements, modifications to the i* model will be performed. Similarly if the CASO model needs to specify additional details or aspects of the i* model, modifications to the CASO model will be made. Once a satisfactory model of the required system has been developed, a requirements specification document is produced.

6   Related Work

Much of the existing work in requirements engineering has focused on modelling and specification languages for the late-phase, and assume that an initial statement of the requirements is always available [11]. The late-phase of requirements engineering focuses on the completeness, consistency, and automated verification of requirements. In contrast, the early phase intends to model and analyse stakeholders' interests and how they might be addressed, or compromised, by means of various system and environment alternatives. However, formal support for the early-phase requirements engineering has been relatively sparse.

Formal Tropos [4] is an intermediate language in which i* models must be defined before an eventual translation into a state machine model on which model checkers can be deployed to verify systems properties. Alternative approach have been proposed for transforming i* models into agent programs in formal agent programming languages such as ConGolog [5]. In their work aspects of the i* model are ignored in the translation process. [8] presents an agent-oriented requirements engineering approach that combines informal i* models

with formal specifications written in the CASL language. Co-evolution of i*
models with 3APL is described in [7] where agent programs are concurrently
maintained and updated. In another notable system called SNet [12], extended
i* diagrams are automatically translated into executable ConGolog programs
in inter-organizational networks in business process management. It also uses
softgoals to model alternates in ConGolog but makes modifications to i* diagrams. One major drawback of most of the existing approaches is that the
agents themselves are limited in the kinds of choices they are able to make during a simulation. During a particular simulation run, the agents simply commit
reactively to the choices required by the design of the agent network and do not
themselves engage in deliberation about what the most appropriate course of
action might be.

7   Conclusion

In this paper, we review an approach to executing i* models by translating these
into set of interacting agents implemented in the CASO language and suggest
how we can perform reasoning with the requirements modelled (both functional
and non-functional) using i* models. In this paper we particularly incorporate
deliberation into the agent design. The methodology allows the requirements engineer to exploit the complementary features of the two frameworks to develop
better models of the application of interest and produce requirements specifications that fulfill the client's goals. A CASO model can be used both as a
requirements analysis tool and as a formal high-level specification for a MAS
that satisfies the requirements because it is possible to produce a high-level,
formal model of the MAS right from the i* diagrams. Our work particularly
focuses on modeling alternatives using softgoals which is different from most
other approaches where softgoals and softgoal dependencies are abstracted out
from SR diagrams before annotations are introduced and softgoal nodes are
dropped from the i* diagrams along with the accompanying contribution links.
We can analyze the system behavior and especially reason with the alternates,
using real-life examples which is otherwise not possible by only looking at the
i* model and CASO agents separately. Using the CASO interpreter, one can
run this high-level model of the system on some sample environment/agent parameters and determine if the behaviour of the program corresponds to the
expected behaviour of the system-to-be. If discrepancies are found, they can
be analyzed and appropriate changes can be made to the original SR diagram.
Due to the tight mapping between SR diagrams and the CASO models, it is
easy to find parts of the SR diagram that are related to specific parts of the
CASO program and vice versa. Using this technique one can specify requirements, define architecture, model behavior as well as perform simulation. Our
future work is to build an automated tool which can help towards achieving this
exercise.

References

1. Chung, L., Nixon, B.A., Yu, E., Mylopoulos, J.: Non-Functional Requirements in Software Engineering. Kluwer Academic Publishers, Dordrecht (1999)
2. Dasgupta, A., Ghose, A.K.: Caso: A framework for dealing with objectives in a constraint-based extension to agentspeak(l). In: Proc. of the 2006 Australasian Computer Science Conference, pp. 121–126 (2006)
3. Dasgupta, A., Krishna, A., Ghose, A.K.: Co-evolution of agent-oriented conceptual models and CASO agent programs. In: IAT 2006. IEEE/WIC/ACM International Conference on Intelligent Agent Technology, Hong Kong, China, pp. 686–689 (2006)
4. Fuxman, A., Kazhamiakin, R., Pistore, M., Roveri, M.: Formal tropos: language and semantics (2003)
5. Giacomo, G.D., Lesperance, Y., Levesque, H.J.: Congolog, a concurrent programming language based on the situation calculus, vol. 121, pp. 109–169 (2000)
6. Hui, B., Liaskos, S., Mylopoulos, J.: Requirements Analysis for Customizable Software Goals-Skills-Preferences Framework. In: RE 2003. International Requirements Engineering Conference, pp. 117–126. IEEE, Los Alamitos (2003)
7. Krishna, A., Guan, Y., Ghose, A.K.: Co-evolution of i* models and 3apl agents. In: QSIC 2006. Sixth International Conference on Quality Software, pp. 117–124 (2006)
8. Lapouchnian, A.: Modeling Mental States in Requirements Engineering – An Agent-Oriented Framework Based on i* and CASL. MSc. Thesis, York Univ., Canada (2004)
9. Rao, A.: Agentspeak(l): BDI agents speak out in a logical computable language. In: Agents Breaking Away: Proc. of the 7th European WS on Modelling Autonomous Agents in a Multi-Agent World, Springer, Heidelberg (1996)
10. Yu, E.: Modelling strategic relationships for process reengineering, Phd. Thesis. Univ. of Toronto, Canada (1995)
11. Yu, E., Liu, L.: Modelling trust for system design using the i* strategic actors framework. In: Falcone, R., Singh, M., Tan, Y.-H. (eds.) Trust in Cyber-societies. LNCS (LNAI), vol. 2246, pp. 175–194. Springer, Heidelberg (2001)
12. Gans, G., Jarke, M., Lakemeyer, G., Schmitz, D.: Deliberation in a metadata-based modeling and simulation environment for inter-organizational networks. In: Information Systems, vol. 30(7), pp. 587–607. Elsevier Science Ltd., Amsterdam (2005)
13. Rao, A.S., Georgeff, M.: BDI Agents: from theory to practice. In: ICMAS 1995. Proc. of First International Conference on Multi-Agent Systems, San Fransisco, pp. 312–319 (1995)
14. Yi, D.W., Kim, S.H., Kim, N.H.: Combined Modeling with Multi-Agent Systems and Simulation: Its Application to Harbor Supply Chain Management. In: Proceedings of Hawaii International Conference on System Sciences, USA (2002)
15. Nuseibeh, B., Easterbrook, S.: Requirements Engineering: a roadmap. In: Proc. of 22nd International Conference on Software Engineering-Future of SE Track, Limerick, Ireland, pp. 35–46 (June 4-11, 2000)

Achieving, Satisficing, and Excelling

Ivan J. Jureta[1], Stéphane Faulkner[1], and Pierre-Yves Schobbens[2]

[1] Information Management Research Unit, University of Namur, Belgium
[2] Institut d'Informatique, University of Namur, Belgium
iju@info.fundp.ac.be, stephane.faulkner@fundp.ac.be, pys@info.fundp.ac.be

Abstract. Definitions of the concepts derived from the goal concept (including functional and nonfunctional goal, hardgoal, and softgoal) used in requirements engineering are discussed, and precise (and, when appropriate, mathematical) definitions are suggested. The concept of satisficing, associated to softgoals is revisited. A softgoal is satisficed when thresholds of some precise criteria are reached. Satisficing does not cover situations in which continual improvement of thresholds is expected. The notion of *excelling* is suggested to cover such cases, along with the concept of *disposition* to represent and reason about excelling.

1 Outline

One motive for representing and reasoning about a system is to precisely understand the purpose thereof and subsequently use this information to identify, analyze, and select among alternative properties and behaviors needed of the system to fulfill its purpose. The *goal* concept stands out among the various abstractions proposed for the representation and reasoning about a system's purpose. It is now accepted that the concept is relevant for the elicitation, elaboration, structuring, specification, analysis, negotiation, documentation, and modification of stakeholders' requirements on a system [1,2,3,4,5,6,7,8,9,10,11,12,13].

Few contributors to the field of requirements engineering (RE) agree on a *precise* definition of the goal concept. Elasticity in definitions may facilitate the basic understanding of goal-based RE frameworks to non-experts: common knowledge substitutes for specialized RE knowledge, thus facilitating learning. Elasticity, however, also involves difficulties in communication, imprecision in intended meaning, and overuse and/or abuse of the terminology.

In response, we study definitions of the goal and its derived concepts, including hardgoal, softgoal, functional goal, and nonfunctional goal. We suggest precise definitions consistent with the literature; when appropriate, definitions are in formal logic. It is usually said that a hardgoal can be achieved, while a softgoal can only be satisficed. We revisit the concept of satisficing, commonly associated to the concept of softgoal. A softgoal is satisficed when thresholds of some precise criteria are reached. We argue that satisficing does not cover situations in which continual improvement of thresholds is preferred. We subsequently suggest the notion of *excelling* to cover such cases, along with the concept of *disposition* to represent and reason about excelling. The contributions of this paper are: (i)

J.-L. Hainaut et al. (Eds.): ER Workshops 2007, LNCS 4802, pp. 286–295, 2007.
© Springer-Verlag Berlin Heidelberg 2007

the set of precise definitions of goal and derived concepts of hardgoal, softgoal, functional goal, and nonfunctional goal; (ii) the notion of excelling intended to complement the concepts of achievement and satisficing in goal-oriented RE; and (iii) the concept of disposition intended for representing and reasoning about requirements to which the notion of excelling applies.

2   Goal Concept in RE Research

System development frameworks include, since the 1970s some form of analysis involving goals [9], among them context analysis, definition study, and participative analysis. Goals have become an essential part of any system's documentation through standards such as e.g., the IEEE-Std-830/1993. There is no established definition for the goal concept: consider Table 1, which lists informal definitions of the goal concept appearing in various goal-oriented RE frameworks. KAOS highlights the nonoperational nature of goals, pointing to the need for taking action to make goals precise by refinement (see, e.g., [2]). Broadly speaking, the KAOS definition is in line with those of Tropos, i^*, GDC, and Lightswitch: a goal designates desirable conditions on the system and/or its environment. Such condititions restrict the set of alternative system and environment states, so that it is appropriate to say that a goal describes desired states. A different conceptualization appears in NFR, where goals are employed for representing nonfunctional requirements, in addition to design decisions, and arguments for or against other goals. We can interpret "design decisions" as restricting potential desired system and environment states. Notions of argument and justification appear in NFR and GBRAM. We have discussed elsewhere [14] the relevance of argumentation and justification for goal-oriented RE, arguing and illustrating that it is more appropriate to maintain the notion of argument separate from the goal concept.

Regarding the use of goals for modeling nonfunctional requirements, two relevant goal taxonomies have been introduced since the seminal contributions in the NFR framework (see, e.g., [9,18] for discussions).[1] *Functional* goals are distinguished from *nonfunctional* ones, and *softgoals* from *hard goals*. *Functional* goals have been used to represent services that the software is expected to deliver (i.e., *what* the software does), whereas *nonfunctional* goals refer to quality requirements that the software needs to satisfy while delivering the services (i.e., *how* the software provides services; e.g., securely, safely, rapidly, etc.). While it is common to equate nonfunctional goals and softgoals (e.g., [1]), it has been subsequently argued that *softgoals* belong to another taxonomy, in which they are distinguished from *hardgoals* [9]. According to the traditional definition, "a softgoal is similar to a (hard) goal except that the criteria for whether a softgoal is achieved are not clear-cut and a priori." [19] The definition used in the REF framework [15] adds details, as shown in Table 1.

While softgoal satisfaction cannot be established in a clear-cut sense [1], the satisfaction of a hardgoal is objective in that it can be established using (formal) verification techniques [2]. In this respect, a hardgoal is said to be *achievable*,

---

[1] This paragraph follows our previous discussions on the subject [18].

Table 1. Informal definitions of the *goal* concept in goal-oriented RE

Framework	Informal definition of the goal (and derived) concepts
KAOS	"A goal is a nonoperational objective to be achieved by the composite system. Nonoperational means that the objective is not formulated in terms of objects and actions available to some agent in the system; in other words, a goal as it is formulated cannot be established through appropriate state transitions under control of one of the agents." [2] "A goal is a desired property about quantities in the environment." [8]
Tropos and i^*	"A goal is a condition or state of affairs in the world that the stakeholders would like to achieve." [3,5,10]
NFR	"Goals [represent] non-functional requirements, design decisions and arguments in support or against other goals." [1,7]
REF	"According to the nature of a goal, a distinction is made between hard goals and soft goals. A goal is classified as hard when its achievement criterion is sharply defined [...]. For a soft goal, instead, it is up to the goal originator, or to an agreement between the involved agents, to decide when the goal is considered to have been achieved [...]. In comparison to hard goals, soft goals can be highly subjective and strictly related to a particular context; they enable the analysts to highlight quality issues [...] from the outset [...]" [15]
GDC	"An enterprise goal is a desired state of affairs that needs to be attained." [16]
GBRAM	"Goals are high level objectives of the business, organization or system. They capture the reasons why a system is needed and guide decisions at various levels within the enterprise." [4]
Lightswitch	"[A] maintenance goal is said to represent a condition that remains constant. [...] [An] achievement goal has definite pre and post-conditions. The pre-condition represents the interpretation that the state of affairs has drifted (or will drift) outside of the threshold associated with the norm [i.e., a variable of the system whose state the system attempts to maintain unchanged as defined by an observer]. The post condition is an interpretation that is within this threshold." [17]

whereas a softgoal is *satisficeable* [1,7,12]. The concept of *satisficing* originates in H. Simon's work [20] in economics: to satisfice is to set a threshold, and accept any achievement above the threshold. In addition to involving satisficing, a softgoal has a subjective component, in that various stakeholders of the system will have different thresholds. We have worked on a more expressive definition of softgoals elsewhere [18], but we did not provide a mathematical definition.

The KAOS framework provides the most precise hardgoal conceptualization: a hardgoal is defined in terms of predicate patterns in a discrete linear temporal first-order logic (see, e.g., [21]).[2] A hardgoal is any one of the following [8]: an

---

[2] In publications on KAOS, what we call hardgoal here is called simply "goal". Note, however, that this conceptualization does not encompass the softgoal concept (which was introduced separately from KAOS): if we know a constraint, written in logic over system histories, we can check at any time if the actual history of the system respects or not the constraint.

achieve hardgoal (pattern: $\phi \Rightarrow \diamond\psi$), a cease hardgoal ($\phi \Rightarrow \diamond\neg\psi$), a maintain hardgoal ($\phi \Rightarrow \square\psi$), an avoid hardgoal ($\phi \Rightarrow \square\neg\psi$). The same conceptualization is adopted in Formal Tropos [11], where patterns are used in the same way to define hardgoals. An informal interpretation of the said conceptualization is that a hardgoal is a constraint over system histories (i.e., behavior over time).

It is clear that the goal concept is intended to be rich in meaning. Instead then of seeking an all encompassing definition, we study derived concepts, obtained by crossing the hardgoal/softgoal and functional/nonfunctional taxonomies; we thus have: (i) *functional hardgoals*, which are hardgoals about what the system should do (e.g., in an email application, such a goal can be: "whenever an e-mail marked as important arrives, the user is informed with a pop-up window and a sound"); (ii) *nonfunctional hardgoals* which describe verifiable criteria for how the system should operate (e.g., "the user should be informed about important e-mail arrival within 1 second of arrival"); (iii) *functional softgoals* describe a subjective requirement of a stakeholder about what the system should do (e.g., "the user should be informed when an e-mail marked as important arrives"); and (iv) *nonfunctional softgoals* which indicate in a subjective and nonverifiable manner how the system should operate (e.g., "the user should be informed rapidly about the arrival of an e-mail marked as important").

In summary, there is no unique definition of goal. One reason for this is that the goal concept is intended to be rich in meaning. Variations in definitions are also due to slightly different uses of the concept in each framework. Whether a goal conceptualization is appropriate depends on how useful is the framework in which it is used. A prescriptive general definition thus seems excluded. It remains, however, of interest to seek a conceptual framework in which the derived concepts mentioned above can be used together, so that the benefits of these complementary concepts can be combined when representing and reasoning about requirements. A precise definition is already available for the hardgoal concept. We can now suggest a common ground for the cited derived concepts.

3   A Common Framework

Consider a toy system that has only two properties, p_1 and p_2. All possible combinations of allowed values for p_1 and p_2 define all possible states of the system. Let S_1, S_2, and S_3 be arbitrary system states, as shown in the bottom part of Figure 1. Assume that measurements are performed on the system in order to evaluate its quality. To perform measurement, we define two metrics d_1 and d_2. To relate what we observe in the system and the values of the metrics, we define mappings M_1, M_2, and M_3 between system states and value combinations of the two metrics. Since some minimum level of quality is expected, we define thresholds on metrics: in Figure 1, $t_1 \equiv d_1 \geq d_1^t$ and $t_2 \equiv d_2 \geq d_2^t$, so that the quality is above the minimal level only when the system is in state S_2 and not in the other two.

Taking the state-based conceptualization of the hardgoal concept, we define a hardgoal $hg_1 \equiv (\top \Rightarrow \square(p_2 = p_2^*))$ as a value p_2^* of the property p_2. hg_1 is

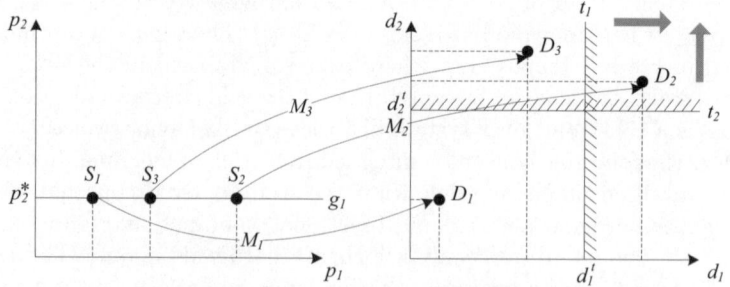

Fig. 1. Arbitrary functional and nonfunctional hardgoals in a toy system

a functional hardgoal, since it says precisely what the system is expected to do (i.e., set property p_2 to value p_2^*). Returning to the informal understanding of the nonfunctional hardgoal given earlier (§2), we see that it cannot be defined over system properties, but on metrics defined for the system. We can thus define two nonfunctional hardgoals in Figure 1: $\tilde{hg}_1 \equiv (d_1 \geq d_1^t)$ and $\tilde{hg}_2 \equiv (d_2 \geq d_2^t)$.

Are there any softgoals in Figure 1? We know that a softgoal is used to model requirements at the earliest stages of an RE process (e.g., [1,9,10,12,18]). Usually, initial requirements, and consequently softgoals are imprecise, subjective, idealistic, and context-specific [18], meaning that we cannot have a softgoal in Figure 1—the figure is already too precise. Consequently, and in line with contributions in the RE field, a softgoal is here understood as an initial, early form of requirements about what the system should do and how "well" it should do it, from which one or more functional and/or nonfunctional hardgoals are extracted during the requirements process. This is appropriate, since we also know that one cannot manage (i.e., assure, control, or improve) what one cannot measure (e.g., [22,23,24,25,26]): if quality-related information contained in softgoals is to be used in decision-making during an RE process, we need to make a softgoal precise, agreed upon by various stakeholders, and realistic—that is, we need to convert it into nonfunctional hardgoals. Same applies for functional softgoals: we require precise, agreed upon, and realistic requirements about what services the system should deliver in order to be able to implement them in later stages of the system development process.

Having established that softgoals appear earlier in an RE process than hardgoals, recall that softgoals are associated to the concept of satisficing. Satisficing is the reason we specified our nonfunctional hardgoals as thresholds only, instead of, e.g., more elaborately stating the desired values of d_1 and d_2. Indeed, nonfunctional hardgoals derived from nonfunctional softgoals serve in RE as criteria for comparing alternative system structures (e.g., [1,7,9,10]). A system structure is chosen over an alternative one if the former dominates the latter over a set of nonfunctional softgoals or the derived nonfunctional hardgoals. In our toy system, we would choose a system structure that is associated to higher values of the two metrics, over one associated with lower values; we would discard structures that are not above both thresholds. Satisficing, while clearly useful and reflecting the inability to identify *the* optimal system structure, does not cover requirements in which continuous improvement is sought. Indeed, satisficing does not go as far as to say what values above a threshold are preferred over other values, also above the threshold. That is, all values are equally desired, provided that they are above the threshold. In many actual cases, we do need to set thresholds, but we need not equally prefer all above-threshold structures. This is the case in particular for adaptable systems based on the agent or services paradigms. We encountered this need in an actual setting: we proposed elsewhere [27] an adaptable system in which above-threshold structures are learned. Therein, a "system structure" corresponds to a composition of web services that allows a service request (coming from a system user and specified in terms of requirements) to be filfiled. To form compositions, a composer web service observes other web services during execution and subsequently selects (for participation in a composition) only those that allow it to obtain more desired values over a given set of metrics. Compositions are revised, so that the quality to which same service requests are fulfilled improves over time. When specifying requirements on such a system, we are clearly not interested only in satisficing—if we did rely on satisficing only, we would not exploit the ability of the system to improve the fulfillment of service requets. We would not exploit the system's ability to adapt. Instead, we need to express that the system both needs to satisfice (so that below-threshold compositions be discarded) and to always improve compositions. We clearly cannot use the notion of satisficing to express requirements on continuous improvement: instead, we use the notion of *excelling* to do so. The limitation of satisficing that we highlight here is not novel: recently, J. L. Pollock proposed the concept of locally global planning, in which "any plan with a positive expected utility is defeasibly acceptable, but only defeasibly. If a better plan is discovered, it should supplant the original one. Satisficing would have us remain content with the original." [28] This discussion brings us to the following position: to use the concept of softgoal grounded in satisficing to express requirements that are associated to the concept of excelling is to extend the softgoal concept too far. We thus propose the concept of *disposition*. A disposition is a preference order defined over goals of the same type; we thus have the following taxonomy for the disposition concept: (i) *hard-functional disposition*, defined over functional hardgoals; (ii) *hard-nonfunctional disposition*, over nonfunctional hardgoals; (iii) *soft-functional disposition*, over functional softgoals; and (iv) *soft-nonfunctional disposition*, over nonfunctional softgoals. An example of a hard-nonfunctional disposition expressed informally is: "the user should be informed about important e-mail arrival within the least time possible" generalizes a preference order in which it is clear that the nonfunctional hardgoal "the user should be informed about important e-mail arrival within 0.5 second of arrival" is preferred to "the user should be informed about important e-mail arrival within 1 second of arrival". Note the following:

- Do not mistake excelling for optimization: the latter applies if the email system is designed so that it always gives the optimal time (i.e., 0 seconds). This is clearly idealistic. Excelling is in a sense optimization over time and given resource boundedness; that is, the email system excels if it reduces time for informing the user at each email arrival compared to the time it needed on the last occasion an email arrived. Excelling applies even if the system does not continually improve (expecting this may be idealistic); what is important for excelling to apply is that, even if, in our example email system, notification time increases, it restablishes and goes down at some later and observable point (i.e., not indefinetly in the future).

– Not always can be a disposition so summarily expressed as in our notification time example for the email application. It may for instance happen that disjoint subsets of metric values are preferred, so that a disposition does not reduce to a formulation of desired direction for metric values. We explore in the remainder a simple notion of disposition mainly because we are introducing the concept here—extensions to its expressivity are of interest in current and future effort.

To express our various types of goals, we start from the multi-sorted first-order version of MITL [29,30], a continuous real-time linear temporal logic. Some predefined sorts (e.g. real numbers) have a fixed interpretation that will be used to express metrics. Our logic starts from Φ, a first-order vocabulary, consisting of predicate and function symbols p, f. They can be declared as flexible (time-dependent) or rigid. As usual, constant symbols are viewed as 0-ary rigid function symbols. Starting with atomic formulas of first-order logic, we form more complex formulas as usual by closing off under truth-functional connectives (i.e., $\wedge$, $\vee$, $\neg$, and $\rightarrow$)[3], temporal operators (i.e., next $\bigcirc$, eventually $\diamond$, always $\Box$, until U and unless W) that can be indexed by a non-singular real-time constraint, existential ($\exists$) and universal ($\forall$) quantification. We denote the resulting language $\mathcal{L}$. We interpret formulas of $\mathcal{L}$ over the structure $\mathcal{T} \equiv \langle \mathcal{D}_\mathcal{S}, \mathcal{S}, \pi, \mathcal{H} \rangle$, where $\mathcal{D}$ gives a domain to interpret each sort, $\mathcal{S}$ is a set of states of the system, π is an interpretation assigning each predicate symbol and function symbol in Φ a predicate or function of the right arity over $\mathcal{D}$, if the symbol is declared rigid. If the symbol is declared flexible, it depends furthermore on the current state. $\mathcal{H}$ is a set of timed state sequences, $\mathcal{S}_0, I_0, \mathcal{S}_1, I_1 \ldots$ i.e. an infinite sequence of *states* and their associated interval of time, representig all possible executions of the system. These intervals I_i must partition the positive reals. To interpret first-order variables, we use a valuation function σ, which, given a variable of sort s returns its value, an element of $\mathcal{D}_f$. Given a structure $\mathcal{T}$, an history h, a time t and a valuation σ, we can associate with every formula of $\mathcal{L}$ a truth value in the usual way. A formula holds in a structure if it yields true for all histories, valuations and times. We call the obtained logic $\mathbb{L}_\mathcal{L}$. We can now give a precise definition of functional hardgoal.

Definition 1. *A functional hardgoal is a formula in $\mathbb{L}_\mathcal{L}$ that restricts the possible histories of a given system only to those desired by system stakeholders.*

To express the nonfunctional hardgoal concept, we use *metrics*. A metric is a rigid function symbol, which will return values in a sort equipped with an order.

---

[3] Usual abbreviations apply, e.g., $(\phi \Rightarrow \psi) \equiv \Box(\phi \rightarrow \psi)$.

Definition 2. *A nonfunctional hardgoal is a formula in $\mathbb{L}_{\mathcal{L}}$ that restricts the values of metrics to those desired by system stakeholders.*

To relate the metrics and the behaviour of the system (recall Figure 1), we also need mappings between functional and nonfunctional hardgoals.

Definition 3. *A hardgoal mapping is a formula in $\mathbb{L}_{\mathcal{L}}$ over one or more functional hardgoals and one or more nonfunctional hardgoals.*

Taking the email application example, the following is a functional hardgoal, and is followed by an equivalent nonfunctional hardgoal:

$$hg \equiv [\forall m : Email \, (arrived(m) \wedge important(m)) \Rightarrow$$
$$\Diamond_{\leq 1sec} \exists \, w : PopupWindow, s : NotifSound \, (display(w) \wedge play(s))]$$

$$\tilde{hg} \equiv [timeToNotification \leq 1sec]$$

We can then define a hardgoal mapping for the above as follows: $hg \equiv \tilde{hg}$.

In contrast, we understand softgoals as expressing dispositions, i.e. preference over goals of the same type. To accommodate dispositions, we extend $\mathbb{L}_{\mathcal{L}}$ in the following way. First, we add a new kind of formulas, *disposition formulas*, such that if ϕ and ψ are formulas of $\mathcal{L}$ then $\phi \succeq_d \psi$ is a disposition formula. We take here the simplest case, in which we do not allow the operator $\succeq_d$ to appear in, e.g., temporal formulas of the language. $\mathcal{L}$ extended with disposition formulas is denoted $\mathcal{L}^{\succeq_d}$. Second, we extend our structures $\mathcal{T}$ to interpret defeasible formulas: we add a function v which maps a real number (informally understood as a utility value) to non-disposition formulae. v is then evaluated as follows: if $\phi \succeq_d \psi$, then $v(\phi) \geq v(\psi)$, which means that ϕ is preferred to ψ.

Following the earlier example, we may have the following hard-nonfunctional disposition:

$$[timeToNotification \leq 0.5sec] \succeq_d [timeToNotification \leq 1sec]$$

We define a disposition in terms of hard and soft disposition as follows:

Definition 4. *A hard disposition is a disposition formula in $\mathbb{L}_{\mathcal{L}^{\succeq_d}}$ between hardgoals.*

Definition 5. *A soft disposition is a preference either between only functional softgoals or between only nonfunctional softgoals.*

4   Conclusions and Future Work

Definitions of the concepts derived from the goal concept (including functional and nonfunctional goal, hardgoal, and softgoal) used in RE are discussed, and precise (and, when appropriate, mathematical) definitions are suggested. The concept of satisficing, associated to softgoals is revisited. A softgoal is satisficed when thresholds of some precise criteria are reached. Satisficing does not cover situations in which continual improvement of thresholds is expected. The notion of *excelling* is suggested to cover such cases, along with the concept of *disposition* to represent and reason about excelling.

Although we have only presented here the simplest notion of disposition, we believe that the paper opens a particularly relevant discussion for goal-oriented RE. We hope it motivates similar efforts to ours in exploring more expressive concepts and techniques for RE. More elaborate expressions of dispositions need to be possible if excelling is to be properly accounted for; these include, e.g., conditional dispositions. We are working on extending the expressivity of the concept and are building a method to use the disposition concept in a systematic manner during the RE process. The method is intended to extend established goal-oriented RE frameworks. Tool support will be explored.

References

1. Mylopoulos, J., Chung, L., Nixon, B.: Representing and using nonfunctional requirements: A process-oriented approach. IEEE Trans. Softw. Eng. 18(6), 483–497 (1992)
2. Dardenne, A., van Lamsweerde, A., Fickas, S.: Goal-directed requirements acquisition. Science of Computer Programming 20 (1993)
3. Yu, E.: Modeling Strategic Relationships for Process Reengineering. PhD thesis, Dept. of Computer Science, University of Toronto (1994)
4. Anton, A.I.: Goal-based requirements analysis. In: Proceedings of the International Conference on Requirements Engineering (1996)
5. Yu, E.: Towards modeling and reasoning support for early requirements engineering. In: Proceedings of the IEEE International Symposium on Requirements Engineering (1997)
6. Roland, C., Souveyet, C., Achour, C.B.: Guiding goal modeling using scenarios. IEEE Transactions on Software Engineering, Special Issue on Scenario Management, 1055–1071 (1998)
7. Chung, L., Nixon, B.A., Yu, E., Mylopoulos, J.: Non-Functional Requirements in Software Engineering. Kluwer Academic Publishers, Dordrecht (1999)
8. Letier, E.: Reasoning about Agents in Goal-Oriented Requirements Engineering. PhD thesis, Dept. d'Ingenierie Informatique, Universite de Louvain (2001)
9. van Lamsweerde, A.: Goal-oriented requirements engineering: A guided tour. In: RE 2001. Proceedings of the International Symposium on Requirements Engineering (2001)
10. Castro, J., Kolp, M., Mylopoulos, J.: Towards requirements-driven information systems engineering: the tropos project. Information Systems 27(6), 365–389 (2002)
11. Fuxman, A., Liu, L., Mylopoulos, J., Pistore, M., Roveri, M., Traverso, P.: Specifying and analyzing early requirements in tropos. Requirements Engineering 9(2), 132–150 (2004)
12. Letier, E., van Lamsweerde, A.: Reasoning about partial goal satisfaction for requirements and design engineering. In: Proceedings of the International Conference on Foundations of Software Engineering (2004)
13. van Lamsweerde, A.: Goal-oriented requirements engineering: A roundtrip from research to practice. In: RE 2004. Proceedings of the International Requirements Engineering Conference (2004)

14. Jureta, I.J., Faulkner, S., Schobbens, P.Y.: Justifying goal models. In: RE 2006. Proceedings of the 14th IEEE International Conference on Requirements Engineering, IEEE Computer Society Press, Los Alamitos (2006)

15. Donzelli, P.: A goal-driven and agent-based requirements engineering framework. Requirements Engineering 9, 16–39 (2004)

16. Kavakli, E.: Goal-Driven Requirements Engineering: Modeling and Guidance. PhD thesis, University of Manchester (1999)

17. Regev, G., Wegeman, A.: Where do goals come from: the underlying principles of goal-oriented requirements engineering. In: RE 2005. Proceedings of the International Requirements Enginering Conference (2005)

18. Jureta, I.J., Faulkner, S., Schobbens, P.Y.: A more expressive softgoal conceptualization for quality requirements analysis. In: Embley, D.W., Olivé, A., Ram, S. (eds.) ER 2006. LNCS, vol. 4215, Springer, Heidelberg (2006)

19. Liu, L., Yu, E.: Designing information systems in social context: a goal and scenario modeling approach. Information Systems 29, 187–203 (2004)

20. Simon, H.: A behavioral model of rational choice. Quarterly Journal of Economics 59, 99–118 (1955)

21. Manna, Z., Pnuelli, A.: The Temporal Logic of Reactive and Concurrent Systems. Springer, Heidelberg (1992)

22. IO for Standardization ISO 8402 Quality management and quality assurance - Vocabulary. International Organization for Standardization (1986)

23. Kitchenham, B., Pfleeger, S.L.: Software quality: The elusive target. IEEE Softw. 13(1), 12–21 (1996)

24. Briand, L.C., Morasca, S., Basili, V.R.: An operational process for goal-driven definition of measures. IEEE Trans. Softw. Eng. 28(12), 1106–1125 (2002)

25. Fenton, N.E., Neil, M.: Software metrics: roadmap. In: ICSE - Future of SE Track, pp. 357–370 (2000)

26. Haag, S., Raja, M., Schkade, L.: Quality function deployment usage in software development. Communications of the ACM 39(1), 41–49 (1996)

27. Jureta, I.J., Faulkner, S., Achbany, Y., Saerens, M.: Dynamic web service composition within a service-oriented architecture. In: ICWS 2007. Proceedings of the International Conference on Web Services (2007)

28. Pollock, J.L.: Thinking about Acting: Logical Foundations for Rational Decision Making. Oxford University Press (Forthcoming)

29. Alur, R., Feder, T., Henzinger, T.: The benefits of relaxing punctuality. In: Proceedings of the Tenth Annual Symposium on Principles of Distributed Computing, pp. 139–152. ACM Press, New York (1991)

30. Henzinger, T., Raskin, J.F., Schobbens, P.Y.: The regular real-time languages. In: Larsen, K.G., Skyum, S., Winskel, G. (eds.) ICALP 1998. LNCS, vol. 1443, pp. 580–591. Springer, Heidelberg (1998)

On the Adequacy of *i** Models for Representing and Analyzing Software Architectures

Gemma Grau and Xavier Franch

Universitat Politècnica de Catalunya
c/ Jordi Girona 1-3, Barcelona E-08034, Spain
{ggrau, franch}@lsi.upc.edu

Abstract. In order to work at the software architecture level, specification languages and analysis techniques are needed. There exist many proposals that serve that purpose, but few of them address architecture and requirements altogether, leaving a gap between both disciplines. Goal-oriented approaches are suitable for bridging this gap because they allow representing architecture-related concepts (components, nodes, files, etc.) and more abstract concepts (goals, non-functional requirements, etc.) by using the same constructs. In this paper we explore the suitability of the *i** goal-oriented approach for representing software architectures. For doing so, we check its properties against the ones suitable for Architecture Description Languages and we define some criteria for solving the unfulfilled aspects in representing the architectures. This paper assumes basic notions on *i**.

1 Introduction

Many researchers have realized that, to obtain the benefits of an architectural focus, software architecture must be provided with its won body of specification languages and analysis techniques [17]. The purpose of Architecture Description Languages (hereafter, ADLs) is to demonstrate the properties of the system at its early stages and minimize the cost of errors. For doing so, ADLs have to provide adequate abstractions and, at the same time, enough level of detail for establishing the properties of interest. However, despite the need for and benefits of specifying non-functional properties, there is a notable lack of support for them in existing ADLs [18].

It is possible to observe that most requirements frameworks provide mechanisms for model analysis in order to inform further decisions but, despite the similarities in both fields, there is a recognized gap between requirements and architectures. According to [12] this gap is mainly due to the different representation of concepts in requirements and in architectures. This leads to the use of new techniques and models for bridging the gap between requirements and architectures. From this point of view, goals are an adequate formalism for representing the concepts on both disciplines with the required level of detail. Because of that, there are many approaches that advocate the use of goal-oriented models [16], and in particular, the *i** notation [20], for representing software architectures (see [13]).

Following this tendency, in [9], [10], we presented SAR*i*M, a Software Architecture Reengineering *i** Method, that aims at assessing the selection of a

J.-L. Hainaut et al. (Eds.): ER Workshops 2007, LNCS 4802, pp. 296–305, 2007.
© Springer-Verlag Berlin Heidelberg 2007

system architecture. SAR*i*M is an evolution of PR*i*M [8] a method for exploring and evaluating process and system alternatives by representing its requirements with *i**. Consequently, SAR*i*M uses the *i** constructs to model the current architecture and, then, supports the exploration of alternative architectures by means of different *i** models, which are evaluated using structural metrics [7]. SAR*i*M copes with the architectures exploration and evaluating process using the *i** constructs for representing architectures and, so, works at the requirements level. For a reliable application of the method, we need to make an adequate use of *i** when modelling the architectures. However, although most of the previously mentioned work uses *i** for modelling software architectures, no consensus has been reached for representing architectural concepts within *i**. As far as we know, only [2] and [5] propose some correspondence between *i** and the agent architectural description language, but they do not address how the use of *i** fulfils the desirable properties for being an ADL.

In order to ensure that *i** is adequate for representing software architecture, we have checked its properties against the desiderata for software architectures presented in [19]. As a result, we have found that *i** accomplishes the principles:

- **Composition.** *i** allows describing a system as a composition of independent components and connections, where components are represented by *i** actors and connections by *i** dependencies.
- **Abstraction.** *i** makes possible to describe the software system at different levels of detail (for instance, early requirements, late requirements, architectural design or detailed design [3]).Thus, the components and their interaction can be described with or without software architecture in a way that clearly prescribes their abstracts roles in a system.
- **Analysis.** There are several proposals on how to analyse *i** models [13], being the most commonly used goal-reasoning techniques [20] and structural metrics [6], [7]. Thus, it is possible to perform rich and varied analyses of architectural descriptions.

However, there still some open aspects that remain open:

- **Reusability.** As it is highlighted in [1], the inherent freedom of the *i** language makes most of the groups working with *i** to use their own criteria and constructs. Thus, despite *i** could make possible to reuse components, connectors, and architectural patterns in different architectural descriptions, the different variations of *i** used and the lack of consensus in representing architectural concepts, could make this reuse difficult. On the other hand, most of the *i** models are build from the scratch and, although some work uses *i** architectural patterns [2], [3], [14] there are no catalogues and directives for reusing them.
- **Configuration and Heterogeneity.** For the same reasons stated above, there is often a lack of prescriptiveness when constructing the models and there exist different works that propose different uses of the constructs, which damages configuration and heterogeneity.

In order to address these issues we have analysed the *i** constructs against the most relevant modelling features proposed in [18] for the classification and comparison of ADLs. In those aspects not fulfilled by *i**, we have used our experience in *i** modelling and architecture modelling, and other work related on *i** [13] to provide a suitable solution. As a result, in section 2, we present a proposal for representing components and connectors by means of actors and dependencies. Then, in section 3 we extend this proposal for representing software architecture configurations and we show how they can be refined through different kinds of models allowing reuse. Finally, in section 4, we present the conclusions and future work.

2   Adapting the *i** Constructs to Software Architectures

In order to adapt the *i** constructs to the needs for representing software architectures, we propose to use the building blocks of the architecture description established in the ADL classification and comparison framework proposed in [18]. The main modelling features are components, connectors, and architectural configurations. For components and connectors, the main elements to take into account are their interface, types, semantics, constraints, evolution, and non-functional properties (see [18] and further sections for a description of the elements). In the following sections we propose a way to adapt the *i** constructs for satisfying these modelling features.

2.1   Using Actors for Modelling Components

A Component in a software architecture is a unit of computation or a data store [18]. In ADLs, components have an interface which contains a set of interaction points (also called *ports*) that allows interacting with the external world. In *i**, actors are the most intuitive way to represent components, being their ports the points where the dependencies are connected to the actor. As we explain in the next section, the different types of dependencies have different architectural meaning and, accordingly, we only consider as *ports* those links related with resource and task dependencies. Following this criteria, an *i** actor may have as many *ports* as needed and, as dependencies are bidirectional, we can distinguish between input *ports* (where the actor is the *depender*) and output *ports* (where the actor is the *dependee*).

The functionality of the components is encapsulated into reusable blocks, which are abstracted by means of component types. Actors can be distinguished by a label with their name. As *i** actors can represent different kinds of entities (i.e., stakeholders, software systems, hardware sensors, etc.) we consider adequate to add an attribute to the actors for indicating their *structural type*. Our proposed initial set of structural types is software, hardware, human, and organization. More structural types can be added if needed, and also, it is possible to further divide each structural type into subtypes, eventually in more than one level. For instance, for software actors we may distinguish commercial off-the-shelf components (COTS), modules, layers, etc.

The semantics of the components define the component behaviour. Semantics are needed to perform analysis, enforce architectural constraints, and ensure consistent mappings. In the semantics of the *i** actors, we have different intentional types that may be used for these purposes by defining the abstraction level of the components:

- A *role* represents a service or group of related services that can be supplied by a certain component. For instance, in Fig. 1, we present the model of a COTS system for a meeting scheduler, where we identify the roles of *Meeting Scheduler*, *Message Delivery*, and *Antivirus*.

- A *position* represents types of components available to cover one or more roles. For instance, in the COTS market we identify that for a Meeting Scheduler System there are packages whose primary purpose is to act as a *Meeting Scheduler* but also provide mailing facilities as a *Message Delivery*. Therefore, we can use a *Communicated Meeting Scheduler* position (see Fig. 1).
- An *agent* represents an specific component that can be integrated into the software architecture. For instance, if in the COTS market there is a specific component that accomplishes the functionality required by the *Communicated Meeting Scheduler* position, the agent (e.g. the ACME Meeting Net in Fig. 1) will occupy this position.

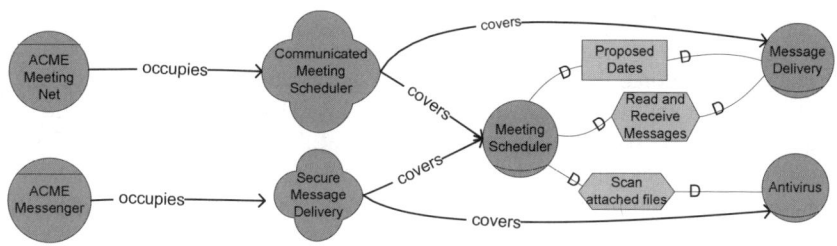

Fig. 1. Excerpt of *i** model for a Meeting Scheduler showing some roles, positions and agents

2.2 Using Dependencies to Represent Connectors

Connectors are architectural building blocks used to model interactions among components and rules that govern those interactions [18]. The interface of the connectors is the set of interactions points between the connector and the components attached to it. In *i** connectors are represented as dependencies where the direction of the dependency indicates the role of each connector, being one the *depender* and the other the *dependee*.

Components communication, coordination, and mediation decisions are encapsulated by the connector types. The *i** dependencies are associated with a *dependum* with a label which indicates the name of the object or concept that is shared among the components. Dependencies have an already associated *structural type* (goal, task, resource and softgoal), which interpretation changes according to the expectative of the *depender* and the *dependee* upon the *dependum*. Thus, the *depender* depends on the *dependee* to bring about a certain state in the world (goal dependency), to attain a goal in a particular way (task dependency), for the availability of a physical or informational entity (resource dependency) or to meet some non-functional requirement (softgoal dependency).

The connector semantics define the high-level model of the connectors' behaviour. In order to enforce the link between requirements and architectures when deciding the kind of a certain dependency, we propose to distinguish two different types of relationships to be represented with the four kinds of *i** dependencies: *intentional*, i.e. what behaviour a component expects from other part of the system, and *architectural*, i.e. how one component communicates with other part of the system. The examples provided correspond to the Meeting Scheduler System represented in Fig. 2(a).

- **Intentional relationships.** They are the ones that involve human or organizational actors as dependee and/or depender. They represent the intentional needs of the actors upon the system as follows:

  - A goal dependencies state functional requirement over the system, e.g. the *Initiator* depends on the *Meeting Scheduler* for *Meeting be scheduled.*
  - Resource dependencies state flow of concepts, and remarkable some type of knowledge, or a concept, relevant for the domain that does not physically exist, e.g. the *Meeting Proposal* concept in the context of the meeting scheduler.
  - Softgoal dependencies state high-level non-functional requirements, which may refer to quality of service, development objectives, or architectural constraints over the components, e.g. the *Antivirus* shall ensure *Easy Configuration* to the *Administrator* actor.

  We remark that we do not model task dependencies at the intentional level, because they enforce that the *depender* provides a prescriptive procedure, and this is not done at this level. The use of the *i** dependencies when modelling early requirements, late requirements and architectural design in TROPOS [3] enforces this decision as they only use goals, softgoals and resources.

- **Architectural relationships.** They are the ones that occur between components of the system. In order to adhere with the architectural concepts, the semantics of the architectural relationships are defined by adapting the six CBSP architectural dimensions proposed in [12] into the *i** framework:

  - Goal dependencies model the elements that that describe system-wide features or features pertinent to a large subset of the system's components or connectors. For instance, the *Meeting Scheduler* depends on the *Message Delivery* for *Messages Received and Read.*
  - Task dependencies model the elements that describe or involve processing components. For instance, in fig. 2 (c), the task dependency *Scan attached files* states a service invocation request to the *Antivirus.*
  - Resource dependencies model the elements that describe or involve data components. For instance, the resource dependency *Proposed Dates* states information interchange between the software actors *Meeting Scheduler* and *Message Delivery.*
  - Softgoal dependencies model the elements that describe or imply data or processing component properties, bus properties or system properties. For instance, the *Meeting Scheduler* depends on the *Message Delivery* for a *Reliable Message Delivery.*

2.3   Constraints, Evolution and Non-functional Properties

Architectural components and connectors also have to represent constraints, evolution, and non-functional properties. Constraints state properties about the system and ensure interaction protocols, intra-connector dependency and enforce usage boundaries. On the other hand, non-functional properties represent requirements for the correct behaviour of the components and connectors. In *i**, both constraints and non-functional properties are represented as softgoal dependencies between the different kinds of actors: human and organization actors constrain the properties of the connectors (software actors) with softgoal dependencies upon them, whilst softgoals dependencies among software actors constrain the properties of the connectors.

In order to allow more formal constraints, the *i** constructs will have to be complemented with more specific work. For instance, [16] allows defining architectural constraints among goals in a formal way. Regarding evolution, the *i** constructs do not provide a way to store the modification of the actor's and connectors properties. However, the work reported in [4] proposes to use actor's inheritance (the *is-a* construct) to support the extension, refinement and redefinition of intentional elements, which can be used to record evolution of the actors (components) and can be extended to record the evolution of their dependencies (connectors).

3   Representing Architectural Configurations in *i**

Architectural configurations are connected graphs of components and connectors that describe the architecture structure [18]. According to [18], characteristics features at the level of architectural configurations are understandability, compositionality, refinement and traceability, heterogeneity, scalability, evolution, dynamisms, constraints, and non-functional properties. Some of these characteristics are directly supported by *i**, whilst others require the adoption on extended work on the field.

The graphical representation of *i** certainly enhances the understandability of the modelled architecture, mainly because the different shapes used and the fact that components and connectors are identified with a textual label, makes it very intuitive. Despite one of the open issues on *i** is how to improve scalability when the models get very big, regarding architectural representations *i** makes possible to work with the components at different levels of detail, i.e. by decomposing a component represented in a high level of detail, into several components and its relationships at lowers levels of details. This use of the *i** models also improves compositionality.

Refinement and traceability deal with the representation of a consistent refinement of the architecture and the traceability of changes. In a similar manner, evolution deals with the how architectures change to reflect and enable evolution of a family of software systems. This features are not directly addressed in the *i** framework, however, we propose to use the semantics provided by the *i** actors to distinguish 3 different models depending on its type. Therefore we may model the architecture at three different levels:

- **Roles model.** Provides a description of the different roles that we may distinguish in the system architecture (i.e., services or groups of related services). Dependencies may be intentional or architectural. However, in the roles model there are no task dependencies between software actors because at this level we do not know the specific architecture of the system neither the protocols that would be used. Consequently, we use goal dependencies to state the services that a software actor requires from another software actor. For instance, in Fig. 2(a) the *Meeting Scheduler* depends on the *Message Delivery* for the goal *Messages Received and Read*, which, in turn, depends on the *Antivirus* for *Attached files scanned*.

- **Position model.** Identifies the different positions that exist in the architecture (i.e. the types of components available that cover one or more of the roles represented in the roles model). Usually this model does not state new dependencies, but, hides dependencies among roles which are covered by the same position. For instance, in Fig. 2(b) the *Communicated Meeting Scheduler* covers the roles of the *Meeting Scheduler* and the *Message Delivery* and so, the dependencies between them remain hidden. We remark that several different instantiations of the roles are possible (e.g. The *Secure Message Delivery* position covers the *Meeting Scheduler* and the *Antivirus* in Fig. 1). This leads to the exploration and modelling of different architectural solutions.

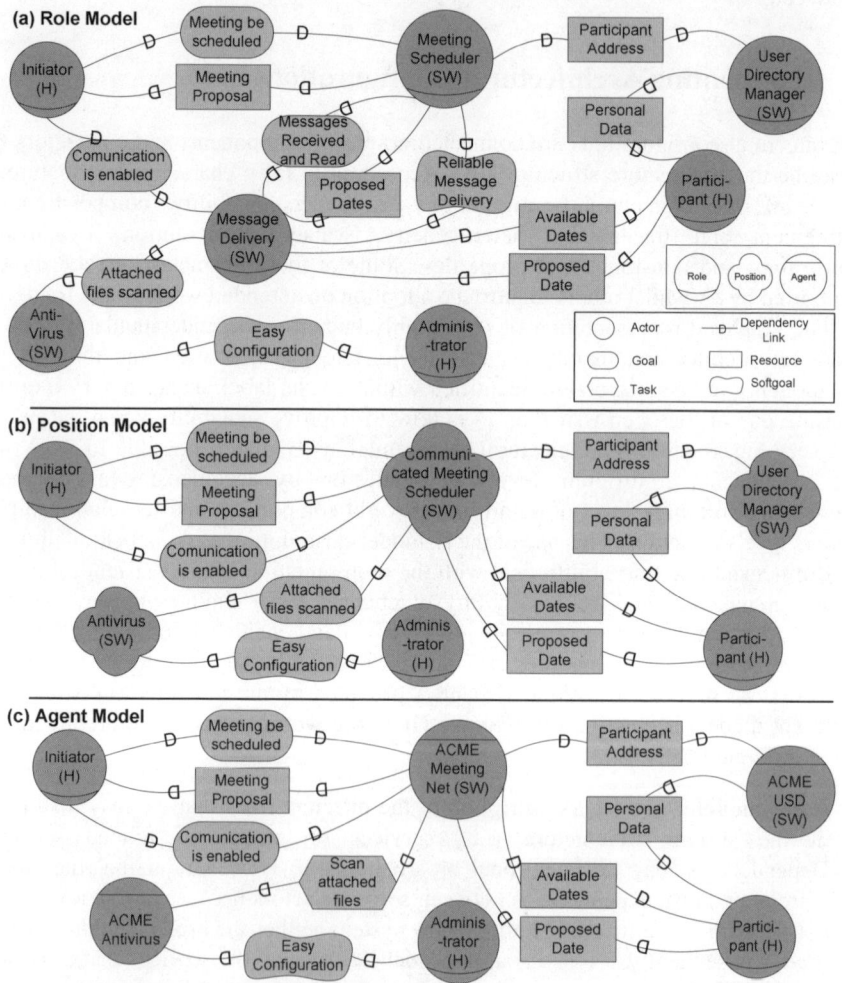

Fig. 2. Role model, position model, and agent model for an excerpt of the Meeting Scheduler

- **Agent model.** States the different agents that exist in the architecture (i.e. specific software component that can be integrated into the software architecture). In this model, dependencies denoting architectural relationships may refine the dependencies that exist on the role model, e.g., in Fig. 2(c) the *ACME Message Delivery* depends on the *ACME Antivirus* for the task *Scan attached files*, as we identify that the communication between those specific software components involves processing this data.

These three levels propose an agent-oriented perspective of the Software Architecture. However, as agent architectures do not have a direct correspondence to some types of software systems, the levels must be used as an abstraction mechanism. More precisely, these three levels provide a structure similar to the 4+1 view proposed in [15], in which the role model would represent the logical and process views, the position model the development view, and the agent model the physical view. A crucial point is that the mapping among models is one-to-many: a role model may have different position models bound, whilst a position model may have different agent models bound.

Another aspect mentioned in [18] for architectural configurations that it is not directly addressed by *i** is heterogeneity. This feature deals with how the language allows developing a software architecture by using pre-existing components and connectors of varying granularity possibly specified in different formal modelling languages, with varying operation system requirements, and supporting different communication protocols. In order to address this issue we propose to enrich the models by adding *attributes* to their modelling elements. Thus, attributes make possible to state characteristics of actors or dependencies. Attributes may be universal (e.g. *priority* in the case of the dependencies) or type-dependant (e.g., *programming language* in the case of software actors or *size* for resource dependencies). Attributes are useful not only to make *i** models more complete, but also for defining metrics and constraints among them.

In previous sections, we have seen that non-functional requirements can be bound to the components and connectors by means of softgoal dependencies that constrain their behaviour. However, at the configuration-level, non-functional requirements deal with the whole system and are used to ensure that the appropriate components and connectors are selected, perform analysis, enforce constraints, map architectural building blocks to processors, and aid in project management. In order to help the evaluation of the components and connectors and inform the analysis, we propose the use of structural metrics to evaluate properties and inform decisions. Structural metrics may depend on different parameters. Consider as an example data privacy. Each resource dependency in a position model identifies a danger, i.e. data interchange among different software components, so a first metric could be defined just as the total number of resource dependencies in the model:

$$DP(M) = \#k: k \in dependencies(M): type(k) = resource$$

This initial definition could be considered too simplistic. Refinements could be to consider the size of this resource:

$$DP(M) = \Sigma k: k \in dependencies(M) \wedge type(k) = resource: size(k)$$

The level of detail depends on how much effort the architect is able to put in the evaluation process and how confident is he/she in the result, for more information we refer to [6], [7].

Non-functional properties aimed to enforce constraints and global constraints over the system cannot be controlled with non-functional properties but, as with components and connectors, it is possible to use some of the extensions (for instance, the formal notation proposed in [16]) in order to represent them and check them.

4   Conclusions and Future Work

In this paper we analyse the *i** framework from an architectural point of view. As *i** allows a certain flexibility in the use of its elements, we have proposed some consensus in order to address architecture modelling. Our four main contributions are: 1) Clarifying the use of the *i** constructs for modelling components and connectors. As this is done by means of actors and dependencies we have provide an architecture-oriented semantics to them in order to help the process; 2) Adding the notion of role model, position model, and agent model in order to help traceability of the architectural representation; 3) Proposing the addition of attributes in the actors and dependencies of the models in order to store information for its analysis; and, 4) Suggesting the use of structural metric to analyse the properties of the final system.

Based on this study, we can conclude that it is possible to use *i**, a framework that includes a requirements language, for modelling architectures using the same constructs. The benefits of this are twofold. On the one hand, it is possible to represent and analyse a design architecture using the same constructs of their requirements model. In the other hand, architectures representation benefits from the notions of goals and softgoals, allowing a better representation of non-functional requirements.

We have to mention that, as shown in [19], it is very difficult to have an ADL that accomplish at its maximum extent all the mentioned features because it becomes too complex. Because of that we know in advance that *i** cannot be complete. Consequently, the modelling features needed and the techniques proposed have to be analysed in order to choose the most adequate. From this starting point, our future work will be directed to use this analysis as the basis for ensuring that the use of *i** in SAR*i*M is adequate for represent, explore and evaluate alternative architectures. We remark that, as we have proposed the addition of attributes on the *i** models (i.e. structural types for the actors, and attributes for all the elements), defining a model becomes laborious and time-consuming activity. However, the use of tool support facilitates this step and, because of that, we will adapt J-PR*i*M [11] to support them as we progress. The research agenda includes: 1) the creation of an ontology of the domain, putting together classical architectural concepts and goal-oriented ones; 2) the definition of a complete and widely applicable catalogue of metrics to be used in different experiences, eventually refining the framework outlined in [7]; 3) the definition and representation of usual architectural patterns in *i**; and, 4) the completion of a validation program, starting by retrospective analysis of existing cases and ending up with industrial case studies.

Acknowledgements. This work has been partially supported by the CICYT programme project TIN2004-07461-C02-01. Gemma Grau work is supported by an UPC research scholarship.

References

1. Ayala, C.P., Cares, C., Carvallo, J.P., Grau, G., Haya, M., Salazar, G., Franch, X., Mayol, E., Quer, C.: A Comparative Analysis of *i**-Based Goal-Oriented Modelling Languages. In: Proceedings of SEKE 2005, pp. 43–50 (2005)
2. Bastos, L.R.D., Castro, J.F.B.: Enhancing Requirements to derive Multi-Agent Architectures. In: Proceedings of WER 2004, pp. 127–139 (2004)
3. Bresciani, P., Perini, A., Giorgini, P., Giunchiglia, F., Mylopoulos, J.: TROPOS: An Agent-Oriented Software Development Methodology. Journal of Autonomous Agents and Multi-Agent Systems 8(3) (2004)
4. Clotet, R., Franch, X., López, L., Marco, J., Seyff, N., Grunbacher, P.: The Meaning of Inheritance in *i**. In: Proceedings of AOIS 2007, pp. 651–666 (2007)
5. Faulkner, S., Kolp, M.: Towards an Agent Architectural Description Language for Information Systems. In: Proceedings of ICEIS 2003, pp. 59–66 (2003)
6. Franch, X.: On the Quantitative Analysis of Agent-Oriented Models. In: Dubois, E., Pohl, K. (eds.) CAiSE 2006. LNCS, vol. 4001, pp. 495–509. Springer, Heidelberg (2006)
7. Franch, X., Grau, G., Quer, C.: A Framework for the Definition of Metrics for Actor-Dependency Models. In: Proceeding of RE 2004, pp. 348–349 (2004)
8. Grau, G., Franch, X., Maiden, N.A.M.: A Goal Based Round-Trip Method for System Development. In: Proceedings of REFSQ 2005, pp. 71–86 (2005)
9. Grau, G., Franch, X.: ReeF: Defining a Customizable Reengineering Framework. In: CAiSE 2007. LNCS, vol. 4495, pp. 485–500. Springer, Heidelberg (2007)
10. Grau, G., Franch, X.: A Goal-Oriented Approach for the Generation and Evaluation of Alternative Architectures. In: Proceedings of ECSA (to appear, 2007)
11. Grau, G., Franch, X., Ávila, S.: J-PRiM: A Java Tool for a Process Reengineering *i** Methodology. In: Proceedings of RE 2006, pp. 352–353 (2006)
12. Grünbacher, P., Egyed, A., Medvidovic, N.: Reconciling software requirements and architectures with intermediate models. Software and Systems Modeling 3(3), 235–253 (2004)
13. The *i** wiki at: http://istar.rwth-aachen.de/ (Last Accessed June 2007)
14. Kolp, M., Giorgini, P., Mylopoulos, J.: Organizational Patterns for Early Requirements Analysis. In: Eder, J., Missikoff, M. (eds.) CAiSE 2003. LNCS, vol. 2681, pp. 617–632. Springer, Heidelberg (2003)
15. Kruchten, P.B.: The 4-1 View Model of Architecture. IEEE Software 12(6), 42–50 (1995)
16. van Lamsweerde, A.: From System Goals to Software Architecture. In: Bernardo, M., Inverardi, P. (eds.) SFM 2003. LNCS, vol. 2804, pp. 25–43. Springer, Heidelberg (2003)
17. Medvidovic, N., Rosenblum, D.S.: Domains of Concern in Software Architectures and Architecture Description Languages. In: Proceedings DSL 1997, pp. 199–212 (1997)
18. Medvidovic, N., Taylor, R.N.: A Classification and Comparison Framework for Software Architecture Description Languages. IEEE Transactions on Software Engineering 26(1), 70–93 (2000)
19. Shaw, M., Garlan, D.: Software Architecture: Perspectives on an Emerging Discipline. Prentice-Hall, Englewood Cliffs (1996)
20. Yu, E.: Modelling Strategic Relationships for Process Reengineering. PhD. thesis, University of Toronto (1995)

Extending Argumentation to Goal-Oriented Requirements Engineering

Ibrahim Habli, Weihang Wu, Katrina Attwood, and Tim Kelly

Department of Computer Science, The University of York, York YO10 5DD
{Ibrahim.Habli,Weihang.Wu,Katrina.Attwood,
Tim.Kelly}@cs.york.ac.uk

Abstract. A key goal in safety-critical system development is to provide assurance that the critical requirements are sufficiently addressed. This goal is typically refined into three sub-goals, namely that the safety requirements are validated, satisfied and traceable. The achievement of these sub-goals is typically communicated by means of a safety argument supported by items of evidence (e.g. testing, review or analysis). In this paper, we explore the relationships between goals, requirements, and arguments. We discuss how argumentation is used to assure the decomposition and traceability of requirements in safety-critical applications. Particularly, we focus on the achievement of goals related to both the requirements artefacts and the underlying requirements process.

1 Introduction

Goal-modelling techniques have long been recognised as an effective support to requirements engineering processes. The work of van Lamsweerde and others [1, 2], for example, has demonstrated a powerful goal-based method to support requirements elicitation, refinement, trade-off analysis and documentation. A goal describes the objective that a system should meet, which may be classified further in terms of functional and quality goals. Goals can be refined hierarchically into sub-goals through various refinement strategies such as AND/OR decomposition, design decisions and domain-specific analysis results. The goal refinement process stops when all identified goals have been satisfied or achieved. Goals may be violated, however, due to unexpected behaviours of a system or its environment. The notion of obstacles [3] or anti-goals [4] has recently been introduced in order to integrate potential violations into goal modelling.

Goal-based techniques have also been used to model the relationships between the system under definition and the environment in which it will operate. For example, the i* technique has been used in early-stage requirements engineering, to capture aspects of user motivation, business and organisational goals in systems characterised by a high degree of human-computer interaction, such as Air Traffic Control [5, 6]. Chung et al [7] have also developed a goal-based framework for clarifying and prioritising non-functional requirements and managing trade-offs between them. At the design level, early work on argument-based design rationale developed a set of

J.-L. Hainaut et al. (Eds.): ER Workshops 2007, LNCS 4802, pp. 306–316, 2007.
© Springer-Verlag Berlin Heidelberg 2007

generic models of design processes in terms of three common elements [8]: issues/questions, positions/options, and arguments/criteria. There have been several applications of design rationale to software engineering. The richest extension to date is the REMAP (REpresentation and MAintenance of Process knowledge) model developed by Ramesh [9], in which the notions of requirements/goals, assumptions, constraints and design objects have been incorporated into the Issue-Based Information System (IBIS) framework.

In the safety-critical system domain, one of the key goals is to provide assurance that the critical requirements are sufficiently addressed. As a minimum, it should be demonstrated that these requirements are validated, satisfied and traceable. Particularly, software standards in the safety domain are shifting towards goal-based approaches where the validation, satisfaction and traceability of safety requirements are the primary goals for demonstrating that a safety-critical software system is acceptably safe. To justify that a software system is acceptably safe, a safety case is typically submitted. A safety case is defined in the UK Defence Standard 00-56 as [10]:

> *"A structured argument, supported by a body of evidence that provides a compelling, comprehensible and valid case that a system is safe for a given application in a given operating environment."*

Underlying the descriptions of the safety case is a view of the safety case consisting of three principal elements: Goals, Argument and Evidence [11]. The argument communicates the relationship between the evidence and goals. Argument without supporting evidence is unfounded, and therefore unconvincing. Evidence without argument is unexplained – it can be unclear that (or how) goals have been satisfied.

In this paper, we explore the relationships between goals, requirements, and arguments based on our experience in the software safety domain. We discuss how argumentation is used to assure the decomposition and traceability of software requirements in safety-critical applications. In particular, we focus on the achievement of goals related to both the *requirements artefacts* and the underlying *requirements engineering process*. To create goal-based arguments, we use the Goal Structuring Notation (GSN) [11]. GSN is a graphical notation for the construction of safety and assurance arguments. Nonetheless, GSN is generic and can be used to structure and present goal-based arguments that demonstrate requirements validation and satisfaction.

The rest of paper is structured as follows. Section 2 presents an overview of GSN and how it can be used to demonstrate how existing requirements are decomposed and managed. Section 3.1 presents an argumentation approach to refining safety goals and anti-goals. 3.2 shifts the discussion towards the assurance of goals related to the underlying requirements process. Section 3.3 focuses on one process aspect, namely requirements traceability. Finally, Section 4 presents a summary and conclusions.

2   The Goal Structuring Notation (GSN)

GSN explicitly represents the individual elements of goal-based arguments (requirements, goals, evidence and context) and (perhaps more significantly) the relationships that exist between these elements (i.e. how individual requirements are supported by specific claims, how claims are supported by evidence and the assumed context that is defined for the argument). The principal symbols of the notation are shown in Fig. 1 (with example instances of each concept).

When the elements of the GSN are linked together in a network they are described as a 'goal structure'. The principal purpose of any goal structure is to show how goals (claims about the system) are successively broken down into sub-goals until a point is reached where claims can be supported by direct reference to available evidence (solutions). As part of this decomposition, using the GSN it is also possible to make clear the argument strategies adopted (e.g. adopting a quantitative or qualitative approach), the rationale for the approach and the context in which goals are stated (e.g. the system scope or the assumed operational role).

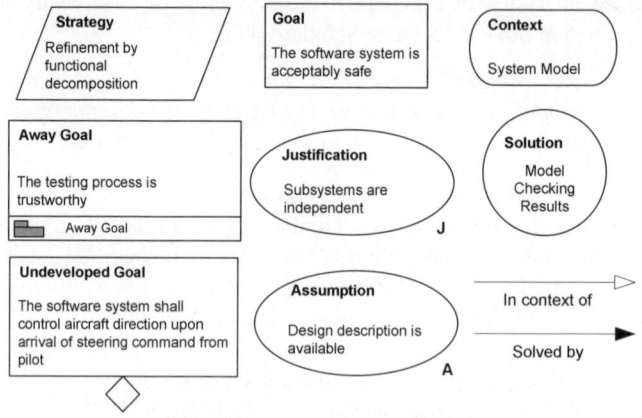

Fig. 1. Elements of the Goal Structuring Notation

The GSN technique adds a rich semantics to this goal-breakdown structure, by forcing the explicit recording of and justification for the strategies used to refine and relate the goals in a semi-formal argument structure. Since both the goals and these refinement strategies hold true only in a defined context (i.e. within the scope of a particular system, or under certain operational conditions), the notation's syntax requires that explicit reference is made to the evidential basis on which the goals and the refinement depend. Within Europe, GSN has been adopted by a growing number of companies within safety-critical industries (such as aerospace, railways and defence) for the presentation of safety arguments within safety cases.

Arguably, GSN can be used in a broader and more general context. It can provide a means for decomposing requirements and recording traceability links between individual claims and sub-claims. These claims are represented as goals, and equate to the requirements statements and specifications. The notation also records the strategies used to decompose the goals. These strategies map to the 'satisfaction argument', in that they seek to provide a basis for the relationship between the goals. As well as presenting a clear record of the goal-decomposition strategy, however, GSN allows this strategy to be validated by the use of an apparatus of justifications and assumptions. These justifications and assumptions are attached to goals and strategies within the structure, as well as by explicit references to artefacts such as system architectural models or contextual information.

3 Modelling Requirements Artefacts and Processes Using GSN

This section presents GSN arguments addressing (1) the refinement of safety goals and anti-goals and (2) the assurance of goals related to the underlying requirements process, with a particular emphasis on requirements traceability.

3.1 Modelling Goals and Anti-goals

Although several different goal formulation techniques exist, it is possible to identify four common elements in the 'goal' concept:

- Artefact: The artefact is the composite system or its parts onto which a goal is applied.
- Context: The context addresses the pre-conditions that a goal refers to and evolves over.
- Stimulus: The stimulus is the trigger condition for the initiation of a goal.
- Response: The response captures the desired properties (i.e., postconditions) that the artefact should hold over time. Quality requirements (e.g., deadline or failure rate) can be specified in this part if they exist.

A GSN goal represents a requirement goal, and can thus be expressed using natural language in the following form:

"The <artefact> shall <respond> upon <stimulus> when <context>"

This goal formulation is consistent with the SEI's quality attribute scenario framework [12]. It can be applicable to both functional and quality goals. As an example, consider a wheel braking system (WBS) on an aircraft [13]. We assume there are a number of top-level system goals that can be stated in terms of aircraft functionality e.g., controlling the aircraft on the ground and safety (Fig. 2). Despite their high level of abstraction, these system goals can be expressed in stimulus-response form. Each functional goal can be decomposed further into a set of sub-goals and these should evolve separately given that they are independent. The goal decomposition may be guided by the decision to use various mechanisms for speed reduction. Goal structures can thus be constructed. Safety goals cannot simply be decomposed via functional goals or system structures; their refinement is based upon the results of deviation analysis and the chosen mitigation. For example, in the WBS, late output from a controller is likely to be safety-significant. A performance goal is thus derived and added into the safety goal structures. Fig. 2 shows a part of the goal structure in which the core functionality of WBS is elicited. All the goals in this structure are expressed using the above form (the context elements '*RefArf_SatisfactionArg*' and '*ReqDev_SWReq*' are addressed separately in the next

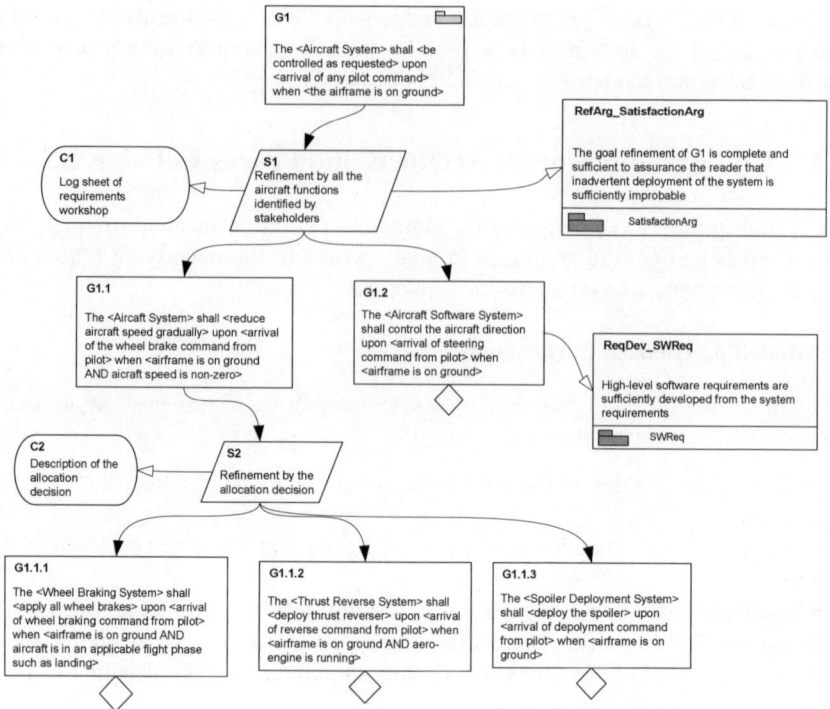

Fig. 2. A top-level goal structure for the WBS example

two sections). The expression language used is a structured natural language, and some expression can be very abstract at this level. For example, both the stimulus and response parts of the top-level goal G1 are very general and need to be refined. This should be acceptable, however, in the early development lifecycle in which many requirements are volatile and unclear.

On the other hand, an anti-goal is the negative correlative of a goal: a condition that, if true, would immediately prevent the system from achieving the corresponding goal. Goals and anti-goals are complementary and thus capture the possible desired and undesired end states of a composite system. A common example of an anti-goal is the loss of a system function where the function is a goal. Nevertheless, the simple negation of a goal in terms of propositional logic cannot guarantee the sufficient completeness of the corresponding anti-goals. A less obvious but perhaps more severe anti-goal would be inadvertent application of that function. Given some goal formulation, it is important to ensure the exhaustiveness of deviations from that goal, at least from the viewpoint of safety. In the safety community, the possible deviations of a system are often characterised in terms of deviation or failure modes. Previous work in York has developed a collection of deviation modes for software systems: SHARD guidewords [14]. We interpret the SHARD modes with respect to the goal formulation in the following Table 1.

By allocating the SHARD modes onto a formulated goal and interpreting them using the above table, we can achieve a high level of confidence in the exhaustiveness of the set of anti-goals elicited. Note that anti-goals do not necessarily have safety implications, although they should always be evaluated with respect to possible safety-related consequences. Let us return to the WBS example. As soon as the system goals of WBS are formulated, the identification of anti-goals can start by considering the SHARD deviations first without information about the elaborated scenarios. In this example, only omission and commission modes are applicable. Table 2 illustrates an anti-goal by negating the context part – wheel braking when the context is not as intended. The definition of the stimulus part is trivial in this case. The anti-goal elicited is expressed at an abstract level.

Table 1. The anti-goal interpretation using SHARD guidewords

SHARD	Anti-Goal Interpretation
Omission	Response part does not hold while stimulus and environment parts hold
Commission	Stimulus or context parts do not hold while response part holds
Timing	Timing constraint specified in the response part is violated while the other parts hold
Value	Value constraint specified in the response part (e.g., accuracy or cost) is violated while the other parts hold

Table 2. An example anti-goal formulation

Portion of Goal	Possible Value
Artefact	WBS
Context	NOT (Airframe is on ground AND aircraft is in landing/taxiing/RTO flight phase)
Stimulus	N/A
Response	All wheel brakes are applied

By expanding the negation operation on the context part using Boolean logic, we can derive a set of well-refined anti-goals: e.g., wheel brakes applied when the aircraft is taking off or when the aircraft is in air (both of these conditions can lead, at worst, to total loss of control). It must be stressed that the expansion here cannot be achieved solely by formal Boolean logic and may need the help of domain experts. For the example of inadvertent wheel braking when the aircraft is taking off, we may need to distinguish further whether the aircraft is taking off before the decision speed V1, as the safety consequences before and after this threshold is passed would be different. Obviously, it would be impossible to identify these two anti-goals by the use of formal logic alone, without the requisite domain knowledge.

When all anti-goals are identified and refined (say, eight anti-goals for the WBS example), they should be linked to the anti-goals of the parent goal of the WBS (i.e., aircraft deceleration) in a bottom up manner, thereby forming an anti-goal structure. The anti-goal structure in the WBS example is shown in Fig. 3. The anti-goal structure should refer to the corresponding functional goal structure. It should be noted that the expression languages used in functional goals, safety goals and anti-goals are slightly different. The functional goals are simply operational and thus

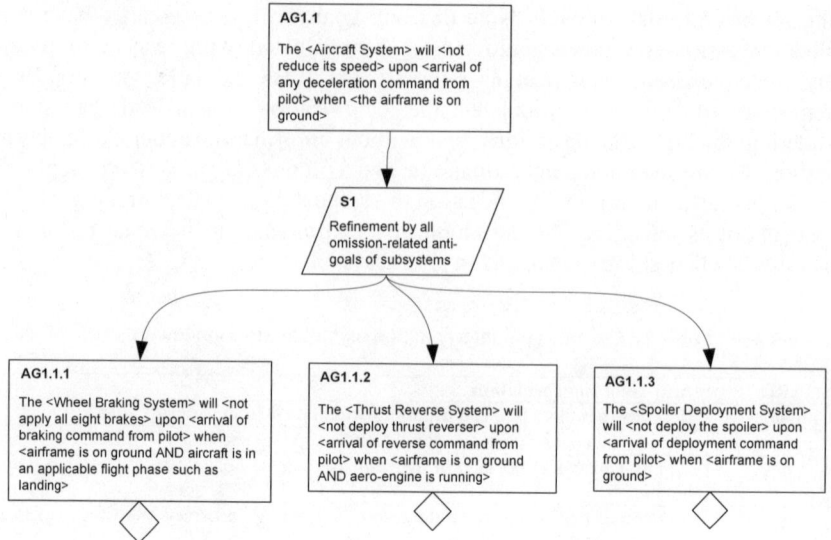

Fig. 3. The anti-goal structure for the WBS example

'shall' statements are suitable; the anti-goals are hypotheses about states and events of the system and thus 'will' statements should be used. The construction of anti-goal structures will prompt the refinement of the goal structure in which decisions need to be made regarding the mitigation of these anti-goals.

3.2 Assuring Goals Concerning the Requirements Process

The previous section has presented an argument supporting the decomposition of requirements goals in isolation from the underlying requirements process, i.e. the level of review, independence, traceability, competency of the requirements analysts and organisational agility. Uncertainties about the trustworthiness of this process may weaken confidence in the allocated and refined requirements. To this end, it is important to satisfy goals about the requirements *process* (i.e. not just the requirements *artefacts*). These goals address attributes such as process completeness, consistency and robustness since flaws in the requirements process may result in flaws in the requirements themselves. Such types of requirements flaws may only be discovered by independent reviews, for example. Factors that need to be addressed by an argument justifying the requirements process include issues such as:

- Is the software requirements team independent from the systems requirements team?
- Is requirements validation carried out using a repeatable and traceable technique?
- Is requirements validation performed on a stable and identifiable version of the allocated system requirements?
- Do the software engineers understand their relationship with the system requirements?

Fig. 4 shows a purely process-based argument that provides evidence addressing goals about the requirements process of the "Aircraft Software System" (G1.2 in Fig. 2). The process goals considered in that argument concern the clarity of the notation, the suitability of the validation methods, the consistency of the configuration, and the competency of the software engineers. Research has shown that domain knowledge is one of the most significant factors in achieving low rates for hazardous failure in safety-critical software development [15]. Validation of the requirements – and the consequent discovery of operational requirements errors – is best performed by domain experts. Therefore, regardless of how well the software requirements are structured, refined and documented, it is of equal importance to show the competency of personnel and it is therefore important that it is included as a separate goal in the process assurance argument. The next section focuses on assuring requirements traceability, a process goal central to all software safety standards.

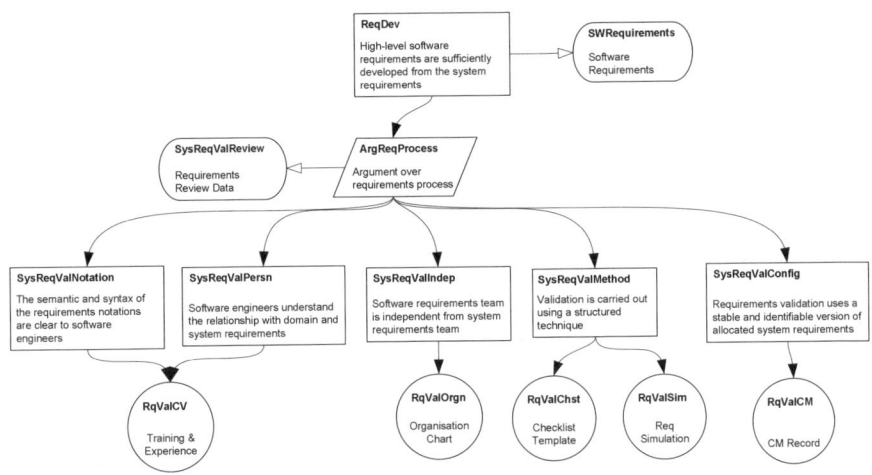

Fig. 4. Software Requirements Validation Argument

3.3 Goal-Based Arguments for Requirements Traceability

Zave and Jackson [16] observe that satisfaction of a requirement (R) can be demonstrated only by a sufficient combination of domain knowledge (K) and specifications (S): S, K $\vdash$ R. Jackson suggests that traceability links between requirements and specifications should be supported by textual 'correctness arguments' which explain how the specifications and domain behaviour combine to provide assurance that the engineered system satisfies the requirement in the application domain [17]. 'Satisfaction arguments' are a development from these 'correctness arguments' and provide assurance that the traceability relationship between requirements at different levels of abstraction is valid within a given application domain [18]. For example, it is essential for the successful reuse of requirements across system families that there is some assurance that the satisfaction relationships remain valid for requirements and specifications in the reuse domain. Failure to observe this principle risks the late, and therefore costly, delivery of erroneous or untenable requirements.

The following example demonstrates how GSN can be used to document satisfaction arguments, and to indicate where changes in design commitments or contexts challenge the satisfaction of the requirements. The decomposition in Fig. 2 does not attempt to justify the decomposition strategy employed. Instead, it contains an 'away goal' reference to a justification claim depicted in Fig. 5 (*Satisfaction Argument*). This goal is the top-level claim of the argument in Fig. 5, which is the satisfaction argument justifying the requirements decomposition. The argument strategy is a two-pronged one: the left-hand side of the goal structure argues that the checks, taken together, are sufficient to satisfy the top-level requirement, while the right-hand side (not fully developed here) argues that all possible failure modes have been considered and are adequately mitigated by the checks. The GSN structure makes clear what evidence is required to demonstrate the satisfaction of the top-level requirement (PSSA).

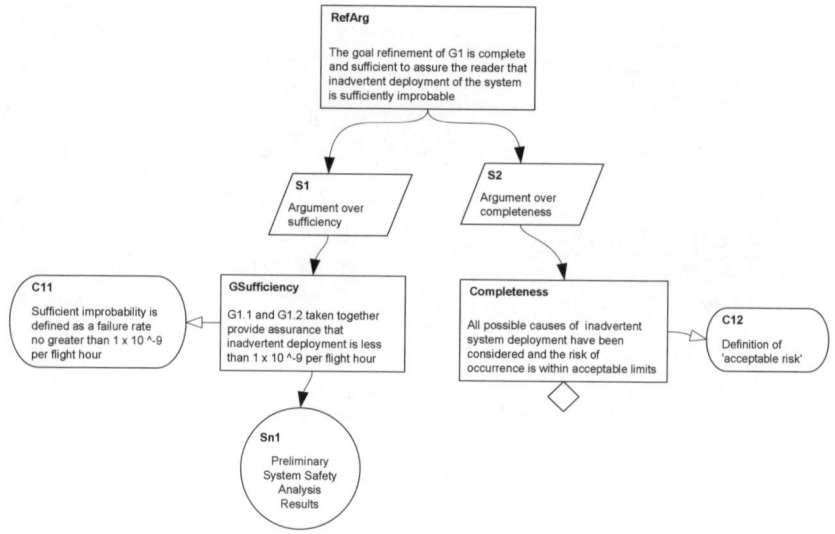

Fig. 5. Satisfaction argument for the refinement of the G1 in Fig. 2

The GSN structure provides a straightforward means for assessing which aspects of the satisfaction relationship are threatened by the design change. GSN satisfaction arguments thus allow for the clear record of domain information and assumptions, and indicate which information sources are required for adequate requirements traceability. The severity of the impact of requirements or contextual change can be assessed by reference to the satisfaction arguments on the requirements decomposition.

5   Summary and Conclusions

In this paper we have discussed how goal-based argumentation can be used to provide assurance for the decomposition and traceability of safety-critical requirements. We have presented an integrated approach that addresses the achievement of goals related to both the requirements artefacts and the underlying requirement process. The key benefit of adopting GSN is that it improves comprehension of the requirements satisfaction and validation argument amongst all of the key project stakeholders (i.e. system developers, safety engineers, independent assessors and certification authorities). In turn, this improves the quality of the debate and discussion amongst the stakeholders and reduces the time taken to reach agreement on the development and analysis approaches being adopted.

References

1. Lamsweerde, A.v., Dardenne, A., Fickas, S.: Goal-directed Requirements Acquisition. Science of Computer Programming 20, 3–50 (1993)
2. Lamsweerde, A.v.: Goal-Oriented Requirements Engineering: A Guided Tour. In: RE 2001. Proceedings of 5th IEEE International Symposium on Requirements Engineering, pp. 249–263. IEEE Computer Society Press, Los Alamitos (2001)
3. Lamsweerde, A.v., Letier, E.: Integrating Obstacles in Goal-Driven Requirements Engineering. In: Proceedings of the 20th International Conference on Software Engineering, pp. 53–62. IEEE Computer Society Press / ACM Press (1998)
4. Lamsweerde, A.v.: Elaborating Security Requirements by Construction of Intentional Anti-Models. In: Proceedings of the 26th International Conference on Software Engineering, pp. 148–157. IEEE Computer Society Press, Los Alamitos (2004)
5. Yu, E.: Towards Modelling and Reasoning Support for Early-Phase Requirements Engineering. In: RE 1997. Proceedings of the Third IEEE International Symposium on Requirements Engineering, Washington D.C., USA, Jan 6-8, 1997, pp. 226–235. IEEE Computer Society Press, Los Alamitos (1997)
6. Maiden, N., Jones, S.: Dependability in RESCUE: A Concurrent Engineering Approach to the Specification of Requirements for Air Traffic Management. In: DSN2004. Proceedings of the Workshop on Interdisciplinary Approaches to Achieving and Analysing System Dependability, Washington D.C., USA, June 29, 2004 (2004)
7. Chung, L., Nixon, B.A., Yu, E., Mylopoulos, J.: Non-Functional Requirements in Software Engineering. Kluwer Academic, Boston, etc. (1999)
8. Shum, S.B.: Design Argumentation as Design Rationale. In: The Encyclopedia of Computer Science and Technology, Marcel Dekker Inc., New York, pp. 95–128 (1996)
9. Ramesh, B., Dhar, V.: Supporting systems development by capturing deliberations during requirements engineering. IEEE Trans. on Software Engineering 18(6), 498–510
10. UK Ministry of Defence, 00-56 Safety Management Requirements for Defence Systems, Part 1: Requirements, Issue 3, UK Ministry of Defence (August 2004)
11. Kelly, T.P.: Arguing Safety - A Systematic Approach to Safety Case Management. DPhil Thesis, University of York, York (1999)
12. Barbacci, M., Ellison, R., Lattanze, A., Stafford, J., Weinstock, C., Wood, W.: Quality Attribute Workshops (QAWs), Third Edition. Technical Report (CMU/SEI-2003-TR-016) Pittsburgh, PA: Software Engineering Institute, Carnegie Mellon University (2003)

13. ARP 4761: Guidelines and Methods for Conducting the Safety Assessment Process on Civil Airborne Systems and Equipment, Society of Automotive Engineers, Inc. (1996)
14. Fenelon, P., McDermid, J., Nicholson, M., Pumfrey, D.: Towards Integrated Safety Analysis and Design. ACM Computing Reviews 2(1), 21–32
15. McDermid, J.A.: Software Safety: Where's The Evidence? In: Proceedings of the Sixth Australian Workshop on Industrial Experience with Safety Critical Systems and Software, Australian Computer Society (2001)
16. Zave, P., Jackson, M.: Four Dark Corners of Requirements Engineering. ACM Transactions on Software Engineering and Methodology 6(1) (1997)
17. Jackson, M.: Problem Frames: Analysing and Structuring Software Development Problems. Addison-Wesley, London (2001)
18. Hull, M., Jackson, K., Dick, J.: Requirements Engineering. Springer, London (2002)

Preface to SeCoGIS 2007

Esteban Zimányi[1] and Michela Bertolotto[2]

[1] Université Libre de Bruxelles, Belgium
[2] University College Dublin, Ireland

The recent advances in remote sensing, positioning systems, and mobile technologies have increased the production, collection, and diffusion of geo-referenced data, thus requiring the rapid development and the wide deployment of multiple geographic information systems(GIS). With the popularity of the World Wide Web and the diversity of GIS on the Internet, geographic information can now be available via personal devices anytime and anywhere. Nowadays, GIS are emerging as a common information infrastructure, which penetrate into more and more aspects of our society, and converge with most areas in the IT scenario, such as office automation, workflow, digital libraries, Web searching, and virtual reality. This has given rise to new challenges in order to accommodate new users requirements for new applications. Conceptual modeling and Semantic Web technologies provide complementary solutions for these issues. They allow eliciting of user requirements and the semantics of the target application domain.

The workshop on Semantic and Conceptual Issues for GIS combines two successful series of workshops, Conceptual Modeling for Geographic Information Systems (CoMoGIS), held in 2004, 2005, and 2006, and Semantic-Based Geographical Information Systems (SeBGIS), held in 2005 and 2006. The objective of joining these workshops is to focus the GIS community on an event in which semantic and conceptual aspects are discussed within a common framework.

SeCoGIS is intended to bring together researchers, developers, users, and practitioners carrying out research and development in GIS to discuss how to integrate conceptual modeling and semantics into current geographic information systems, and how this will benefit the end users. The workshop provides a forum for original research contributions and practical experiences of conceptual modeling and Semantic Web technologies for GIS, fostering interdisciplinary discussions in all aspects of these two fields, and will highlight future trends in this area.

Three or four experts from the same or a closely related discipline as the authors reviewed each of the 21 submissions. We would like to sincerely thank the Program Committee and the additional experts who realized an excellent work in carefully reviewing the papers: they made a strong contribution to the quality of the workshop. Ten high-quality papers were chosen, making a successful workshop.

The papers were presented in four sessions. Session 1 focused on moving objects. Session 2 hosted the invited keynote speaker of the workshop, Gary Hunter. Session 3 dealt with advances in conceptual modeling for GIS. Finally, Session 4 covered issues related to integrity constraints and approximate reasoning.

Modeling Historical and Future Spatio-temporal Relationships of Moving Objects in Databases

Reasey Praing and Markus Schneider*

University of Florida, Department of Computer & Information Science & Engineering,
Gainesville, FL 32611, USA
{rpraing, mschneid}@cise.ufl.edu

Abstract. Moving object databases have recently found large interest in the database community, geographical information systems, and geosciences. So far, moving object models have focused exclusively either on past or future movements. The so-called *Balloon Model* is the first attempt to treat past and future movements of moving objects in a consistent and integrated manner. Based on a new set of spatio-temporal data types and operations as well as former research, in this paper, we propose a design of *spatio-temporal predicates* on Balloon objects. These predicates explore the topological relationships of Balloon objects over time and take into account both past and future movements. We present new kinds of queries to demonstrate the use of these predicates.

1 Introduction

Research in spatio-temporal databases have recently received a lot of interest in many disciplines such as robotics, mobile computing, and geographical science. Much of the research in this field focuses on defining data models for continuously changing geometric objects typically known as *moving objects*. While most previous approaches assume that we have precise knowledge about the objects' movements (i.e., historical movements), some recent research considers the problem of predicting the future movements of moving objects in specific environments. The lack of a generic data model for the future movements and the separation of past and future movement models for moving objects have fueled a new development in this field. Recently, we have proposed a new moving object model called *balloon model* which considers both the past and the future movements of moving objects while preserving their temporal consistency.

With this new model, a new set of moving data types and operations for balloon objects are introduced. At the same time, new kinds of interesting queries can be posed. For instance, assuming that hurricane Katrina is making its way across the Gulf of Mexico, we can ask a question like "What is the chance that hurricane Katrina will make land fall in New Orleans?" Although this question can be answered by examining the projected path of the hurricane which is modeled as a balloon object, some other questions require the use of a spatio-temporal relationship or predicate between balloon objects. For example, we can post a query "List all airplanes that will potentially cross

---

* This work was partially supported by the National Science Foundation under grant number NSF-CAREER-IIS-0347574.

J.-L. Hainaut et al. (Eds.): ER Workshops 2007, LNCS 4802, pp. 318–327, 2007.

the projected path of hurricane Katrina." Here, we can model each airplane as a *balloon point* consisting of a moving point past and a moving point prediction since an airplane may have a defined route. We can also model the hurricane as a balloon point but with a moving region prediction since a hurricane does not have a defined route. The answer to this query requires the evaluation of the *balloon predicate* "potentially cross" between each airplane and the hurricane. Even though in the past, spatio-temporal predicates have been defined between moving objects with precisely known movements, spatio-temporal predicates between balloon objects taking into account the uncertainty of future relationships have not been defined.

The goal of this paper is to propose a spatio-temporal predicate model for defining *balloon predicates* between balloon objects. Since balloon objects are defined as a *temporal composition* of the past and future movements of moving objects, a balloon predicate between two balloon objects is defined based on a temporal composition of the spatio-temporal predicates between their past and future movements. With this model, users can specify balloon predicates and use them in queries.

We present the paper by first providing a brief overview of the existing moving object model and spatio-temporal predicate (STP) model for moving objects in Section 2. We then give a description of the balloon model in Section 3 as a foundation for defining our STPs for balloon objects in Section 4. Having defined the predicate model, we explore the querying possibilities of balloon predicates in Section 5. Finally, in Section 6, we draw some conclusions and discuss future work.

2   Related Work

Within the past decade, there have been several major developments of STP models. Among these developments, STP models for moving objects have received wide-spread interest in both application and research directions. Each of these models is defined on the basis of a specific moving object data model which only considers either the relationship between the past movements or future movements of moving objects.

2.1   A Spatio-temporal Predicate Model for Past Movements

The STP model presented in [4] characterizes the developments of topological relationship between moving objects which are defined by a moving object model proposed in [7,2,1]. To simplify our discussion, we refer to these models as the traditional STP model and traditional moving object (MO) model respectively. The traditional MO model defines a moving object as a function from time to space. For an arbitrary data type α, the corresponding moving type of α is a function $\tau(\alpha)$ that provides the mapping from the temporal domain to α, i.e., $\tau(\alpha) = time \rightarrow \alpha$. By instantiating this definition with spatial data types like *point*, *line*, and *region*, moving data types such as moving point (*mpoint*), moving line (*mline*), and moving region (*mregion*) are obtained. Figure 1(a) and 1(b) illustrate a moving point and a moving region respectively.

The spatio-temporal predicate $stp(\tau(\alpha), \tau(\beta))$ between two moving objects of type $\tau(\alpha)$ and $\tau(\beta)$ is defined by the traditional STP model as a sequence of alternating spatial predicates that only hold for a period of time (period predicates) and those that

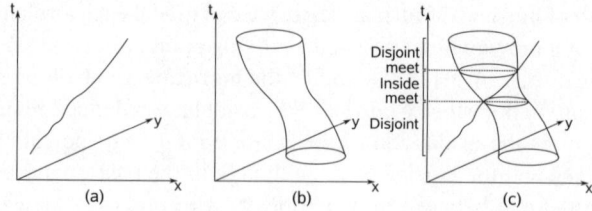

Fig. 1. Examples of a moving point (a), a moving region (b), and a crossing situation (c)

can hold for an instant of time (instant predicates). For example, the predicate *Cross* between a moving point and a moving region is defined as $Cross := Disjoint \triangleright meet \triangleright Inside \triangleright meet \triangleright Disjoint$ where *Disjoint* and *Inside* are period predicates, *meet* is an instant predicate, and the symbol $\triangleright$ signifies a development or a change of relationship. Figure 1(c) depicts the spatio-temporal predicate *cross*. From another perspective, a spatio-temporal predicate between two moving objects is a *temporal composition* of period and instant predicates between the two objects. This view of a spatio-temporal predicate is made obvious in later works such as the Query-By-Trace [5] model which focuses on visual specification of spatio-temporal predicates for use in query. Although the traditional STP model is able to effectively capture the development of relationship over time, it is defined based on the traditional moving object model which assumes that we have precise knowledge about the movements of moving objects, i.e.past movements. The aspect of uncertainty such as the relationship between future movement predictions of moving objects is not part of the model. In reality, the ability to model and query such relationship is of great importance to many applications. In our balloon predicate model, we make use of this traditional STP model to characterize the relationship between the past and future movements of balloon objects.

2.2 A Spatio-temporal Predicate Model for Future Movements

With regard to the future movements of moving objects, the STP model presented in [6] captures the uncertainty aspect of the future relationship between a moving object, more specifically a moving point, and a static region. The future movement of a moving point is defined by a future *motion plan* or *trajectory* and a threshold value signifying an acceptable deviation of the actual movement from the trajectory. The application of a threshold around a future trajectory creates a *trajectory volume* which represents the set of all possible future motion curves. A STP is then defined based on the relationship between the *spatial projection* of such a trajectory volume and a static region. Depending on this relationship, the uncertainty of a future STP can be captured and represented by using any combination of the prefixes *sometimes, always, possibly,* and *definitely*. While this model is able to model future STPs to a certain extent, it is limited to only those relationships between a moving point and a static region. In contrast, our balloon predicate model is a generic model which supports both the past and future relationships between any combination of balloon object types, thus any combination of moving object types.

3   Foundation Data Model for Balloon Predicates

To motivate the approach for representing moving objects, consider a scenario of modeling a hurricane. It is desirable to be able to model this hurricane as a single moving object consisting of its past movement up until its current state and continued by its future prediction. Due to the fact that existing moving object models only support either the past or a restricted type of future movements of moving objects, there is a need for a generic data model which can handle both the past and future movements of moving objects. A first step in this direction is made with the FuMMO model presented in [8] which emphasizes the separation between prediction methods and future data models and offers a generic data model for future movements of moving objects. Based on this idea and the need for a unique and consistent representation of moving objects, the balloon model is proposed in [9] as a generic data model for supporting both the past and future movements of moving objects.

3.1   The Balloon Model

In this model, a balloon object is defined as a temporal composition of the past component and the future component of the object. The connection point of this composition is known as the object's *present* denoted by t_p and represents the instant of the latest known state of the object. The past temporal domain *time$_h$* of the past component is defined as $time_h = (-\infty, t_p]$. Similarly, the future temporal domain *time$_f$* of the future component is defined as $time_f = (t_p, +\infty)$. Each of these components is defined as a traditional moving object with a specific continuity property. To capture the uncertainty aspect of future movements, the future component also consists of a *moving confidence distribution* indicating the confidence level of each point of the future movement. Whereas the past component can be of type *mpoint*, *mline*, or *mregion*, the future component can be of type *fpoint*, *fline*, or *fregion* each consisting of a moving geometry and a moving confidence distribution. The separation of the moving geometry and the moving confidence distribution in the future component allows us to use a traditional moving data type to represent this moving geometry. It turns out that not all combinations of moving object types are valid. For example, it is not possible to use a moving point to represent the potential future extent of a moving region. If this were possible, it would mean that a region can evolve to collapse its dimension into a single point in the future. This proves to be impossible based on the definition of the continuity of movements described in [9]. In fact, the future component must be based on a moving object whose spatial dimension is greater than or equal to that of the past component. Hence, only six valid combinations exist which translate to six balloon data types. Let α and β be a spatial data type. In general, a balloon data type is defined by a type constructor $\Omega(\alpha, \beta) = \tau(\alpha) \times \varphi(\beta)$ where $\tau(\alpha)$ represents the past component (see Section 2.1) and $\varphi(\beta)$ models the future component including the moving confidence distribution.

3.2   Data Model Considerations for Balloon Predicates

Defining the relationship between uncertain movements of moving objects is a very complex task. For instance, consider a prediction of an airplane that crosses a prediction of a hurricane. It is not necessary that the airplane will always cross the hurricane;

it may only get close to or touch the actual hurricane even though its prediction crosses the hurricane's prediction. However, there is a chance that the airplane would cross the hurricane as well. The quantification of this chance depends on a complex calculation of the moving confidence distributions of both objects. Since these distributions can be represented in many forms, e.g., using probabilistic or fuzzy concepts, considering this quantification as part of a predicate model here proves to be extremely complex. As a first step in an attempt to model spatio-temporal predicates between balloon objects, we therefore present the model in its most simplest and understandable form as possible. Thus, in this paper, we consider only the existence of a chance that a relationship can occur between uncertain movements instead of modeling the quantification of this chance, which can be done in further research. For this reason, we can focus only on the moving geometry of the future movement and replace $\varphi(\beta)$ by simply $\tau(\beta)$. Therefore, the balloon data type constructor can be written as $\Omega(\alpha,\beta) = \tau(\alpha) \times \tau(\beta)$. Based on this type constructor, the six balloon data types can be instantiated as follows.

$balloon\_pp = mpoint \times mpoint$ $balloon\_ll = mline \times mline$
$balloon\_pl = mpoint \times mline$ $balloon\_lr = mline \times mregion$
$balloon\_pr = mpoint \times mregion$ $balloon\_rr = mregion \times mregion$

For a balloon object $b = (b_p, b_f) \in \Omega(\alpha,\beta)$, the first moving object b_p, called the *past part*, describes the past movement of b. The second moving object b_f, called the *future part*, describes the collection of future potential movements of b, that is, it is the geometry of a future prediction of potential positions or extent of the balloon object. Figure 2(a) and 2(b) illustrate a *balloon_pp* and a *balloon_pr* object respectively.

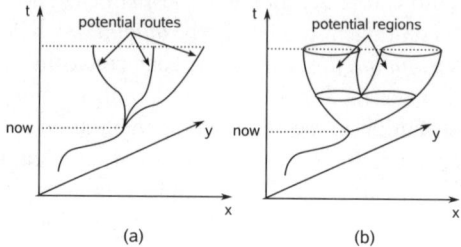

Fig. 2. Examples of a *balloon_pp* object (a) and a *balloon_pr* object (b)

4   Predicates on Balloon Objects

In this section, we define balloon predicates and explore their properties. After describing our general mechanism for defining balloon predicates in Section 4.1, we then discuss how such a predicate can be specified using traditional STPs in Section 4.2. Finally, we determine the canonical collection of balloon predicates in Section 4.3.

4.1   General Mechanism for Balloon Predicates

The approach we present here is based on two main goals. The first goal is to develop a formalism that works independently of the data types to which it is applied. It is desired

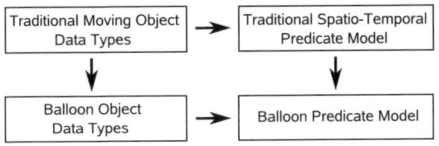

Fig. 3. Relations between traditional moving object data model and balloon data model

that the formalism can be equally applied to any pair of balloon objects irrespective of their data types. The second goal concerns the importance of making use of existing definitions of traditional STPs. Since balloon objects, as described in Section 3, are constructed based on traditional moving objects. It is only consistent to let balloon predicates be constructed from traditional STPs. With this goal, we can benefit from both theoretical and implementation advantages such that the formalism and implementation of balloon predicates can make use of the existing work done for traditional moving object data model. Figure 3 shows the relationships between traditional moving object data model and balloon object data model.

The general method we propose characterizes balloon predicates based on the idea that as two spatial objects move over time, the relationship between them may also change over time. By specifying this changing relationship as a predicate, we can ask a true/false question of whether or not such a changing relationship occurs. Thus, we can define a balloon predicate as a function from balloon objects to Boolean (that is, bool).

Definition 1. A balloon predicate is a function of the form $\Omega(\alpha_1, \beta_1) \times \Omega(\alpha_2, \beta_2) \rightarrow$ *bool* for $\alpha_1, \beta_1, \alpha_2, \beta_2 \in \{point, line, region\}$.

The change of relationship over time between two balloon objects indicates that there is a sequence of relationships that hold at different times. This suggests that a balloon predicate can also be modeled as a development (sequence) of spatial and spatio-temporal predicates. Due to the fact that a balloon object consists of a past part followed by a future part, the specification of a balloon predicate must also take into account the uncertainty of future relationships between the objects. To do this, let us first explore how relationships between balloon objects can be modeled.

Each balloon object has a defined present state at its present instant t_p which separate the past part and the future part. Between two balloon objects $A = (A_p, A_f)$ and $B = (B_p, B_f)$, A's present instant may either be earlier, at the same time, or later than B's present instant. In each of these scenarios, certain sequences of spatio-temporal relationships are possible between the parts of A and B. Here, we are only interested in the relationship between a part of A and another part of B whose temporal domains overlap since, in this case, the two parts may be defined on the same period of time. Figure 4 illustrates all the possible related pairs for each scenario between parts of A and B.

Although there are four possible types of relationships between all parts of balloon objects, it turns out that in any case, there are at most three types of relationships that may exist between parts of any two balloon objects. These include *past/past*, *past/future* or *future/past*, and *future/future* relationships. The past/future and future/past relationships cannot exist at the same time due to the temporal composition between the past and future parts of a balloon object.

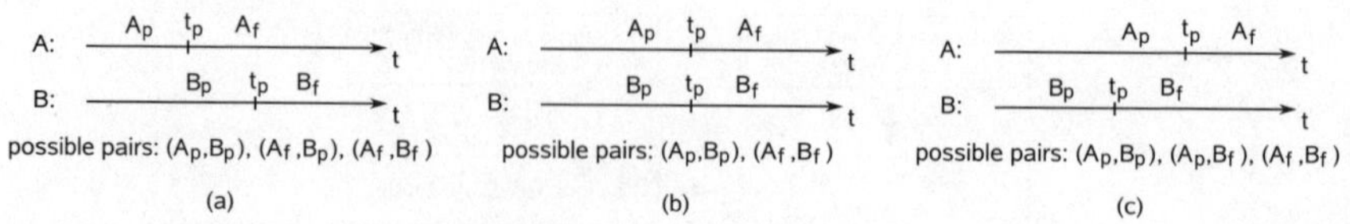

A: Ap tp Af → t   A: Ap tp Af → t   A: Ap tp Af → t
B: Bp tp Bf → t   B: Bp tp Bf → t   B: Bp tp Bf → t

possible pairs: (A_p,B_p), (A_f,B_p), (A_f,B_f) possible pairs: (A_p,B_p), (A_f,B_f) possible pairs: (A_p,B_p), (A_p,B_f), (A_f,B_f)

(a)   (b)   (c)

Fig. 4. Possible relationships between parts of balloon objects A and B when A's present instant is earlier (a), at the same time (b), or later (c) than that of B's

4.2 Specification Based on Traditional STPs

If we observe more closely, all the relationships between the parts of two balloon objects that may exist in a scenario form a development such that the entire relationship between the two balloon objects can be seen as a sequence of these relationships between their parts. For example, consider an airplane represented by a *balloon_pp* object $P = (P_p, P_f)$ and a hurricane represented by a *balloon_pr* object $R = (R_p, R_f)$ (Figure 5). In the past, P has been disjoint from R's path as well as part of R's predicted future. However, the future route of P crosses the predicted future of R.

The relationship between P and R can be described as a development or sequence of uncertain spatial and spatio-temporal predicates which hold at different times, i.e., $Disjoint_u \triangleright meet_u \triangleright Inside_u \triangleright meet_u \triangleright Disjoint_u$ (the subscript u indicates uncertain predicates). However, these spatial and spatio-temporal predicates may represent relationships between different parts of the balloon objects. For instance, the first $Disjoint$ predicate is actually a temporal composition of three different types of disjointedness between the corresponding parts of P and R, i.e., $Disjoint(P_p, R_p) \triangleright Disjoint_u(P_p, R_f) \triangleright Disjoint_u(P_f, R_f)$. The rest of the predicates represent relationships between the future parts of both objects. Hence, we can expand the original sequence as $Disjoint(P_p, R_p) \triangleright Disjoint_u(P_p, R_f) \triangleright Disjoint_u(P_f, R_f) \triangleright meet_u(P_f, R_f) \triangleright Inside_u(P_f, R_f) \triangleright meet_u(P_f, R_f) \triangleright Disjoint_u(P_f, R_f)$. In this sequence, the subsequence $Disjoint_u(P_f, R_f) \triangleright meet_u(P_f, R_f) \triangleright Inside_u(P_f, R_f) \triangleright meet_u(P_f, R_f) \triangleright Disjoint_u(P_f, R_f)$ can be represented by an STP $Cross_u(P_f, R_f)$, as shown in Section 2.1, since they are applied to the same future parts of the two balloon objects. Thus, we have $Disjoint(P_p, R_p) \triangleright Disjoint_u(P_p, R_f) \triangleright Cross_u(P_f, R_f)$.

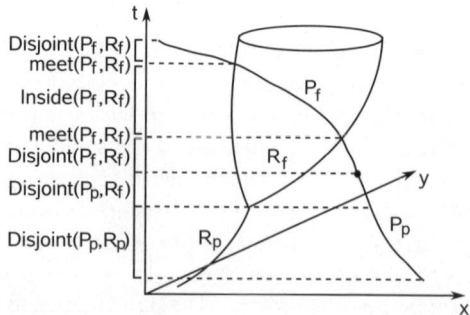

Fig. 5. A future crossing situation between a *balloon_pp* object P and a *balloon_pr* object R

As a result, we are left with a sequence of three STPs each applied to different combination pairs of parts of the balloon objects. This example illustrates that balloon predicates can be appropriately modeled by sequences of three STPs between the related parts of balloon objects. Hence, we can specify balloon predicates based on the traditional STPs as follows:

Definition 2. Let P and R be two balloon objects of type $\Omega(\alpha_1, \beta_1)$ and $\Omega(\alpha_2, \beta_2)$ respectively. A balloon predicate between P and R is a sequence of spatio-temporal predicates: $\langle stp(\tau(\alpha_1), \tau(\alpha_2)), (stp(\tau(\alpha_1), \tau(\beta_2)) | stp(\tau(\beta_1), \tau(\alpha_2))), stp(\tau(\beta_1), \tau(\beta_2)) \rangle$.

We consider an STP between two moving objects to be meaningful if and only if there exists a period of time for which both objects are defined. Hence, each element of the above sequence is meaningful only if the relationship between the corresponding parts is meaningful. The predicate of the first element in the sequence represents an interaction that *did occur*. The first and second alternative predicates of the second element in the sequence represents an interaction that *might occur* and *may occur* respectively. These predicate options reflect the constraint described in Section 4.1 which dictates that the two predicates cannot exist at the same time. The predicate of the third element in the sequence denotes an interaction that *probably will occur*. The combinations of multiple of these interactions represents a more complex relationship between balloon objects. For example, an interaction that *did occur* in the past and *probably will occur* in the future can indicate that it *probably always occurs*. Table 1 shows an example of assigning a meaningful prefix to the name for each pairwise combination between these interactions. Other combinations with larger number of interactions also exist, but it is usually not obvious to name these relationships. Here are some examples of balloon predicates:

$$
\begin{aligned}
did\_cross &:= \langle Cross(\tau(\alpha_1), \tau(\alpha_2)) \rangle \\
probably\_will\_cross &:= \langle Cross_u(\tau(\beta_1), \tau(\beta_2)) \rangle \\
may\_have\_been\_disjoint &:= \langle Disjoint_u(\tau(\alpha_1), \tau(\beta_2)), Disjoint_u(\tau(\beta_1), \tau(\beta_2)) \rangle \\
probably\_always\_inside &:= \langle Inside(\tau(\alpha_1), \tau(\alpha_2)), Inside_u(\tau(\beta_1), \tau(\beta_2)) \rangle
\end{aligned}
$$

4.3 Canonical Collection of Balloon Predicates

Having defined a model for balloon predicates, we can now search for a canonical collection of balloon predicates. The use of traditional STPs in the definition of balloon predicates suggests that the canonical collection of balloon predicates can be expressed in terms of the canonical collection of traditional STPs, which is provided in [4]. Another important factor that affects the canonical collection is whether dependencies exist

Table 1. Assigning naming prefixes to pairwise combinations of interactions

	did	might	may	probably will
did	-	might have	may have	probably always
might	might have	-	-	may have been
may	may have	-	-	probably will have
probably will	probably always	may have been	probably will have	-

Table 2. Number of balloon predicates between *balloon_pp*, *balloon_pr*, and *balloon_rr* objects

	balloon_pp	*balloon_pr*	*balloon_rr*
balloon_pp	4,394	14,924	43,904
balloon_pr	14,924	1,600,144	136,996,944
balloon_rr	43,904	136,996,944	21,237,972,784

between the three elements of the sequence. More specifically, we need to investigate whether the existence of a STP as an element of the sequence can prevent or restrict another STP from representing another element of the sequence.

According to [4], the dependency between STPs, which are parts of a continuous development, is expressed using a *development graph*. This graph describes all the possible developments of STPs which correspond to continuous topological changes of moving objects. For example, if a moving point is inside a moving region, it must meet the boundary of the moving region before it can be disjoint from the region. This constraint relies on the continuity of the moving point. If we allow discontinuity such as a period of unknown movement as in the case of the balloon model, then such constraint cannot be applied. Although the past part and the future part of a balloon object cannot temporally overlap each other, it is possible that they can be separated by a period of unknown movement. Further, there can also be periods of unknown movement within the past or the future part of a balloon object. Due to the possible discontinuity of balloon objects, we can deduce that each element of the predicate sequence, which is a STP between the parts of two balloon objects, are independent of each other. Thus, all the combinations of the STPs involved are possible. This means that the canonical collection of balloon predicates can be determined solely based on the canonical collections of the traditional STPs involved. As provided in [4], there are 13 distinct temporal evolutions between two moving points without repetitions, 28 between a moving point and a moving region, and 2,198 between two moving regions. With this information, we can determine, for example, the number of distinct, non-repetitive balloon predicates between two *balloon_pp* objects to be $13 \times (13 + 13) \times 13 = 4,394$. Each of the three numbers of the multiplication represents the number of distinct STPs for each element of the sequence. Similarly, we can determine the number of balloon predicates between all type combinations of *balloon_pp*, *balloon_pr*, and *balloon_rr* as shown in Table 2 below. Since the numbers of STPs that involve moving line objects are not available in [4], we omit those balloon predicates that involve balloon objects which are based on moving line objects.

5   Using Balloon Predicates in Queries

Due to the large numbers of possible balloon predicates, it is essential to consider all the different ways in which these predicates can be used in queries. Either the user is provided with a small, application-specific set of balloon predicates, or the user is allowed to construct balloon predicates according to his needs. Several mechanisms are available which enable users to specify and use balloon predicates in queries. One solution to this problem is to use the *spatio-temporal query language* (STQL) [3] to support textual specifications of balloon predicates in queries. Assuming that all necessary data for hurricanes and airplanes are available in a database. We can define, for example, a balloon predicate *potentially_cross* and use it in a query as follows:

```
DEFINE potentially_cross AS <Cross(future/future)>;
SELECT flights.id FROM flights, hurricanes
WHERE flights.route potentially_cross hurricanes.path;
```

Another solution is to use the *visual query language* [5]. By extending this concept to allow a graphical specification of a sequence of developments, we can use this mechanism to specify balloon predicates and use them in queries.

6   Conclusions

Modeling spatio-temporal predicates between spatio-temporal objects with uncertainty movements is a very challenging task. In this paper, we take a first step in this direction by first focusing on modeling the spatio-temporal relationships between the geometries of precise and uncertain movements of spatial objects. In doing so, we show that relationships between balloon objects are composed of a sequence of certain and uncertain relationships between their parts. With the understanding of the existence of these relationships, the next step is to consider the quantification of the chances in which such relationships can occur.

References

1. Erwig, M., Güting, R.H., Schneider, M., Vazirgiannis, M.: Abstract and Discrete Modeling of Spatio-Temporal Data Types. In: ACM GIS. ACM Symp. on Geographic Information Systems, pp. 131–136. ACM Press, New York (1998)
2. Erwig, M., Güting, R.H., Schneider, M., Vazirgiannis, M.: Spatio-Temporal Data Types: An Approach to Modeling and Querying Moving Objects in Databases. GeoInformatica 3(3), 265–291 (1999)
3. Erwig, M., Schneider, M.: Developments in Spatio-Temporal Query Languages. In: IEEE Int. Workshop on Spatio-Temporal Data Models and Languages, pp. 441–449. IEEE Computer Society Press, Los Alamitos (1999)
4. Erwig, M., Schneider, M.: Spatio-Temporal Predicates. IEEE Trans. on Knowledge and Data Engineering (TKDE) 14(4), 1–42 (2002)
5. Erwig, M., Erwig, M., Schneider, M.: Query-By-Trace: Visual Predicate Specification in Spatio-Temporal Databases. In: The Fifth Working Conference on Visual Database Systems, vol. 168, pp. 199–218 (2000)
6. Hinrichs, K., Trajcevski, G., Wolfson, O., Chamberlain, S.: Managing Uncertainty in Moving Objects Databases. ACM Trans. on Database Systems (TODS) 29, 463–507 (2004)
7. Güting, R.H., Böhlen, M.H., Erwig, M., Jensen, C.S., Lorentzos, N.A., Schneider, M., Vazirgiannis, M.: A Foundation for Representing and Querying Moving Objects. ACM Trans. on Database Systems (TODS) 25(1), 881–901 (2000)
8. Praing, R., Schneider, M.: A Universal Abstract Model for Future Movements of Moving Objects. In: AGILE International Conference on Geographical Information Systems (2007)
9. Praing, R., Schneider, M.: The Balloon Model: Supporting Historical and Future Movements of Moving Objects. Technical report, CISE Department, University of Florida (2007)

Towards a Semantic Spatial Model for Pedestrian Indoor Navigation

Edgar-Philipp Stoffel, Bernhard Lorenz, and Hans Jürgen Ohlbach

University of Munich, Oettingenstr. 67, 80538 Munich, Germany
{stoffel,lorenz,ohlbach}@pms.ifi.lmu.de
http://www.pms.ifi.lmu.de/

Abstract. This paper presents a graph-based spatial model which can serve as a reference for guiding *pedestrians* inside buildings. We describe a systematic approach to construct the model from geometric data. In excess of the well-known topological relations, the model accounts for two important aspects of pedestrian navigation: firstly, *visibility* within spatial areas and, secondly, generating *route descriptions*. An algorithm is proposed which partitions spatial regions according to visibility criteria. It can handle simple polygons as encountered in floor plans. The model is structured hierarchically - each of its elements corresponds to a certain domain concept ('room', 'door', 'floor' etc.) and can be annotated with meta information. This is useful for applications in which such information have to be evaluated.

1 Introduction and Related Work

Pedestrian guidance within buildings [6] differs from customary navigation based on networks. This is due to the following reasons:

1. Features like roads or railways can be commonly modelled by *one-dimensional* elements in a clearly defined network. In contrast, features of a building do not fit so nicely into this schema – a building exhibits a nested, *three-dimensional* structure. There are several ways to overlay a **floor plan** with a graph [5]. It is arguable which representation to choose: e.g. should a corridor be mapped to a node, a chain of connected nodes, or rather to an edge?
2. Pedestrians can roam *freely* between the interior boundaries of buildings. Their movement is less restricted than those of vehicles (which are often bound to their networks, e.g. trains). However, spatial orientation and visibility play a vital role for **human wayfinding** [16]: An *L-shaped* room, for example, cannot be perceived as a whole without moving around the corner. Although the primary concern of navigation is to determine a shortest (fastest) path, one should not underrate the importance of comprehensible **route descriptions** [13] – they tell us *how* to follow a path in a region.

Floor plans document the interior layout of a building in terms of boundaries. Particularly in computational geometry [3] and robot motion planning [2], so-called roadmap methods derive detailed navigational graphs from geometric information. At the other end of the spectrum, there are symbolic models which can

J.-L. Hainaut et al. (Eds.): ER Workshops 2007, LNCS 4802, pp. 328–337, 2007.

be classified [1] into topological models like Region Adjacency Graphs [10,14] or Cell and Portal Graphs [11] and hierarchical models [8,12,15]. All these abstract away from geometric details; they merely represent qualitative spatial relations between regions [4]. However, as argued by Hu et al. [8], if the relations are too abstract or coarse, they are impractical – they cannot model reachability among regions via different entry and exit points. Finally, cognitive models of indoor spaces [16,17] are interesting insofar as they cater for the representation of space from the perspective of humans. They are well-suited for generating route descriptions. What is missing is an elegant way to couple these different aspects into one coherent model, e.g. a high-level topological representation which can be refined to reveal the inner structure of a region if required.

The main contribution of this paper is the formalisation of a spatial model which can handle different levels of abstraction and provides a basis for the generation of route instructions. We describe a systematic approach to construct the model from floor plan data. It is worth noting that the notion of visibility is embedded into the model.

The paper is structured as follows: in Sect. 2.1 and 2.2, we formally define the basic elements of the model based on a floor plan's geometry. Sect. 2.3 explains the extesion to a hierarchy. The algorithm presented in Sect. 2.4 refines the hierarchy based on visibility criteria. Sect. 2.5 motivates the potential of annotating the model with meta information.

2   The Proposed Model

We assume that a set of floor plans of the environment are available in a vector-based format (these data originate e.g. from a CAD application). The geometric data structure consists of a (planar) **mesh of polygons**; each polygon encloses a spatial region R. In the following we use the terms polygon and spatial region interchangeably, although we are aware of the subtle difference [4] that a polygon is only a representation of the boundary ∂R of a spatial region.

2.1   Spatial Regions

Definition 1 (Spatial Region). *A **spatial region** $R := C_1C_2..C_{z\geq3}$ is geometrically represented as a list of corners $C_{i\in1..z}$ (indices $i > z$ are calculated **modulo** z). Corners are ordered **counter-clockwise (ccw)**. Each corner takes up a position $C_i.pos = (x, y)$ in an underlying reference system. Two consecutive corners fix a **boundary line** $bLine(i) = C_iC_{i+1}$. Boundary lines do not intersect elsewhere: $bLine(i) \cap bLine(j \neq i \pm 1) = \emptyset$. **interior**$(R) :=$ all points p with odd crossing number (number of times a ray starting from p intersects boundary lines of R).*

Two spatial regions R_p, R_q **touch** [4] if $\exists i, j$, $bl_p := bLine(i)$ in R_p, $bl_q := bLine(j)$ in R_q: $bl_p = bl_q$, i.e. they share at least one boundary line. A mesh of polygons is characterised more precisely as a **cell complex** (cf. Plümer et al. [14]) by the following axioms:

Condition 1 (pairwise disjoint). *Spatial regions may **not overlap**, but **touch**:* $\forall bl_p \in R_p, bl_q \in R_q: bl_p = bl_q \vee bl_p \cap bl_q = \emptyset \vee bl_p \cap bl_q = \{C \| C \in R_p \cap R_q\}$.

Condition 2 (jointly exhaustive). *All n spatial regions **make up** the environment to be modelled:* $\bigcup_{i \in 1..n} R_i = FloorPlan$.

Whereas convex polygons are not critical for route descriptions, we have to treat **non-convex polygons** specially (see Sect. 2.4). This is simply because humans can't see what is beyond a corner.

Definition 2 (Concave Corner, Chain). *A corner X enclosing an internal angle greater than $180°(\pi)$ is called **concave**. A non-convex polygon has at least one concave corner. **Concave chains** are maximum sequences $X_i X_{i+1}..X_{i+k}$ of consecutive concave corners.*

2.2 Boundary Nodes

For the purpose of navigation, we have to define connectivity among spatial regions. Even though two adjoining spatial regions are touching, they can still be physically separated by walls or other divisions (loges in a theatre, platforms on a station). Access and egress are only possible through specific points on their shared boundary (i.e. **doors** and other openings), called **boundary nodes**. In the literature these elements are also referred to as *gateways* [17] or *exits* [8]. A prototypical setting with two boundary nodes is depicted in Fig. 1.

Definition 3 (Boundary Node). *A **boundary node** $B := (id, t, R_p, R_q, w, \Omega)$ is a waypoint for a path between two adjoining spatial regions R_p, R_q: $\exists a, b, bl: bl = bLine(a)$ in $R_p \wedge bl = bLine(b)$ in $R_q \wedge B$ in bl. It has exactly one type t and a **unique** identifier id. Furthermore, w denotes its total width and Ω the angle/orientation perpendicular to its boundary line bl.*

Remarks:

1. Boundary nodes are key points for navigation. Their type t is a conceptual representation of the underlying architectural feature, e.g. of a door, a window or an opening.
2. Two spatial regions can have more than one boundary node in common (see Fig. 1), thus their connection has a certain multiplicity. The *unique* identifier id guarantees that each boundary node can be distinguished as a separate entity.

A boundary node can be regarded, from a **dual** perspective, as a transition relation (or connection) between regions. In Region Adjacency Graphs [12,14] or Cell and Portal Graphs [11], they are represented as *multi*edges (remark 2) between the nodes R_p and R_q.

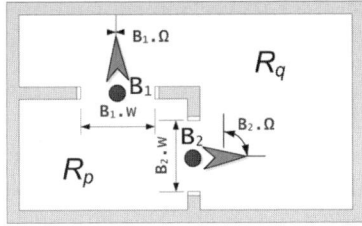

Fig. 1. Two Exemplary Boundary Nodes

2.3 Hierarchical Graphs

Apart from coinciding boundaries, there is another fundamental relation between spatial regions: they can be nested. Premises are inherently organised into constituent floors, sections, rooms, and so forth. Therefore it makes sense to define a relation for containment. This relation can be used to extend the aforementioned flat graphs into **hierarchical graphs** [1,12,15]:

Definition 4 (Child Relation ◁). *A spatial region R_c is **child** of another spatial region R_p, denoted as $R_c \vartriangleleft R_p$, if $\exists(C \in R_c) : C \in interior(R_p)$ and $\forall(R_a \neq R_p) : [C \in interior(R_a) \Rightarrow \exists(C_p \in R_p) : C_p \in interior(R_a)]$. R_p, R_a are called **parent** and **ancestor** region, respectively.*

Remarks:

1. According to axioms 1 and 2, the first condition of Def. 4 implies that the other corners $C_{other} \neq C$ of R_c lie either in the interior of R_p or on a boundary line shared by R_p and R_c. The second condition guarantees that R_p is indeed the minimal region containing R_c - there is no other region R_a between R_p and R_c that also contains R_c (illustrated in Fig. 2).
2. A further consequence is that only one region R_p can be parent of the region R_c. However, an important question arises: How can, for example, rooms R_c be represented which belong to a floor R_{p1} and a wing R_{p2} at the same time? There is a way to fit them into the model: one can simply substitute R_{p1} and R_{p2} by the three regions $R_{p1} \setminus R_{p2}$, $R_{p1} \cap R_{p2}$, and $R_{p2} \setminus R_{p1}$. This

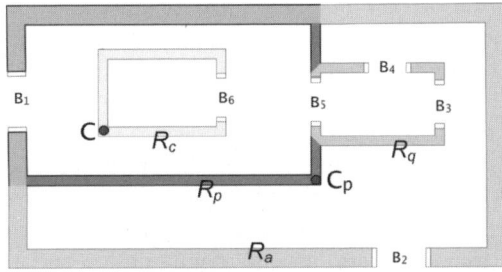
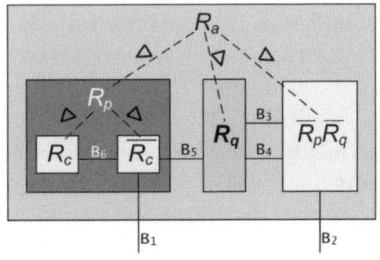

Fig. 2. An Example for a Hierarchy

method works in general for n overlapping parent regions. They are replaced by at most $2^n - 1$ non-overlapping regions.

Definition 5 (Region Graph). $G_R := (N_R, E_R, t)$ *is the* **region graph** *of a spatial region R. The type t of a region graph represents concepts like floors, sections, rooms, etc. All region graphs G_Q with $Q \lhd R$ are nodes in N_R. Comprised in the edge set E_R are the boundary nodes of R as well as those of all child regions Q.* **Local edges** *of G_R are boundary nodes between $Q_1, Q_2 \in N_R$. Otherwise, if one $Q_{i \in \{1,2\}} \notin N_R$ they are* **interface edges** *of G_R.*

One can easily see that the relation $\lhd$ is anti-symmetric and irreflexive. It thus defines a partial ordering on spatial regions. This ordering represents the multiple levels of the hierarchy, from coarse- to fine-granular spatial regions:

Definition 6 (Level, Root, Leaf)
level$(G_{R_r}) := 0$ *if $\nexists R_p : R_r \lhd R_p$. Such R_r are called* **roots**.
level$(G_{R_c}) := 1 + level(G_{R_p})$ *if $R_c \lhd R_p$, i.e. $G_{R_c} \in N_{R_p}$.*
All R_l with $\nexists R_c : R_c \lhd R_l$ are **leaves**. *They lie at the bottom of the hierarchy.*

Pragmatic considerations speak in favour of using hierarchical graphs as underlying navigation model: Experiments conducted e.g. by Jing et al. [9] and Shekhar et al. [18] have shown a gain in the processing time of shortest path queries. Besides, humans can rather make sense of hierarchical structures than of coordinates returned by a positioning system since the former are qualitative. The hierarchy is always coined by the floor plans. However, it requires some reorganisation with respect to *reachability* between child regions:

- $G_{Q_i} \in N_R$ may *not* be *mutually* reachable by a sequence of boundary nodes $B_N \in E_R$. For instance, two rooms Q_1, Q_2 in fully separated sections of a floor R can only be reached via another floor. It makes sense to split G_R into its **connected components**, each becoming a new region graph.
- The removal of one **articulation edge** from E_R (say B_5 in Fig. 2 is locked), leads to having two connected components. Knowing this, one can split G_R in advance at all articulation edges.

2.4   Partitioning Algorithm and Navigation Process

Hierarchical graphs reflect a building's topology. Nevertheless, their resolution is too coarse for navigation: The interior of *non-convex* leaf regions may be complex so that several route instructions are necessary (e.g. for a door around two corners). The path between two boundary nodes (e.g. B_1 and B_2 in Fig. 1) in a leaf region (R_q) is not always the line of sight, unless the region is convex.

   Visibility depends on the shape of a leaf region. In the following, we present an algorithm which partitions leaf regions according to visibility criteria: The principal idea is to connect corners in a non-convex leaf region in such a way that they partition the region into non-overlapping **convex sub-regions** (see Fig. 3). The partitioning is not arbitrary (we could use any triangulation then), but concave corners play a major role. The actual process of partitioning is described in the main algorithm:

```
 1  List<Polygon> convexPartitioning(Polygon p)
    {   List<Polygon> subPolys = new List<Polygon>(); /* store sub-polygons */
 3      for (ConcaveCorner r in p.concaveCorners())                /* ccw list */
        {   ConcaveCorner rNext = p.nextConcaveCorner(r);    /* ccw from r */
 5          if (rNext == null)  /* (1) the only concave corner of p is r */
                subPolys.add(matchOneConcaveCorner(p,r));
 7          else                          /* (2) more than one concave corner in p */
            {   if (p.index(r) + 1 == p.index(rNext)) continue;
 9                  /* (2.1) r, rNext in concave chain: skip one iteration */
                Polygon sub =
11                  p.createSubpolygon(range(p.index(r), p.index(rNext)));
                    /* puts all corners from r to rNext in sub, rNext points to r */
13              if (not(sub.isConvex())) //(2.2) match concave in sub with corner
                {   List<Polygon> subSplits = convexPartitioning(sub);
15                  if (not(p.nextConcaveCorner(rNext) == r))
                        subSplits.remove(polygons with bLine rNext to r);
17                  subSplits.remove(polygons with concave corner of p inside);
                    subPolys.add(subSplits); }
19              else            /* (2.3) sub-polygon is convex */
                {   if (no concave corner of p inside sub)
21                  {   p = p.splitSubpolygon(sub);
                        /* bypass all corners between r, rNext */
23                      subPolys.add(sub);    } } } }
        p.updateConcaveCorners();    /* some can fall away now */
25      if (not(p.isConvex()))          /* still non-convex */
            subPolys.add(convexPartitioning(p));    /* recursive call on p */
27      return subPolys;    }
```

In order to cut off a sub-polygon *sub* (line 10,11), the algorithm tries to connect each concave corner r with its next concave corner $rNext$ in ccw (line 4). If this succeeds, a new boundary line is created between r and $rNext$. In case r is the only concave corner (line 5,6), it has to be matched to (ideally) one or two convex corners (see next listing). If r and $rNext$ are already connected (line 8), the algorithm skips r and proceeds to $rNext$.

The polygon *sub* is only cut off if it is convex and no other concave corners of p are enclosed in its interior (line 19-23). As a counterexample, the cut $r_2 r_3$ on the left of Fig. 3 would enclose r_1. However, assuming a cut is allowed, all corners between r and $rNext$ are removed from p (line 21).

If *sub* is non-convex (line 13-18), it implies that r and/or $rNext$ are also concave in *sub*. The algorithm is called recursively on *sub*. The result of this call is *subSplits*, a convex partitioning of *sub*. All polygons of *subSplits* with a concave corner of p inside have to be removed (line 17), also the polygon with the *bLine* from r to $rNext$ if this cut wouldn't be done by $rNext$ (line 15,16). The rest of the polygons in *subSplits* are then cut off from p (line 18). After one complete tour around the polygon p, it might have become smaller due to some successful cuts. The concave corners of p are updated (line 24); if there are still concave corners in p, the algorithm is called recursively (line 25,26).

The partitioning of a polygon with only *one* concave corner (line 6) is illustrated on the right hand side of Fig. 3 by the description of cut 2, as well as in the following listing:

```
 1  List<Polygon> matchOneConcaveCorner(Polygon p, ConcaveCorner r)
    {   int ix = p.index(r); List<Polygon> subPolys = new List<Polygon>();
 3      List<Corner> Cmatches = all corners c of p with
            c leftOf bLine(ix-1) and c leftOf bLine(ix);            /* LL: ideal
                */
```

```
5        if (Cmatches == null)   /* can't reduce with one ideal cut (LL) */
         {   Cmatches = corners c,d of p such that d = p.nextCorner(c) and
7            c rightOf bLine(ix-1) and c leftOf bLine(ix)   /* RL */
             and d leftOf bLine(ix-1) and d rightOf bLine(ix);  }   /* LR */
9        for (Corner c in Cmatches)
         {   Polygon sub = p.createSubpolygon(range(p.index(r),p.index(c)));
11           sub.addBoundaryNodeBetween(c, r);
             if (no concave corner of p inside sub)
13           {   p = p.splitSubpolygon(sub);
                 subPolys.add(sub); } }
15       return subPolys; }
```

A corner in the area 'LL' would be ideal to connect to (line 4), since the concave corner would become convex. But if there is none, one can pick the last corner in the area 'RL' (line 7) and the first one in the area 'LR' (line 8).

On the basis of the partitioning, one can define a navigational graph for representing also **paths between boundary nodes**. This is exemplified, too, in Fig. 3: paths are indicated by the small dashed lines. Each sub-region is convex, so all boundary nodes in the *same* sub-region are *per se* mutually visible. They can be directly connected by a path whose distance is simply the Euclidean distance. For route descriptions, one can make use of the orientations $B.\Omega$ encoded in boundary nodes and divide the space into front, left and right. This is working also in cases where boundary nodes lie on the same boundary line (e.g. B_5 and B_4 on the right hand side of Fig. 3): With the angles enclosed between the path $B_5 B_4$ and $B_5.\Omega$ (resp. $B_4.\Omega$) one can find out that a left turn is required starting at B_5. A route description such as "Turn left [at B_5] and move along the wall until you reach the first door [B_4] on your left" can be obtained.

Boundary nodes in *different* sub-regions can be directly connected by a path if they intersect *all* cutting lines between their sub-regions (on the left hand side of Fig. 3: $B_1 B_2$ with cut 5, $B_3 B_7$ with cuts 2 and 4). Otherwise, they either lie outside the considered leaf region L (left hand side: $B_7 B_6$), or intersect some boundary line of another sub-region of L (left hand side: $B_2 B_8$, $B_4 B_3$). This means, in any case, that they are not mutually visible in L. However, they can be connected by a chain of paths which additionally run through points, e.g. the centres, of the cutting lines between the involved sub-regions. The centre of cut 4 (left hand) is added as an intermediary point along the path from B_2 to B_8. When connecting B_3 with B_4, the center of cut 2 is superfluous because the

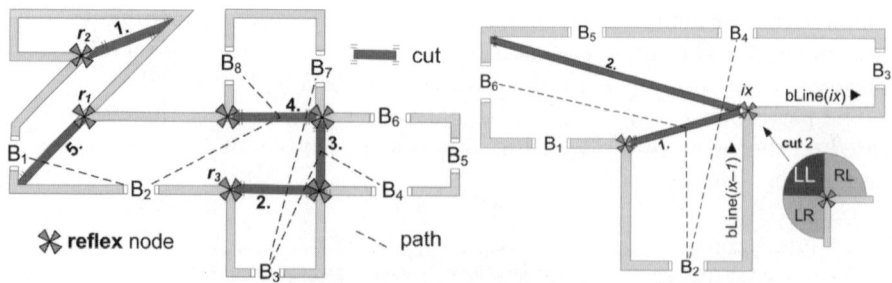

Fig. 3. Applying the Partitioning Algorithm

path between B_3 and the centre of cut 3 intersects cut 2 – this means that using only cut 3 is sufficient.

2.5   Evaluation of Constraints

Although the focus of this paper is on the spatial model, it is worth noting that the model can be annotated with meta information. Especially the types t of boundary nodes and region graphs could contain further attributes, e.g. in form of a list of key-value-pairs. This could be very useful in practise, for applications which require a more detailed processing of *context* information. The notion of distance could be understood in a variety of different ways, depending on the **semantics** [19,20] of the application and its context (encoded in these attributes). Consider the following examples:

- Doors (boundary nodes) can be locked or, more general, access requires *authorisation* (key, card, biometric scan etc.).
- Admission of entry can be limited in *time*, e.g. opening hours of an office.
- Certain sections of a public building (all interface edges into a region graph) may be *restricted* in access ("staff only", high-security wings, laboratories).
- Special exits and base level windows can be used for *emergencies*.

The examples from above can be modelled as **boolean** (hard) **constraints** of the form $\bigwedge^* (attr = value \vee attr \in valuePartition)$ on boundary nodes and/or region graphs of a certain type. After evaluation, such a constraint yields a truth value. Boolean constraints can, hence, be used to determine under which conditions motion is physically possible ('can') or admitted ('may') in the environment [7]. The environment can be filtered only for the relevant parts which fulfil these binary constraints *before* the actual navigation process. A rich indoor model should take these kinds of constraints into account, but not exclusively. *Person-related* properties like roles, privileges, or preferences also have a significant impact on navigation:

- Imagine a person inside a building, pushing a pram. She intends to get from the ground floor to an upper floor. This person opts for a path with an `elevator` (in case the pram fits in), deliberately accepting a detour.
- In the same building a second person on business has an appointment in an office. Say it is on the *second floor*. Rather than waiting for the elevator, this person uses the `staircase` in order to arrive timely. Now let us assume a slightly modified situation: The appointment takes place on the *ninth floor*. In this case the person may instead be willing to use the elevator.

Although in both situations, the topology of the building is exactly the same, there are *two* interpretations of distance. The *personal* context of the wayfinder matters. The trade-off described in these situations can be modelled by **soft constraints**: $\forall path \in G_R[of\ type\ t] : path.cost = [time_{wait} +]\ bonus/penalty * path.time$. They alter the costs of traversing certain regions of the environment (e.g. $t = stairs \rightarrow penalty = 4$) in favour of others (e.g. $t = elevator \rightarrow time_{wait} = 20$).

3   Conclusion

In this paper we presented a hierarchically structured model of an indoor environment which accounts for different entry and exit points of regions. Defined upon concrete geometries, the model is not abstract but can be implemented and provided with real data from floor plans. Furthermore, we presented an algorithm which partitions regions according to visibility criteria, so that route descriptions can be given for their interior. It would be interesting to study deeper the ties between the spatial entities in the model and their linguistic counterparts for route descriptions. Another point for research is the further development of constraints: Semantic Web technologies are appealing for their specification (annotation of maps in a wiki-like style), and especially their processing. It would be also worthwhile to examine inhowfar constraint processing and hierarchical planning could be intertwined.

Acknowledgements

This research has been co-funded by the European Commission and by the Swiss Federal Office for Education and Science within the 6th Framework Programme project REWERSE number 506779 (cf. http://rewerse.net).

References

1. Becker, C., Dürr, F.: On location models for ubiquitous computing. Personal Ubiquitous Comput. 9(1), 20–31 (2005)
2. Choset, H., Burdick, J.: Sensor-Based Exploration: The Hierarchical Generalized Voronoi Graph. International Journal of Robotics Research 19(2), 96–125 (2000)
3. de Berg, M., van Kreveld, M., Overmars, M., Schwarzkopf, O.: Visibility Graphs: Finding the Shortest Route. In: Computational Geometry: Algorithms and Applications, ch. 15, pp. 307–317. Springer, Heidelberg (2000)
4. Egenhofer, M.J., Franzosa, R.D.: Point Set Topological Relations. International Journal of Geographical Information Systems 5, 161–174 (1991)
5. Franz, G., Mallot, H., Wiener, J.: Graph-based Models of Space in Architecture and Cognitive Science - a Comparative Analysis. In: Proceedings of the 17th International Conference on Systems Research, Informatics and Cybernetics, pp. 30–38 (2005)
6. Gilliéron, P.-Y., Merminod, B.: Personal Navigation System for Indoor Applications. In: Proceedings of the 11th IAIN World Congress on Smart Navigation, Systems and Services, Berlin (2003)
7. Hendricks, M.D., Egenhofer, M.J., Hornsby, K.: Structuring a Wayfinder's Dynamic Space-Time Environment. In: Kuhn, W., Worboys, M.F., Timpf, S. (eds.) COSIT 2003. LNCS, vol. 2825, pp. 75–92. Springer, Heidelberg (2003)
8. Hu, H., Lee, D.L.: Semantic Location Modeling for Location Navigation in Mobile Environment. In: Proceedings of the 5th IEEE International Conference on Mobile Data Management, pp. 52–61. IEEE Computer Society, Los Alamitos (2004)
9. Jing, N., Huang, Y.-W., Rundensteiner, E.A.: Hierarchical Encoded Path Views for Path Query Processing: An Optimal Model and Its Performance Evaluation. Knowledge and Data Engineering 10(3), 409–432 (1998)

10. Kuipers, B., Modayil, J., Beeson, P., MacMahon, M., Savelli, F.: Local metrical and global topological maps in the Hybrid Spatial Semantic Hierarchy. In: Proceedings of the IEEE International Conference on Robotics and Automation. ICRA, vol. 5, pp. 4845–4851. IEEE Computer Society, Los Alamitos (2004)
11. Lefebvre, S., Hornus, S.: Automatic cell-and-portal decomposition. Technical Report 4898, INRIA (July 2003)
12. Lorenz, B., Ohlbach, H.J., Stoffel, E.-P.: A Hybrid Spatial Model for Representing Indoor Environments. In: Carswell, J.D., Tezuka, T. (eds.) W2GIS 2006. LNCS, vol. 4295, pp. 102–112. Springer, Heidelberg (2006)
13. Ohlbach, H.J., Rosner, M., Lorenz, B., Stoffel, E.-P.: NL Navigation Commands from Indoor WLAN fingerprinting position data. In: REWERSE Deliverable A1-D7 (2006), http://rewerse.net
14. Plümer, L., Gröger, G.: Nested Maps – a Formal, Provably Correct Object Model for Spatial Aggregates. In: Proceedings of the 4th ACM International Workshop on Advances in Geographic Information Systems, pp. 76–83. ACM Press, New York (1996)
15. Poulovassilis, A., Levene, M.: A Nested-Graph Model for the Representation and Manipulation of Complex Objects. ACM Trans. Inf. Syst. 12(1), 35–68 (1994)
16. Raubal, M., Worboys, M.: A Formal Model of the Process of Wayfinding in Built Environments. In: Freksa, C., Mark, D.M. (eds.) COSIT 1999. LNCS, vol. 1661, pp. 381–401. Springer, Heidelberg (1999)
17. Rüetschi, U.-J., Timpf, S.: Using Image Schemata to Represent Meaningful Spatial Configurations. In: Meersman, R., Tari, Z., Herrero, P. (eds.) OTM 2005 Workshops. LNCS, vol. 3762, pp. 1047–1055. Springer, Heidelberg (2005)
18. Shekhar, S., Fetterer, A., Goyal, B.: Materialization Trade-Offs in Hierarchical Shortest Path Algorithms. In: Symposium on Large Spatial Databases, pp. 94–111 (1997)
19. Stahl, C., Haupert, J.: Taking Location Modelling to New Levels: A Map Modelling Toolkit for Intelligent Environments. In: Hazas, M., Krumm, J., Strang, T. (eds.) LoCA 2006. LNCS, vol. 3987, pp. 74–85. Springer, Heidelberg (2006)
20. Tsetsos, V., Anagnostopoulos, C., Kikiras, P., Hadjiefthymiades, S.: Semantically enriched navigation for indoor environments. International Journal of Web and Grid Services 2(4), 453–478 (2006)

Modeling Collaborative Semantics with a Geographic Recommender

Christoph Schlieder

Laboratory for Semantic Information Technology
Bamberg University, Germany
christoph.schlieder@wiai.uni-bamberg.de

In the Semantic Web paradigm, geo-ontologies are closely related to geospatial information communities. Each community comes with its own ontology which is modeled, most frequently, within the framework of description logics. The paper questions a central assumption underlying this approach, namely that communities (and ontologies) are defined by crisp semantic boundaries. The idea of a semantic boundary contrasts sharply with the notion of a community of data producers/consumers that characterizes Web 2.0 applications. Well-known examples are GPS-trail libraries for hikers and bikers or image libraries of places of touristic interest. In these applications, conceptualizations are created as folksonomies by voluntary contributors who associate georeferenced objects (e.g. trails, images) with semantic tags. We argue that the resulting folksonomy can not be considered an ontology in the sense of Semantic Web technology. However, we propose a novel approach for modeling the collaborative semantics of geographic folksonomies. This approach is based on multi-object tagging, that is, the analysis of tags that users assign to composite objects, e.g. a group of photographs.

Subject areas: Spatio-temporal ontologies and reasoning, Definition, extraction, elicitation, and capture of spatial semantics.

1 Introduction

Communities of voluntary contributors have been successful at building extensive collections of georeferenced data. Libraries of mountain biking trails recorded with handheld GPS receivers constitute typical examples (Morris et al., 2004). The idea of community-based mapping has been extended to features that, unlike biking trails, belong to the core topographic data set, most notably to road networks[1]. Generally, the rationale behind such low-quality remapping efforts is that of creating a data set which comes with a public domain license. Another type of application that is capable of attracting contributors in great numbers are libraries of georeferenced photographs. For instance, in the United Kingdom and the Republic of Ireland, amateur photographers collect images of geographic features aiming at documenting each square kilometer of the British Isles with at least one photograph[2].

---

[1] For example: Open Street Map Project (http://www.openstreetmap.org/)
[2] Geograph project: http://www.geograph.org.uk/

J.-L. Hainaut et al. (Eds.): ER Workshops 2007, LNCS 4802, pp. 338–347, 2007.

Most forms of collaborative acquisition of georeferenced data include the acquisition of metadata. Spatio-temporal coverage (Where and when was the image taken, the trail recorded?) and content type (What objects are depicted? What type of trail is it?) constitute indispensable parts of content-related metadata sets. *Social tagging* is probably the simplest way to obtain this metadata from a community. Data producers categorize their contribution by assigning the key words ("tags") that they think best characterize the spatio-temporal coverage and the content type. The resulting tag vocabulary is called a *folksonomy*.

When O'Reilly (2005) described his vision of the Web's future, folksonomies were already widely used and he could present them as a typical tool of the Social Web, or, as he preferred to called it, the Web 2.0. From the point of view of conceptual modeling, the Social Web with its emphasis on collaborative metadata acquisition (Miller, 2005) and the Semantic Web with its emphasis on ontological engineering (Fensel et al., 2003) stress complementary aspects. This is especially true for GI processing where the OGC has always been very clear about the relationship between ontologies and geospatial information communities. The concept of *information community* is given a concise definition in the OGC white paper by Gould and Hecht (2001): "An information community is a group of people who share a common geospatial feature data dictionary (including definitions of feature relationships) and a common metadata schema."

We argue that the most remarkable difference between folksonomies and ontologies is not that of different degrees of formalization but the fact that they address different use cases. Folksonomies handle the modeling of semantics before the emergence of information communities and before crisp semantic boundaries have been established whereas the ontology languages of the Semantic Web (OWL, and to a lesser extent RDF) address the modeling of semantics after that phase. To support that view, we introduce a novel approach for modeling the collaborative semantics of geographic folksonomies.

The paper makes two main contributions: firstly, it bases semantic analysis on multi-object tagging instead of single-object tagging, and secondly, it introduces a new method for analyzing geospatial information communities by exploiting data about user-to-user similarity from a geographic recommender. *Multi-object tagging* is a form of tagging in which not individual items but groups of items are associated with a tag. For instance, a slide show consisting of photographs of tourist sites could constitute a multi-object and would be tagged with a single place concept such as "Southern France". The rest of the paper explores the semantics of multi-object tagging and is organized as follows.

First, we discuss proposals that have been made to formalize single-object tags (section 2). We show that in many cases a judgment about typicality can be recovered from multi-object tags and analyze consequences for the semantics of place concepts (section 3). We connect multi-object tagging to the problem of multi-object recommendation, that is, the task to assist users in selecting a multi-object based on his or her previous choices for such composite objects. Using a recommender system for composite images (patchwork postcards) as an example, we illustrate how to obtain data about multi-object tagging. Finally, we argue that for multi-object tagging, a key computational concept of recommending, namely user similarity, should be understood as a measure that indicates the extent to which two users share a common conceptualization. We show that as a consequence, the community of taggers is not going to be divided by crisp semantic boundaries (section 4). The paper concludes with a discussion of related work (section 5).

2   The Meaning of Tags

The Geograph project that was mentioned in the introduction provides a good example for the standard approach to folksonomies: *single-object tagging*. Information about the tags and their frequency can be retrieved directly from the project's web interface. In the beginning of May 2007, the folksonomy consisted of 2784 categories. It is revealing to sort them in order of decreasing frequency of use. The resulting list starts with the most popular tags, "Church", "Farmland", "Farm", and ends with tags that have been used only once: "Windmill stump", "luminous object in space (Sun)", "Penstock". It is known that folksonomies tend to show an exponential tag frequency distribution (Guy and Tonkin, 2006). In other words, tag frequency follows a power law: 36% of the Geograph tags are used only once, a further 24% appear just 2-5 times in the collection of 422.895 images whereas the most frequent tag "Church" is used 17.360 times. Not surprisingly, the Geograph tag set also shows other properties characteristic of folksonomies. A tag comes in different inflections ("house" vs. "houses"). Synonyms ("mansion" vs. "manor house"), homonyms ("bank" vs. "bank"), and hypernyms ("geological feature" vs. "rock outcrop") appear as tags. And last, but not least, some images are not categorized correctly.

Irrespective of these problems, the Geograph folksonomy may be considered a "shared understanding … in a given subject area" as the frequently quoted definition of ontology by Ushold and Gruninger (1996) requires. We could expect, for instance, the tag set to be disjoint from that of medical folksonomies. However, the understanding that the community of Geograph users has agreed upon is rather limited as it ignores all semantic relations between tags. The ontologies of the Semantic Web, in contrast, are formalized by specifying relations between concepts in form of role restrictions in description logic formulas. This means that while folksonomies can be considered ontologies in the sense of the above defintion, they definitely do not qualify as ontologies in the sense of Semantic Web technology.

Some Web 2.0 applications permit *user tagging* which means that not just the original contributor but also other users may categorize a piece of data. User tagged data items generally receive more than one category tag. We argue that it is user tagging with its potential of semantic conflict which makes folksonomies an interesting alternative to the Semantic Web approach to conceptual modeling. Gruber (2005) pointed out that tagging should not be seen as a binary relation *tagging(object, tag)*. Semantic analysis should instead consider more complex relations all of which have at least three arguments: *tagging(object, tag, user)*. In the following, we restrict our attention to Social Web applications capable of recording and evaluating information about user tags corresponding to this ternary tagging relation.

Generally, Social Web applications process tags in order to obtain information about the correct categorization of objects. A typical question asked in that context is: which of the many tags given to *image23.jpg* are the correct ones? We propose to study the converse problem which consists in processing tags to learn about the semantics of the categories. We ask questions such as: which images in the collection have been considered images of historical buildings? This approach amounts to recover from tag data the characterization of a concept by the set of its instances.

However, object categorization and concept characterization are not fully symmetric problems. A single user tag can provide a correct categorization for an object and a few tags may be sufficient to inform us about all categorizations relevant in a given problem domain. This is certainly not true of the *concept characterization problem* because the set of instances is not finite for most relevant concepts. An interesting variant of the concept characterization problem arises if we move from single-object tagging to multi-object tagging.

3 Multi-object Tagging

Probably the most frequent type of aggregated objects that users of Social Web applications encounter are collections of (atomic) objects which result from a selection process in the problem domain. The 50 images that have been chosen to illustrate a mobile travel guide to Paris or the 7 roundtrips that a geographic recommender system suggests making during a Weekend stay in Florence constitute *collections of selected objects*. Such collections are not restricted to the digital world. A bestselling book describes 1,000 places to see before you die[3] and the printed patchwork postcard in Fig. 1 features a selection of 8 images that convey a multi-faceted view of the City of Bamberg.

Collections of selected images frequently come with a tagging that informs about the selection criterion. This type of tag can be interpreted as a caption that applies to all images of the collection. Actually, this is exactly the purpose of the place name caption "Bamberg" on the printed patchwork postcard in Fig. 1. Collections of selected images that come with a caption or are tagged otherwise constitute examples of *multi-object tagging*. Formally, we can still use the ternary tagging relation but with a collection as first argument: *tagging({object1, ... , objectN}, tag, user)*. This generalizes the description of tagging given by Gruber (2005) since single-object tagging can be treated as the special case of a one object collection.

We argue that multi-object tagging shows a semantic characteristic that is missing in single-object tagging, namely that of a selection process which chooses objects that are (1) typical instances of the concept and, at the same time, (2) show the variability of the instances. A patchwork postcard with the caption "London" featuring 8 images of Big Ben would satisfy criterion (1) but not (2). As tagging of multi-images is still rare in today's image sharing platforms and data on user-tagged multi-images are not available, we take a look at printed patchwork postcards to support our argument.

The sample consists of all patchwork postcards that are currently available of Bamberg, Germany. They show the historical center of the town, a listed UNESCO World Heritage Site. What makes this example particularly interesting is the fact that Bamberg has been listed as architectural ensemble and not because of a single building of outstanding value. Since there are many buildings of equal historical interest, the postcards could show a random sample of the city's built heritage with all

[3] Schultz, P. (2003). 1,000 Places To See Before You Die, Workman Publishing Company.

Fig. 1. Printed patchwork postcard and its digital counterpart generated by the Tripost system

historical buildings having about the same probability to appear on a postcard. There would be no typicality effect (criterion 1) in that case. On the other hand, if some historical site appears on almost all patchwork postcards we would assume that the image selection process is guided by some notion of typicality. Similarly, a variability effect (criterion 2) would be present if there are many images appearing on only very few, if any, other postcards. 15 patchwork postcards from Bamberg have been collected with a total of 108 images. However, only 24 different historical sites are depicted. Their frequency distribution is shown in Fig. 2 which reads as follows: half of the sites (12 of 24) appear on 1 to 3 postcards, a quarter (6 of 24) on 4 to 6 postcards, ..., and just 1 site appears on 13 to 15 postcards. We find the frequency drop off that is characteristic of the power law, and consequently a first piece of evidence for both, the typicality and the variability of the image selection process.

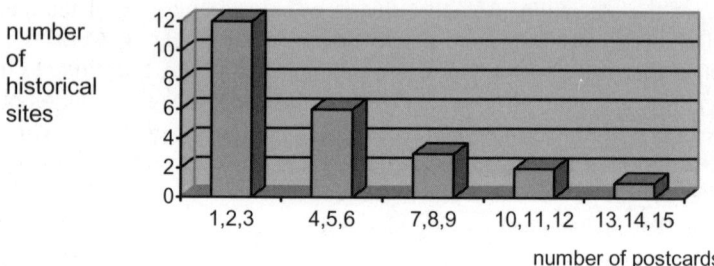

Fig. 2. Typicality of associations between a place name and a geographic object

We have seen that data about multi-object tagging can be useful to analyze the different meanings of a place concept. But from what sources can we obtain such data? We suggest to collect multi-object taggings as a side-activity of running a recommender service for multi-objects as described in the next section.

4 Collaborative Semantics

Classical recommender systems solve an information filtering task (Burke, 2002). They suggest data sets, typically text documents describing some product or service, that seem likely to be relevant to the user based upon previous choices that this user has made. A widely known recommending approach is the item-to-item similarity approach used by Amazon to recommend books based on books one has bought before (Linden et al., 2003). It is easy to verify that the recommendations are based on item-to-item similarity rather than user-to-user similarity. With a travel guide to Bali in the shopping cart, the recommender suggest buying further travel guides to Bali and not, for that matter, books on Hinduism or surfing. A *geographic recommender system* recommends items from a library of georeferenced objects. In multi-object recommendation, collections of selected items are suggested to the user. A geographic multi-object recommender could make suggestions for a list of cities to visit or a slide show of images illustrating a certain place.

The latter use case is that of the Tripod Geographic Recommender. This recommender is currently implemented as part of the Tripod project which is concerned with the automatic generation of captions for georeferenced images including multi-images[4]. Fig. 3 shows the screenshot of the e-mail postcard application Tripost that takes advantage of the Tripod recommending service to automatically suggest patchwork postcards based on postcards that the user has previously built manually by selecting images from a photo sharing collection.

Data about the tagging of multi-images with place name tags is easily obtained from Tripost submissions to the recommender. Users select a fixed number of images to be shown on the postcard and provide a place name caption. You may use Tripost to generate e-mail postcards of your home town – never one like the other – without that you have to scan the collection for interesting new photographs. Also, you may want Tripost to send you a postcard from a location unknown to you which you specify by a place name. If you submitted, for instance postcards of Bamberg similar to the printed postcard shown in Fig. 1 and asked for a postcard from Florence you will

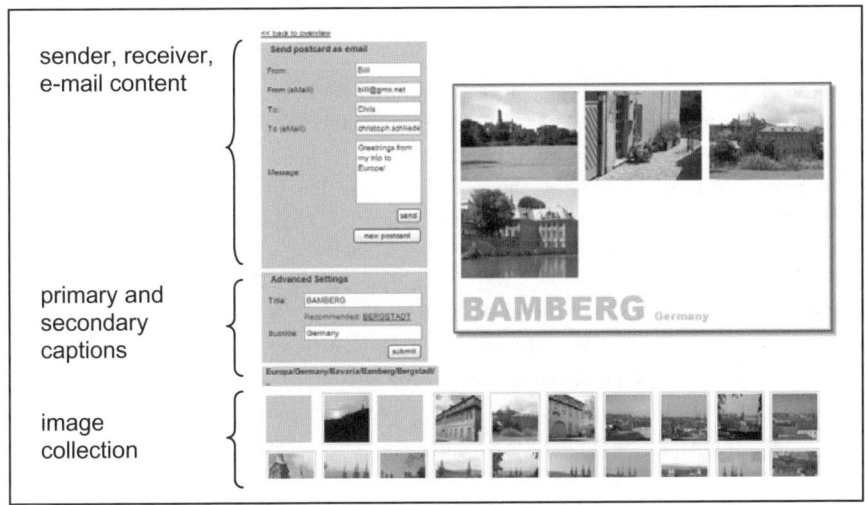

Fig. 3. Screenshot of TRIPOST, a service based on the Tripod Geographic Recommender

---

[4] Tripod – Automating caption creation: http://tripod.shef.ac.uk

probably get a postcard showing built heritage sites. Only probably, because this depends on what selections other users have made for images of Florence. Obviously, user-to-user similarity is more of a help in this case than item-to-item silimarity.

GroupLens is generally considered the first system that used a collaborative filtering approach for generating recommendations based on user-to-user similarity (Resnick et al., 1994). The authors of that landmark system characterize the basic assumption underlying their approach as "the heuristics that people who agreed in the past, are likely to agree again". We transfer that idea from the temporal to the spatial domain – GroupLens made recommendations for news based on ratings of old news. Our spatial heuristic consists in assuming that people who agreed on the qualities of some places, are likely to agree on the qualities of other places too.

We consider a very simple example that helps to explain how a recommender based on user-to-user similarity can help to model the semantics of place concepts. The example data is summarized in Tab. 1. There are 4 place concepts (Antwerpen, Bamberg, Cardiff, Dublin) each of which comes with a collection of exactly 6 images numbered 1 to 6. Furthermore, 6 users (Anna, ..., Franz) have submitted patchwork postcards, each showing 3 images, to the recommender with an appropriate place caption. The table entry Antwerpen-Anna, for instance, contains the information that user Anna has chosen image 1, 3, and 6 from the Antwerpen collection to represent that city:

tagging({imageA1.jpg,imageA3.jpg,imageA6.jpg}, images of Antwerpen, Anna).

The standard data table of a recommender is indexed by users (columns) and objects (rows). Since our goal consists in modeling the semantics of place concepts, we need to rewrite the standard table and index rows by concepts. The objects, that is, the images, now appear as table entries. User-to-user similarity compares users based on their previous selections. In our case, the similarity between two users, say, Bill and Emma, is based on the entries of Tab. 1.

Different similarity measures might be applied such as the variant of the Tversky measure proposed by Rodriguez and Egenhofer (2004) for the comparision of geospatial feature sets. To keep the example as simple as possible, we use the Tanimoto measure for similarity. Let A be the set of images that user a associates with place concept *Place* and B the set of images that user b associates with *Place*. We measure the similarity of the semantics of *Place* specified by A and the semantics of Place specified by B through the expression $sim(A,B) = |A \cap B| / |A \cup B|$. In other words, the more images two patchwork postcards with the same place name caption share, the more similar is the place name semantics that they convey: $sim(\{2,4,6\}, \{4,5,6\} = 2/3 \approx 0.66$. Again for the sake of simplicity, similarity values are computed for entries that are non-void for both users and averaged: $sim(Anna, Bill) = 1/2 \cdot (sim(\{1,2,6\},\{2,4,6\}) + sim(\{2,4,5\},\{2,5,6\})) = 1/2 \cdot (2/3 + 2/3) \approx 0.66$ (Tab. 2).

User-to-user similarity measures how similar two users are in their choices of images representing place concepts. We interpret this as a measure for a shared understanding between these users. This interpretation has interesting consequences for the notion of information community. Which are the users that share the understanding of

Table 1. Results of multi-object tagging by users of a geographic recommender

	Anna	Bill	Clio	Don	Emma	Franz
Antwerpen	1, 3, 6		2, 4, 5	1, 4, 6		2, 3, 4
Bamberg		2, 3, 5	1, 4, 6	2, 4, 6	1, 4, 5	1, 2, 3
Cardiff	1, 2, 6	2, 4, 6	3, 5, 6	1, 3, 6	4, 5, 6	1, 2, 6
Dublin	2, 4, 5	2, 5, 6	1, 3, 5		1, 2, 3	2, 5, 6

Table 2. User-to-user similarity resulting from multi-object tagging

	Anna	Bill	Clio	Don	Emma	Franz
Anna		.66	.22	.66	.33	.66
Bill	.66		.22	.33	.44	.77
Clio	.22	.22		.55	.66	.42
Don	.66	.33	.55		.33	.55
Emma	.33	.44	.66	.33		.33
Franz	.66	.77	.42	.55	.33	

place concepts of user Bill? There is no clear-cut answer because there are degrees of similarity. Most similar to Bill is Franz, then comes Anna, then Emma etc. An obvious way to define the "information community" of a user u consists in taking the set of the *k-nearest neighbors of u*. The boundaries of these communities are no longer crisp as k may take different values. Two communities that are disjoint when k corresponds to 5 % of the users may well overlap when k is chosen to correspond to 20 % of the users.

Another central property of information communities, as they are known from the Semantic Web, is lost. For "communities" induced by tagging, user u may be a member of the k-community of v, without that v is a member of the k-community of u. This can be seen from the 3-nearest neighbor communities in our example that are depicted in Fig. 4. Anna is a member of the community of Franz and vice versa, however, Franz is a member of the community of Clio without that Clio is a member of Franz' community. It is the breaking down of the symmetry of mutual membership that leads us to question whether the notion of information community really makes sense in the context of tagging. The structuring of the users according to similarity seems rather to precede the emergence of information communities in the sense of the Semantic Web.

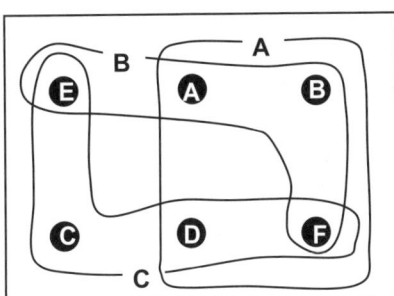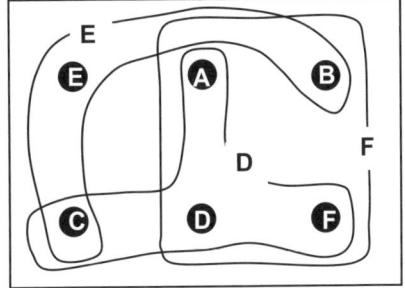

Fig. 4. Communities for $k=3$ of Anna, Bill, and Clio (left); Don, Emma, and Franz (right)

If this is true then we should think about a novel term to designate groups of users characterized by similarity.

5 Related Work and Discussion

We have presented an approach to collaborative semantics that is based on multi-object tagging and defines "information communities" for folksonomies in terms of user-to-user similarity measurement. We illustrated the approach with an analysis of the semantics of place concepts that are used as tags to describe groups of images.

Our analysis shares the ternary interpretation of tagging with Gruber (2005) who points out that folksonomies and ontologies should not be seen in opposition because "there is increasing value in applying Semantic Web technologies to the Social Web". However, we generalized the interpretation to include multi-object tagging. To our knowledge, the main idea presented in this paper, namely to analyze geospatial information communities by exploiting data about user-to-user similarity from a geographic recommender, has not been explored before.

Recently, Bishr and Kuhn (2007) have joined Gruber (2005) in rejecting the binary interpretation of tagging by providing arguments from a GI processing perspective. They suggest the use of typed tags (what-tags, where-tags, etc.) to enhance the semantic structure of folksonomies. With a different emphasis, Matyas (2007) has used description logics to implement data quality checks for the collaborative acquisition of geospatial data. While we agree with the main conclusion of these authors – sophisticated Social Web applications will have to make use of the knowledge representation technologies developed for the Semantic Web – we also see that the two approaches to semantics differ radically on the technological, the formal, and probably also on the philosophical level of analysis.

In conceptual modeling, folksonomies are often perceived as the poor men's ontology. Community-based mapping projects seem to provide evidence for that view. Their spatial data accuracy and the quality of the specification of their feature semantics are often inferior to that of commercial products. However, by adopting this perspective, one would miss the real strength of folksonomies which lies in their ability to handle use cases where information communities have not yet emerged. Web 2.0 applications illustrate that data sharing happens long before it is possible to identify "a group of people who share a common geospatial feature data dictionary (including definitions of feature relationships)", as Gould and Hecht (2001) formulated in their definition of information community. We might therefore think of folksonomies as a way of modeling "pre-community semantics". In that perspective, the collaborative metadata acquisition methods of the Social Web and the ontological engineering methods of the Semantic Web appear truly complementary.

Acknowledgements

The author gratefully acknowledges support by the European Commission who funded parts of this research within the Tripod project under contract number

IST-FP6-045335. He also whishes to thank Christian Matyas who implemented the Tripost service for interesting discussions about the Tripod Geographic Recommender.

References

Bishr, M., Kuhn, W.: Geospatial Information Bottom-Up: A Matter of Trust and Semantics. In: Proc. AGILE Int. Conf. on Geographic Information Science, Aalborg University (2007)

Burke, R.: Hybrid recommender systems: Survey and experiments. User Modeling and User-Adapted Interaction 12, 331–370 (2002)

Fensel, D., Wahlster, W., Lieberman, H., Hendler, J.: Spinning the Semantic Web: Bringing the World Wide Web to Its Full Potential. MIT Press, Cambridge (2003)

Hecht, G.: A Framework for Geospatial and Statistical Information Integration. Open GIS Consortium White Paper (2001)

Guy, M., Tonkin, E.: Folksonomies: Tidying up Tags? In: D-Lib Magazine, vol. 12(1) (2006), http://www.dlib.org/dlib/january06/guy/01guy.html

Gruber, T.: Ontology of Folksonomy: A Mash-up of Apples and Oranges. In: MTSR 2005. Keynote at the First Online Conference on Metadata and Semantics Research (2005), http://tomgruber.org/writing/ontology-of-folksonomy.htm

Linden, G., Smith, B., York, J.: Amazon.com Recommendations: Item-to-Item Collaborative Filtering. IEEE Internet Computing, 76–80 (February 2003)

Matyas, S.: Collaborative Spatial Data Acquisition. In: Proc. AGILE Int. Conf. on Geographic Information Science, Aalborg University (2007)

Miller, P.: Web 2.0: building the new library, Ariadne 45 (2005), published on http://www.ariadne.ac.uk/issue45/miller/intro.html

Morris, S., Morris, A., Barnard, K.: Digital trail libraries. In: JCDL 2004. Proc. Joint Conference on Digital Libraries, pp. 63–71. ACM Press, New York (2004)

O'Reilly, T.: What Is Web 2.0: Design Patterns and Business Models for the Next Generation of Software (2005), published by O'Reilly Media on http://www.oreillynet.com/pub/a/oreilly/tim/news/2005/09/30/what-is-web-20.html

Resnick, P., Iacovou, N., Suchak, M., Bergstrom, P., Riedl, J.: GroupLens: An Open Architecture for collaborative Filtering of News. In: Proc. Conf. Computer-supported cooperative work, pp. 175–186. IEEE Press, Los Alamitos (1994)

Rodríguez, A., Egenhofer, M.: Comparing geospatial entity classes: an asymmetric and context-dependent similarity measure. International Journal of Geographical Information Science 18(3), 229–256 (2004)

Uschold, M., Gruninger, M.: Ontologies: principles, methods, and applications. Knowledge Engineering Review 11, 93–155 (1996)

Dynamically Traveling Web Service Clustering Based on Spatial and Temporal Aspects

Yang Hao, Chen Junliang, Meng Xiangwu, and Qiu Bingyu

State Key laboratory of Networking and Switching Technology
Beijing University of Posts & Telecommunications,Beijing 100876, China
fifayh@gmail.com, chjl@bupt.edu.cn, mengxw@bupt.edu.cn, bingyuqiu@gmail.com

Abstract. Traditional web service orchestration is always implemented by business process work flow or I/O matching, just taking What services provide and How to provide. But this method will invoke some problems including too many composite results, inefficiency, and useless service composition without taking account of When and Where services are provided. Based on divide-and-conquer rule, this paper provides a clustering algorithm based on spatial & temporal aspects which are used to describe spatial and temporal attributes of services, so When and Where are provided for service composition. After composite services cluster by cluster, the whole complex service is created and is sure to satisfy the time and space constraints. Testing cases illustrate that complex traveling services can be created and is optimized to meet users' requests.

Keywords: Service Orchestration, Spatio & Temporal Aspects, Clustering algorithm.

1 Introduction

The W3C definition of Web services as "··· a software system designed to support intraoperative machine-to machine interaction over a network. It has an interface described in a machine-processable format(specifically WSDL). Other systems interact with the Web service in a manner prescribed by its description using SOAP messages, typically conveyed using HTTP with an XML serialization in conjunction with other Web related standards".

The encoding and protocol stack typically used for creating and publishing Web services including the following technologies: SOAP, WSDL and UDDI. SOAP is a protocol defining the exchange of messages containing Web service requests and response. WSDL is the standard means for expressing Web service description. UDDI is the directory services schema commonly used to register and discover Web services.

A Web service should be used when an application builder wishes to expose some reactive operation expressible as a programmatic function with or without parameters that may or may not return a response. This is essentially a remote method invocation using an XML encoded message over HTTP.

Two languages for web service composition have emerged: (1) Business Process Execution Language(WS-BPEL)[1], developed by BEA, IBM, Microsoft

J.-L. Hainaut et al. (Eds.): ER Workshops 2007, LNCS 4802, pp. 348–357, 2007.

and Siebel, was submitted to the WS-BPEL Technical Committee at OASIS,(2) Choreography Description Language(WS-CDL)[2], developed by the Web Service Choreography Working Group at W3C. The goal of these languages is to glue Web Services together in a process-oriented way[3]. WS-BPEL and WS-CDL are both real-world Business Process Management System[4], and they address how organizations can identify, model, develop, deploy, and manage their business processes, consisting of a set of logically related tasks. As the correct business rules produce a business outcome, "order-to-cash" is an example of a business process.

However, complex services composition based on WS-BPEL and WS-CDL is only answer "What Web services provide" and "How Web services are provided", and is not enough without time and space constraints. For example, "I want to get a pizza and then go to the cinema", just based on process logic, the web services logic is (1) find a pizza store (2) order the pizza (3) get the pizza and pay for it (4)find a cinema (5) buy a ticket and enjoy the film. But when it is implemented by tradition web service composition, many bugs maybe emerge. Firstly, if there are several pizza stores and cinemas, there will be too many results and time costing for calculation maybe unacceptable. Secondly, it is possible that the chosen pizza store is in the east of city, while the cinema is in the west, so this service composition is unacceptable too. Last, if the chosen pizza store is open from 11 A.M to 5 P.M, while the film starting time is 10 A.M, the service composition is also failed, even if it satisfies the business logic. The causes of these errors are traditional web services orchestration WS-BPEL and WS-CDL only take account of what SP(abbreviation for Service Provider) provide and how they provide, ignoring where and when. But where and when issues are also important in complex services orchestration system.

This paper provides a solution, which implements traditional web service with an interface with space and time aspects, which is used to answer "When Web services are provided" and "Where Web services are provided". Then clustering the services into several clusters according to these aspects. After orchestrating services cluster by cluster, one complex service is created. The rest of the paper is organized as follows. In the next section, we describe an object-oriented spatial-temporal data model. Section3 outlines how to choose the cluster size and vector space for clustering, then analyze some clustering algorithms. We discuss in detail traveling web services orchestration taking time and space into account in section 4. Section 5 describes our prototype and examines the results comparing with results from traditional service orchestration. The final section makes a conclusion and discusses future work.

2   Spatio and Temporal Aspects

2.1   Creating Spatio and Temporal Aspects Based on AOP

Traditional solution of spatial-temporal data model is using database to describe the spatial data and provides a sequence of time stamps to track these data over time[5].This method is mainly used in geo-raster. In this model, all data are labeled with a validated time stamp, and with the time stamp sequence, a series of spatial data are saved. This method is easy to be implemented and query operations such as insert, search, update spatial data at accurate time are also easy to do. But there are complete copies of spatial date for every time slice, the redundancy in the time axis is huge. In addition time-costing is large if we want to get some spatial object status between time A and time B. And time interval must be some constant, two data sets which are described with not equal time interval will not be merged. For example, preserving the newspaper from each Monday to Friday can't go along with ordering The Times magazine from January to September.

Aspect-Oriented programming is the modularization of crosscutting concerns. After defining common functionality in one place, you can declaratively define how and where this functionality is applied without having to modify the class to which you are applying the new features. Cross-cutting concerns now be modularized into special objects called aspects.

This has two benefits. For the first, the logic for each concern is now in one place, as opposed to being scattered all over the code base. For another, our service modules are now cleaner since they contain code for their core functionality and secondary concerns have been moved to aspects.

Distributed coordinated filters can be defined as aspects, then each object is provided with an composition filter approach. Related services can only be executed when service description satisfies both service peers filtering policies.

In our applications, we extend traditional cross-cutting, which are always including "Security" and "Transaction", with "Spatial Issues" and "Temporal Issues", in Figure1.

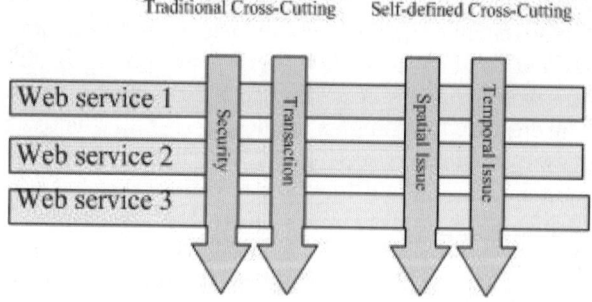

Fig. 1. Aop Model for Traveling Service Composition

2.2   Three-Tiered Spatio-temporal Data Model for Services

To describe where to provide the Web service and when to provide the service, beyond what is the service, we propose three-tiered service class diagram taking space, time and service attribute into account in figure 2.

Time class is a composite class of weekday, year, month, date, and hour in figure3. And end year is used to distinguish dead services from active ones. For

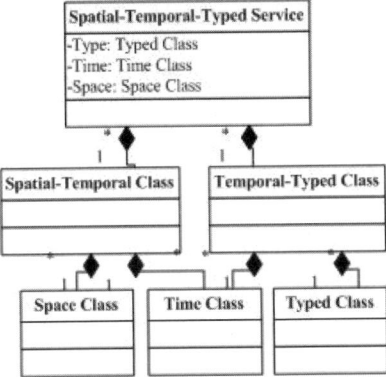

Fig. 2. Three-tiered Spatio-temporal Data Model for Services

example, service A, which was available from Monday to Friday every week in 2000-2002, is described with a time class: set(year(2000-2002), weekday(Monday, Tuesday, Wednesday, Thursday, Friday)); while service B is available everyday from 1980, then it is pictured as time class : set(year(1980-)).

Space class is based on traditional Node-Arc-Polygon, and also can be 2 dimensions, 3 dimensions, For simplicity in service description, space is usually described in 2 dimensions.

Attribute class is based on the service content classification, such as restaurant, hotel, hospital, and so on.

Composite pattern[10] for spatio-temporal class is used to resolve the object's address and time relation without considering its content. Another composite pattern for temporal-typed class is used to describe attributed service with its time character.

Finally spatio-temporal-typed service is the complex service which is composed of space class, time class, attribute class, and then we can implement complex service logic to create complex service composition. Whole class diagram is shown in Figure3. Transport service is spatio-temporal arc service, hotel service is spatio-temporal node service, and traveling service is spatio-temporal-typed service, And this object-oriented service orchestration logic is based on clustering on spatio-temporal data model in next section.

3   Clustering on Spatio-temporal Data Model

From the preceding section, services are object-oriented with time and space attributes. Then how to composite so many services with space and time logic To decrease the algorithm complexity, spatial clustering is used to divide services into several sets, and naturally orchestration with subsets is easy to conquer based on time sequence.

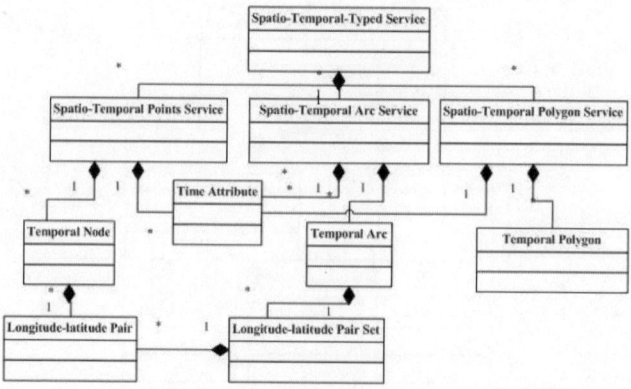

Fig. 3. Spatio-temporal-typed class diagram

3.1   Getting Cluster Size from Temporal Metric

Temporal metrics are closely related with action. To get appropriate metric for traveling service, cycled approach is used. The process is like below:

Step 1. Define an initial temporal metric M. For traveling service composition, day can be the choice. M = 1day;
Step 2. Get cluster size S, which equals Total time T_Σ divided by M. $S = T_\Sigma/M$;
Step 3. Cluster services and get the QoS(Quality of Service);
Step 4. If satisfied, the program is over, else M = M / 2, goto Step 2;

3.2   Clustering Service Objects Based on Spatial Relations

Clustering can be considered the most important unsupervised learning problem; so, as every other problem of this kind, it deals with finding a structure in a collection of unlabeled data. A loose definition of clustering could be "The process of organizing objects into groups whose members are similar in some way". A cluster is therefore a collection of objects which are "similar" between them and are "dissimilar" to the objects belonging to other clusters.

Mature clustering algorithms includes K-means[6], Fuzzy C-means[7], Hierarchical[8], Mixture of Gaussians[9]. K-means is exclusive and algorithm complexity $O(n)$, Fuzzy C-means is overlapping clustering algorithm. Hierarchical is also exclusive with algorithm complexity is $O(n^2)$, and lastly Mixture of Gaussians is based on probability. For simplicity and efficiency, we choose K-means as our clustering algorithm.

K-means is one of the simplest unsupervised learning algorithms that solve the well known clustering problem. The procedure follows a simple and easy way to classify a given data set through a certain number of clusters (assume k clusters) fixed a priori. The main idea is to define k centroids, one for each cluster. These centroids shoud be placed in a cunning way because of different location causes different result. The next step is to take each point belonging to a given data set and associate it to the nearest centroid. When no point is pending, the first step is completed and an early groupage is done. At this point we need to re-calculate k new centroids as barycenters of the clusters resulting from the previous step. After we have these k new centroids, a new binding has to be done between the same data set points and the nearest new centroid. A loop has been generated. As a result of this loop we may notice that the k centroids change their location step by step until no more changes are done. In other words centroids do not move any more.

4 Traveling Service Composition Based on Spatio-temporal Clustering

4.1 EBNF for User Request

For a traveling service request, we defined EBNF of user request below:

- Source Address?
- Destination Address+
- Transport($Plane \mid Train \mid Bus$)
- Total traveling time+
- Cost?
- (Place of interests?)+
- Hotel Level? Favorite Hotel?
- Restaurant type ? Favorite Restaurant?

4.2 Web Service Orchestration

Traveling service creation sequence is pictured in Figure4. And orchestration steps are:

Step 1. send source address, destination address, transport type(bus, train or plain) to book the ticket.

Step 2. send total traveling time and destination address to spatio-temporal clustering system, including time metric, if not existing one, use the default value as one day. Get all places of interests of destination address and calculate the cluster size by dividing total time by time metric. Then cluster places of interests based on spatial data and send them back to the orchestration service.

Step 3. Choose the first batch of places of interests by rank, and carefully make sure the summary of time less than maximum traveling time in one time metric.

Step 4. If hotel location chosen differs every day, calculate the centroids of step3's cluster and find the hotel nearby that point, else choose the hotel as a stable point for the first time, and skip over for next clusters.

Step 5. Do as step4, book the restaurant nearby the individual cluster centroid, which type meet user's request.

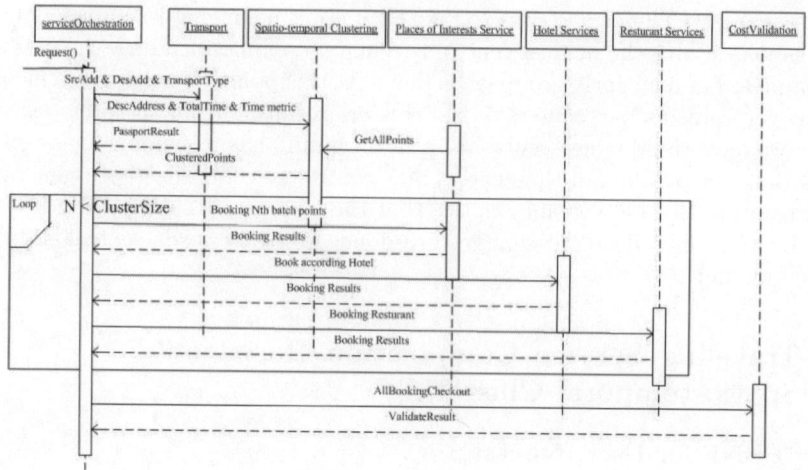

Fig. 4. Service Orchestration Sequence Diagram

Step 6. Go to step 3 for remaining clusters.
Step 7. Validate the cost, and adjust scene number. If user's cost is below the
          least of service composition, throw an exception to user to report the
          cost is not enough for the traveling.

4.3 Service Orchestration Based on Spatial Issues, Temporal Issues, Security, and Transaction Aspects

Filters in our platform are divided individually as outgoing filter and incoming
filter. Before giving the two filters definition, we define two command in our
nodes. Vertical Command is issued when an event is invoking services in local
node. Horizontal Commands are carried out if some event in node m invoking
some services in other node, and then each service that requires node-to-node
interactions must provide a Slice to serve Horizontal Commands.

When a Vertical Command is issued, each service filter processes the command
and does something, and this sequence of filter is named as Outgoing Filter. Then
when service request is coming from another node, each service filters in local
node are invoked sequently and also do something, the sequence of service filter
is named as incoming Filter.

An example about Vertical Command, Horizontal Command, outgoing filter
aspects and incoming filter aspects are described in Figure 5.

5 Implementation in PingGu Traveling System

5.1 PingGu Traveling System Overview

PingGu traveling system is a traveling project based on service oriented architecture, There are fifteen types services, nearly 150 services, including interest

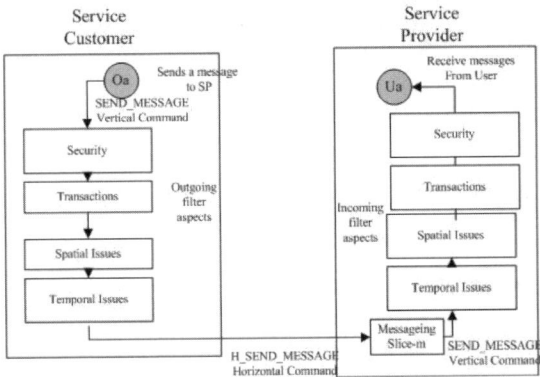

Fig. 5. Incoming Filter and Outgoing Filter based on Spatial Issues, Temporal Issues, Security and Transactions

points, hotels, restaurants, passport agencies, hospital, petrol station, and so on. All service are implemented by web service described by WSDL[WSDL]. For service orchestration, we defined a mapping between service objects and WSDLs, and all objects implement the spatio-temporal-typed interface for providing longitude, latitude, existing time, available time and service attribute.

5.2   Function Measurements

User request: From Shanghai to Pinggu, 5 days, budget costs below 4000RMB. From the request, we can get input parameters as follows:

$$srcAddress = Shanghai$$
$$destAddress\ = Pinggu$$
$$Totaltime\ = 5(days)$$
$$costMaximum = 4000RMB$$

5.3   Traveling Orchestration Results and Analysis

Clustering results are pictured in figure6 and Service Composition results are pictured in figure7. The traveling line created by our service is described in table1.

5.4   Comparing with Traditional BPEL Service Composition Results

For algorithm complexity, divide-and-conquer can decrease algorithm complexity from $O(n)$ to $O(\log(n))$[11]. Services composition based on clustering has more efficiency than traditional service composition algorithm.

Taking account of time aspect, condition such as service composition disordering or overrunning will not occur .

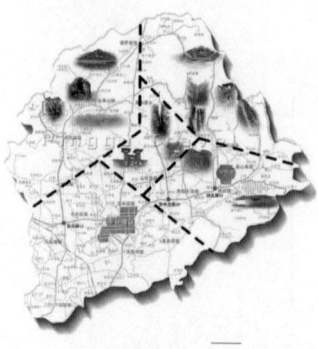   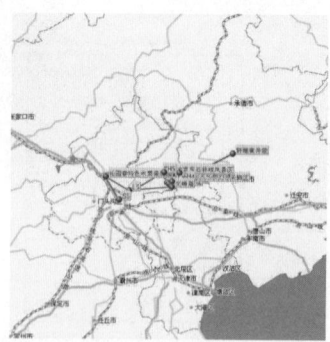

Fig. 6. Traveling Arrangement
for Five Days

Fig. 7. Traveling Arrangement
for Five Days

Table 1. Traveling Line Created by Service Composition

	Beauty Spots	Restaurant	Hotel
1st day	JingHai Lake JingDong Gorge	YuYang Hotel	
2nd day	JingDong Cave HuDongShui Park	DongXiaoXinYue	
3rd day	FeiLongGu ShiLinXia Beauty Spot	BiHai Manor	PanFeng Hotel
4th day	YaHuanShan Park TaoJinGu Beauty Spot	ZiYueWenTao	
5th day	ShangZhai Meseum	NanHua Restaurant	

For space issues, through clustering places with latitude and longitude, traveling composition, which includes place A in east of city and then followed by place B in west of city, will not exist.

6   Conclusion and Future Work

Traveling service orchestration based on spatio-temporal data model and clustering, successfully resolve service composition, not only answer questions about what and how, but also where and when, which is especially important for some complex service composition.

For future works, two directions exist. For the first, many complex service domains are taking into account and how to build a common middleware to erase the complexity of cross aspects. Another, how to provide an easy utility for user to change the service composition, change the important rank of time, space, costs, and change the ratio of restaurant, scenic places, and hotels.

Acknowledgment

This project is supported by the National Natural Science Foundation of China under Grant No. 60432010 and National 973 project(Universal trustworthy Network and pervasive Services) under Grant No 2007CB307100.

References

1. Khalaf, R., Keller, A., Leymann, F.: Business processes for web services: principles and applications. IBM Systems Journal 45(2), 425–446 (2006), http://www.research.ibm.com/journal/sj/452/khalaf.html
2. WS-CDL, Web Services Choreography Description Language Version 1.0, [online], http://www.w3.org/TR/2004/WD-ws-cdl-10-20040427/
3. Newcomer, E., Lomow, G.: Understanding SOA with Web Services, pp. 280–282. Addison Wesley, Reading (2004)
4. Majed, M.: BPMS, Business process management-major challenges, Business Process Management, 411–412 (2002)
5. Rojas-Vega, E., Kemp, Z.: An Object-Oriented Data Model for Spatio-Temporal Data[A]. In: Proc of the Ninth Annual Symposium on Geographic Information Systems[C], Vancouver, Canada (1995)
6. MacQueen, J.B.: Some Methods for classification and Analysis of Multivariate Observations. In: Proceedings of 5-th Berkeley Symposium on Mathematical Statistics and Probability, Berkeley, vol. 1, pp. 281–297. University of California Press (1996)
7. Bezdek, J.C.: Pattern Recognition with Fuzzy Objective Function Algoritms. Plenum Press, New York (1981)
8. Johnson, S.C.: Hierarchical Clustering Schemes. Psychometrika 2, 241–254 (1976)
9. Dempster, A.P., Laird, N.M., Rubin, D.B.: Maximum Likelihood from Incomplete Data via theEM algorithm. Journal of the Royal Statistical Society, Series B 39(1), 1–38 (1977)
10. Gamma, E., Helm, R., Johnson, R., Vlissides, J.: Design Patterns, Elements of Reusable Object-Oriented Software, pp. 163–174. Pearson Education (1995)
11. Corrmen, T.H: Introduction To Algorithms, pp. 257–355. The Mit Press, Cambridge (2002)

A Graph-Oriented Model and Query Language for Events

Miguel Romero[1] and M. Andrea Rodríguez[2]

[1] Universidad del Bío-Bío, Chile
mromero@ubiobio.cl
[2] Universidad de Concepción, Chile
andrea@udec.cl

Abstract. Different applications can be modeled by sets of interrelated events, such as, sequence of natural disaster events, events in the urban development of a city, and innovation process trajectory in a manufacturing industry. This work is motivated by limitations of current database systems for modeling and querying about interrelated events. It proposes a graph-oriented model and query language for events that extend current graph data models by incorporating temporal and spatial settings of events and different levels of granularity for event representation. The model and language are implemented in an extensible database system.

1 Introduction

Events are present in diverse areas of engineering and science. Examples are simulation, monitoring, business process, knowledge representation, environmental modeling, and active database systems [19, 11]. Two complementary views of events in information systems are: (1) events are triggering conditions rather than being stored and queried as traditional data, such as events in active databases [19] and in reactive Web applications and services [5], and (2) events are subjects of data representation and query processing, such as events in spatio-temporal databases [18].

In this work, we consider events in the context of spatio-temporal databases, where events establish something that happens at a specific location and time instant or interval, but then it is gone. Considering events as subjects of representation leads us to model events as entities in database systems. Unlike entities in traditional database systems, however, the relationships between events, such as temporal order, spatial connectivity or causality, play an important role for querying events.

Consider as an example a database that handles natural disasters in the world. Among these disasters, the greatest earthquake whose ground motion strength has been measured, recorded, and verified occurred in 1960 in the South of Chile, rating 9.5 on the moment magnitude scale. This earthquake was not a single event, but a set of events including several ground motions, tsunamis, landslides, and massive flows of earthen debris and rock. Figure 1 illustrates some of these events and their interrelations.

J.-L. Hainaut et al. (Eds.): ER Workshops 2007, LNCS 4802, pp. 358–367, 2007.

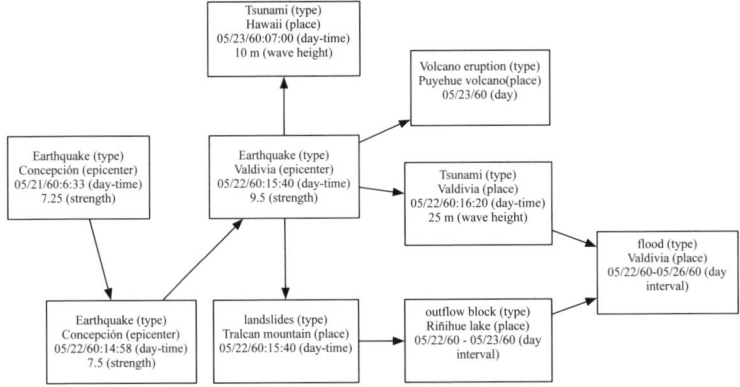

Fig. 1. Events related to the 1960 Chilean Earthquake

In this example, we have several *atomic events* (single events) of different types, and therefore, with different attributes. Note that the temporal setting of events is sometimes expressed by date and time, and other times, by intervals in terms of days. So, we need to handle different levels of granularity in the specification of the temporal setting of events. These events together are part of the *composite event* known as the *1960 Chilean Earthquake*, and are interrelated by causality relations, which could also have some attributes or be of different types, such as necessary and sufficient conditions. Then, a graph-oriented representation of such events lets us obtain the sequence of events and not just the set of events that occurs between two time instants.

In this work, we extend models and query languages for graphs to represent events in two different ways: (1) we explicitly identify nodes in the graphs as events with their particular properties (temporal and spatial settings) and (2) we model events and their interrelations at different levels of granularity. The main contribution of the work is to address the modeling, representation, and query specification of events from a different perspective, where the focus is on querying events rather than changes in objects' states. In this paper we present the graph-based model and language for events and describe briefly their implementation in the extensible database SECONDO [10].

The organization of the paper is as follows. Section 2 discusses the semantics and use of events. Section 3 describes the graph-oriented model for events in algebraic terms, that is, it describes types and operations over these types. Section 1 presents the query language which is followed by its implementation in section 5. Conclusions and future research directions are given in section 6.

2   Events and Related Work

There is no clear consensus in the meaning of the term *event*. Events are also referred as *occurrents*, as opposite to *continuants or objects* [17] [8]. Occurrents happen or occur and are then gone. From a semantic point of view, interesting distinctions are made depending on whether or not an occurrent is initiated and terminated by human or no human agents. In information systems, events are considered as a way that an agent classifies patterns of changes that are useful or relevant for an application [2]. They are used to specify reasons why changes in objects' states occur [16,18]. Depending on the type of events, they can occur at a time instant or last for a time interval. A single event that occurs at a time instant or lasts for a time interval is *atomic*, whereas a *composite or complex* event lasts for a time interval and is defined by an aggregation of events, atomic and composite events [4,19].

In spatio-temporal databases, tree basic concepts have been used to model applications in this domain: objects, events, and processes [7]. They have spatial and temporal locations, also called settings [17], and can be related by means of involvement and participant relations [8].

The different formalizations of events in logic model the reality with the goal of doing some kind of reasoning about changes and time-varying attributes of objects. Examples of such formalizations are Event Calculus [12,15] and Situation Calculus [14]. These formalisms allow us not only knowing the origin of changes, but also omitting those changes that can be reconstructed from the associated events [3,13]. The formalization of events serve as a reference for query languages, but unfortunately for our interest, they have been focused on querying objects' states rather than querying events and events' interrelations.

Some studies in distributed computer systems consider events as messages over networks that allow the interaction between applications or between applications and users [4]. Unlike these studies, we follow a database perspective for modeling and querying events, where events are facts with their own properties, but more importantly, with their interrelations. Although previous works on events have not discussed a specific query language for events in the context of database systems, they have outlined several operators concerning events' relationships [19], mainly addressing temporal, multiple occurrences, and sufficient and necessary conditions between events.

3   A Graph-Oriented Model for Events

In this section we introduce the use of a graph-oriented approach to modeling and querying events. We define an event graph as a directed and acyclic graph of the form $EG = [V, E]$, where V is a set of events, and E is a set of relationships between events such that $uv \in E$ iff $u, v \in V$. For space restrictions we will not include a full algebraic specification of data types.

At an abstract level, the system is composed of a set of *object classes* (of different *types* or *schemas*). Each object class has a set of *objects* with a number of *attributes* with values drawn from certain *domain* or *data types*. We introduce the following signature [9] that describes the types in the system:

$$\rightarrow \text{BASE } int, real, string, bool$$
$$\rightarrow \text{SPATIAL } point, line, polygon, region$$
$$string \rightarrow \text{INSTANT } instant$$
$$string \times BASE \rightarrow \text{LABEL } intLabel, stringLabel, realLabel$$
$$INSTANT \times INSTANT \rightarrow \text{INTERVAL } interval$$
$$INTERVAL \times SPATIAL \rightarrow \text{ATOMIC\_EVENT } atomic\_event$$
$$ATOMIC\_EVENT \times EVENT\_GRAPH \rightarrow \text{COMPOSITE\_EVENT } composite\_event$$
$$LABEL^+ \times (\text{ATOMIC\_EVENT} \cup \text{COMPOSITE\_EVENT}) \rightarrow \text{EVENT } event$$
$$EVENT \times EVENT \rightarrow \text{ATOMIC\_EDGE } atomic\_edge$$
$$COMPOSITE\_EVENT \times COMPOSITE\_EVENT \times$$
$$EDGE^\star \rightarrow \text{COMPOSITE\_EDGE } composite\_edge$$
$$LABEL^+ \times (\text{ATOMIC\_EDGE} \cup \text{COMPOSITE\_EDGE}) \rightarrow \text{EDGE } edge$$
$$EVENT^+ \times EDGE^+ \rightarrow \text{PATH } path$$
$$EVENT^\star \times EDGE^\star \rightarrow \text{EVENT\_GRAPH } event\_graph$$

In this declaration, capital letters represent *kinds, sorts* or set of types, *italics* represents type constructors, and s^+ represents a list of one or more operands of kind s. We have also extended the basic concepts in [9] with an additional constructor for *sets*, which we denoted by $s^\star$ for operands of kind s, whose specification is left out of the paper. Type constructors *event* and *edge* apply to types *atomic_event* or *composite_event*, and *atomic_edge* or *composite_edge*, respectively.

Spatial Type. Type *spatial* represents location and shape. We consider the following basic spatial data types: *points* in $\mathbb{R}^m$ defined as m-tuples of real numbers $[x_1, \ldots, x_m]$, polylines and polygons defined as lists of points $< point, \ldots, point >$, and regions defined as sets of polygons $\{polygon, \ldots, polygon\}$. Formally, $A_{point} \triangleq \mathbb{R}^m \cup \{\bot\}$, $A_{polyline} \triangleq \{(p_1, \ldots, p_n) | p \in \mathbb{R}^m\} \cup \{\bot\}$, and $A_{polygone} \triangleq \{(p_1, \ldots, p_n) | p \in \mathbb{R}^m, p_1 = p_n\} \cup \{\bot\}$. We include an undefined value $\{\bot\}$ to represent cases when the spatial setting is unknown. At this moment, we have only used topological relations between regions and points, where the relation between unknown and any spatial type value is also unknown.

Instant Type. Type *instant* is a temporal point represented by date (yy-mm-dd) and time (hh:mm:ss.s). Instants are considered as strings with their carrier set defined by $A_{instant} \triangleq string \cup \{\bot\}$. The value $\bot$ indicates that the instant is unknown. In addition, we can define an instant based on a day-based granularity, in which case, time is set with the default value 00:00:00.0. Likewise, a month-based granularity defines an instant where the day is set to the default value 1. Using a parameter string to create instants allows us to set the time of events posteriorly to their occurrence. The operations over instants are presented in the following table:

Operation signature	Description
$<, \leq, =, \neq, \geq, > : instant \times instant \rightarrow bool$	relational operations
year, month, day, hour, minute : $instant \rightarrow int$	return part of an instant representation
second : $instant \rightarrow real$	return the seconds of an instant representation

Label Type. Type *label* represents a tuple with the name and value of an attribute. The carrier set of *label* is $A_{label} \triangleq \{[s, v] | s \in A_{string}, v \in A_\alpha\}$, with A_α being the carrier set of a base type. Operations on labels return their values.

Interval Type. Type *interval* represents a range limited by two time instants (starting and ending time instants) and on which a total order exists. The carrier set of *interval* is $A_{interval} \triangleq \{[x,y]|x,y \in A_{instant}, x \leq y \vee x = \bot \vee y = \bot\}$. An interval $[x, \bot]$ represents an interval that starts at x and has an unknown ending instant. The temporal relation involving an interval with an unknown bounding instant is also unknown. This definition could be further extended by handling events with open time intervals, such as the case of an event that has not finished yet. We leave this extension as future work and concentrate here on historical events. The operations over intervals are presented in the following table:

Operation signature	Description
$\tau : interval \times interval \rightarrow bool$	check a temporal relation between intervals, with τ being one of thirteen temporal relations in [1]
start, end : $interval \rightarrow instant$	return the starting or ending time instant.

Event Type. Type *event* is defined by a list of labels, time interval, and spatial setting in case of an *atomic event*, and by an atomic event plus an event graph in case of a *composite event*. The carrier set of *event(atomic_event)* is $A_{event(atomic\_event)} \triangleq \{[l^+, t, s]|l \in A_{label}, t \in A_{interval}, s \in A_{spatial}\} \cup \bot$ and the carrier set of *event(composite_event)* is $A_{event(composite\_event)} \triangleq \{[ae, eg]|ae \in A_{event(atomic\_event)}, eg \in A_{event\_graph}\} \cup \bot$. By default, events have a unique identifier in the system, which is assigned automatically to new event. In the specification, this identifier is one of the labels of the type event. The operations on events are presented in the following table:

Operation signature	Description
id : $event \rightarrow real$	get id of an event
get_interval : $event \rightarrow interval$	get interval of an event
is_atomic : $event \rightarrow bool$	check whether or not an event is atomic
get_graph : $event \rightarrow event\_graph$	get the event_graph that composes an event
str_label_value : $event \times string \rightarrow string$	get a string label of an event
int_label_value : $event \times string \rightarrow int$	get an int label of an event
real_label_value : $event \times string \rightarrow real$	get a real label of an event
is_in : $event \times event \rightarrow bool$	check if the events are related by an edge, with the first event as source and second event as target
is_out : $event \times event \rightarrow bool$	check if the events are related by an edge, with the first event as target and second event as source

Edge Type. Type *edge* defines the relationship between atomic or composite events, distinguishing between *atomic* or *composite* edges. An *atomic edge* is defined by a list of labels and two events. A *composite edge* is defined by a list of labels, two composite events and a set of edges. Thus, this is a recursive definition of edges. The carrier set of *edge(atomic_edge)* is $A_{edge(atomic\_edge)} \triangleq \{(l^+, e_1, e_2)|l \in A_{label}, e_1, e_2 \in A_{event}\} \cup \bot$ and of *edge(composite_edge)* is $A_{(composite\_edge)} \triangleq \{(l^+, e_1, e_2, e*) |l \in A_{label}, e_1, e_2 \in A_{composite\_event}, e \in (A_{edge(atomic\_edge)} \cup A_{edge(composite\_edge)})\} \cup \bot$. The operations on edges are presented in the following table:

Operation signature	Description
source, target : $edge \rightarrow event$	get the incident events
is_atomic : $edge \rightarrow bool$	check whether or not an edge is atomic
edges : $edge \rightarrow edge^*$	get the set of edges of a composite edge
str_label_value : $edge \times string \rightarrow string$	get a string label of an event
int_label_value : $edge \times string \rightarrow int$	get a int label of an event
real_label_value : $edge \times string \rightarrow real$	get a real label of an event

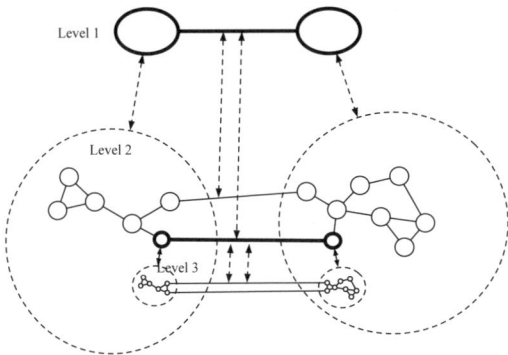

Fig. 2. Edges at multiple levels of granularity

To clarify the difference between atomic and composite edges, consider the example in Figure 2. In this figure, composite events and edges are represented with thicker lines.

Event_Graph Type. Type *event_graph* represents graphs of events interrelated. The carrier set of *event_graph* is $A_{event\_graph} \triangleq \{[V,E]|V \subseteq A_{event}, V \subseteq A_{edge}, \forall e \in E, e.source() \in V, e.target() \in V\} \cup \bot$. The operations over event graphs are presented in the following table:

Operation signature	Description
union, intersection, difference:	
event_graph × *event_graph* → *event_graph*	graph union, intersection and difference
get_edge : *event_graph* × *event* × *event* → *edge*	get the edge with specified identifier
get_event : *event_graph* × *real* → *event*	get the event with specified identifier
get_interval : *event_graph* → *interval*	get interval involving all events
is_empty : *event_graph* → *bool*	check if the event_graph is empty
add, update : *event_graph* × *event* → *event_graph*	add or update an event to an event_graph
add, update : *event_graph* × *edge* → *event_graph*	add or update an edge to an event_graph
project_over_event :	return an event graph where events include only
event_graph × *string*⁺ → *event_graph*	labels in the list of strings
project_over_edge :	return an event graph where edges include only
event_graph × *string*⁺ → *event_graph*	labels in the list of strings
select_over_event :	return an event graph where events' labels
event_graph × (*event* → *bool*) → *event_graph*	satisfy a boolean condition
select_over_edge :	return an event graph where edges' labels
event_graph × (*edge* → *bool*) → *event_graph*	satisfy a boolean condition

In the selection operators, a value of type function is used to define the boolean condition, whose notation follows the syntax in [9] and is an adaptation of Lambda Calculus. In this notation, a value of type $s_1 \times \cdots \times s_n \to s$ is expressed by **fun**$(x_1 : s_1, \ldots, x_n : s_n)$ t, where t is a term of sort s with free variables $x_1, \ldots, x_n$. Using this type function a query that retrieves an event graph whose events contain a label with name "S" and value less than 100 would be expressed by *select_over_event*$(g, \mathbf{fun}(x : event)int\_label\_value(x, S) < 100)$.

Path Type. Type *path* defines the connecting links between two events. The carrier set of *path* is $A_{path} \triangleq \{(v_1, \ldots, v_n) \cup (e_1, \ldots, e_{n-1})|\forall v_i, e_j, 1 \leq i \leq n, 1 \leq$

$j \le n-1, v_i \in A_{event}, e_j \in A_{edge}, e_j.source() = v_j, e_j.target() = v_{j+1}\} \cup \perp$. The operations over paths are presented in the following table:

Operation signature	Description
first, last : $path \to event$	return the first and last event in a path
nextEdge, prevEdge : $path \times event \to edge$	return the next or previous edge of an event in a path
nextEvent, prevEvent : $path \times edge \to event$	return the target or source event of an edge in a path

4   Event Query Language (EQL)

EQL is a language for querying event graphs using the data model given in Section 3. It is basically a language that specifies algebra expressions. A more well-known language such as the SQL-like OQL (object Query Language [6]) can also be defined, but it requires an additional translator to map SQL-like statements to EQL or to the algebra expressions of SECONDO. In Table 1 we present the most relevant syntactic elements of the language. We left out the specification of some elementary operators and object references (identifier) due to lack of space.

A query applies to event graphs, paths, events, and edges, all being objects in the database or formulas that return values of one of these types. A clause <graph query> allows us to apply projection or selection operators over an

Table 1. Partial BNF specification of EQL

```
<query>::= query {<graph formula> | <path formula> | <event formula> | <edge formula>}
<graph formula>::= {<graph query> | <graph expression> | <graph object>}
<graph query>::=<graph formula> do
                [proyect_each_event(<label name> [{, <label name>}...])]
                [proyect_each_edge(<label name> [{, <label name>}...])]
                [select_each_event( fun(<var name>:event) <boolean formula>)]
                [select_each_edge(fun(<var name>:edge) <boolean formula>)]
done
<graph expression>::= {<graph formula> union <graph formula>
              | <graph formula> intersection <graph formula>
              | <graph formula> difference <graph formula>
              | get_event_graph(<event formula>)}
<path formula>::=<path object> | paths(<graph formula>)
<event formula>::=<event object> | <event out operator>
<edge formula>::=<edge object> | <edge out operator>
<boolean formula>::=<boolean object> | not(<boolean formula>)
              | <boolean formula> and <boolean formula>
              | <boolean formula> or <boolean formula>
              | <allen formula>
              | <relational formula>
<interval formula>::=<interval object> | <interval out operator>
<allen formula>::=<interval formula><allen operator><interval formula>
<allen operator>::=equals | diferent | before | after | meets | metby | finishes |
                finishedby | overlaps | overlappedby |starts | startedby | during | contains
<relational formula>::=<instant formula><relational operator><instant formula>
              | <base formula><relational operator><base formula>
<base formula>::=<base object> | <base out operador>
<instant formula>::=<instant object> | <instant out operador>
<base out operator>::= /* operators of the system than return a value of type base*/
<instant out operator>::= /* operators of the system than return a value of type instant*/
<interval out operator>::= /* operators of the system than return a value of type interval*/
<event out operator>::= /* operators of the system than return a value of type event*/
<edge out operator>::= /* operators of the system than return a value of type edge*/
```

event graph or a query that returns a graph. For example, the following is a
query that obtains all events occurred before 2005, filtering the labels of events
to only include labels *type* and *place*.

```
query get_event_graph(evento1) do
          project_each_event(type, place)
          select_each_event(fun(ev: event) get_interval(ev) before new_interval(2005))
done
```

The term **query** is used only once, even if we apply two graph formulas
connected by graph operators. For example:

```
query get_event_graph(evento1) do
          project_each_event(type, place)
          select_each_event(fun(ev: event) get_interval(ev) before new_interval(2005))
done
difference
get_event_graph(evento1) do
          select_each_edge(fun(ed:edge) int_label_value(ed, likelihood)> 0.8)) done
```

5   Implementation

We have implemented the data model and query language over SECONDO,
an extensible database system [10]. To do so, all types and their corresponding
operators where implemented as classes in Java and linked to the kernel provided
by SECONDO. In addition, we have extended the user interface of SECONDO
to visualize events and their interrelations as graphs.

To show its implementation and use, we created a database about earthquakes
in Chile. Figure 3 shows part of the event graphs associated with the 1960 Chilean
Earthquake.

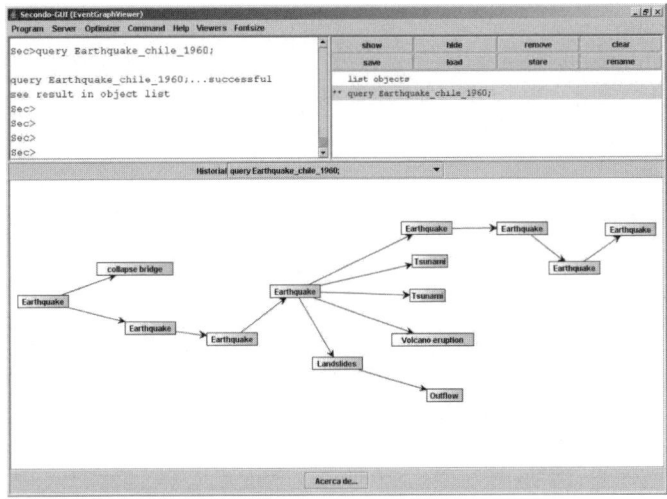

Fig. 3. User interface of the event-graph model in SECONDO

A query over this database in the EQL language could be to retrieve the events of type *earthquake* with a ground motion strength above 7.5 and that cause events of type *tsunami*. In addition, we could want to project the labels of events to filter only label *place*. The syntax of this query is as follows, where g denotes the event graph "Earthquake_chile_1960":

```
query g do
    select_each_edge[fun(e:edge) g get_event[target(e)] str_label_value["Type" sj] ==("Tsunami" sj)]
    select_each_event[fun(v:event) (v str_label_value["Tipo" sj] == ("Earthquake" sj))
                and (v real_label_value["Strength" sj]>= 7.5)]
        project_each_event["Place" sj]
done;
```

Note that in this query we used variables sj next to strings due to implementation constraints to translate strings from Java to SECONDO.

6   Conclusions

In this paper we proposed a graph-based approach to modeling events and their interrelations. We discussed the need of making explicit the relationships between events, and we have extended current graph data models to include temporal and spatial settings and to manage different levels of granularity for representing events and their relationships. As future work, we would like to fully implement the model in the extensible database system, including most relevant spatial operators. The event query language needs further development. First, it can be redefined following a syntax closely to standard query languages and, secondly, it should incorporate additional operators to exploit the representation of complex events and multiple granularities. There are also research issues to address concerning reasoning with temporal granularity and unknown temporal or spatial settings.

Acknowledgment. This work has been partially funded by Conicyt-Chile, under grant FONDECYT 1050944. In addition, Miguel Romero has been partially funded by CYTED-Spain under grant COMPETISOFT 506AC0287.

References

1. Allen, J.F.: Maintaining knowledge about temporal intervals. Commun. ACM 26(11), 832–843 (1983)
2. Allen, J.F., Ferguson, G.: Actions and events in interval temporal logic. J. Log. Comput. 4(5), 531–579 (1994)
3. Bertino, E., Ferrari, E., Guerrini, G.: An approach to model and query event-based temporal data. In: TIME, pp. 122–131 (1998)
4. Bry, F., Eckert, M., Patranjan, P.-L.: Querying composite events for reactivity on the web. In: Shen, H.T., Li, J., Li, M., Ni, J., Wang, W. (eds.) APWeb 2006 Workshops. LNCS, vol. 3842, pp. 38–47. Springer, Heidelberg (2006)
5. Bry, F, Eckert, M., Patranjan, P.-L.: Reactivity on the web: Paradigms and applications of the language xchange. J. Web Eng. 5(1), 3–24 (2006)

6. Cattell, R.G.G.: The Object Database Standard: ODMG-93 (Release 1.2). Morgan Kaufmann, San Francisco (1996)
7. Galton, A., Worboys, M.F.: Processes and events in dynamic geo-networks. In: Rodríguez, M.A., Cruz, I., Levashkin, S., Egenhofer, M.J. (eds.) GeoS 2005. LNCS, vol. 3799, pp. 45–59. Springer, Heidelberg (2005)
8. Grenon, P., Smith: SNAP and SPAN: Towards dynamic spatial ontology. Spatial Cognition and Computation 4(1), 69–103 (2004)
9. Güting, R.H.: Second-order signature: A tool for specifying data models, query processing, and optimization. In: Buneman, P., Jajodia, S. (eds.) Proceedings of the 1993 ACM SIGMOD International Conference on Management of Data, Washington, D.C, May 26-28, 1993, pp. 277–286. ACM Press, New York (1993)
10. Güting, R.H., de Almeida, V.T., Ansorge, D., Behr, T., Ding, Z., Höse, T., Hoffmann, F., Spiekermann, M., Telle, U.: Secondo: An extensible dbms platform for research prototyping and teaching. In: ICDE, pp. 1115–1116. IEEE Computer Society, Los Alamitos (2005)
11. Hlupic, V., Robinson, S.: Business process modelling and analysis using discrete-event simulation. In: Winter Simulation Conference, pp. 1363–1370 (1998)
12. Kowalski, R.: Database updates in the event calculus. Journal of Logic Programming 12(162), 121–146 (1992)
13. Luckham, D.C., Vera, J.: An event-based architecture definition language. IEEE Trans. Software Eng. 21(9), 717–734 (1995)
14. McCarthy, J., Hayes, P.: Machine Intelligence. In: Some Philosophical Problems from the Standpoint of Artificial Intelligence, pp. 463–502. Edinburg University Press (1969)
15. Miller, R., Shanahan, M.: The event calculus in classical logic - alternative axiomatisations. Lonköping Electronic Articles in Computer and Information Science 4(16), 1–30 (1999)
16. Nørvåg, K.: Issues in transaction-time temporal object database systems. J. Database Manag. 12(4), 40–51 (2001)
17. Worboys, M., Hornsby, K.: Geographic Information Science. In: Egenhofer, M.J., Freksa, C., Miller, H.J. (eds.) GIScience 2004. LNCS, vol. 3234, pp. 327–343. Springer, Heidelberg (2004)
18. Worboys, M.F.: Event-oriented approaches to geographic phenomena. International Journal of Geographical Information Science 19(1), 1–28 (2005)
19. Zimmer, D., Unland, R.: On the semantics of complex events in active database management systems. In: ICDE, pp. 392–399. IEEE Computer Society, Los Alamitos (1999)

PLR Partitions: A Conceptual Model of Maps

Mark McKenney and Markus Schneider*

University of Florida, Department of Computer and Information Sciences
{mm7,mschneid}@cise.ufl.edu

Abstract. The traditional spatial data models model space in terms of points, lines, and regions. These models employ a *disjoint dimension model* in which a spatial object can only consist of a zero, one, or two-dimensional point set. However, such models cannot adequately represent spatial reality. For example, consider a river network that contains both rivers and lakes. Intuitively, this river network is a single object made up of one-dimensional components (the river segments), and two-dimensional components (the lakes). Typically, collection types are used to represent such an object, but they present new problems. In this paper, we propose the *PLR partition* model that is able to model space in the form of a *map geometry* that can contain point, line, and region features within the same object. This model solves the problems associated with the traditional spatial data models.

1 Introduction

Spatially oriented disciplines such as spatial databases, geographical information systems (GIS), digital cartography, CAD, geocomputation, geoinformatics, artificial intelligence, computer vision, image databases, robotics, and cognitive science are fundamentally affected by notions of space and the properties of the models used to define space for these fields. In many cases, the expressiveness and the limitations of the formal spatial model underlying applications in these fields directly impacts the abilities of systems built upon them. Thus, the development of new, more powerful data models has far reaching implications.

Spatial data models attempt to model space using geometric constructs to represent real-world spatial features. For example, the spatial data types defined in most spatial data models consist of points, lines, and regions, which are used to model zero-, one-, and two-dimensional features, respectively. We denote such models as *traditional spatial models*. However, the design of spatial models based on these types reflects fundamental assumptions about space that limit their generality. Specifically, we identify three limitations of models that represent space in this manner. First, traditional spatial models are *disjoint dimension models*, meaning that spatial objects are defined based on their dimensionality (that is, points, lines, and regions). A common problem associated with such models arises with the use of spatial operations. For example, consider the two

---

* This work was partially supported by the National Science Foundation under grant number NSF-CAREER-IIS-0347574.

J.-L. Hainaut et al. (Eds.): ER Workshops 2007, LNCS 4802, pp. 368–377, 2007.

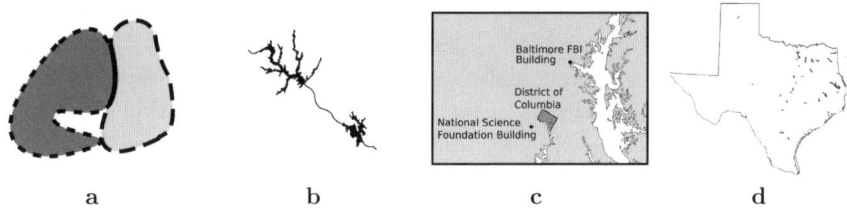

| a | b | c | d |

Fig. 1. Scenes that require multiple points, lines, or regions for representation

regions in Figure 1a. In this case, the intersection of the regions does not result in a region; however, the spatial intersection between regions is typically defined such that a new region is returned. In practice, this situation is handled by defining specific intersection operations that take two regions and return a point, a line, or a region, respectively. The assumption is that the user will indicate what type of object they wish to compute by choosing the appropriate operation. However, the intersection of the regions in this case results in both a point and a line. It is clear that defining multiple intersection operations does not address the fundamental problem. In order to compute this information, the user must compute two different spatial intersection operations. The end result is that no matter which operation is chosen, information about the intersection may be lost. We denote this problem the *dimension reduction problem* for spatial operations.

A second problem that follows from the disjoint dimension model of spatial data types is that the model implicitly imposes the restriction that an object must contain features of only a single dimension. In practice, this constraint causes current spatial data models to have limitations in their abilities to represent geographic reality. For instance, consider Figure 1b, and Figure 1c. These figures each depict an object containing components of multiple dimension: a river system containing lakes and the river feeding them (**b**), and United States Federal Government properties including the District of Columbia, the Baltimore FBI building, and the National Science Foundation building (**c**). Each of these figures contain what can intuitively be modeled as a single object containing features of multiple dimensions. These objects cannot be modeled by a single spatial object using current spatial models. For instance, assuming that the river in **b** is to be represented as a line, then either the disjoint one-dimensional sections must be grouped into an object, or a representative path of the river through the two-dimensional objects (lakes) must be arbitrarily chosen. Similarly in **c**, current spatial models will not allow a single object to represent both the two-dimensional area of the District of Columbia along with the points indicating the federal buildings, even though they are all federal properties. We term this inability to model components of multiple dimensions of a single object the *dimension represention problem* of spatial object models.

Finally, traditional spatial models do not allow multiple components of an object to be identified separately from each other. For instance, if a region has two faces, both of those faces belong to the region and cannot be identified separately. Thus, two faces in a region cannot meet along a common boundary since they will be indistinguishable from their union. A common consequence of this problem occurs when modelling bodies of water within a state (Figure 1d). A lake in the interior of a state is often modeled as a hole in the state, even though the lake belongs to the state (the state patrols the lake, builds structures that extend into the lake, etc.). The only choices for representation are to not represent that lake in the region representing the state (which is geographically inaccurate) or to model it as a hole (which does not indicate the state's ownership of the lake). In either case, a tradeoff occurs when modelling geographic or political reality. We term this situation the *feature restriction problem*, because it arises from fundamental restrictions of traditional spatial models.

In this paper, we address the shortcomings of traditional spatial models by introducing a new spatail representation model, called the point-line-region (PLR) partition model, that does not suffer from the dimension reduction problem, the dimension representation problem, or the feature restriction problem. We achieve this by defining a PLR partition as a fundamental unit of spatial representation. A PLR partition is a type of map geometry that can contain multiple features such that each feature can be zero-, one-, or two-dimensional, thus solving the dimension representation problem. We define each feature in a PLR partition to be associated with a *label*. The use of labels allows for PLR partitions to model thematic information of arbitrary complexity, and also provides a solution to the feature restriction problem. For example, two region features in a map geometry can meet along a one-dimensional boundary, but be differentiated based on their labels. Finally, because a PLR partition can model features of multiple dimensions, multiple types of a single spatial operation, such as intersection, need not be defined, thus solving the dimension reduction problem. Furthermore, we design PLR partitions in such a way that the geometries of point, line, and region features in a PLR partition correspond to the definitions of points, lines, and regions used in traditional spatial models. This allows us to leverage the vast amount of knowledge built upon traditional spatial models, such as topological relationships, to be directly applied to the features of map geometries. Thus, our model is conceptually intuitive to people familiar with traditional spatial models, but overcomes the limitations associated with them.

In Section 2 we review relevant related work to map geometries and spatial data models. We then define the type for PLR partitions in Section 3. Finally, in Section 4, we draw some conclusions.

2   Related Work

Research into spatial data types tends to focus on the disjoint dimension models of spatial data. Specifically, early work dealt with simple points, lines, and regions. Increased application requirements and a lack of closure properties of the simple spatial types lead to the development of the complex spatial types: complex points (a single point object consisting of a collection of simple points), complex lines (which can represent networks such as river systems), and complex regions that are made up of multiple faces and holes (that is, a region representing Italy, its islands, and the hole representing the Vatican) [1]. These types are defined based on concepts from point set topology, which allow the identification of the interior, exterior, and boundary of the spatial objects.

The idea of a *map* or *map geometry* as a data type has received significant attention. In [2,3,4,5], a map is not defined as a data type itself, but as a collection of spatial regions that satisfy some topological constraints. This approach is similar to the geometry collection concept put forth by the Open Geospatial Consortium. The drawback to this model is that a map is not regarded as a data type, but a collection of other types. Thus, it is unclear in many cases how constraints between participating spatial objects can be enforced and how type closure under spatial operations can be achieved. Other approaches to defining maps have focused on *raster* or *tessellation* approaches [6,7]. However, such approaches are not general enough for our purposes in the sense that the geometries of maps in these models are restricted to the tessellation scheme in use. In [8], the authors consider a map geometry to be a planar subdivision; however, they do not discuss how a planar subdivision should be modeled except to say that data structures such as winged edge or quad edge structures should be used. In [9] map geometries are represented as planar graphs. However, the plane graph, as defined, is not able to model thematic properties of the map.

The work that comes closest to ours is presented in [10]. The authors of this paper define an abstract, mathematical data model that formally describes the type of *spatial partitions*. A spatial partition is able to model multiple labeled region features together in a single object. However, this model is unable to represent line and point features.

3   The PLR Partition Model

Intuitively, we define a PLR partition as a map geometry containing point, line, and region features such that each feature is associated with a label. For example, a single PLR partition can represent a map of a country that depicts the country's provinces, cities, and roads, complete with their names. PLR partitions are also able to represent the example map geometries shown in Figure 1.

Spatial objects have traditionally been defined by partitioning the plane into point sets that identify the different parts of an object. For example, lines and regions have been defined such that they partition the plane into point sets containing the interior, boundary, and exterior points of the objects. We define PLR partitions in a similar manner by partitioning the plane into point sets. However, we cannot simply define a PLR partition as a set theoretic partition of $\mathbb{R}^2$. In general, a set partition is defined as a complete decomposition of a set S into non-empty subsets $\{S_i \in S | i \in I\}$ called *blocks* such that the union of the blocks is equal to the original set and the blocks are pairwise disjoint. We cannot model PLR partitions as a set partition of the plane (that is, partition $\mathbb{R}^2$ by a function $\pi : \mathbb{R}^2 \to I$) for two reasons: I does not contain semantically relevant values, and we must impose constraints on blocks in order for the features in a PLR partition to correspond to the traditional spatial types. Therefore, we proceed in three steps: we first define a mapping from points in the plane to blocks identified by semantically relevant labels. We then provide mathematical notations that we use to identify labeled blocks, and we then identify blocks and define the properties of these blocks that must hold in order for such a mapping to be a valid PLR partition. For the remainder of this paper, we use the terms *PLR partition* and *partition* interchangeably.

3.1   Definition of a Spatial Mapping

One of the goals of our PLR partition definition is that it associates labels with spatial features in partitions. Intuitively, this is similar to marking or coloring different portions of a partition. In general, arbitrary identifying values, called *labels*, that model thematic information of any complexity should be able to be assigned to different points, lines, or regions within a partition so that components of partitions can be identified by their label. Thus, the set A of labels used to mark point, line, and region features in a partition determines the type of the PLR partition. We make no assumptions as to the structure or contents of a specific label. For example, a PLR partition showing the countries of Canada, the US, and Mexico such that each country is labeled with its name can have a type $A = \{Canada, US, Mexico\}$. Any area in a PLR partition that is not specifically labeled in a partition is given the $\bot$ label. Thus, in the PLR partition of Canada, the US, and Mexico, the area shown that does not belong to those countries is labeled with $\bot$ and the type of the partition is $A = \{Canada, US, Mexico, \bot\}$. All points with the $\bot$ label are considered to be in the exterior of a partition.

We cannot simply map the plane to the elements of a type A because the boundaries between two features cannot be identified as belonging to one feature or the other. Instead, shared boundaries between two features belong to both features and are thus labeled with the labels of both features. Therefore, a PLR partition of type A is defined by a *spatial mapping* that maps points in the plane to elements of the power set of A. Points in the plane that belong solely to region features are mapped to an element $e \in 2^A$ that correspond to an element in A (that is, $e = \{l\}$ with $l \in A$). For example, the region $R2$ in Figure 2 consists of all points mapped to the label $\{R2\}$. If we consider the exterior of a partition to be an unbounded region, then points and lines always intersect at least one region. For example, line $L1$ in Figure 2 intersects region $R1$ and the exterior of the partition. Therefore, each point that belongs to a line feature also belongs to a region feature, and must carry the labels of both. This is also true for point features. Thus, boundaries between features are mapped to the labels of all features that surround the boundary, and intersecting features take the labels of all participating features. It follows that a PLR partition of type A is defined by a function $\pi : \mathbb{R}^2 \rightarrow 2^A$ that maps points to the power set of the set of labels that defines the type of the partition. For example, Figure 2a depicts a scene where each feature is annotated with a name. Figure 2b shows the result of the spatial mapping that corresponds to this scene.

As was mentioned previously, any points in a PLR partition that are not explicitly labeled are considered to be in the *exterior* of the partition and are

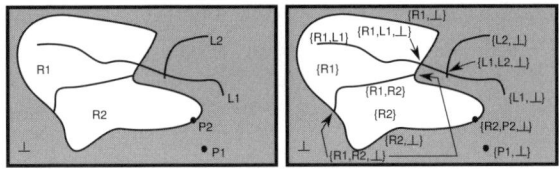

Fig. 2. A scene (a), and its associated spatial mapping (b)

labeled with the *undefined element* or *empty label* $\perp_A \in A$. If no ambiguities arise, we sometimes omit the type subscript and simply use $\perp$. We define a *spatial mapping* as the following:

Definition 1. *A* spatial mapping *of type A is a total function* $\pi : \mathbb{R}^2 \to 2^A$.

Therefore, a spatial mapping partitions the plane into blocks of identically labeled points. However, blocks defined in this way are not restrictive enough for our purposes. Spatial data types are typically defined not merely as point sets, but are in some sense regular; that is, regions are not allowed to have cuts or punctures, etc. The topological concept of regular open sets as shown in [11] models this well. Therefore, we choose to model regions as two-dimensional, regular open point sets, and lines as one-dimensional, regular open point sets (points are trivial). This implies that a given spatial mapping may not necessarily represent a valid PLR partition. Therefore, we must define the properties a spatial mapping must have if it is supposed to represent a PLR partition. In order to do this, we must first provide a summary of the mathematical notation we use throughout the remainder of the paper.

3.2 Notations

A treatment of regular open point sets can be found in [11]. Let (X, T) be a topological space[1] with topology $T \subseteq 2^X$, and let $S \subseteq X$. The *interior* of S, denoted by S°, is defined as the union of all open sets that are contained in S. The *closure* of S, denoted by $\overline{S}$ is defined as the intersection of all closed sets that contain S. The *exterior* of S is given by $S^- := (X - S)^\circ$, and the *boundary* or *frontier* of S is defined as $\partial S := \overline{S} \cap \overline{X - S}$. An open set is *regular* if $A = \overline{A}^\circ$. The type of regular open sets is closed under intersection. In this paper, we deal with topological space $\mathbb{R}^2$.

The application of a function $f : A \to B$ to a set of values $S \subseteq A$ is defined as $f(S) := \{f(x) | x \in S\} \subseteq B$. For example, given a spatial mapping π of type A, the function $\pi(\{(1, 2), (2, 3)\})$ would return the set of labels of the given points. In some cases we know that $f(S)$ returns a singleton set, in which case we write $f[S]$ to to denote the single element, that is $f(S) = \{y\} \to f[S] = y$. For doubly

---

[1] In a topological space, the following three axioms hold [12]: (i) $U, V \in T \to U \cap V \in T$, (ii) $S \subseteq T \to \bigcup_{U \in S} U \in T$, and (iii) $X \in T, \emptyset \in T$. The elements of T are called *open sets*, their complements in X are called *closed sets*, and the elements of X are called *points*.

nested singleton sets, we use $f[[S]]$ similarly. The inverse function $f^{-1} : B \to 2^A$ of f is defined $f^{-1}(y) := \{x \in S | f(x) = y\}$. Given a spatial mapping π of type A and label $l \in A$, the inverse function $\pi^{-1}(l)$ returns the block of points with the label l. It is important to note that f^{-1} is a total function and that f^{-1} applied to a set yields a set of sets.

3.3 Defining PLR Partitions

In this section, we show how to identify each feature in a spatial mapping. We then provide constraints on these features in order to define the type of PLR partitions. However, identifying features in a spatial mapping is somewhat non-intuitive. Consider region $R1$ in Figure 2b and the portion of line $L1$ that intersects it. The points that make up $R1$ are associated with one of two blocks, the block labeled $\{R1\}$ and the block labeled $\{R1, L1\}$. To find the point set consisting of all points that define the region $R1$, we cannot simply use the function $\pi^{-1}(\{R1\})$, since it returns a point set containing a cut (the block labeled $\{R1, L1\}$ will not be returned). We make the observation that although points in $R1$ have one of two different labels, $\{R1\}$ is a subset of both labels. Therefore, we say that the label $\{R1\}$ is the *discriminant label* of the region, since it can be used to distinguish all points that make up that region. We now need a method to identify and retrieve blocks from a spatial mapping based on discriminant labels.

In order to identify the different features of a partition based on their discriminant labels, we take advantage of the property of regular open sets in the plane that every point that is a member of a regular open set is contained in a *neighborhood* of points that are also members of that set. We define the neighborhood of a given point $p = (p_x, p_y)$ in two dimensions as the set of points forming a circle around p with an infinitesimally small radius r. In order to retrieve the neighborhood of a point p, we define the neighborhood operation which returns the set of points contained in the neighborhood of p:

$$N := \mathbb{R}^2 \to 2^{\mathbb{R}^2}$$
$$N(p) = \{(x,y) | (x - p_x)^2 + (y - p_y)^2 = (r)^2\}$$

We can now determine if a point in a partition belongs to a point, line, or region feature based on the properties of that point's neighborhood. For example, a point p that belongs solely to a region (as opposed to the intersection of a region and a line) will have the same label as every point in its neighborhood. In other words, the label of p will be equivalent to the union of the labels of all points in $N(p)$. The predicate *isBasicRegion* takes a point and returns a value of *true* if it belongs solely to a region feature, and *false* otherwise. For instance, all points in the interior of region $R2$ in Figure 2b are basic region points.

$$isBasicRegion := \mathbb{R}^2 \to \mathbb{B}$$
$$isBasicRegion(p) = (\pi(p) = \bigcup_{q \in N(p)} \pi(q))$$

In order to determine if a point p belongs to a region in general (and not a boundary between regions) regardless of how many line and point features also include that point, we must determine that all points in the neighborhood of p that happen to be identified as basic region points have the same label (that is, the set containing the labels of all basic region points in $N(p)$ has a cardinality equal to 1). If p is on a boundary between regions, then its neighborhood will contain basic region points from each differently labeled region. For a spatial mapping π of type A, we also define the *discriminant spatial mapping* for regions $\pi_r : \mathbb{R}^2 \to 2^A$ that maps a point to its discriminant region label. For instance, a point p that lies on the intersection of region $R1$ and line $L1$ in Figure 2a is a region point but not a basic region point, and $\pi_r(p) = \{R1\}$.

$$isRegion := \mathbb{R}^2 \to \mathbb{B}$$
$$isRegion(p) = |\{\pi(q)|q \in N(p) \ \wedge \ isBasicRegion(q) \ \wedge \ \pi(q) \subseteq \pi(p)\}| = 1$$
$$\pi_r : \mathbb{R}^2 \to 2^A$$
$$\pi_r(p) = \{\pi[q]|q \in N(p) \ \wedge \ isBasicRegion(q) \ \wedge \ isRegion(p)\}$$

A region boundary point is characterized as belonging to multiple regions; therefore, a point p lies on a region boundary if at least two points in its neighborhood are basic region points from different regions (that is, the set of basic region labels belonging to points in $N(p)$ must contain at least two elements). We define the predicate *isRegionBound* to test if a point lies on a region boundary, and the discriminant spatial mapping for region boundaries below. For example, point p labeled $\{R2, P2, \perp\}$ in Figure 2 is a region boundary because $N(p)$ contains basic region points from both $R2$ and the exterior region $\perp$, and $\pi_{rb}(p) = \{R2, \perp\}$.

$$isRegionBound := \mathbb{R}^2 \to \mathbb{B}$$
$$isRegionBound(p) = |\{\pi(q)|q \in N(p) \ \wedge \ isBasicRegion(q) \ \wedge \ \pi(q) \subseteq \pi(p)\}| > 1$$
$$\pi_{rb} : \mathbb{R}^2 \to 2^A$$
$$\pi_{rb}(p) = \{\pi[q]|q \in N(p) \ \wedge \ isBasicRegion(q) \ \wedge \ isRegionBound(p)\}$$

Point features are unique in that the neighborhood of a point feature will not contain the discriminant label of the point. Therefore, to determine if point p is a point feature, we compute the difference of each label in $N(p)$ with the label of p. The resulting label will either be the discriminant label of p, if it is a point feature, or the empty set if p is not a point feature:

$$isPoint := \mathbb{R}^2 \to \mathbb{B}$$
$$isPoint(p) = |\pi(p) - \bigcup_{q \in N(p)} \pi(q)| > 0$$
$$\pi_p : \mathbb{R}^2 \to 2^A$$
$$\pi_p(p) = \pi(p) - \bigcup_{q \in N(p)} \pi(q)$$

To identify line features in a spatial mapping, we must first disregard the discrimant labels for point and region features. We define the label stripping function S that takes a point and returns the label of that point with any point, region, and region boundary discrimant labels removed.

$$S := \mathbb{R}^2 \to 2^A$$
$$S(p) = \pi(p) - \pi_r(p) - \pi_{rb}(p) - \pi_p(p)$$

Lines present a problem due to the fact that the interior of two lines can intersect along a line or at a point. Thus, we must treat lines differently than regions or points. We can determine that a point p is part of the interior of a line if there are at least two points, q and s, in its neighborhood such that the stripped labels for those points are identical. However, if two line interiors intersect at a point, the discriminant spatial mapping must return two discriminant labels for that point, one for each line. Furthermore, the inverse of the discriminant spatial mapping must return the point in question if given either of the discriminant labels for the intersecting lines. An example of such a scenario is the intersection of lines $L1$ and $L2$ in Figure 2b.

$$isLine := \mathbb{R}^2 \to \mathbb{B}$$
$$isLine(p) = \exists q, s \in N(p) | q \neq s \ \wedge \ S(q) = S(s) \ \wedge \ S(q) \neq \varnothing \ \wedge \ \pi(q) \subseteq \pi(p)$$
$$\pi_l : \mathbb{R}^2 \to \{2^A\}$$
$$\pi_l(p) = \{S(q) | q, s \in N(p) \ \wedge \ isLine(p) \ \wedge \ S(q) = S(s) \ \wedge \ q \neq s \ \wedge \ S(q) \neq \varnothing\}$$
$$\pi_l^{-1} := 2^A \to 2^{\mathbb{R}^2}$$
$$\pi_l^{-1}(Y) = \{x \in \mathbb{R}^2 | Y \in \pi_l(x)\}$$

A point p is on the boundary of a line if the line extends in only one direction from p; therefore, if there exists a point q in the neighborhood of p such that $S(q) \subseteq S(p)$ and no other point in p's neighborhood contains the same stripped label as q, then point p is a boundary of a line.

$$isLineBoundary := \mathbb{R}^2 \to \mathbb{B}$$
$$isLineBoundary(p) = \exists q \in N(p) | \pi(q) \subseteq \pi(p) \ \wedge \ S(q) \neq \varnothing$$
$$\wedge \ (\nexists s \in N(p) | q \neq s \ \wedge \ S(q) = S(s))$$
$$\pi_{lb} : \mathbb{R}^2 \to 2^A$$
$$\pi_{lb}(p) = \{S[q] | (\nexists s \in N(p) | q \neq s \ \wedge \ S(q) = S(s)) \ \wedge \ isLineBoundary(p)$$
$$\wedge \ S(q) \neq \varnothing\}$$

Given a spatial mapping, we can now identify its features:

Definition 2. Let π be a spatial mapping of type A
    (i) $\rho(\pi) := \pi_r^{-1}(2^A)$ (regions)
    (ii) $\omega_\rho(\pi) := \pi_{rb}^{-1}(2^A)$ (region borders)
    (iii) $\lambda(\pi) := \pi_l^{-1}(2^A)$ (lines)
    (iv) $\omega_\lambda(\pi) := \pi_{lb}^{-1}(2^A)$ (line borders)
    (v) $\varphi(\pi) := \pi_p^{-1}(2^A)$ (points)

Finally, a spatial mapping is a PLR partition if its lines and regions are regular, open point sets, and the borders between features are labeled with the union of the labels of all features sharing the border.

Definition 3. A *PLR partition* is a spatial mapping π of type A with:
    (i) $\forall r \in \rho(\pi) : r = \overline{r}^\circ$
    (ii) $\forall l \in \lambda(\pi) : l = \overline{l}^\circ$
    (iii) $\forall b \in \omega_\rho : \pi_{rb}[b] = \{\pi_r[[r]] | r \in \rho(\pi) \ \wedge \ b \subseteq \partial r\}$
    (iv) $\forall b \in \omega_\lambda : \pi_{lb}[b] = \{\pi_l[[l]] | l \in \lambda(\pi) \ \wedge \ b \subseteq \partial l\}$

4   Conclusions and Future Work

In this paper, we have defined the PLR partition model which is able to overcome the dimension reduction, dimension representation, and feature restriction problems associated with traditional spatial data models. PLR partitions are able to represent map geometries containing point, line, and region features along with the thematic information associated with each feature. Furthermore, the individual features within a PLR partition are defined as open point sets so that they directly correspond to the traditional spatial types of complex points, lines, and regions. Thus, existing concepts defined on the traditional spatial types can be directly applied to the individual features in a PLR partition.

Future work includes formally defining operations over PLR partitions and investigating an implementation model for use in spatial systems. We also plan to explore the use of PLR partitions as a mechanism to support spatial query processing in a spatial database system. Furthermore, it is unclear if traditional spatial querying mechanisms are adequate to query PLR partitions. Therefore, we plan to investigate SQL extensions that can support PLR partition queries.

References

1. Schneider, M., Behr, T.: Topological Relationships between Complex Spatial Objects. ACM Trans. on Database Systems (TODS) 31(1), 39–81 (2006)
2. Güting, R.H.: Geo-relational algebra: A model and query language for geometric database systems. In: Schmidt, J.W., Missikoff, M., Ceri, S. (eds.) EDBT 1988. LNCS, vol. 303, pp. 506–527. Springer, Heidelberg (1988)
3. Güting, R.H., Schneider, M.: Realm-Based Spatial Data Types: The ROSE Algebra. VLDB Journal 4, 100–143 (1995)
4. Huang, Z., Svensson, P., Hauska, H.: Solving spatial analysis problems with geosal, a spatial query language. In: Proceedings of the 6th Int. Working Conf. on Scientific and Statistical Database Management, Institut f. Wissenschaftliches Rechnen Eidgenoessische Technische Hochschule Zürich, pp. 1–17 (1992)
5. Voisard, A., David, B.: Mapping conceptual geographic models onto DBMS data models. Technical Report TR-97-005, Berkeley, CA (1997)
6. Ledoux, H., Gold, C.: A Voronoi-Based Map Algebra. In: Int. Symp. on Spatial Data Handling (2006)
7. Tomlin, C.D.: Geographic Information Systems and Cartographic Modelling. Prentice-Hall, Englewood Cliffs (1990)
8. Filho, W.C., de Figueiredo, L.H., Gattass, M., Carvalho, P.C.: A topological data structure for hierarchical planar subdivisions. In: 4th SIAM Conference on Geometric Design (1995)
9. Floriani, L.D., Marzano, P., Puppo, E.: Spatial queries and data models. In: Campari, I., Frank, A.U. (eds.) COSIT 1993. LNCS, vol. N. 716, pp. 113–138. Springer, Heidelberg (1993)
10. Erwig, M., Schneider, M.: Partition and Conquer. In: Frank, A.U. (ed.) COSIT 1997. LNCS, vol. 1329, pp. 389–408. Springer, Heidelberg (1997)
11. Tilove, R.B.: Set Membership Classification: A Unified Approach to Geometric Intersection Problems. IEEE Trans. on Computers C-29, 874–883 (1980)
12. Dugundi, J.: Topology. Allyn and Bacon (1966)

A Conceptual Framework to Support Semantic Interoperability of Geospatial Datacubes

Tarek Sboui[1,2], Yvan Bédard[1,2], Jean Brodeur[3,1], and Thierry Badard[1]

[1] Department of Geomatic Sciences and Centre for Research in Geomatics, Université Laval,
Quebec, Qc, G1K 7P4, Canada
[2] NSERC Industrial Research Chair in Geospatial Databases for Decision-Support
[3] Natural Resources Canada, CIT, 2144-010 King West St, Sherbrooke, Qc, J1J 2E8, Canada
Tarek.Sboui.1@ulaval.ca, Yvan.Bedard@scg.ulaval.ca,
brodeur@nrcan.gc.ca, Thierry.Badard@scg.ulaval.ca

Abstract. Today, we observe a wide use of geospatial databases that are implemented in many forms (e.g. transactional centralized systems, distributed databases, multidimensional datacubes). Among those possibilities, the multidimensional datacube is more appropriate to support interactive analysis and to guide the organization's strategic decisions, especially when different epochs and levels of information granularity are involved. However, one may need to use several geospatial multidimensional datacubes which may be heterogeneous in design or content. Overcoming the heterogeneity problems in a manner that is transparent to users has been the principal aim of interoperability for the last fifteen years. In spite of successful initiatives and widespread use of standards, today's solutions do not address yet geospatial datacubes. This paper aims at describing the interoperability of geospatial datacubes, defining the semantic heterogeneity problems that may occur when using different geospatial datacubes, and proposing a conceptual framework to support semantic interoperability of these datacubes.

Keywords: Geospatial datacubes, interoperability, semantic heterogeneity, ontology, context.

1 Introduction

In the last decades, we have witnessed an exponential increase in the amount of data available from multiple sources. Additionally, there have been important innovations in geospatial[1] information technology, especially in decision support systems, geographic knowledge discovery and interoperability. Data warehouses are being considered as efficient components of decision support systems. They are usually structured as datacubes, i.e. according to the multidimensional paradigm defined in the field of Business Intelligence (BI). This paradigm enables making strategic decisions by supporting the user's mental model of data. It allows users to navigate

---

[1] The term "Geospatial" is used in this paper to refer to spatial information for objects located on the Earth (e.g. spatial reference systems, relationships between objects, constraints, etc.).

J.-L. Hainaut et al. (Eds.): ER Workshops 2007, LNCS 4802, pp. 378–387, 2007.
© Springer-Verlag Berlin Heidelberg 2007

aggregated and summarized data according to a set of dimensions with different hierarchy [9], [4], [16], [15]. Multidimensional databases (hereafter called datacubes) are becoming widely used in the geographic field [4], [16], [15]. Geospatial datacubes contain spatial, temporal and thematic data which may differ in format and content (e.g. geospatial characteristics such as location and geometry, levels of abstraction and meanings) resulting in heterogeneous geospatial datacubes. The heterogeneity of geospatial datacubes presents a major obstacle when people need to make strategic decisions or discover spatio-temporal trends using information located in different geospatial datacubes. For example, a health organization, willing to analyze the risk of the West Nil virus on population, may use two geospatial datacubes; one containing data related to forest stand and the other containing data related to the population density. Both geospatial datacubes are modeled differently (e.g. different conceptualizations, purposes, etc.) and developed using different techniques.

Interoperability has been widely recognized as an efficient paradigm for joining heterogeneous systems to facilitate an efficient exchange of information [5], [6], [10] [13]. It aims at resolving technical, organizational, and semantic heterogeneities between various systems in many fields (information management, engineering technologies, etc.). However, resolution of semantic heterogeneity still remains a challenge for enabling interoperability among databases [13], [6].

Although the interoperability of information systems and more especially in the geographic information realm has attracted the attention of many researchers [8], [5], [13], [6], there have not been many works on the semantic interoperability among datacubes. Also, no work on semantic interoperability of geospatial datacubes has been found in the literature. The purpose of this paper is to discuss the need for interoperating geospatial datacubes and to propose a conceptual framework to support the semantic interoperability among geospatial datacubes.

In the next section, we present the need for interoperating geospatial datacubes, we define the interoperability between geospatial datacubes, and we classify the semantic heterogeneity problems that may occur in these datacubes. In section 3, we present a conceptual framework to overcome the semantic heterogeneity in geospatial datacubes. We conclude and present further works in section 4.

2   Interoperability and Datacubes

2.1   Interoperability Between Information Systems

Interoperability has been described in various ways. It has been generally defined as the ability of heterogeneous systems to communicate and exchange information and applications in an accurate and effective manner [5], [6], [14].

Geospatial interoperability is considered here as the ability of information systems to a) communicate all kinds of spatial information about the Earth and about the objects and phenomena on, above, and below its surface, and b) cooperatively run applications capable of manipulating such information [17]. Brodeur and Bédard defined the interoperability as a communication between different agents who use their common background to establish a common understanding of reality [7]. Brodie proposed an elegant definition of geographical databases interoperability: "Two geographical databases X and Y can interoperate if X can send requests for services R to Y on a mutual understanding of R by X and Y, and Y can return responses S to X based on a mutual understanding of S as responses to R by X and Y" [8], [5].

Typically, there are three levels of interoperability among geographic information systems: technical, semantic, and institutional [10]. The technical level aims to resolve the differences in format, languages and user interfaces. The semantic interoperability aims to provide a mutual understanding of different meanings. For example, the term forest may have more than one meaning: 1) "a defined area of land formerly set aside in England as a royal hunting ground" or 2) "a land that is covered with trees and shrubs" [1]. Finally, at the institutional level, the interoperability aims at unifying the organization process. For spatial information systems, we include the consideration about object's geometry in the semantic level since geometry is not inherent to objects but defined according to the needs of a given application (for example, a polygon representing a building may correspond to the roof and may be measured using photogrammetry for a given application, while it may correspond to the foundations and measured using land surveying for another application).

2.2 Geospatial Datacubes

Data warehouses are being considered efficient components of decision support systems. They may be structured as datacubes, i.e. according to the multidimensional paradigm, and implemented using database management systems (DBMS) with a multidimensional server, directly with a multidimensional server, or with hybrid models. From an On-Line Analytical Processing point of view (OLAP), these different architectures are known as Relational OLAP (ROLAP), Multidimensional OLAP (MOLAP) and Hybrid OLAP (HOLAP) [4], [9], [15], [16].

A datacube is composed of a set of measures aggregated according to a set of dimensions with different levels of granularity. Both dimensions and measures of a geospatial datacube may contain geospatial components [4]. Geospatial datacubes support the user's mental model of the data and enables him to make strategic decisions [2], [4], [16]. In fact, they enable users to interactively navigate through different levels of granularity so; he can get a global picture of a phenomenon, and can get more insight into that phenomenon detailed information. Moreover, geospatial datacubes contains spatial component (e.g. street address, geographic coordinates, map coordinates) which allow the visualization of phenomena and, hence, help users to extract insights that can be helpful to understand these phenomena [2].

2.3 Need for Interoperability Between Geospatial Datacubes

In many situations, we may need to use several heterogeneous geospatial datacubes. We group these situations into three categories:

1. *A simultaneous and rapid navigation through different datacubes*: Users from different disciplines may need to access and navigate simultaneously through heterogeneous geospatial datacubes. Navigating separately through each datacube would be an arduous work for users, since they likely need to make extra efforts to manually resolve the problems of heterogeneity between datacubes (e.g. comparing the meaning of concepts, establishing a mapping between them, etc.). The principal aim of interoperability is to automatically overcome such differences and, hence,

can considerably facilitate the navigation task. Interoperability of geospatial datacubes would enable, for example, to create a temporary multidimensional structure (i.e. structure without data records) that is connected to the different dimensions of geospatial datacubes. Such structure is useful especially in emergency situations when users need to rapidly navigate through data stored in the geospatial datacubes without preoccupying themselves with the problems of heterogeneity. An example of a situation is a natural disaster that affects adjacent countries. In such situation, we may need to navigate different geospatial datacubes, developed in these countries, in order to get the right information and respond quickly at different levels (local, provincial, federal, etc.) or at different domains (geographic, political, etc.). In such cases, interoperability of geospatial datacubes is crucial to prevent catastrophic losses.

2. *A rapid insertion of data in a datacube*: While data in datacubes are usually collected from legacy systems, they can be imported from other heterogeneous datacubes [4]. We may need to rapidly insert new data in a geospatial datacube from other datacubes. For example, we may need to rapidly insert new data in a geospatial datacube which contains data about the construction of winter bridges, from two other geospatial datacubes; one of them contains data about traffics. The other cube contains data about lakes.

3. *An interactive and rapid analysis of phenomena changes*: In order to analyze phenomena changes (such as the forest stand evolution), we need to compare data describing these phenomena at different epochs. We may need to compare data stored in geospatial datacubes build also at different epochs. Interoperating geospatial datacubes would enable to interactively compare data and analyze phenomena changes. For example, we may need to rapidly compare forest stand changes following an environmental disaster. Interoperability of geospatial datacubes would enable to detect changes in the wood volume.

However, data in datacubes are usually collected from different source systems. One may ask "why shouldn't we just interoperate the source systems rather than datacubes?" There are three reasons for interoperating geospatial datacubes:

1. We possibly have no more access to data source systems from which we created the datacubes due to multiple reasons including administration policies (such as security reason).

2. We need to use long-period data (i.e. historic data) that usually exist only in datacubes. In fact, in source systems, long-period data are usually modified or replaced by new data and then destroyed or archived, whereas datacubes keep historic data for strategic decision-making purposes [4]. So if we need to reuse such data, we have no choice but to consider datacubes.

3. In the context of decision-making, interoperating datacubes is more efficient than interoperating source systems. In fact, within a geospatial datacube, contrary to source systems, possible aggregations of measures for all possible combinations of members are pre-calculated using different operators (e.g. mathematical, metric such as distance and area, and topological relations such as disjoint, interior intersection, etc.) [16]. These aggregations usually require an arduous work for geospatial datacubes developers (e.g. defining procedures for aggregation, defining a new spatial layer as an aggregation of others). Reusing datacubes means that we don't need to define such aggregations from scratch, thus saving time and money.

2.4 Defining the Interoperability Among Geospatial Datacubes

Inspired from the definition provided by Brodie [8], we define *the interoperability among two geospatial datacubes C1 and C2 as the ability of C1 to request a service in a manner that can be understood by C2, and the ability of C2 to respond to that request in a manner that can be understood by C1. The request and response are conducted automatically.* Services could include:

- importing/exporting data contained in dimensions or facts;
- comparing a dimension/fact against another;
- getting information about the dimensions or the facts (e.g. language used);
- taking into account a concept evolution (e.g. meaning or format changes); or
- adapting the meaning of a concept when the context changes.

While at first sight, there may not seem to have differences between semantic interoperability of transactional databases and that of datacubes, the latter particularly stresses the importance of dealing with the semantics of aggregation and generalization relationships, the semantics of summarizing methods and algorithms, the semantics of summarizability conditions and the semantics of cross-tabulations for every level of details and every member of the datacube dimensions. Although one could do it using traditional transactional solutions, the efficiency can be improved with an enriched framework that explicitly supports datacubes.

We are especially interested in the semantic interoperability which still remains the main issue for the geographic information systems communities [5], [13]. Accordingly, the remaining part of this paper focuses specifically on semantic interoperability of geospatial datacubes. Such interoperability aims at overcoming semantic heterogeneity problems in geospatial datacubes.

2.5 Semantic Heterogeneity Conflicts in Geospatial Datacubes

Semantic heterogeneity represents a major challenge for enabling interoperability among information systems [5], [6]. It occurs when there are differences in the meaning or the interpretation of the related concepts (i.e. concepts having a similarity, generalization, or specialization relationship). For geospatial databases, the semantic heterogeneity may come from the difference in the abstraction of concepts (for example, the concept forest may be represented, with different geometries, as vegetations, as trees, or as wooded areas) and databases schemas [6]. These problems appear more difficult to deal with when working with geospatial datacubes since concepts in these databases are typically more complex (dimensions with different levels, measures to analyze, etc.). In order to support geospatial datacubes interoperability, we first propose a classification of semantic heterogeneity types that may occur between different geospatial datacubes.

1. Cube-to-Cube heterogeneity
- *Cube-to-Cube context conflicts* arise when two cubes are created or used in different contexts (e.g. different purposes, languages, etc.).
2. Dimension-to-Dimension heterogeneity
- *Dimension-to-Dimension meaning conflicts* arise when related dimensions of different datacubes have mismatched meanings (e.g. different names, definitions,

geospatial – i.e. difference in geometry or in temporality – or thematic properties) For example, a bridge may be represented with a line or a polygon. These problems may also occur when unrelated dimensions are named equivalently.

- *Dimension-to-Dimension context conflicts* arise when related dimensions of datacubes have been defined and used in different contexts.
- *Dimension-to-Dimension hierarchy conflicts* arise when related dimensions of different datacubes have different hierarchies.
- *Dimension-to-Dimension level differences* arise when dimensions of datacubes have different levels.

3. Fact-to-Fact heterogeneity

- *Fact-to-Fact meaning conflicts* arise when related facts of datacubes have different meanings (e.g. different names or definitions).
- *Fact-to-Fact context conflicts* arise when related facts of datacubes have been defined and used regarding different contexts.

4. Measure-to- Measure heterogeneity

- *Measure-to-Measure meaning conflicts* occur when related measures have different meanings (e.g. different names or scales).
- *Measure-to-Measure context conflicts* occur when related measures have different contexts.

In order to overcome these conflicts, we defined a conceptual framework based on human communication, ontology, context and multidimensional structure.

3   A Conceptual Framework to Support Semantic Interoperability Between Geospatial Datacubes

3.1   Human Communication, Ontology, Context, and Semantic Interoperability of Geospatial Datacubes

The conceptual framework presented in this paper is based on human communication, ontology and context. The human communication process constitutes an improved form of interoperability [6]. It is based on agreements between a set of agents to use a shared vocabulary (i.e. ontology) and on specific circumstances under which the vocabulary can be used (i.e. context). An ontology is a set of related concepts and a set of assumptions about the intended meaning of these concepts in a given domain or application [12], [11]. Ontologies play an important role in enabling semantic interoperability between agents by providing them a common understanding of the reality. Ontology of geospatial datacubes would include definitions, assumptions, and properties of the datacubes concepts. The properties include non-spatial and spatial aspects (i.e. geometries and graphical representations).

Context is any information that surrounds and facilitates the interpretation of concepts. Ontologies contain some elements of context which are usually defined in the assumptions (for example, fountains should be represented with a point). However, other elements of context may vary from a specific use to another (for example, the security measures vary from a province to another) and are normally not included in ontologies. In order to support semantic interoperability of geospatial datacubes, we aim to explicitly identify all possible elements of context related to geospatial datacube. To guide context definition, we identify four context levels: *Goal Context* level which defines the purpose for which the geospatial datacubes will be used (such as population density), the *Domain Context* level which contains the context elements of the domain (such as forest stand), the *Dataset Context* level which consists of elements related to dataset of geospatial datacubes (such as the specification used to describe concepts), and *Concept Context* level which includes the characteristics of dimensions, or measures of datacubes (role, properties, etc).

3.2 A Theoretical Framework for Geospatial Datacube Interoperability

Some research works on interoperability were based on the communication process between people [7], [18]. Brodeur and Bédard defined a conceptual framework for spatial data interoperability based on human communication [7]. They considered the context of a given concept as a) the meaning of this concept and b) its dependencies with other concepts. We build our work on their point of view; however, we define a richer model that explicitly represents not only elements extracted from datacubes metadata, but infers new context information from these elements. We define an agent communication framework which is based on a *Context Agent* that helps agents, representing geospatial datacubes (called *Datacubes Agents*), to appropriately interpret information exchanged between them (see Fig.1). We generate two ontologies for each *Datacube Agent*: one from the datacube model and another from metadata related to that model. The *Context Agent* defines and explicitly represents the context elements using ontologies, and stores them in what will form the context knowledge base that will be available for *Datacubes Agents,* to reason about and interpret the concepts of their geospatial datacubes.

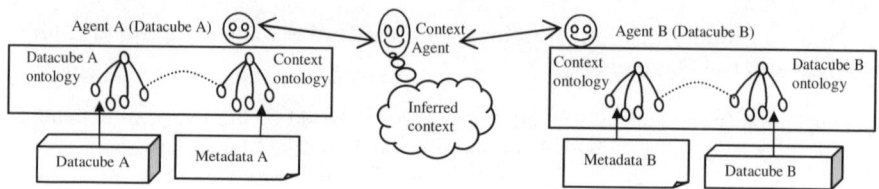

Fig. 1. A communication between agents representing geospatial datacubes

For example, *Agent A* (representing a geospatial cube developed in Ontario using English) communicates the following information to *Agent B* (representing a geospatial cube developed in Quebec using French): "The river *R*, represented with a line, intersects the forest *F*". The context elements of this model are:

- Goal Context: building bridges.
- Domain Context: geographic domain, civil engineering (large scale map).
- Dataset Context: The language used is *English*. The modeling language is UML.
- Concept Context: Time of the concept definition was 1995. The place is Ontario.

The *Context Agent* helps *Agent B* to interpret the data within its context (i.e. figure out what *Agent A* means by the concept *river*). In other words; will *river* be interpreted as *rivière* or *fleuve* (in French)? In order to make the appropriate interpretation, agents would reason about the semantic of the concept *river*, including its definition, its geometry and its context. We defined a model to reason about the semantics of the concepts in order to facilitate their interpretation and hence support the interoperability between geospatial datacubes.

3.3   A Model for Semantic Interoperability of Geospatial Datacubes

We believe that, in order to overcome the semantic heterogeneity of geospatial cubes, we should reason about their semantic. We define semantic regarding the elements of ontologies (i.e. concepts, definitions, assumption, properties such as thematic, geometric, graphic and temporal aspects) and the elements of context of geospatial datacubes concepts (e.g. language, techniques used to define spatial objects, etc.). Both ontology and context elements define the semantic characteristics of geospatial data cubes concepts. In order to guide the reasoning about the concepts semantics, and inspired by the VUEL concept (View Element) [3], we introduce a model that is based on multidimensional structure called *SemEL* (i.e. Semantic Element) where ontology and context represent the facts (see Fig.2). This model enables to explicitly represent the meaning and to define a relevant interpretation of a concept regarding the ontology and the context in which it has been defined and used. The ontology model has five dimensions (i.e. definitions, assumptions, geometries, time and graphical representations) and a fact table that has the ontology description of datacubes concepts (*Ont_Desc*) as its unique measure. *Ont_Desc* will contain textual definition, geometry, graphical and temporal properties, as well as axioms. The context model is defined according to four dimensions (i.e. Goal Context, Domain Context, Dataset Context, and Concept Context) and a fact table that has the description of context (*Context_Desc*) as its unique measure.

Since it is based on multidimensional structure, *SemEL* enables to rapidly navigate from one level to another and from dimension to another and apply reasoning capabilities (e.g. inference) to draw conclusions based on relations between semantic elements (i.e. ontology and context elements). For example, if the term *Forest* was used in England's royal context, then by inference, this term can be interpreted as a "hunting ground". More specifically, the model would allow to:

− provide the appropriate meaning of a concept (i.e. the concept defined in the specific context, represented with a specific geometry, a specific graphic, in a specific date and according a predefined assumption). For instance, the meaning of the concept *river* can be determined by 1) its definition within a general ontology: "Natural stream of water that flows in a channel" [1], and a general assumption specifying that it flows into the sea, 2) its geometry: *line*, 3) its graphic: *blue* and 4) its context elements: *English* as the language used, agriculture as the domain in which the concept is used, etc. Consequently, the appropriate meaning of the concept *river* in *French* would be *fleuve*.
− facilitate the conversion of concepts semantics. That is, navigating through different levels of dimensions, we can change the semantic characteristics of each concept and define the impact of that change on the interpretation of this concept.
− analyze phenomena changes by facilitating the comparison of different semantic elements of the same phenomenon. In fact, *SemEL* helps to rapidly navigate through different dimensions and compare different measures of a given

phenomenon (i.e. *Ont_Desc* and *Context_Desc*) and infer what changes have affected that phenomenon. For example, if an assumption, specifying that "people can easily walk", was added to the semantic of the concept *forest*, we can conclude that the forest has been managed for hiking.

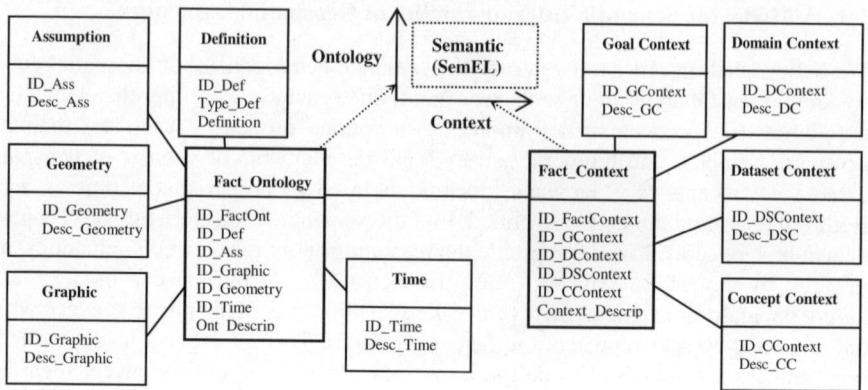

Fig. 2. A representation of the Semantic Element concept (*SemEL*)

4   Conclusion

In this paper we described the interoperability of geospatial datacubes, and we proposed a conceptual framework to overcome semantic heterogeneity problems when using geospatial datacubes. The framework is based on human communication, ontology, context, and the multidimensional structure. We defined a communication model which is based on agents representing geospatial datacubes (called *Datacubes Agents*) and a mediator agent (called *Context Agent*). The *Context Agent* helps the datacubes agents to appropriately interpret information exchanged between them. The interpretation is supported by a model which is based on the multidimensional paradigm (i.e. *SemEL*). The semantic of concepts will be discussed regarding the dimensions of this model (i.e. elements of both ontology and context).

Further work is required to refine *SemEL* and define a mapping between two different multidimensional models. Then we would implement our framework.

Acknowledgments. We wish to acknowledge the contribution of the NSERC Industrial Research Chair in Geospatial Databases for Decision Support.

References

1. Answers, http://www.answers.com/ (Last visted September 20, 2006)
2. Bédard, Y., Rivest, S., Proulx, M.J.: Spatial on-line analytical processing (SOLAP): concepts, architectures and solutions from a geomatics engineering perspective. In: Koncillia, W.R (ed.) Data warehouses and OLAP: concepts, architectures and solutions (2005)

3. Bédard, Y., Bernier, E.: Supporting Multiple Representations with Spatial View Management and the Concept of "VUEL". In: Joint Workshop on Multi-Scale Representations of Spatial Data, ISPRS WG IV/3, ICA Com. on Map Generalization (2002)
4. Bédard, Y., Merrett, T., Han, J.: Fundamentals of spatial data warehousing for geographic knowledge discovery. In: Miller, H.J., Han, J. (eds.) Geographic Data Mining and Knowledge Discovery (2001)
5. Bishr, Y.: Overcoming the semantic and other barriers to GIS interoperability. Int. J. Geographical Information science 12, 299–314 (1998)
6. Brodeur, J.: Interopérabilité des données géospatiales: élaboration du concept de proximité géosémantique. Ph.D. Dissertation. Université Laval (2004)
7. Brodeur, J., Bédard, Y.: Geosemantic Proximity, a Component of geospatial Data Interoperability. In: Internat. Workshop, Semantics of Enterprise Integration, ACM Conference on OOPSLA, pp. 14–18 (2001)
8. Brodie, M.L.: The promise of distributed computing and the challenge of legacy Information systems. In: Proceedings of the IFIP WG2'6 Database Semantics Conference on Interoperable Database Systems (1992)
9. Codd, E.F., Codd, S.B., Salley, C.T.: Providing OLAP (On-Line Analytical Processing) to User- Analysts: An IT Mandate. Hyperion white papers, p. 20 (1993)
10. Goodchild, M.F., Egenhofer, M.J., Fegeas, R.: Interoperating GIS. Report of a specialist meeting held under the auspices of the Varenius project panel on computational implementations of geographic concepts. Santa Barbara, California (December 5-6, 1997)
11. Guarino, N.: Formal Ontologies and Information Systems. In: Guarino, N. (ed.) Proe. of FOIS 1998, pp. 3–15. IOS Press, Amsterdam (1998)
12. Gruber, T.R.: Toward principles for the design of ontologies used for knowledge sharing. Original paper presented at the International Workshop on Formal Ontology (1993)
13. Harvey, F., Kuhn, W., Pundt, H., Bishr, Y., Riedemann, C.: Semantic Interoperability: A Central Issue for Sharing Geographic Information Annals of Regional Science. Special Issue on Geo-spatial Data Sharing and Standardization, pp. 213–232 (1999)
14. ISO/IEC 2382. Information technology – Vocabulary – Part 1: Fundamental terms (1993)
15. Malinowski, E., Zimanyi, E.: Representing spatiality in a conceptual multidimensional model. In: Proceedings of ACM int. workshop on Geographic information systems, ACM Press, New York (2004)
16. Rivest, S., Bédard, Y., Proulx, M., Nadeau, M., Hubert, F., Pastor, J.: SOLAP Technology: Merging Business Intelligence with Geospatial Technology for Interactive Spatio-Temporal Exploration and Analysis of Data. J. ISPRS Advances in spatio-temporal analysis and representation, 17–33 (2005)
17. Roehrig, J.: Information Interoperability for River Basin Management. J. Technology Resource Management and Development - Scientific Contributions for Sustainable Development 2 (2002)
18. Xhu, Z., Lee, Y.C.: Semantic Heterogeneity of Geodata. In: Proceedings of ISPRS Commission IV Symposium (2002)

On Languages for the Specification of Integrity Constraints in Spatial Conceptual Models

Mehrdad Salehi[1,2], Yvan Bédard[1,2], Mir Abolfazl Mostafavi[2], and Jean Brodeur[2,3]

[1] Canada NSERC Industrial Research Chair in Geospatial Databases for Decision Support
[2] Center for Research in Geomatics, Department of Geomatics Sciences, Laval University, Quebec City, Canada, G1K 7P4
[3] Center for Topographic Information, Natural Resources Canada, Sherbrooke, Canada, J1J 2E8
mehrdad.salehi.1@ulaval.ca, yvan.bedard@scg.ulaval.ca,
mir-abolfazl.mostafavi@scg.ulaval.ca, brodeur@nrcan.gc.ca

Abstract. This paper intends to classify, review, and compare the existing languages for the specification of spatial integrity constraints at the conceptual level. We classify these languages into natural, visual, first-order logic, and hybrid and review their syntax and semantics. We compare these language categories based on expressiveness and pragmatics qualities. The results of this study show that controlled natural languages and natural hybrid languages with pictograms are good candidates for expressing spatial integrity constraints at the conceptual level of spatial databases. At the end, we identify new research challenges that should be addressed in this domain in order to facilitate spatial integrity constraint specification and representation.

1 Introduction

A database design process typically starts with a *conceptual model* and moves towards development based on an *implementation model*. A Conceptual model concerns clients' views of the data of interest. In this paper, we consider its abstraction level the same as the Computation Independent Model (CIM) in the Object Management Group (OMG)'s Model-Driven Architecture (MDA) [1]. An implementation model, however, serves as a developer's view of the data for the implementation on a given family of platforms (e.g., transactional DBMS, OLAP servers) or for a specific software package. These respectively correspond to the Platform Independent Model (PIM) and Platform Specific Model (PSM) in MDA. A conceptual model includes two components: (1) *a schema* which shows how the data are structured, and (2) a *data dictionary* which includes additional information about the data [2].

Typically, integrity constraints (ICs) are defined at the CIM level with the client's point of view, and then translated for the PIM and PSM levels. Integrity Constraints describe semantically valid states of the data and preserve database integrity [3]. It is emphasized that ICs convey important semantic information about the database application domain [4]. Hence, it is necessary to specify ICs at the conceptual model

J.-L. Hainaut et al. (Eds.): ER Workshops 2007, LNCS 4802, pp. 388–397, 2007.

of applications (i.e., CIM) and to make them amenable to implementation via the PIM and PSM. Each of these levels may require a different language to express these ICs. In this paper, we are interested in the languages used for the definition of ICs in spatial conceptual level.

In spatial conceptual models, additional types of ICs, known as Spatial ICs (SICs), must be specified. Examples of these ICs are *topological ICs* restricting topological relationships, e.g., two building do not *overlap* and *non-topological ICs* restricting non-topological relationships (such as metric), e.g., the *distance* between a residential area and a gas station must be more than *30 meters*.

Different kinds of languages for the specification of SICs are proposed in the literature. Despite the existence of research surveys on modeling approaches for spatial database applications (e.g., [5],[6]), there is a gap concerning the investigation of the proposed languages for expressing SICs. This paper intends to review and classify the state-of-the-art SICs specification languages at the conceptual level. In addition, it compares the languages and identifies their strengths and weaknesses which contributes to developing future IC specification languages, e.g., for spatial multidimensional models (datacubes).

To this end, Section 2 reviews, classifies, and discusses the state-of-the-art languages for the specification of SICs. Section 3 compares these categories of languages. Final conclusions and future research requirements are given in Section 4.

2 Classifying and Reviewing the Languages

Various kinds of languages have been developed for the specification of SICs at different abstraction levels. For example, an ontology-driven language named Semantic Web Rule Language (SWRL) is used by [7] for expressing SICs within ontologies. At the implementation level, languages like Spatial SQL [8] are proposed. However, in this paper we focus on the languages used at the *conceptual level* within the spatial database community, i.e., natural languages [9] [10], a visual language [11], a hybrid natural language [12], spatial OCL [13], spatio-temporal modeling languages [14] [4], and first-order logic [15]. In spite of the capability of spatio-temporal conceptual modeling languages to express SICs directly in the conceptual schema, more specific languages, called Integrity Constraint Specification Languages (ICSL) have been introduced. These languages specify SICs in a data dictionary since they cannot be expressed efficiently in a schema. We classify all these possibilities into four main categories, i.e., natural languages, visual languages, first-order logic language, and hybrid languages.

2.1 Natural Languages

People use natural languages (e.g., English and French) for their daily communications. These languages are the simplest and easiest languages for the clients to specify SICs. We classify natural languages into two categories: (1) *Free natural language*, and (2) *Controlled natural language* as described below.

2.1.1 Free Natural Languages
A free natural language is a natural language without implication of any limit on syntax and semantics of the language. *Syntax* of a language describes the way

language's symbols should be combined to create well-formed sentences. However, *semantics* reveals the meaning of syntactically correct strings in a language [16]. SICs can be defined using a free natural language. For example, in English: "A road does not cross a building" is the simplest form of specifying SICs. Free natural languages support a rich vocabulary. However, they are sometimes ambiguous or used too loosely. In such languages, several words may bear the same semantics and a word may have several meanings depending on the context. Nevertheless, they remain today's most widely used language for expressing SICs.

2.1.2 Controlled Natural Languages

In order to overcome the ambiguity of free natural languages, controlled natural languages have been proposed. Controlled natural languages are subsets of natural languages whose syntax and semantics have been restricted [17].

For the specification of topological ICs, [10] proposes a controlled natural language in a form similar to predicates. Within this language, the syntax of a topological IC consists of entity class 1, a topological relationship between entity classes characterized by extended 9-intersection model [18] (e.g., inside, cross, join), entity class 2, and a multiplicity quantifier which can be one of the followings: forbidden, at least *n* times, at most *n* times, and exactly *n* times:

```
(Entity class1, Relation, Entity class2, Quantifier)
```

For instance, the following SIC in free natural language "Road cannot cross a Building" is specified using this controlled natural language as follows:

```
(Road, Cross, Building, forbidden)
```

Similarly, [9] introduces a controlled natural language for the definition of topological ICs. In this language, the syntax of a topological IC is composed of object class 1, one of the eight topological relationships *equal, disjoint, intersect, touch, cross, within, contain, overlap* (of ISO19107:2003 – Geographic information – Spatial schema) extended by three notions *tangent, border,* and *strict,* object class 2, and cardinality of the relationship. For example, in this language, "each road segment should be end-to-end connected to at least one or at most two road segments" is expressed as:

```
Road Segment Touch-Tangent Road Segment [1,2]
```

Where [-,-] is the cardinality of the topological IC indicating minimum and maximum number of instances of the two classes defined in the relationship.

2.2 Visual Languages

A visual language employs graphical and image notations to communicate the information with the user. Various visual languages for different purposes have been proposed; for a survey of these languages refer to [19]. In this section, by visual languages, we mean the languages that use only graphical or image notations. We will study hybrid languages, which combine visual and other languages, in Section 2.4.

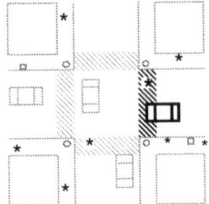

Fig. 1. A visual language which highlights SICs: "cars and people cannot be inside a crosswalk simultaneously" [11]

For the specification of SICs, [11] present a model and a visual language based on depicted pictures. The pictures show unacceptable database states, termed "constraint pictures". For instance, Figure 1 expresses topological IC "cars and people cannot be in a crosswalk simultaneously". In this figure, people, cars, and crosswalks are represented by asterisks, small boxes, and dashed lines respectively.

The aim of visual languages is to create an easy to use and perceive language [20]. However, representing all the information of SICs using a visual language necessitates clients to learn the meaning of every visual construct of the language to be able to use it, and to understand very well the very specific context of it usage, which requires a big effort. In spite of such efforts, several ambiguities and unintended meanings can emerge from such representations.

2.3 First-Order Logic Language

First-order logic language is a common formal language for representing ICs in databases and knowledge bases. First-order logic allows quantification on objects, i.e., the first-order entities exist in the real world.

Hadzilacos and Tryfona [15] describe a formal model with first-order logic language for expressing topological ICs at the conceptual level. In this language, the syntax of a topological sentence is built of atomic topological formulae with negation ($\neg$), conjunction ($\wedge$), disjunction ($\vee$), and universal ($\forall$) and existential ($\exists$) quantification. Atomic topological formulae consist of topological relations between objects based on binary topological relationships [21], geometric operators over objects, and comparison between attributes of objects. As an example, topological IC "roads and buildings are disjoint" in first-order logic is:

$$\forall(r,b)[\text{disjoint}(r,b)]$$

Where r and b stand for "road" and "building" respectively, and *disjoint* is a topological relationship.

First-order language supports precise semantics and syntax which avoids error in interpretation. Nevertheless, understanding this language requires a mathematical background. Clients do not necessarily have such a background.

2.4 Hybrid Languages

A number of languages used for the specification of SICs are not purely natural, visual, or logical and can be best described as a combination of them. We call such languages *hybrid languages*. Depending on the dominant component of a language, a hybrid language can be *visual hybrid* or *natural hybrid*.

2.4.1 Visual Hybrid Languages

The main part of visual hybrid languages consists of visual symbols, which are enriched by a natural language. As a limited number of visual constructs are easy to perceive [20], visual hybrid languages combine this advantage with the richness of natural languages. Unified Modeling Language (UML) is such a language.

As previously stated, a conceptual model consists of conceptual schema and data dictionary. We did not find any visual hybrid ICSL for defining SICs in data dictionary. However, spatio-temporal conceptual modeling languages, such as Perceptory [14] and MADS [4], are among visual hybrid languages that can express a number of ICs (e.g., topological and synchronization constraints) in a conceptual schema. Most of these modeling languages provide spatial and temporal data types or object stereotypes with associated icons. While some of these languages provide a number of spatial and temporal topological relationships (e.g., disjoint, overlap, starts, finish), others leave it up to the client to select the most appropriate relationships based on their field vocabulary (e.g., homemade 3D set of relationships, spatio-temporal relationships). These constructs can be employed for defining SICs.

For instance, a conceptual schema for a road network application is likely to consist of object classes such as "route" and "roundabout" classes, their attributes, and the relationships between classes, e.g., each roundabout is crossed by at least one route. This schema relationship "cross" between "roundabout" and "route" expresses a SIC. The expression of this SIC using a visual hybrid modeling language Perceptory [14] is shown in Figure 2. In this figure, "route" class and "roundabout" class hold geometric attribute of type line and polygon. The SIC between two classes is expressed by the relationship between two classes.

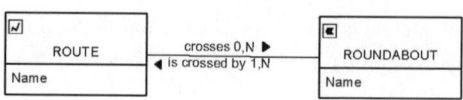

Fig. 2. SIC using visual hybrid modeling language Perceptory at the conceptual schema

The elements of schema are necessary for maintaining database consistency. However, they cannot express all the possible SICs (e.g., the minimum allowable distance between two entities) as simply, easily, and shortly as ICSL which are designed specially for this purpose. Moreover, SICs are detailed information and may end up over hundreds of times the number of object classes in a schema [14]. Hence, systematically representing them in a conceptual schema contradicts the rule that conceptual schema should be concise enough to remain readable. Thus, visual hybrid modeling languages are not sufficient to be used alone for defining all SICs in a practical way.

2.4.2 Natural Hybrid Languages

A natural hybrid language consists of a natural language as well as pictograms or symbols, but the natural language part is dominant.

2.4.2.1 Natural Hybrid Language with Pictograms. A natural hybrid language with pictograms consists of a natural language which is enriched by pictograms [22]. In the spatial database community, pictograms are intuitive symbols employed for the representation of objects' geometries. Normand [12] proposes a natural hybrid language with a limited number of pictograms to overcome the limitations of visual hybrid modeling languages in the specification of SICs. This language can specify all types of SICs between object classes, including the SICs that depend on specific attribute values or on objects' geometries (when alternate or multiple geometries are involved). In this language, class names, topological relationships, and cardinality of relationships are specified by controlled natural language. Additionally, pictograms ⊡⟋⊡ etc., stand for point, line, polygon, and more complex geometries (see [14] for all supported cases) of the classes. The language uses a table for each object class to express its SICs with regards to all other object classes (the class name is at the top of the table). The table consists of six columns: "Dimension Object 1" shows the geometry of object 1 (there maybe several in case of complex geometries), "Operator" describes logical operator (AND, OR, XOR) when more than one spatial relationship is possible between two classes, "Relations" addresses the spatial relationship between two classes based on 9 intersection model [23] or on metric constraints (e.g., maximum distance), "Cardinality" expresses the minimum and maximum occurrences of this relationship, "Object 2" represents the name of object class 2 (there may be several), and "Dimension Object 2" holds the geometry of object 2 (there may also be several in case of complex geometries for object 2). This method shows all its power when complex SICs or several SICs are involved for an object class. For instance, Figure 3 shows a simple SIC between "Canal" and "Watercourse" classes which support multiple representation for "Watercourse" class.

Canal

Dimension Object 1	Operator	Relations	Cardinality	Object 2	Dimension Object 2
⟋	OR	Disjoint	-	Watercourse	⟋
		Adjacent	1-2		
	OR	Disjoint	-		⊡
		Adjacent	1-2		

Fig. 3. Specification of a topological IC using Normand's method [12]

Figure 3 describes that a linear canal must be disjoint from a linear watercourse or must be adjacent to one or at last two linear watercourses. In addition, a linear canal must be disjoint from a polygonal watercourse or must be adjacent to one or at last two polygonal watercourses.

2.4.2.2 Natural hybrid language with symbols. It is possible to add a number of specific symbols to a controlled natural language to make it more abstract than a natural language for knowledge representation. Object Constraint Language (OCL), which is used along UML and developed by the OMG, is such a language. This language allows expressing constraints and queries on UML class diagrams. OCL is a formal language designed to overcome the ambiguity of natural languages and the disadvantage of formal languages which are usable only for individuals with a mathematical background [24].

In order to precisely define topological ICs along with UML for conceptual modeling of spatial databases, [13] proposes a spatial extension of OCL. The authors in [13] add new basic geometric primitives, i.e. `point`, `polyline`, and `polygon`, generalized by `BasicGeoType` to the OCL meta-model. Furthermore, for describing topological ICs based on 9 intersection method [23], the following new OCL operators: `disjoint`, `meet`, `contains`, `covers`, `inside`, `coveredBy`, `equal`, and `overlap` are defined and added to OCL. For specifying the SIC "all buildings are disjoint from all roads" one can write:

```
context Building inv:
Road.allInstances()->forAll(R|R.geometry->Disjoint self.geometry))
```

Where "context" specifies the model entity for which the OCL expression is defined; `allInstances` and `forAll` are universal and existential operators respectively. "R" refers to one instance of Road class. The keyword `self` returns an instance of the class declared in the "context." Although less abstract than the mathematically formal languages, it remains a language with a low-level of readability from the clients' point of view.

3   Comparing Languages for the Specification of SICs

Inspired by the work of [16], [25], and [26], by 20 years of research and experimentation of our team in real projects, and by the final objective of our research, we analyzed the languages discussed in Section 2 by focusing on the two aspects: expressiveness and pragmatics.

Expressiveness takes into consideration semantic quality (the degree of correspondence between the concepts that a language supports and the meaning of SICs in the real world), syntactic quality (the degree to which rules govern the structure of expressions), richness (the capability to express clearly the needed elements of SICs), and inherence (the precision of a language to be straight to the point and focus on essential aspects of SICs).

Pragmatics, in the context of conceptual modeling, refers to two qualities. First, it refers to the usability of the language by those who must express the domain-related constraints that are needed for the system under development, and also who must validate the database designer's understanding of these constraints, that is the clients (i.e., the end-users of the system, not the computer). Second, pragmatics also refers to the facility to translate (by the database designer when necessary) these ICs in a technical language amenable to computerization. In our context of expressing ICs for conceptual modeling, the former pragmatic quality has priority over the latter since experience has taught us that we have higher chances of success when we ask the database designer to translate users requirements into technical languages for implementation than when we ask the clients to learn technical languages to express and validate themselves their database requirements. In other words, we consider that a good pragmatic quality lets the client focus on the ICs, not on learning a new language (N.B. the client's involvement into database development is a *ad hoc* activity taking place only once or very few times in a career). Using OMG's MDA vocabulary, we are targeting the CIM and PIM levels rather than the PSM level.

As this research is in progress, in this paper we compare the categories of the languages instead of their instances. In addition, since we are not aiming at finding "the best" language (if such a thing exists) but rather at summarizing potential avenues for our research on SICs for spatial datacubes, we focused on expressiveness and pragmatics as two general indicators rather than considering a larger series of detailed criteria for a very precise comparison. Thus, this research is not an end *per se* but a mean to better attack ICSL for spatial datacubes.

Table 1 shows the evaluation results of the languages' categories with respect to expressiveness and pragmatics, and ranks them by "good", "medium", and "weak". The values in this table represent our opinion resulting from a study of the literature and of 20 years of experience in several industrial projects involving spatial databases, spatio-temporal databases, and spatio-temporal datacubes in diverse fields and implementations with very diverse technologies.

Table 1. Our opinion about the categories of languages for the specification of SICs

Language		*Expressiveness*	*Pragmatics*
Natural	Free	Medium	Good
	Controlled	Good	Good
Visual		Weak	Medium
First-Order Logic		Good	Weak
Visual Hybrid		Good	Medium
Natural Hybrid	With Pictograms	Good	Good
	With Symbols	Good	Medium

Considering the above table, controlled natural languages and natural hybrid languages with pictograms seem more suitable ICSL for the conceptual level. Although visual hybrid languages could be considered good candidates, the existing proposed visual hybrid languages typically cope with SICs in the conceptual schema. Consequently, due to the large number of ICs and their level of complexity, [14] suggested defining ICs in data dictionary in a complementary manner to those defined in the schema in order to prevent the schemas from becoming too complex and unreadable. Thus, conceptual modeling languages can be used to a certain degree if completed by detailed SICs described in another language.

Finally, the pragmatics values change according to the target audience. For example, when moving from conceptual to implementation modeling, different suggestions come out in such cases. Therefore, one will typically have to deal with more than one ICSL to completely cover the development process from needs analysis to design, implementation, operation, and updating of a database.

4 Conclusions and Future Research

In this paper, we categorized the languages for the specification of SICs at the conceptual level into: natural (free and controlled), visual, first-order logic, and hybrid (visual, natural with pictograms, and natural with symbols) languages. We discussed on state-of-the-art languages and compared the categories of languages based on their expressiveness and pragmatic qualities. The results indicate that controlled natural languages and natural hybrid languages with pictograms are the best choices as ICSL at the conceptual level. One may also expect to see a different result for the implementation level and have to deal with more than one ICSL.

The existing spatial ICSL presented in this paper have been developed for transactional databases. However, to specify ICs of spatial datacubes, additional syntax and semantics concerning multidimensional conceptual models should be taken into account [27]. Our future research is focused on developing ICSLs for spatial multidimensional databases at the conceptual level. This ICSL can be an extension to a discussed language in this paper. Therefore, our next step is to compare into more details the quality of the proposed languages' instances.

Acknowledgments. The authors would like to acknowledge the financial support from the Canada NSERC Industrial Research Chair in Geospatial Databases for Decision Support. We are also grateful to the three anonymous reviewers for their useful comments on this paper.

References

1. Miller, J., Mukerji, J.: MDA Guide Version 1.0, OMG Document (2003)
2. Brodeur, J., Bédard, Y., Proulx, M.J.: Modeling Geospatial Application Databases using UML-based Repositories Aligned with International Standards in Geomatics. In: ACMGIS. Proceedings of Eighth ACM Symposium on Advances in Geographic Information Systems, Washington, D.C, pp. 39–46. ACM Press, New York (2000)
3. Godfrey, P., Grant, J., Gryz, J., Minker, J.: Integrity Constraints: Semantics and Applications. In: Logics for Databases and Information Systems, pp. 265–306. Kluwer, Dordrecht (1997)
4. Parent, C., Spaccapietra, S., Zimányi, E.: Conceptual Modeling for Traditional and Spatio-temporal Applications: The MADS Approach, p. 466. Springer, Heidelberg (2006)
5. Miralles, A.: Ingénierie des Modèles pour les Applications Environnementales. Ph.D. Thesis, Université de Montpellier II (2006)
6. Parent, C.: A Framework for Characterizing Spatio-temporal Data Models. In: Advances in Multimedia and Databases for the New Century, pp. 89–97. World Scientific, Singapore (2000)
7. Mas, S., Fei, W., Wolfgang, R.: Using Ontologies for Integrity Constraint Definition. ISSDQ05. Beijing, China (2005)
8. Egenhofer, M.J.: Spatial SQL: A Query and Presentation Language. IEEE Transactions on Knowledge and Data Engineering 6(1), 86–95 (1994)
9. Vallières, S., Brodeur, J., Pilon, D.: Spatial Integrity Constraints: A Tool for Improving the Internal Quality of Spatial Data. In: Devillers, R., Jeansoulin, R. (eds.) Fundamentals of Spatial Data Quality, Great Britain, ISTE Ltd., pp. 161–177 (2006)
10. Ubeda, T., Egenhofer, M.J.: Topological Error Correcting in GIS. In: Scholl, M.O., Voisard, A. (eds.) SSD 1997. LNCS, vol. 1262, pp. 283–297. Springer, Heidelberg (1997)
11. Pizano, A., Klinger, A., Cardenas, A.: Specification of Spatial Integrity Constraints in Pictorial Databases. Computer 22(12), 59–71 (1989)

12. Normand, P.: Modélisation des contraintes d'intégrité spatiale, théorie et exemples d'applications, M.Sc. Thesis, Laval University, Quebec, Canada (1999)

13. Kang, M., Pinet, F., Schneider, M., Chanet, J., Vigier, F.: How to Design Geographic Databases? Specific UML Profile and Spatial OCL Applied to Wireless Ad Hoc Networks. In: Proceedings of 7th AGILE Conference on Geographic Information Science (2004)

14. Bédard, Y., Larrivée, S., Proulx, M.J., Nadeau, M.: Modeling Geospatial Databases with Plug-Ins for Visual Languages: A Pragmatic Approach and the Impacts of 16 Years of Research and Experimentations on Perceptory. In: Wang, S., Tanaka, K., Zhou, S., Ling, T.-W., Guan, J., Yang, D.-q., Grandi, F., Mangina, E.E., Song, I.-Y., Mayr, H.C. (eds.) CoMoGIS 2004. LNCS, vol. 3289, pp. 17–30. Springer, Heidelberg (2004)

15. Hadzilacos, T., Tryfona, N.: A Model for Expressing Topological Integrity Constraints in Geographic Databases. In: Frank, A.U., Formentini, U., Campari, I. (eds.) Theories and Methods of Spatio-Temporal Reasoning in Geographic Space. LNCS, vol. 639, pp. 252–268. Springer, Heidelberg (1992)

16. Slonneger, K., Barry, L.K.: Formal Syntax and Semantics of Programming Languages: A Laboratory Based Approach, p. 637. Addison-Wesley, Reading (1995)

17. Schwitter, R.: Controlled Natural Language as Interface Language to the Semantic Web. In: 2nd Indian International Conference on Artificial Intelligence, pp. 1699–1718 (2005)

18. Clementini, E., Di Felice, P.: A Comparison Method for Representing Topological Relationships, Information Systems, vol. 80, pp. 1–31 (1994)

19. Chang, S.: Visual languages: A Tutorial and Survey. Software 4(1), 29–39 (1987)

20. Boursier, P., Mainguenaud, M.: Langages de Requêtes Spatiales: SQL Étendu vs. Langage Visuel vs. Hypermédias. SIGAS 2(1), 37–51 (1992)

21. Egenhofer, M., Herring, J.: Pint-set topological spatial relations. IJGIS 5(2), 133–152

22. Bédard, Y., Larrivee, S.: Spatial Database Modeling With Pictrogrammic Languages. In: Shekhar, S., Xiong, H. (eds.) Encyclopedia of GIS, p. 14. Springer, NY (in press)

23. Egenhofer, M., Herring, J.: Characterizing Binary Topological Relations between Regions, Lines, and Points in Geographic Databases. NCGIA Technical Report 94-1 (1994)

24. UML 2.0 OCL Specification, p. 170 (2003)

25. Lindland, O.I., Guttorm, S., Solvberg, A.: Understanding Quality in Conceptual Modeling. IEEE Software 11(2), 42–49 (1994)

26. Teeuw, W.B., Van den Berg, H.: On the Quality of Conceptual Models. In: Behavioral Models and Design Transformations: Issues in Conceptual Modeling, ER 1997 (1997)

27. Salehi, M., Bédard, Y., Mostafavi, M.A., Brodeur, J.: Towards Integrity Constraints of Spatial Datacubes, ISSDQ 2007, the Netherlands (2007)

Approximate Queries by Relaxing Structural Constraints in GIS

Arianna D'Ulizia, Fernando Ferri, and Patrizia Grifoni

IRPPS-CNR, via Nizza 128, 00198 Roma, Italy
{arianna.dulizia, fernando.ferri, patrizia.grifoni}@irpps.cnr.it

Abstract. In the Geographic Information System (GIS) domain the evaluation of similarity among geographical objects plays an important role. This paper proposes an approach based on semantic and structural similarities to provide more flexible matches between the query search condition expressed by the user and the possible answers provided by the system. The relaxation model considers with different weights the semantic similarity of geographical concepts, that is evaluated by adopting the *information content* approach, and the structural similarity of the attributes and types of geographical objects that is inspired by the *maximum weighted matching* problem in bipartite graphs. The aim of the proposed methodology is to relax structural query constraints, in order to obtain meaningful answers for imprecise or missing data.

Keywords: Geographical information systems, structural similarity, semantic similarity, approximate queries.

1 Introduction

Geographical information systems (GIS) have recently received considerable attention due to the widespread use of their applications in many different fields, such as location-based services, engineering applications and so on. These systems usually have to manage a large amount of data and have to be used by people that are generally non-computer scientists. Visual GIS query languages allows to express geographical queries by using a drawing and consequently they are most suitable for such kind of users, also due to their ease of use, user friendliness and to the fact that there is no need to know the language textual syntax. In spite of these advantages, the use of Visual GIS query languages can lead to unexpected results. In fact, the user query could involve geographical concepts with no instances in the database because of both the user inexperience and the lack of knowledge of the database structure and contents. As a consequence the user obtains empty answers and he/she has to reformulate the pictorial query several times. Approximate queries provide a solution to this problem. The use of a similarity model facilitates the identification of objects that are conceptually close but not identical, and therefore supports information retrieval by allowing the user to receive not only the exact answers but also similar answers obtained by relaxing some of the constraints. In the case of queries on a geographical database, constraints can be classified as three main types: spatial,

J.-L. Hainaut et al. (Eds.): ER Workshops 2007, LNCS 4802, pp. 398–408, 2007.

semantic, and structural. Spatial constraints refer to the spatial relationships existing between geographical objects, semantic constraints refer to the concepts represented by the geographical objects and structural constraints refer to the internal characteristics of geographical objects.

In this paper we propose an approach to relax semantic and structural constraints of a geographical query. This work is an extension of a previous work [1] in which we propose a method to evaluate structural similarity between the attributes of geographical objects. The relaxation model we present in this paper considers with different weights the structural similarity of the attributes and types of geographical objects and the semantic similarity of geographical concepts, that is evaluated by adopting the *information content* approach. To validate the approach we performed usability tests and we show the results we obtained.

The paper is organized as follows. Section 2 examines some important studies from the literature. Section 3 briefly introduces the geographical pictorial query language GeoPQL and its main peculiarities. Section 4 explains the similarity model and the evaluation of semantic and structural similarities between geographical objects. Section 5 describes the usability tests we performed to validate the proposed approach. Finally, Section 6 provides some conclusions and future works.

2   Related Research

The need to consider geographical queries involving unavailable information requires the evaluation of similarity. When a query search condition cannot be matched in the database because a geographical concept is missing, users would rather receive approximate answers through the use of a similar concept than no information at all. As a consequence, similarity is an important, widely used concept that plays a significant role in GIS, as it supports the identification of geographical objects that are conceptually close, but not equal. The problem of how to evaluate structural similarity has been addressed by various authors in different fields and from different points of view.

Information retrieval is one of the most common field in which structural similarity is studied. In this context this kind of similarity refers to the evaluation of distances for estimating the similarity between tree structures in terms of the hierarchical relationship of their nodes. In particular, methods for evaluating structurally similar XML documents, modeled as ordered tree, are explored. In [2] the problem of automatically mapping semantically similar but syntactically different elements in an XML document using the inherently available structural information is addressed. In order to match two nodes in two different hierarchies, the paper proposes to map the two hierarchies into a common Multi-dimensional space through a data analysis method, called Multi-dimensional scaling, for discovering the underlying spatial structure of the elements of the hierarchy from the distance information among them. Another paper that focuses on the identification of structurally similar XML document is [3]. It presents an approach to calculate trees distances and define a structural distance metric to estimate the structural similarity between hierarchies. This approach uses a dynamic programming algorithm which is close to Chawathe's algorithm.

Another interesting field dealing with the computation of structural similarity is the ontology matching. In this context, structural similarity refers to a measure of distances among concepts based on the similarity of their structures in term of attributes, types and values. In [4] a proposal for evaluating concept similarity in an ontology management system, called SymOntos, is presented. The authors consider four notions of similarity: *tentative similarity*, declared in the concept definition for determining a preliminary similarity degree with other concepts; *flat structural similarity*, that considers the concepts structure (attributes, values) and evaluates the similarity of every concept; *hierarchical structural similarity*, that pertains to concept pairs which are hierarchically related; *concept similarity*, obtained by combining the structural and the tentative similarity. Moreover, a method that allows similarity evaluation among concepts is derived on the basis of the previous definitions.

The discussion of the related works shows that, even if the evaluation of the structural similarity has been faced by different fields, there is not an approach to evaluate this kind of similarity between geographical objects in order to provide approximate queries in a GIS domain, as intended in our work.

3   GeoPQL

In this section we describe the main features of the visual GIS query language, named GeoPQL, that we use to formulate geographical queries.

The Geographical Pictorial Query Language (GeoPQL, [5]) allows the user to draw pictorial geographical queries and obtain the objects selected in the geographical database, which are the result of the query target. To achieve that, this language is based on various concepts that represent also the peculiarities of the language.

First of all, GeoPQL allows to specify queries using *Symbolic Graphical Objects* (*SGOs*) that have the appearance of the three classic types of shapes: point, polyline and polygon. In literature SGOs are usually referred to as icons, symbols, feature, etc. A formal description of this data structure used by GeoPQL is given below.

Definition 1. An SGO is a 4-tuple $\psi = <$id, objclass, Σ, Λ $>$ where:

- id is the SGO identifier. The id is the code assigned in order to identify each SGO in queries in a project. A GeoPQL project is formed by a set of queries and the reference to the selected database for querying.
- objclass is the class name iconized by ψ;
- Σ represents the attributes to which the user can assign a set of values; this allows selection to be made among the classes of objects or their instances, iconized by ψ. The attributes of a class correspond to the attributes of the layer of the database represented by the class. In a query, the user can express the set of desired values for an attribute by an expression involving boolean and other operators, value, references to other attributes of the same and of other objects and classes.
- Λ is the ordered set of coordinate pairs (h, v), which defines the spatial extent and position of the SGO with respect to the coordinate reference system of the working area. Using Λ it is possible to determine spatial relationships among symbolic graphical objects.

Secondly, GeoPQL algebra consists of 12 operators. It needs to explicitly express only three operators: Geo-any (to allow any relationship between a pair of SGO to be considered valid), Geo-distance (to express the numeric value which specifies the constraint linked to the operator), and Geo-alias (to allow the same SGO to be duplicated in order to express the OR operator). All other operators, that are Geo-union, Geo-difference, Geo-disjunction, Geo-touching, Geo-inclusion, Geo-crossing, Geo-pass-through, Geo-overlapping and Geo-equality, are applied without being expressing, as they are automatically deduced by the query's pictorial representation.

Another peculiarity is that GeoPQL interfaces with ESRI's GIS ArcView®. The geographical query drawn by the user is translated in an eXtended SQL language, called XSQL. The textual query is continuously updated during the drawing phase, and it follows modifications, deletions and shifting of the pictorial query. At the end of the drawing phase the XSQL query is translated into ArcView® and executed on ArcMap® (the geographical database of ArcView®). The user thus obtains a layer on which the geographical objects that satisfy the query are stored and visualized.

Figure 1 shows the main window of GeoPQL, in which there is an area for the pictorial query formulation, an area for the XSQL query visualization and an area for the visualization of the relationships between SGOs.

The query formulated in the pictorial area is "Find all the Towns which are PASSED THROUGH by a Railway AND which INCLUDE a Church".

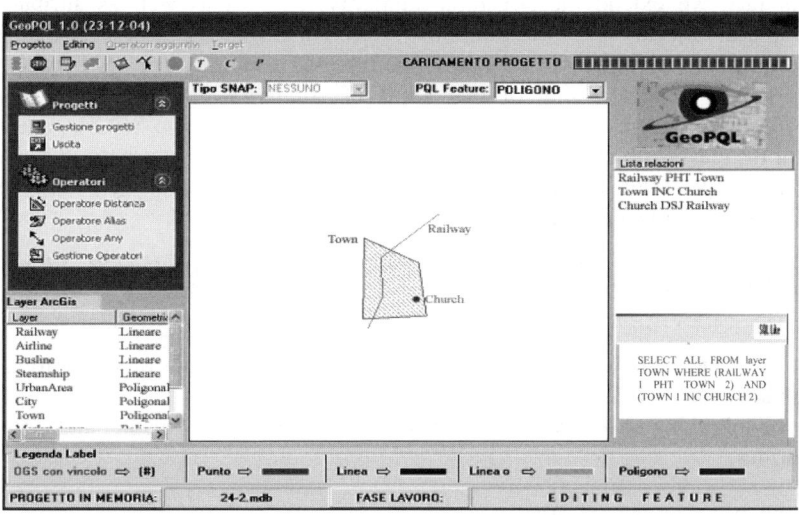

Fig. 1. Formulation of a pictorial query by GeoPQL

In the following we suppose that the geographical database of ArcView® consists of the geographical concepts and the relative attributes given below.

Line (number:int, time_path:int, dep_station:string, dest_station:string)
Railway (number:int,local:boolean,goods:boolean,passengers:int,
        dep_railway_station:string, dest_railway_station:string)

Airline(number:int,company:string,travelers:int,dep_air_terminal:string,
dest_air_terminal:string)
Busline(number:int,local:boolean,passengers:int,dep_bus_terminal:string,
dest_bus_terminal:string)
Steamship (number:int,passengers:int,dep_port:string,dest_port:string)
Urban_area (region:string,country:string)
City (name:string,region:string,country:string,numpeople:int)
Town (name:string,region:string,country:string,inhabitants:int)
Market_town (name:string,region:string,country:string,market_day:string)
Building (address:address,height:int,width:int,floors:int)
Church (name:string,location:address,foundation:date)
Cathedral (name:string, location:address, style:string, foundation:date).

4 The Similarity Model

The aim of the proposed similarity model is to identify SGOs that are conceptually and structurally close in order to not receive empty answers when the user query involves an SGO with no instances in the database but to receive approximate answers by replacing the missing SGO by the most similar one which has instances in the database. This approach can be considered a refinement of the work introduced in [1]. In fact, we extend the structural similarity to types as well as attributes of geographical objects.

Starting from definition 1, we can define the total similarity *TotSIM* between two SGOs as the weighted sum of two kinds of similarity: the *semantic similarity* *SemSIM*, that performs a conceptual comparison between the two SGO ψ_1 and ψ_2 in term of the classes objclass$_1$ and objclass$_2$ they belong to, and the *structural similarity* *StrSIM*, that performs a comparison between ψ_1 and ψ_2 in term of their attributes Σ_1 and Σ_2, the types of Σ_1 and Σ_2.

$$TotSIM\,(\Psi_1, \Psi_2) = \frac{w_{sem} * SemSIM\,(\Psi_1, \Psi_2) + w_{str} * StrSIM\,(\Psi_1, \Psi_2)}{w_{sem} + w_{str}} \quad (1)$$

where w_{sem} and w_{str} are the weights associated with the two kinds of similarity and by default are equal to 0.5.

In the following of this section we explain how to evaluate semantic and structural similarities.

4.1 The Semantic Similarity

Semantic similarity in a taxonomy is determined according to the *information content* approach proposed by Lin [6], originally introduced by Resnik [7].

The method's starting assumption is that SGOs represent concepts, which are organized according to a taxonomy. Note that, according to definition 1, each SGO has an objclass that represent the class name it belongs to, and consequently this class name is the geographical concept represented by the SGO. Furthermore, each concept is associated with a weight, standing for the *probability* of the given concept. In the literature, probabilities are estimated according to the *frequencies* of concepts. Given a concept *c*, the *probability p(c)* is defined as follows:

$$p(c) = \frac{freq(c)}{M} \qquad (2)$$

where *freq(c)* is the *frequency* of the concept *c* estimated using noun frequencies from large text corpora, such as the *Brown Corpus of American English* [8], and *M* is the total number of observed instances of nouns in the corpus. In this paper probabilities have been defined according to *SemCor* project [9], which labels subsections of Brown Corpus to senses in the *WordNet* lexicon [10], but any other lexical database for the English language can be used as well. According to *SemCor*, the total number of observed instances of nouns is *88,312*.

The association of probabilities with the concepts of the taxonomy allows us to introduce the notion of a *weighted taxonomy*. For instance, the *WordNet* weighted taxonomy of the concepts presented in the geographical database of ArcView® is shown in Figure 2.

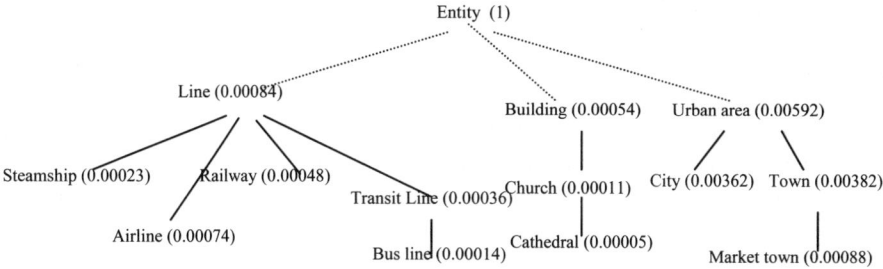

Fig. 2. Fragment of WordNet weighted taxonomy

After introducing the weighted taxonomy of SGOs, the information content of a concept *c* is defined as *-log p(c)*, that is, as the probability of a concept increases, its informativeness decreases, therefore the more abstract a concept, the lower its information content.

Therefore, according to [6], the *information content similarity (ics)* of concepts is essentially defined as the maximum information content shared by the concepts divided by the information content of the compared concepts. In our approach, we follow this definition to introduce the semantic similarity between two SGOs ψ_1 and ψ_2.

$$SemSIM(\Psi_1, \Psi_2) = 2 * \frac{\log p(sharedconcept)}{\log p(objclass_1) + \log p(objclass_2)} \qquad (3)$$

where *sharedconcept* is the least upper bound of the concepts $objclass_1$ and $objclass_2$ in the weighted taxonomy.

For instance, consider the SGOs City and Town. According to the probabilities shown in Figure 2, the following holds:

$SemSIM(City, Town) = 2 \log p(Urban\_area) / (\log p(City) + \log p(Town)) =$
$2 * 7.39965 / (8.10839 + 8.03372) = 0.92$

since Urban_area, which is also the least upper bound of the pair, provides the maximum information content shared by City and Town.

4.2 The Structural Similarity

The evaluation of structural similarity between two SGOs requires the comparison between: Σ_1 and Σ_2, that are the sets of attributes of ψ_1 and ψ_2 respectively, the cardinalities of Σ_1 and Σ_2 and finally the types of Σ_1 and Σ_2.

$$StrSIM(\Psi_1, \Psi_2) = w_1 * Sim(\Sigma_1, \Sigma_2) + w_2 * Sim(\|\Sigma_1\|, \|\Sigma_2\|) +$$
$$+ w_3 * Sim(\Sigma_1 type, \Sigma_2 type) \tag{4}$$

where w_1, w_2, and w_3 are the weights to balance the function, that by default are equal to 0.33.

In the following we show the approach we propose to evaluate each contribution of this formula, starting from the similarity between the set of attributes Σ_1 and Σ_2.

The method to evaluate the similarity between the set of attributes Σ_1 and Σ_2, is inspired by the *maximum weighted matching* problem in bipartite graphs [11]. This problem, starting from an undirected graph in which each edge has a weight associated with it, aims at finding a matching, that is a subset of the edges such that no two edges share a vertex, such that it has the maximum total weight.

In our approach, we aim at finding a matching between pairs of attributes of two SGOs that maximizes the sum of the corresponding information content similarities (ics). To achieve that, we consider two SGOs, namely ψ_1 and ψ_2, and let Σ_1 and Σ_2 be the related sets of attributes. Then, we consider the Cartesian product between the two sets of attributes and we select all the sets of pairs of attributes such that there are no two pairs of attributes sharing an element. We identify each of such sets as $S_{a,b}$. Therefore we have:

$$S_{a,b} = \{(a_1, b_1), \ldots, (a_n, b_n) \in \Sigma_1 \times \Sigma_2 \ / a_k \neq a_h \wedge b_k \neq b_h, \forall 1 \leq h, k \leq n\} \tag{5}$$
$$S = \{S_{a,b}\}$$

Then, for each selected set $S_{a,b}$ of pairs we consider the sum of the information content similarity (*ics*) of the attributes of each pair of the set. Finally, the set of pairs of attributes that maximizes this sum is chosen, and the maximal sum corresponds to the similarity $Sim(\Sigma_1, \Sigma_2)$. The maximal sum has to be divided for the maximal cardinality between Σ_1 and Σ_2 that we indicate as $\|\Sigma_{max}\|$.

$$Sim(\Sigma_1, \Sigma_2) = \frac{1}{\|\Sigma_{max}\|} \max_{S_{a,b} \in S} \sum_{\langle i, j \rangle \in S_{a,b}} ics(i, j) \tag{6}$$

For instance, suppose that *Railway* and *Airline* have the following sets of attributes:

Railway(number:int, local:boolean, goods:boolean, dep_railway_station:string, dest_railway_station:string, passengers:int)

Airline(number:int, company:string, travelers:int, dep_air_terminal:string, dest_air_terminal:string)

For calculating the similarity between the attributes of these two SGOs we must calculate the *ics* among all pairs of attributes that belong to the same weighted hierarchy. For instance, suppose that R*ailway_station* and *Air_terminal* belong to the weighted taxonomy shown in Figure 3.

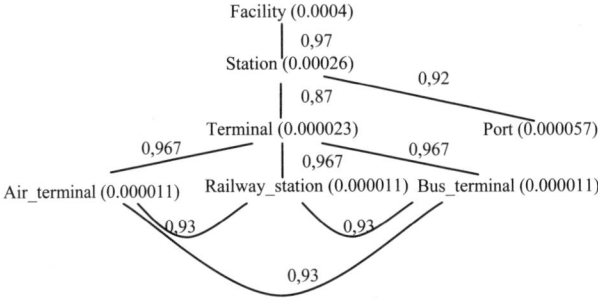

Fig. 3. Fragment of WordNet weighted taxonomy for attributes

The set of pairs of attributes of *Railway* and *Airline* that maximizes the sum of the related *ics* is the following:

{*(number,number)*,*(passengers,travelers)*,*(dep_railway_station, dep_air_terminal)*, *(dest_railway_station, dest_air_terminal)*}

since, by assuming that *passengers* and *travelers* are synonyms, we have:

ics(number,number)=1
ics(passengers,travelers)=1
ics(dep_railway_station, dep_air_terminal) = 0.93
ics(dest_railway_station, dest_air_terminal) = 0.93
Therefore, the similarity between the attributes of *Railway* and *Airline* is:
Sim(Railway,Airline) = 1/6 * [1+1+0.93+0.93] = 0.643.

To evaluate the similarity between the cardinalities of Σ_1 and Σ_2, as the they are numeric values, we can consider the difference between these values, as shown in formula (7).

$$Sim(\|\Sigma_1\|,\|\Sigma_2\|) = \|\Sigma_1\| - \|\Sigma_2\| \qquad (7)$$

The method for the evaluation of the similarity between the types of Σ_1 and Σ_2 is quite similar to the method for the attributes. In fact, we aim at finding a matching between pairs of types that maximizes the sum of the pairs. To achieve that, we consider two SGOs, namely ψ_1 and ψ_2, and let Σ_1 and Σ_2 be the related sets of attributes, and T_1 and T_2 be the sets of types associated with Σ_1 and Σ_2. Then, we consider the Cartesian product between the two sets of attributes and we select all the sets of pairs of attributes such that there are no two pairs of attributes sharing an element and the attributes have the same type. We identify each of such sets as $T_{a,b}$. Therefore we have:

$$T = \{T_{a,b}\} = \left\{ \begin{array}{l} (a_1,b_1),....,(a_n,b_n) \in \Sigma_1 \times \Sigma_2 \;/\; a_k \neq a_h \wedge b_k \neq b_h, \forall 1 \leq h,k \leq n \\ and(t_{a_j} = t_{b_j}) \forall 1 \leq j \leq n \end{array} \right\} \tag{8}$$

Then, the set of pairs of attributes with maximal cardinality is considered, and the maximal cardinality divided for the maximal cardinality between Σ_1 and Σ_2, that we indicate as $|\Sigma_{max}|$, corresponds to the similarity $Sim(\Sigma_1 type, \Sigma_2 type)$.

$$Sim(\Sigma_1 type, \Sigma_2 type) = \frac{\left\| T_{a,b} \right\|_{max}}{\left\| \Sigma_{max} \right\|} \tag{9}$$

5 Experiments

To validate the proposed approach, we have performed a set of experiments among the research staff of our Institute (CNR-IRPPS).

The total number of people involved in the usability tests is twenty. We divided these people in two groups: group 1 is composed of members of the computer science research staff, with special skill with database, and group 2 is composed of members of the social science research staff, without any skill with databases.

To perform the experiments we predispose a PC workstation for each member of the research staff with GeoPQL software available. The GeoPQL interface allows the user to specify if he/she wants to formulate an exact query or an approximate query. In this way we can perform a comparison between two systems that allow to formulate queries on a geographical database with and without the possibility to receive similar answers.

The text consisted of formulating a set of fifteen pictorial geographical queries, expressed in natural language, by using first the modality for formulating an exact query and then the modality for formulating an approximate query.

The criteria we used for evaluating the user behavior are the following:

-    Number of repetition for formulating each query;
-    Satisfaction of the user.

The second criteria consists in a final questionnaire in which we ask the user to grade the performance of the application.

The results of the experiments shows that the approximate queries are more useful for inexperienced user, without any skill for databases, rather than for expert users, even if there is an higher satisfaction for the approximate queries rather than the exact one also for the expert users. Figure 4 illustrates average values for the number of repetitions and the satisfaction of the two modalities (exact and approximate) for each group of user.

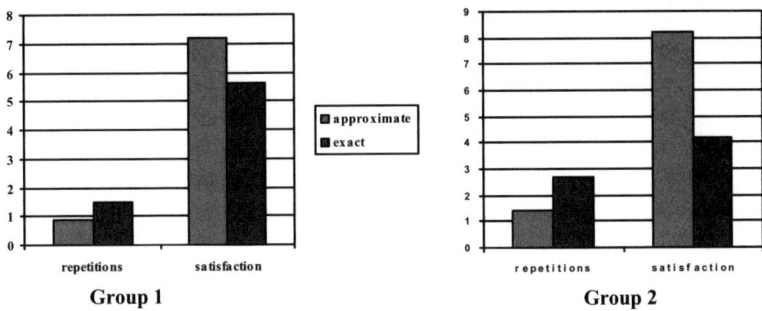

Fig. 4. Experiment results

6   Conclusions

This paper proposed an approach to evaluate semantic and structural similarities to enable the approximation of query answers in geographical information systems.

When a query involves information on a missing concept or a concept with no instances in the geographical database, the system provides approximate answers by evaluating the similarity of the missing concept and the other concepts which are present in the database and considering as acceptable queries in which the missing concept is replaced by a similar one.

The experiments we performed demonstrate the usefulness of the approximate queries especially for the inexperienced users.

References

1. D'Ulizia, A., Ferri, F., Formica, A., Grifoni, P., Rafanelli, M.: Structural similarity in geographical queries to improve query answering. In: ASIIS 2007. ACM Symposium Applied Computing, Track on Advances in Spatial and Image-based Information Systems, Seoul, Korea, March 11-15, pp. 19–23. ACM Press, New York (2007)
2. Candan, K.S., Kim, J.W., Liu, H., Suvarna, R.: Structure-based mining of hierarchical media data, meta-data, and ontologies. In: 5th Workshop on Multimedia Data Mining, Seattle, WA (August 2004)
3. Dalamagas, T., Cheng, T., Winkel, K.J., Sellis, T.K.: Clustering XML Documents by Structure. In: Vouros, G.A., Panayiotopoulos, T. (eds.) SETN 2004. LNCS (LNAI), vol. 3025, pp. 112–121. Springer, Heidelberg (2004)
4. Formica, A., Missikoff, M.: Concept Similarity in SymOntos: an Enterprise Ontology Management Tool. The Computer Journal 45(6), 583–594 (2002)
5. Ferri, F., Rafanelli, M.: GeoPQL: a Geographical Pictorial Query Language that resolves ambiguities in query interpretation. In: Spaccapietra, S., Zimányi, E. (eds.) Journal on Data Semantics III. LNCS, vol. 3534, pp. 50–80. Springer, Heidelberg (2005)
6. Lin, D.: An Information-Theoretic Definition of Similarity. In: ICML 1998. Proceedings of 15th International Conference on Machine Learning, Madison, WI, pp. 296–304 (1998)

7. Resnik, P.: Using information content to evaluate semantic similarity in a taxonomy. In: IJCAI 1995. Proceedings of the 14th International Joint Conference on Artificial Intelligence, San Francisco, pp. 448–453 (1995)
8. Francis, W.N., Kucera, H.: Frequency Analysis of English Usage. Houghton Mifflin, Boston (1982)
9. Fellbaum, C.: A Semantic Network of English: the Mother of all WordNets. Computers and the Humanities 32, 209–220 (1998)
10. WordNet 2.1: A lexical database for the English language (2005), http://www. cogsci.princeton.edu/cgi-bin/webwn
11. Galil, Z.: Efficient algorithms for finding maximum matching in graphs. ACM Computing Surveys 18, 23–38 (1986)

Ensuring the Semantic Correctness of Complex Regions*

Mark McKenney, Alejandro Pauly, Reasey Praing, and Markus Schneider

University of Florida
Department of Computer and Information Sciences and Engineering
{mm7,apauly,rpraing,mschneid}@cise.ufl.edu

Abstract. Ensuring the semantic and topological correctness of spatial data is an important requirement in geographical information systems and spatial database systems in order to preserve spatial data quality and enable correct operation execution. Spatial objects like complex regions are usually represented as an ordered sequence of segments (sequence view) to support and ease the computation of spatial operations by means of plane sweep algorithms. The semantic correctness of such a sequence is usually simply assumed but is not easy to see. In this paper, we present a novel and efficient algorithm to discover the cyclic structure and check for the semantic correctness of the sequence representation of a complex region by determining its cyclic structures (component view) in terms of multiple faces possibly containing holes. The algorithm producing the component view is also interesting for object construction, manipulation, and visualization.

1 Introduction

The study of spatial objects and spatial operations has received widespread attention in varied fields such as computational geometry, spatial databases, geographic information science (GIS), computer-aided design, computer vision, and computer graphics. A large amount of the research these areas has focused on the representation and manipulation of spatial objects for use in spatial systems. Within this research, the type of complex regions has received significant attention due to their ability to model many aspects of geographic reality. A complex region a is two-dimensional spatial object consisting of multiple *faces*, each of which can contain a number of *holes*.

Because of the varied uses of complex regions in spatial systems (as well as other complex data types such as points and lines), two forms of representation have emerged to store and manipulate them. The first form of representation is the *sequential view* of a complex region, which treats a complex region as an *ordered sequence* of *segments*. This view has become popular because it is used as an input and output format for many spatial algorithms, specifically

---

* This work was partially supported by the National Science Foundation under grant number NSF-CAREER-IIS-0347574.

J.-L. Hainaut et al. (Eds.): ER Workshops 2007, LNCS 4802, pp. 409–418, 2007.

plane sweep algorithms, used to implement spatial operations and topological predicates. The main drawback to this view is that the structural components of a complex region (i.e., the faces and holes) are not explicitly represented. The second form of representation for complex regions is the *component view*, in which the segments of the region are grouped by the structural components of the region; in other words, the segments that form a hole or face are grouped together. The component view is typically used to create, manipulate, or visualize a complex region, because its cyclic structure (i.e., the holes and faces) is known and can be utilized. For example, if a user wants to remove a face from a region, then the segments that make up that face, as well as the segments that make up any holes in that face, must be identified and removed. Such information is explicitly encoded in the component view.

The study of spatial data quality has received significant attention in the field of spatial data. An important problem in this field is to determine if a given spatial object is valid in the sense that it conforms to the spatial object's type definition. In the case of complex regions, we need to know if the object defined by a set of segments is semantically and topologically correct. If a complex region is defined based on the component view, then the cyclic structure of the region is known. Therefore, this can be validated, and the topological constraints that the faces and holes must satisfy can be checked. However, given a region represented in the sequential view, the cyclic structure must be computed explicitly. Currently, there are no known algorithms to compute this information in an efficient manner.

At this point it is important to note that the two views of region representation can be merged into a hybrid view in which segments are ordered sequentially and annotated with cyclic information. Thus, this view allows input to spatial operations and explicitly encodes cyclic information so that semantic correctness can be checked. However, such a view cannot be maintained through spatial operations. In other words, even if the cyclic structure of two regions is known, the cyclic structure of their intersection cannot be computed based on the known structure of the original regions. This means that even though the input to an intersection operation is two regions in the hybrid view, the output is a region in the sequential view; thus the cyclic structure must still be computed. This holds for the intersection, union, and difference operations between regions.

The main contribution of this paper is an efficient algorithm that takes a region represented in the sequential view as input and returns the region in a component view as output. Thus, given a region in the sequential view, we are able to discover its cyclic structure. As we mentioned before, the need to *validate* a region, or ensure that it is semantically and topologically correct, is an important concern in spatial data management. In addition to finding the cyclic structure of a region, our algorithm simultaneously checks that the region is semantically and topologically correct. Therefore, this algorithm eliminates the need to store a region in the component view because regions can quickly be validated based on our algorithm, and the component view can be quickly computed. This allows regions to be stored in the sequential view, which is more compact than the hybrid view and is typically easier to manage on disk than the component view.

In Section 2, we present existing work related to our problem. Section 3 introduces the formal definition of regions upon which our problem solution is based. Our algorithm for deriving the component view of regions is presented in Section 4. Section 5 describes the time complexity of the algorithm. Finally, in Section 6 we provide conclusions and outline future work.

2   Related Work

To the authors' knowledge, no previous publication includes a plausible solution to the problem we have introduced in the previous section. Instead, plenty of literature exists that explores problems for which solutions can provide us with important insight for solving our problem.

The well known plane sweep algorithmic technique for geometric intersection problems, original to Shamos and Hoey [9] and also popularly employed by Bentley and Ottmann [4] as well as many others [7,3], serves as a basis for our solution. The plane sweep model proves useful in identifying properties of the segments that are critical for efficiently solving our problem.

Related problems of arrangements of lines and segments are presented in [1,2,5]. Such an arrangement consists of a partitioning of the space such that the lines or segments do not cross the boundaries of the partitions. Computation of planar maps to detect polygons from sets of segments are studied in [6]. The authors provide a solution to the problem of detecting the polygons formed by a set of intersecting segments. That is, each polygon detected is defined by three or more pairwise intersections of segments. This problem is fundamentally different to our problem in that (1) there are no restrictions on the polygons that are detected, (2) in our problem we must consider holes and outer cycles separately, and (3) all holes and outer cycles must adhere to the definition of complex regions.

3   Complex Regions

In this section, we are interested in the implementation model of the region data type. To simplify our discussion, we define the implementation data type of complex regions based on segments. Then, we modify this definition such that it is based on halfsegments, which allows processing of region objects in spatial predicates and operations. In order to define a complex region, we first define the required concepts such as points, segments, polygons, and faces. Due to space limitations, we can only provide an informal definition of these concepts. For a formal definition, see [8].

The type *point* includes all single, two dimensional points. We assume that the equality "=" relation and the lexicographic order relation "<" are defined between any two points. The type *segment* incorporates all straight lines bounded by two endpoints p and q. We make use of the predicates *disjoint, collinear* :

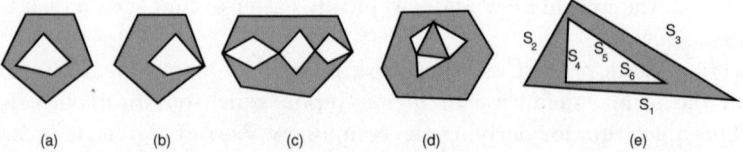

Fig. 1. Regions (a) and (b) consist of a single face. Regions (c) and (d) each have two faces. Region (e) is a single face and is annotated to identify its individual segments.

$segment \times segment \rightarrow bool$ to determine whether two segments are disjoint or are located on the same infinite lines, respectively. A function $len : segment \rightarrow real$ computes the length of a segment. These operations are required to define the order relation of halfsegments which are defined later. Using segments, a simple polygon can be implemented as a connected sequence of segments that forms a single cycle. We say that two simple polygons are *edge-disjoint* if their interiors are disjoint and they possibly share single boundary points but not boundary segments. A face is a simple polygon possibly containing a set of edge-disjoint holes, which are simple polygons, such that these holes do not collectively separate the interior of the face. A complex region is a set of edge-disjoint faces. Figure 1 shows some example regions.

Although the formal definition of *region* in [8] ensures uniqueness of representation for complex regions, in some cases, it is not obvious if a set of segments forms a hole cycle, or if the segments should be part of an outer cycle of a face. Furthermore, certain configurations of faces and holes form non-intuitive scenarios. For example, Figure 1(b) forms a face containing a hole, and not a single face that happens to meet itself at a point. Figure 1(c) depicts two faces that meet at four points, but contain no holes. Figure 1(d) shows two faces, the larger one containing a hole. The uniqueness of representation of complex regions is critical because given a sequence of segments, there is exactly one valid semantic interpretation of the cyclic structure of the region it represents. Our algorithm in Section 4 correctly interprets any valid region described as a sequence of segments.

Spatial operation implementations between regions based on the plane sweep algorithm require input to be a region encoded not as a sequence of segments, but as a sequence of halfsegments. We define the type $halfsegment = \{(s, d) | s \in segment, d \in bool\}$. A halfsegment is a hybrid between a point and a segment since it has features of both geometric structures. For a halfsegment $h = (s, d)$, if d is true (false), the smaller (greater) endpoint of s is the *dominating point* of h, and h is called a *left* (*right*) *halfsegment*. Hence, each segment s is mapped to two halfsegments $(s, true)$ and $(s, false)$. Furthermore, halfsegments are typically annotated with an *interior-above* flag, which indicates whether the interior of the region lies above or below the halfsegment. In addition to the use of halfsegments, the representation of a region object requires an order relation on halfsegments. Let dp be a function which yields the dominating point of a halfsegment. For two distinct halfsegments $h_1 = (s_1, d_1)$ and $h_2 = (s_2, d_2)$ with a common endpoint p, let α be the enclosed angle such that $0° < \alpha \leq 180°$. Let a predicate $rot(h_1, h_2)$ be true if, and only if, h_1 can be rotated around p through α to overlap h_2 in counterclockwise direction. We define a complete order on halfsegments as:

$$
\begin{aligned}
& h_1 < h_2 \Leftrightarrow & \\
& dp(h_1) < dp(h_2) \vee & (1) \\
& (dp(h_1) = dp(h_2) \wedge ((\neg d_1 \wedge d_2) \vee & (2a) \\
& \qquad\qquad\qquad\qquad (d_1 = d_2 \wedge rot(h_1, h_2)) \vee & (2b) \\
& \qquad\qquad\qquad\qquad (d_1 = d_2 \wedge collinear(s_1, s_2) \wedge len(s_1) < len(s_2)))) & (3)
\end{aligned}
$$

Since a segment can be substituted by two halfsegments, a region object can be implemented as an ordered sequence (array) of halfsegments. As an example, Figure 1(e) shows a complex region object (with a single face containing a hole) whose segments are labeled s_i. Let $h_i^l = (s_i, true)$ and $h_i^r = (s_i, false)$ denote the left and right halfsegments of a segment s_i respectively. The order sequence of halfsegments for this complex region is $\langle h_1^l, h_2^l, h_6^l, h_4^l, h_4^r, h_5^l, h_2^r, h_3^l, h_5^r, h_6^r, h_3^r, h_1^r \rangle$.

4 Computing the Cyclic Structure of Complex Regions

We assume that the input to our algorithm is a sequence of ordered halfsegments. If the input represents a region, the algorithm returns the region with its cyclic structure information; otherwise, the algorithm exits with an error message indicating the input sequence does not form a semantically correct region. We begin by providing a high level overview of the algorithm and then present the algorithm and provide a discussion of its details.

In general terms, the algorithm must identify all cycles present in the halfsegment sequence, and classify each cycle as either an outer cycle or a hole cycle of a particular face. To accomplish this, each halfsegment is *visited* once by the algorithm. Note that due to the definition of the type *region*, each segment belongs to exactly one cycle. When a halfsegment is visited, the algorithm marks the halfsegment indicating to which face and cycle it belongs, and whether that cycle is an outer cycle or a hole cycle. The algorithm does not alter the input when marking halfsegments, rather a parallel array to the input sequence is used to represent the cycle information. The algorithm visits halfsegments by stepping through the input list sequentially.

The first halfsegment in the input sequence will always be part of the outer cycle of a face, due to the definition of complex regions and the halfsegment ordering defined previously. Therefore, it can be visited and marked correctly. Once a halfsegment has been visited, it is possible to visit and correctly mark all other halfsegments in the cycle that it belongs to in a procedure which we denote as the *cycle walk*. Thus, all halfsegments that form the cycle to which the first halfsegment in the input sequence belongs are then visited. The algorithm then begins stepping through the remaining halfsegments. The next unvisited halfsegment encountered will be part of a new cycle. The algorithm then visits this new halfsegment. The algorithm can deduce whether this halfsegment is an outer cycle of a new face or a hole in an existing face by examining where the halfsegment lies in relation to already known cycles. To determine this, we use a plane sweep algorithm to step through the halfsegments. Thus, we can take advantage of the plane sweep status structure to find whether or not the current halfsegment lies in the interior of a previously visited face. Once the new halfsegment is visited, we perform a cycle walk from it. Then, the algorithm continues stepping through the input list until it reaches another unvisited halfsegment, visits it, and repeats this procedure. The algorithm is shown in Algorithm 1.

To properly describe the algorithm outlined in Algorithm 1, we introduce several notations. The function $info(h)$ for a given halfsegment h returns its cyclic information, that is, its *owning* cycle, and if part of a hole, its owning face. A cycle *owns* a halfsegment if the halfsegment is part of the boundary of the cycle, and a face owns a hole if the hole is inside the face. We define the function $NewCycle(h)$ to annotate h with a unique identifier for a new cycle. Let f be a halfsegment belonging to an outer cycle of a face. The function $Owns(h, f)$ annotates the halfsegment h to indicate that it belongs to a hole in the face that owns f. Finally, we employ the function $Visit(p)$ to mark a point p as having been visited. The function $Visited(p)$ is used to verify is point p was marked as visited already. Points are only marked as visited when a halfsegment with dominating point p has been visited during the cycle walk. We mark points as being visited in order to identify the special case of a hole cycle that meets the outer cycle of a face at a point. The function $Visited(h)$ is used to verify if halfsegment h has been visited already. A halfsegment has been visited if it has been annotated with face/hole information. For a halfsegment h, we can directly compute its corresponding right (left) halfsegment h_b, which we call its *brother* by switching its boolean flag indicating which end point is dominant. We define the next halfsegment in the cycle to which h belongs as h_+ such that the dominating endpoint of h_b is equal to the dominating point $dp(h_+)$ and $h_+ \neq h_b$ and h_+ is the first halfsegment encountered when rotating h_b clockwise (in an outer cycle) or counter-clockwise (in a hole cycle) around its dominating point. The previous halfsegment in the cycle is similarly defined as h_-.

4.1   Classifying Outer and Hole Cycles

By using a sweep line, the algorithm steps through the halfsegment sequence to find the smallest unannotated halfsegment h, create a new cycle for this halfsegment, and mark its dominating point as visited (line 2-4). At this point, the algorithm needs to determine whether h belongs to a hole cycle (line 5) or an outer cycle (line 9). If a cycle is identified as a hole cycle, the outer cycle to which it belongs must also be identified (line 6-7), and the cycle must be walked using counter-clockwise adjacency of halfsegments (line 8). Recall that the plane sweep algorithm maintains the sweep line status structure, which is a ordered list of *active* segments, such that it provides a consistent view of all halfsegments that currently intersect the sweep line, up to the current *event* (the addition or removal of a halfsegment). By examining the halfsegment directly below a halfsegment h in the sweep line status, we can determine whether h is a part of an outer cycle or a hole cycle of an existing face. In other words, if halfsegment p is directly below halfsegment h in the sweep line status structure and the interior-

---

Algorithm 1: The algorithm for deriving the component view of a region

---

Input: Sequence of unannotated halfsegments H
Output: Sequence H with fully annotated halfsegments

```
 1  while not end of sweep do
 2  │   Advance sweep line to h. h is the left-most halfsegment yet to be annotated;
 3  │   Using sweep line status, determine h as part of an outer cycle or a hole cycle;
 4  │   NewCycle(h); Visit(dp(h));
 5  │   if h belongs to a hole then
 6  │   │   Using sweep line status, retrieve halfsegment f from its owning outer cycle;
 7  │   │   Owns(h, f);
 8  │   │   Set cycle walk mode to use counter-clockwise adjacency;
 9  │   else
10  │   │   Set cycle walk mode to use clockwise adjacency;
11  │   end
    │   /* Begin walking the cycle */
12  │   c ← h₊;
13  │   while c ≠ h do
14  │   │   if Visited(dp(c)) then
15  │   │   │   q ← c; c ← c₋; NewCycle(c); Owns(c, h);
16  │   │   │   while dp(c) ≠ dp(q) do
    │   │   │   │   /* Trace back anchored hole */
17  │   │   │   │   info(c₋) ← info(c); c ← c₋;
18  │   │   │   end
19  │   │   else
20  │   │   │   info(c) ← info(h); Visit(dp(c)); c ← c₊;
21  │   │   end
22  │   end
23  end
```

above flag of p is set to *true*, it follows that h is either in the interior of the cycle to which p belongs, or h is part of the cycle to which p belongs. Recall that as soon as a halfsegment is classified as being apart of a hole or face, the cycle to which it belongs is walked (Section 4.2) and all other halfsegments in that cycle are marked accordingly (lines 12-22). Therefore, if a halfsegment belongs to the same cycle as any halfsegment that has been previously encountered by the sweep line, it is already known to which face and/or hole cycle it belongs. Furthermore, all halfsegments that are less than a given halfsegment in halfsegment order have already been classified. Therefore, we can determine if an unmarked halfsegment belongs to a hole or outer cycle by examining the halfsegment immediately below it in the sweep line status structure.

From the definition of a face, the outer cycle of a face of a region always covers (encloses) all of its hole cycles. This means that the smallest halfsegment of this face is always a part of the outer cycle. This is also true for the entire region object where the smallest halfsegment in the ordered sequence is always a part of the first outer cycle of the first face. Furthermore, due to the order relation of halfsegments and the cyclic structure of a polygon, the smallest halfsegment of a face will always be a left halfsegment with the interior of the face situated above it. Thus, when we process this halfsegment, we set its interior-above flag to indicate this fact. Since we have classified this cycle as an outer cycle, we can walk the cycle and set the interior-above flag for all halfsegments of this cycle. For example, Figure 2(a) illustrates the case where the smallest halfsegment of the sequence is processed and the cycle is classified as an outer cycle.

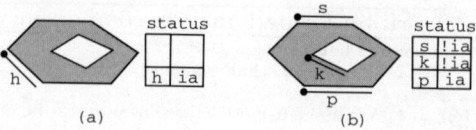

Fig. 2. Processing the smallest halfsegment h of the sequence (a) and k of a cycle (b)

Once the first outer cycle of a face in a region has been processed, we continue to process halfsegments that have not yet been classified based on the plane sweep status structure. Figure 2(b) shows an example. Here, we add/remove visited halfsegments into/from the sweep line status in sequence ordered up to the smallest unvisited halfsegment k. This halfsegment *must* be the start of a new cycle that we must now classify. We know k is the start of a new cycle because all halfsegments of an existing cycle that include a halfsegment j such that $j < k$ must have been marked as visited by the walking process. Once we reach this new cycle represented by its starting halfsegment k, we add this halfsegment into the sweep line status. We classify the type of cycle k belongs to by examining the interior-above flag of the halfsegment p (its predecessor) which was already visited and sits immediately below k in the sweep line status structure. If the predecessor indicates that the interior of the face is above it (the interior-above attribute of p is set to true), then k lies in the interior of the cycle to which p belongs; thus, k must be part of a hole cycle and the interior of the face to which k belongs must lie below k. If the interior-above flag of p indicates that the interior of the face to which p belongs is below p, then the current halfsegment k must be part of an outer cycle of a new face. In case that there is no predecessor, then the current halfsegment must be a part of an outer cycle of a new face, because it does not lie in the interior of any other face's outer cycle. Once the cycle is classified as either an outer cycle of a new face or a hole cycle of an existing face, the cycle walking procedure is carried out to determine all halfsegments that belong to the cycle.

4.2   Walking Cycles

In general terms, we use the phrase *walking a cycle* to indicate the traversal of a cycle such that each halfsegment that forms the cycle is visited. Furthermore, the halfsegments in such a traversal are visited in the order in which they appear in the cycle. In other words, given a halfsegment h, all halfsegments in the cycle to which h belongs are found by repeatedly finding h_+ until the the original halfsegment is encountered again. For example, when walking the outer cycle of the region in Figure 1e in clockwise order beginning from S_1, the halfsegments would be encountered in the order $h_1^l, h_1^r, h_3^r, h_3^l, h_2^r, h_2^l$. The two main challenges to this portion of the algorithm are (i) to identify cycles correctly such that they correspond to the unique representation of a region as stated in the definition of complex regions, and (ii) to achieve this efficiently. In this section we show how to satisfy the first challenge. Time complexity is discussed in the next section.

When a halfsegment h is encountered by the algorithm that has not yet been classified, it is classified as belonging to a hole or outer cycle in line 5. If h belongs to an outer cycle, then the cycle walk portion of the algorithm in lines 12-22 is executed. Due to the halfsegment ordering and the definition of regions, the smallest unvisited halfsegment in the input sequence that the plane sweep encounters is always a left halfsegment of an outer cycle of a face and the interior of that face always lies above the halfsegment. If we rotate h_b clockwise around its dominating point, it will intersect the interior of the face. Thus, the first halfsegment encountered when rotating h_b clockwise around its dominating point will be part of the outer cycle of the same faces (except for a special case discussed below) and will be h_+. We know this to be true because if we find h_+ in this fashion and it turns out to be part of another face, then two faces would intersect, which is prohibited by the definition of complex regions. It follows that each successive halfsegment in the outer cycle can be found by rotating the brother of the current halfsegment clockwise around its dominating point because the location of the interior relative to the halfsegment can always be deduced based on the previous halfsegment encountered in the cycle walk.

One special case occurs when walking outer cycles: the existence of a hole in a face that meets the outer cycle at a point (see Figure 1b). When walking an outer cycle that contains such a hole, the halfsegments that form the hole will be classified as being part of the outer cycle using the procedure just described. In order to remedy this, we mark each point that is a dominating point of a halfsegment encountered during the cycle walk (line 20). Each time we find a new halfsegment that is part of an outer cycle, we first check if its dominating point has been visited yet (line 14). If it has been visited, then we know that we have encountered that point before, and a hole cycle that meets the outer cycle must have been discovered. When this happens, we loop backwards over the cycle until we find the halfsegment whose dominating point has been visited twice (lines 15-18). The halfsegments forming the hole are then marked as such. The remainder of the outer cycle is then walked.

Walking a hole is identical to walking an outer cycle, except that a counter-clockwise rotation from h_b is used to find h_+. A counter-clockwise rotation is required because the interior of the face is intersected by h_b when rotating h_b around its dominating point. When walking holes, the special case exists that two holes may meet at a point. Thus, we employ the same strategy to detect this case as we did with the special case of a hole meeting a face (lines 15-18).

5 Complexity

The classification component of our algorithm requires the use of a plane sweep algorithm over n halfsegments; thus, a complexity of $O(n \log(n))$ is required. However, because a segment intersection indicates an invalid region, we do not need to compute or report intersections, and thus do not require an output sensitive algorithm. The cycle walk algorithm requires that we locate a half-segment based on its dominating point in a list of halfsegments. Because the list of halfsegments is ordered, it is possible to employ a searching technique to locate halfsegments quickly. If the list of halfsegments is implemented as an array, we can simply use a binary search. Because each halfsegment is searched for one time in this stage of the algorithm, the cycle walk can be computed in $O(n \log(n))$ time for n halfsegments. Furthermore, once a halfsegment has been classified, it will at most be visited one additional time by the sweep line portion of the algorithm. Finally, it is possible to keep track of the number of times a point has been visited in an ordered array. Thus, we can rely on a binary search to find points in order to mark them as visited. The number of points is bounded by the number of halfsegments, thus, this has complexity at most $O(n \log(n))$. Therefore, the algorithm has a worst case time complexity of $O(n \log(n))$.

6   Conclusions

In this paper we have introduced an $O(n \log(n))$ algorithm for computing the component view of a complex region originally represented by a sequence of half-segments. Furthermore, the algorithm checks the semantic correctness of region objects since it will not be able to compute the component view of a halfsegment sequence that does not represent a region. The algorithm was successfully implemented as part of a spatial algebra that was embedded into an existing database management system for the purpose of managing spatial data.

References

1. Amato, N.M., Goodrich, M.T., Ramos, E.A.: Computing Faces in Segment and Simplex Arrangements. In: ACM Symposium on Theory of Computing, pp. 672–682. ACM Press, New York (1995)
2. Asano, T., Guibas, L.J., Tokuyama, T.: Walking on an Arrangement Topologically. In: ACM Annual Symp. on Computational Geometry, pp. 297–306. ACM Press, New York (1991)
3. Balaban, I.J.: An Optimal Algorithm for Finding Segments Intersections. In: ACM Annual Symp. on Computational Geometry, pp. 211–219. ACM Press, New York (1995)
4. Bentley, J.L., Ottmann, T.: Algorithms for Reporting and Counting Geometric Intersections. IEEE Trans. on Computers C-28, 643–647 (1979)
5. Edelsbrunner, H., Guibas, L.J., Sharir, M.: The Complexity of Many Faces in Arrangements of Lines of Segments. In: ACM Annual Symp. on Computational Geometry, pp. 44–55. ACM Press, New York (1988)
6. Ferreira, A., Fonseca, M.J., Jorge, J.A.: Polygon Detection from a Set of Lines. In: Encontro Portugues de Computacao Grafica, pp. 159–162 (2003)
7. Nievergelt, J., Preparata, F.P.: Plane-Sweep Algorithms for Intersecting Geometric Figures. Communications of the ACM (CACM) 25, 739–747 (1982)
8. Schneider, M., Behr, T.: Topological Relationships between Complex Spatial Objects. ACM Trans. on Database Systems (TODS) 31, 39–81 (2006)
9. Shamos, M., Hoey, D.: Geometric Intersection Problems. In: IEEE Symp. on Foundations of Computer Science, IEEE Computer Society Press, Los Alamitos (1976)

Author Index

Printing: Mercedes-Druck, Berlin
Binding: Stein+Lehmann, Berlin

Lecture Notes in Computer Science

Sublibrary 3: Information Systems and Application, incl. Internet/Web and HCI

For information about Vols. 1– 4317
please contact your bookseller or Springer